T0387065

ALBERT THE GREAT AND HIS ARABIC SOURCES

PHILOSOPHY IN THE ABRAHAMIC TRADITIONS OF THE MIDDLE AGES

TEXTS AND STUDIES IN INTERPRETATION AND INFLUENCE AMONG PHILOSOPHICAL THINKERS OF THE MEDIEVAL ARABIC, LATIN, AND HEBREW TRADITIONS

VOLUME 5

# Albert the Great and his Arabic Sources

*Medieval Science between Inheritance and Emergence*

*Edited by*
KATJA KRAUSE AND RICHARD C. TAYLOR

BREPOLS

British Library Cataloguing in Publication Data
A catalogue record for this book is available from the British Library



D/2024/0095/23
ISBN 978-2-503-60937-9
eISBN 978-2-503-60938-6
DOI 10.1484/M.PATMA-EB.5.135807

Printed in the EU on acid-free paper.

# Table of Contents

# Acknowledgements

The idea for *Albert the Great and his Arabic Sources: Medieval Science between Inheritance and Emergence* originated in a richly productive workshop held at the De Wulf-Mansion Centre for Ancient, Medieval, and Renaissance Philosophy and Katholieke Universiteit Leuven, Belgium, in June 2012: 'Translation and Transformation in Philosophy: Albert between Aquinas and "the Arabs"', organized by Richard C. Taylor and Andrea Robiglio. We warmly thank the participants in that event for their intellectual generosity, and especially Andrea Robiglio for his invaluable contribution to our long-term project.

Katja Krause and Richard C. Taylor's work on the book was made possible by the Max Planck Institute for the History of Science, Berlin. At Katja Krause's Max Planck Research Group 'Experience in the Premodern Sciences of Soul and Body, ca. 800–1650', Kate Sturge coordinated the manuscript project and worked tirelessly on the style editing; Fabio Di Gregorio, Max Steinwandel, and Anina Woischnig helped to prepare the bibliographies. The Open Access publication of this volume was kindly funded by the MPIWG Open Access Monograph Publishing Fund.

No collection of this scale can come to fruition without the enormous communal commitment, creativity, and attention to detail of all the authors. The editors also had the privilege of Steven Harvey's steady scholarly and personal encouragement throughout the process.

In the project's later stages, the volume benefited greatly from the expertise of the staff at Brepols, especially Guy Carney and Haris Tsartsali, and the indexing skills of Kellyann Wolfe. Finally, the editorial board of the Brepols series 'Philosophy in the Abrahamic Traditions of the Middle Ages', led for this volume by Thérèse-Anne Druart with delicacy and skill, was what made this book possible. We particularly thank Jules Janssens for his kind and erudite engagement.

Last but not least, the editors thank their families for seemingly infinite patience and support. In particular, Katja Krause would like to thank Frank and Elisabeth, and her mother, Vera Krause; Richard C. Taylor would like to thank his wife Carolyn.

KATJA KRAUSE AND RICHARD C. TAYLOR

# Chapter 1. Introduction. Albert's Philosophical *scientia*

*Origins, Geneses, Emergences*

***Wagner.***
*Ach Gott! die Kunſt iſt lang;*
*Und kurz iſt unſer Leben.*
*Mir wird, bey meinem kritiſchen Beſtreben,*
*Doch oft um Kopf und Buſen bang'.*
*Wie ſchwer ſind nicht die Mittel zu erwerben,*
*Durch die man zu den Quellen ſteigt!*
*Und eh' man nur den halben Weg erreicht,*
*Muß wohl ein armer Teufel ſterben.*

***Fauſt.***
*Das Pergament, iſt das der heil'ge Bronnen,*
*Woraus ein Trunk den Durſt auf ewig ſtillt?*
*Erquickung haſt du nicht gewonnen,*
*Wenn ſie dir nicht aus eigner Seele quillt.*[1]

Albert the Great (*c.* 1200–80) was one of the great philosophers, if not the greatest, among the thirteenth-century Scholastics. Yet he has been under-appreciated by modern scholars, who tend to focus on his far more famous student, Thomas

1 Goethe, *Faust: Ein Fragment*, p. 15. The English translation (Goethe, *Faust*, trans. Taylor, pp. 48–49) runs as follows. 'WAGNER: Ah, God! but Art is long, | And Life, alas! is fleeting. | And oft, with zeal my critic-duties meeting, | In head and breast there's something wrong. | How hard it is to compass the assistance | Whereby one rises to the source! | And, haply, ere one travels half the course | Must the poor devil quit existence. FAUST: Is parchment, then, the holy fount before thee, | A draught wherefrom thy thirst forever slakes? | No true refreshment can restore thee, | Save what from thine own soul spontaneous breaks'.

*Albert the Great and his Arabic Sources*, ed. by Katja Krause and Richard C. Taylor, Philosophy in the Abrahamic Traditions of the Middle Ages, 5 (Turnhout: Brepols, 2024), pp. 9–42
BREPOLS PUBLISHERS 10.1484/M.PATMA-EB.5.136481

Aquinas (*c.* 1225–1274). This is especially true for the English-speaking world.[2] Lamentable as the situation is, it has begun to change — thanks in great part to studies by the illustrious scholars who have contributed to this volume. *Albert the Great and his Arabic Sources: Medieval Science between Inheritance and Emergence* aims to continue this trend by examining one major aspect of Albert's philosophy: his use of the Arabic sources available to him at the time.

For present-day historiography of philosophy, the 'source' is a destination in itself. It symbolizes the ideal point of origin and appears to be the fountainhead of historical truth, or at least the most reliable witness to the originator's proper intention. But was that what the Arabic sources symbolized for Albert? And what do they symbolize for us in this book?

Albert himself did not and could not walk the path *ad fontes* in our sense of the term. Unlike us, he was not heir to the Renaissance, the theology of Martin Luther, the Romantic critical historiography of Johann Gustav Droysen, or the objective historiography of Leopold von Ranke. Albert saw *fontes* as things in nature: springs, the heart, or intellectual material. The Arabic Peripatetic 'sources', as we call them, were, in Albert's eyes, material to be read, interpreted, and used mostly on an equal footing with the texts of Aristotle. He saw their usefulness as dependent on the context:

> It should be known as a consequence that Augustine ought to be trusted to a greater extent than the philosophers in matters concerning faith and morals, if there is disagreement. As far as medicine is concerned, however, I would trust Galen or Hippocrates to a greater extent, and speaking of the natures of things, I trust Aristotle more or another who is experienced in the natures of things.[3]

The other experienced natural philosophers besides Aristotle whom Albert trusted regarding 'the natures of things' are easily identified as philosophers hailing from Arabic-speaking lands — and Aristotle himself was known to Albert in part through the mediation of the Arabic-speaking Peripatetic philosophers. As will become clear in the contributions presented in this volume, the most important of these thinkers for Albert's purposes were Avicenna and Averroes, followed by Alfarabi, Algazel, Avempace, and Maimonides.

At the time when Albert was completing his early anthropological treatise *De homine* (1240–42), he was teaching on the *Sentences*, and in these lectures directly referred to more than a dozen Arabic-speaking figures — not only

---

2 The books currently available in English on Albert are Resnick and Kitchell, *Albertus Magnus and the World of Nature*; Blankenhorn, *The Mystery of Union with God*; O'Meara, *Albert the Great*; Resnick, *A Companion to Albert the Great*; Vost, *St Albert the Great*; Cunningham, *Reclaiming Moral Agency*; Bonin, *Creation as Emanation*; Weisheipl, *Albertus Magnus and the Sciences*; and see the special issue Wallace, 'Albertus Magnus'.

3 Albertus Magnus, *Commentarii in II Sententiarum*, d. 13C, a. 2, ed. by Borgnet, p. 247a: 'Unde sciendum, quod Augustino in his quae sunt de fide et moribus plusquam Philosophis credendum est, si dissentiunt. Sed si de medicina loqueretur, plus ego crederem Galeno, vel Hipocrati: et si de naturis rerum loquatur, credo Aristoteli plus vel alii experto in rerum naturis'.

Peripatetic philosophers, but also thinkers best known as experts in medicine, astronomy, or mathematics — whose works were available to him in Latin: Alfraganus (al-Farghānī, d. after 861 CE), Alkindus/Alkindi (al-Kindī, d. 873), Iohannitius (Ḥunayn ibn Isḥāq, d. 873), Constabulus (Qusṭā ibn Lūqā, d. 912), Albategnius (al-Battānī, d. 929), Ysaac Iudaeus (Isḥāq ibn Sulaymān al-Isrāʾīlī, d. *c.* 955), Alfarabius/Alfarabi (al-Fārābī, d. 970), Avicenna (Ibn Sīnā, d. 1037), Algazel (al-Ghazālī, d. 1111), Avempace (Ibn Bājja, d. 1138, indirectly through Averroes), Alpetragius (al-Biṭrūjī, fl. 1185–92), Averroes (Ibn Rushd, d. 1198), Rabbi Moyses (Maimonides, Mūsā ibn Maymūn, d. 1204), and the anonymous *Liber de causis*. One Jewish thinker writing in Arabic is conspicuously absent from the sources woven into Albert's *De homine*: Avicebron (Ibn Gabirol, d. 1058) made his debut in Albert's commentary on the *Sentences*, Book I.[4]

Whether Albert counted some of the experts he mentions in the passage we have quoted under the category of medicine and others under that of natural philosophy can be determined only in part, on the basis of similar pronouncements later in his philosophical corpus.[5] However, the passage does indicate Albert's intellectual concern to select sources relevant for particular disciplines, and his considerable skill in discerning the specific expertise that each source offered. He notes in his *Metaphysica*:

> Therefore, let this be the end of this disputation [on Aristotle's *Metaphysica*], in which I have said nothing according to my own opinion, but everything said is in accordance with the positions of the Peripatetics. And whoever wishes to examine [*probare*] this, let them read their books diligently, and praise or blame not me, but them.[6]

For Albert, therefore, his Arabic sources were not points of destination. They were points of departure. They were authorities to be trusted in their value of

4 Albertus Magnus, *Super I librum Sententiarum*, XXV, d. 24, A, art. 2, ed. by Borgnet, p. 609a: 'Item, Philosophus in libro Fontis vitae: Primum quod recipit a primo, est recipiendo duo: quia recipiens et receptum: ergo cum unitas creata recipiat esse suum a primo, ipsa erit duo, et non unum'. For a recent discussion of Albert's use of Avicebron in his entire subsequent oeuvre, see Miteva, 'The Reception of Ibn Gabirol's *Fons vitae* in Albertus Magnus'.

5 Examples can be found at Albertus Magnus, *Super Ethica*, III.13, ed. by Kübel, p. 207, vv. 4–16: 'circa delectationes tactus in dictis duabus partibus corporis est temperantia, quia istae sunt validissimae et in eis maxime opus est principali virtute refrenante. Quare autem istae delectationes sint validissimae, causa potest assignari secundum naturalem et secundum theologum et secundum ethicum. Secundum naturalem, quia Avicenna et Constantinus dicunt, quod quia per huiusmodi partes coniunguntur in nobis ea quae sunt ad conservationem naturae in specie vel individuo, ideo posuit in eis natura maximum delectamentum, ut sollicitetur animal circa huiusmodi et non negligatur salus naturae'; Albertus Magnus, *Quaestiones super De animalibus*, XII.17, ed. by Filthaut, p. 231, v. 31: 'Isaac in Dietis'=Isaac Israeli, *De dietis universalibus*. See Jacquart, 'La place d'Isaac Israeli'.

6 Albertus Magnus, *Metaphysica*, I.13, tr. 2, c. 4, ed. by Geyer, p. 599, vv. 61–66: 'Hic igitur sit finis disputationis istius in qua non dixi aliquid secundum opinionem meam propriam, sed omnia dicta sunt secundum positiones Peripateticorum. Et qui hoc voluerit probare, diligenter legat libros eorum, et non me, sed illos laudet vel reprehendat'.

truth, but always within disciplinary limits. They were stewards of philosophical positions that he, with his own erudition and synthetic capacity, could bring into the Latin world. They proposed views that he voiced to his peers and students by balancing contradictory accounts and presenting them, more often than not, as a single Peripatetic voice. Truth, certainty, and comprehensiveness were the epistemic values that Albert cherished dearly, and the Peripatetic positions helped him to put these values into practice.

Albert's discourse on these positions took place in Paris in 1242, in Cologne in 1252, in Orvieto and Viterbo in 1261, and in Würzburg in 1264, to name only some of the many locations and periods where he worked. He did not pursue that discourse — as we mostly do today — as a comparison or dispute between two parchments, of which the 'source' parchment presents an idea's point of origin and his own parchment records it. Albert debated with his sources not by reporting, representing, or preserving their content to the letter, but by conveying its meaning afresh in his own times, in harmony with other sources, replacing the errors, faults, and blunders he spotted, and adding new information or even new books to the corpus so as to achieve comprehensive truth and certainty as goals of his *scientia*.

The metaphor of the *source*, we suggest in this volume, stands not for the correct transmission of information alone, but equally for a 'loss of continuity as the emerging current meets and traverses the terrain', as Christopher Wood has aptly noted. 'The uneven, ramifying flow of water symbolizes the relaying of messages forward in time. The liberated water seeks level ground, forms channels, splits into streams'.[7] What, then, were the messages that Albert meant to relay into the future as his Arabic sources met and traversed the terrain of the Latin medieval world?

## The Intrinsic Value of Philosophy

Our book follows the current of Albert's scientific creativity from the early 1240s to the late 1260s and asks how he drew on the Arabic sources he had at hand at any given time. In twelve detailed case studies, it investigates how Albert tackled particular research questions within the philosophical programme that he built up over those years in Paris, Cologne, Worms, Agnani, Regensburg, Viterbo, Orvieto, Würzburg, and Strasbourg.

We take this chronological approach to our book because we view the unfolding of Albert's *scientia* as the *explanans* of the ways in which he used his Arabic sources, and not vice versa.[8] Albert chose how to read, what to select, and which

7 Wood, 'Source and Trace', p. 6.

8 The predominant model of interpretation in the history of philosophy does the opposite of our approach and makes the historical source texts of any given (medieval) thinker the *explanans* of their philosophical argumentation and knowledge. There are many epistemic problems with this model of

way to order these sources in light of the views on *scientia* that he himself held. This autonomous engagement with his sources, we argue, stabilized the content of the body of knowledge found in the sources in many meaningful ways. In other ways, however, it changed that content to accord with the 'images of knowledge' that were dominant in Albert's Latin context.[9]

Images of knowledge, as the historian of science and philosophy Yehuda Elkana explained, are 'beliefs held about the task of science'. They depend on the time and culture of the historical actor at stake and determine which problems are chosen for study.[10] This volume reads Albert's use of his Arabic sources as a way of harnessing concepts, discourses, and bodies of knowledge to the purposes of his own philosophy. Without the sources as material and instrumental ingredients, he could never have realized or even pursued those purposes. In no case, as our studies show, was Albert's use of the Arabic sources a simple 'reconstruction' of the material[11] — neither was this the intention of his scientific practices, even though he makes claims that look like it, as we saw in the *Metaphysica* passage above. Rather, his philosophy was a unique composition that he shaped out of an inherited body of knowledge in his own engagement with novel interpretations or doctrines not always in accord with the doctrines of his sources.

Amidst the variety of this book's contributions, several themes related to inheritance — and its consequence, emergence — recur, but one in particular stands out. This is Albert's self-imposed mission of asserting the role of philosophical *scientia* as an intrinsically valuable activity in the Latin world of the Scholastic academy. Philosophical *scientia*, he proposes, has value in enabling the search for *proximate* causes (instead of remote ones) and, through that search, the perfection of one's own human nature: *homo inquantum homo solus intellectus*.[12]

This is the primary context in which we place Albert's use of his Arabic sources, and what we take to be the *explanans* of the ways in which Albert gave these sources new epistemic meanings, identities, and roles. It implies, too, that Albert was clearly moving in a different direction from the efforts of some of his Latin contemporaries — among them towering figures such as Bonaventure of Bagnoregio, Roger Bacon, and Thomas Aquinas — whose interests lay, each in his

---

interpretation, but the major one is that it ascribes far more agency to texts than to people. The model also subscribes to an impoverished and reductionist causal history, entailing that the whole meaning an author inscribes into a philosophical text is *either* already contained in his source texts *or* presents us with novel ideas — ideas that we, as historians of philosophy, can then excavate from the text in front of us. Yet historical texts are not simply presentations of ideas; they are representations of the lived worlds of located people. It is these lived worlds — those of the source texts and those of the recipients — that we wish, at least in some of their facets, to investigate in our book.

9 The notion of change in line with new images of knowledge is Yehuda Elkana's. Elkana, 'A Programmatic Attempt at an Anthropology of Knowledge'.

10 Ibid.

11 Canguilhem, *Ideology and Rationality in the Life Sciences*, p. 2.

12 This finds its origin in Aristotle, *Nicomachean Ethics* IX.4, 1166a17. For Albert, however, this axiom of the human final causality takes centre stage in his natural philosophy as well. See also below.

own idiosyncratic way, in establishing theology as the ultimately decisive *scientia*. Their reductionist models did not impress Albert, as is evidenced in the writings he produced after the mid-1240s. His unique vision was to give a place, indeed intellectual freedom, to the rationality that he discovered in philosophy and the robustness of the logical, epistemological, and psychological foundations upon which it rested.

This idea of pursuing philosophical *scientia gratia scientiae* — as one may call Albert's will to liberation here — did not occur to him overnight. It dawned on him gradually, in a hard-earned process of intellectual labour through which he simultaneously acquired comprehensive knowledge of his sources and ordered it in his own ways. This labour began in the 1240s, when Albert naturalized Arabic-Peripatetic anthropology in the second part of his *Summa de creaturis*, the *De homine*. It came to its first autonomous fruition in his first commentary on the *Ethica Nicomachea*, the *Super Ethica*, written around 1250–52. And it culminated in the conclusion of his scientific system with a commentary on the *Politics*, written soon after 1264.

It is striking that it was the 'practical sciences' (*philosophia moralis*) — ethics and politics — which historically framed Albert's erection of his very own edifice of philosophical *scientia*. The practical sciences may have offered Albert a call to human action in the emergent medieval cities. For Albert, however, such action meant predominantly an action performed on the self — a self that, despite usually standing within the civic sphere and law, was nonetheless called to perfection.[13] This is because, in line with Aristotelian ideals, Albert regarded philosophical ethics as a prerequisite for politics for the man of education. Accordingly, we find both chronological and substantive overlap between Albert's writing on practical sciences and his composition of a full-fledged philosophical *scientia*, his assimilation of all philosophical knowledge available at the time, and, most importantly, the development of his very own intellect in the process. That third endeavour was Albert's instantiation of the ethical programme he found in his Arabic sources.[14]

This self-perfective aspect of Albert's scientific activity is momentous. The philosophical system he set out was not meant to be an objective, detached, self-standing one that could then be studied in the same way — objectively, neutrally, and independently of the scientist as its subject. Quite the contrary: his philosophical system was what made Albert, and all who followed his path, human in the full sense, as *homo inquantum homo solus intellectus*. It is only within this subject-centred perspective (which is not to be confused with the contemporary curtailment of 'subjectivity') that Albert's philosophy as a whole and his use of the Arabic sources can be understood.[15]

---

13 See, for instance, Cassirer, *The Philosophy of the Enlightenment*.

14 Krause and Anzulewicz, 'Albert the Great's *Interpretatio*'.

15 On this matter, see also Krause, 'Albertus Magnus zur Philosophie und Theologie'. These thoughts are developed in more detail in Krause's forthcoming book *Albert the Great*.

In Albert's sights, we argue, was the search for a philosophical *scientia* in service of the human scientist who studied it, and that system had itself to follow certain ideals. Albert's programme was saturated with Aristotle's ideal of truth, the ideal of certainty he inherited from the Latin translations of Aristotle's works, and the ideal of human intellectual perfection obtainable in this life (through the comprehensive study of philosophical *scientia*) and in the afterlife (through contemplating separate substances) that was transmitted to him from his Arabic Peripatetic sources.[16] In addition to the specific thematic legacies that are charted in each chapter of this book, it was these three larger ideals that guided, directed, and focused Albert's practices of engaging with his sources. No less than Albert's programme-building practices, those ideals have their own history in his intellectual and Scholastic activity.

The challenge that Albert explicitly issued was to erect a new scientific programme built on a philosophical procedure rather than a theological one.[17] This was a programme that would, in Albert's idealizing view of his intended audience, suit the specific needs of the thirteenth-century Latin world.[18] His way of meeting his own challenge was to utilize the transmitted ideas of his Peripatetic sources.

In reality, however, Albert's construction of a new scientific programme was no mere utilization. It was an unprecedented, originative, and deliberate response to multiple inheritances from ancient Greek *sophia*, *technē*, and *epistēmē* and their counterparts in the Arabic language. It revolutionized the Latin practice of *scientia*. The case studies in this volume offer a magnifying glass through which to discern the meticulousness, the colossal memory, and the acquaintance with the inherited knowledge that Albert brought to the task of constructing his philosophical programme. By focusing on particular doctrines that he developed in conversation with his Arabic sources, we can identify the building blocks of Albert's scientific practice and programme.

Still, there is a more fundamental question that needs to be asked before we can begin to grasp Albert's philosophical *scientia*: the question of the historical conditions. How did Albert distance himself from his own Latin tradition by legitimizing philosophical *scientia*'s independence at the outset of his scholarly career?

---

16 On the afterlife, see Albertus Magnus, *De natura et origine animae*, tr. 2, cc. 13–17, ed. by Geyer, p. 37, v. 61–p. 44, v. 23.

17 See Krause and Wietecha, 'Albert the Great on Negative-Mystical Theology'.

18 Albertus Magnus, *Physica*, I.1.1, ed. by Hossfeld, p. 1, vv. 9–22: 'Intentio nostra in scientia naturali est satisfacere pro nostra possibilitate fratribus ordinis nostri nos rogantibus ex pluribus iam praecedentibus annis, ut talem librum de physicis eis componeremus, in quo et scientiam naturalem perfectam haberent et ex quo libros Aristotelis competenter intelligere possent. Ad quod opus licet non sufficientes nos reputemus, tamen precibus fratrum deesse non valentes, postquam multotiens abnuimus, tandem annuimus et suscepimus devicti precibus aliquorum ad laudem primo dei omnipotentis, qui fons est sapientiae et naturae sator et institutor et rector, et ad utilitatem fratrum et per consequens omnium in eo legentium et desiderantium adipisci scientiam naturalem'.

Let us briefly take a step back from Albert to contemplate, at least in extremely broad contours, the wider backdrop of his system. Across disciplines as varied as theology, medicine, and philosophy, thirteenth-century Europe saw an unprecedented rise in ancient Greek and Arabic Peripatetic forms of knowledge at its newly founded educational institutions in Bologna, Paris, Oxford, Cambridge, and elsewhere.[19] Yet for the first forty years or so of that century, few figures had the erudition required to take on the challenge of scientific reform, rearrange the curricula of the arts faculties in accordance with the new forms of knowledge, and propose a practical systematization to guide those students bright enough to embark on the most complex intellectual activities being undertaken at the time.

The new legacies inherited from the Arabic-speaking sphere were thus initially slotted, by default, into the older curricula of the *artes liberales* and theological *sapientia*, as James Weisheipl, for instance, has eloquently argued.[20] Moreover, William of Auxerre, Philip the Chancellor, William of Auvergne, Alexander of Hales, Jean de la Rochelle, Odo Rigaldus, and William of Middleton — to name just a few leading theologians — all integrated Aristotelian concepts and discourses into their existing theological questions and answers.[21]

This certainly led to a more robust defence of theological doctrines, but it also occasioned some perturbance and in some cases even chaos within what had been a well-organized network of concepts and premises. Any 'newly discovered component brings disquiet, if not disorder, into the entire system', as Yehuda Elkana put it, 'if one has not already developed a new architecture into which it fits'.[22] Ultimately, the more new components, the more perturbance, and the more need for a new architecture. This, we believe, is one reason why the route of integration was not the one that Albert the Great took. His was a path of naturalization, of taking the epistemic commitments entailed in his Peripatetic sources seriously and following them through.[23] He thus elevated the pursuit of philosophical *scientia* from something purely instrumental to something with a truly intrinsic value.[24] In so doing, Albert greatly impacted Latin education, research, and institutions for at least four centuries to come. But how did he reach this point of naturalizing philosophy as something with its very own intrinsic value?

---

19 See, for instance, Brungs, Mudroch, and Schulthess, 'Institutionelle Voraussetzungen'.

20 Weisheipl, 'Classification of the Sciences in Medieval Thought'.

21 See, for instance, Suarez-Nani, 'Die theologische Fakultät Paris in der ersten Hälfte des Jahrhunderts', with references to secondary literature on pp. 619–22; Putallaz, 'Die Ersten Franziskaner', with references to secondary literature on pp. 622–27.

22 Elkana, *Leben in Kontexten/Life in Contexts*, p. 81.

23 For the particular use of the concept 'naturalization', see Sabra, 'The Appropriation and Subsequent Naturalization of Greek Science in Medieval Islam'.

24 See Honnefelder, 'Einleitung', p. 21.

## Modalizing *Scientia*

Our answer to that question in this volume turns on the development of Albert's work before his *Super Ethica* and the 'theological rationality', as we wish to call it, that he developed there. We cannot detail this entire development in our short introduction, both for reasons of space and because much of it has been lost to history. We do, however, wish to pinpoint some aspects that we deem most important.

Awareness of the immediate 'historical conditions *under* which' and the closest 'means *with* which' Albert naturalized philosophy's intrinsic value in his own way is the most important point in this respect,[25] because it is these that contextualize the genesis of Albert's system by way of its proximate history. They contextualize not by shedding light on the material, instrumental, or even accidental components of the early body of knowledge that Albert penned, but by shedding light on the *formal* source of that body of knowledge: Albert's own scientific practices as they arose in the space and time he inhabited.

As a result, these conditions and means *explain* how the autonomy of Albert's philosophical *scientia* took shape. Such a historization of epistemology, as historian of science Hans-Jörg Rheinberger has written, implies subjecting any 'theory of knowledge to an empirical-historical regime, grasping its object as itself historically variable, not based in some transcendental presupposition or a priori norm'.[26] What better place to start, then, than by subjecting Albert's evolving view of *scientia* to the regime of historical epistemology and following its transformation into an autonomous philosophical programme.

If we had to pick out the most critical moment in the scientific activities of Albert, it is surely the composition of his commentary on Peter Lombard's *Sentences* (1242–45). Insightful scholarship has long regarded this commentary as the main witness to a new beginning — a beginning at which Albert conventionalizes *sapientia cum affectu* as the modality of his *scientia theologiae*.[27] Some of its inspiration derived from structural considerations concerning the divisions of *scientiae*. Albert knew very well that an all-encompassing *scientia*, as theology was for him, could not be separated from the *scientia* of philosophy by divorcing particular aspects in subject matter from others. If the subject matter failed to be all-encompassing, then so did the *scientia* of theology.

But Albert also knew, most likely from his acquaintance with the translation of Avicenna's *Summa de convenientia et differentia subiectorum* in Dominicus Gundissalinus's *De divisione philosophiae*, that two different *scientiae* could also be separated by distinguishing between the different aspects entailed by overlapping

25 Thus the fundamental historiographical questions proposed by Hans-Jörg Rheinberger. Rheinberger, *On Historicizing Epistemology*, p. 2.

26 Ibid., p. 3.

27 See Anzulewicz, 'The Systematic Theology of Albert the Great'.

subject matters.[28] Medicine and natural philosophy, in the example given by the Persian philosopher Avicenna (Ibn Sīnā, 980–1037), overlap in subject matter, but each studies that subject matter under a different aspect: medicine under the aspect of health and disease, natural philosophy under the aspect of nature or essence.[29]

Albert's pronouncement that theology's all-encompassing subject matter is to be studied under the aspect (*secundum rationem determinatam*) of the final causality of the beatifying end (*finis beatificans*), which he considers to be operant in theology, looks like a similar move. Whatever is studied in theology is studied under the aspect of this final causality of the beatifying end. God as the objective *finis beatificans* thus belongs to theology most intimately as its subject matter, but created reality and human actions also do so insofar they are conducive to attaining that end.

Albert's effortless combination of theology's universal subject matter under the aspect of its teleology also reflected back on the type of *scientia* that theology was for him. The eternal end of theology, viewed as an objective and a subjective end, could not be a merely speculative truth; it had to be an affective truth (*veritas affectiva beatificans*). The union reached with God as objective, eternal end and good bestowed delight on the saint as his subjective, eternal end and good, and as a direct result of his beatific knowledge. For Albert, the object of theology — truth in accordance with piety (*veritas secundum pietatem*) — was thus inseparable from truth and the good. Unlike philosophy, therefore, theology had the potential to perfect both the human intellect and human affect, in a unified fashion. In that sense, it was God-like, and it led to this end by its very own principles of faith and meritorious actions.

By proposing this holistic and unifying modality of theology, Albert also accomplished yet another clear distinction between theology and philosophy. He began to distinguish the different rationalities of philosophy and theology more clearly — a project he brought to its initial climax when he lectured at the new Dominican *studium generale* in Cologne between 1248 and 1254.

It was there, in front of students including Thomas Aquinas, Giles of Lessines, and Ulrich of Strasbourg, that Albert realized the limitations of systematic theology for pursuing the end of the *scientia* of theology. Systematic theology, worked out as a branch of theology in the *Sentences*, is capable only of defending faith argumentatively and establishing its truths exhortatively.[30] Negative theology,

---

28 Dominicus Gundissalinus, *De divisione philosophiae*, ed. and trans. Fidora and Werner, pp. 237–52. See also Janssens, 'Le *De Divisione philosophiae* de Gundissalinus', especially pp. 561–62.

29 In other words, the physician adds a per se accident to body as the more general subject matter of natural philosophy. See Strobino, *Avicenna's Theory of Science*, pp. 114–18.

30 Albertus Magnus, *Super I librum Sententiarum*, d. 1, c. 1, ed. by Borgnet, p. 15, vv. 23–42: 'Quod concedimus dicentes quod habitus eius lumen fidei est. Instrumentum autem duplex secundum duplicem finem doctrinae et artis, qui duplex finis praemissus est in auctoritate Apostoli: scilicet exhortari in doctrina sana et contradicentes revincere. Et quoad exhortationem habet quadruplicem expositionem: scilicet historialem, allegoricam, moralem et anagogiam, quorum modorum numerus

which Albert found in Dionysius's works and especially his *Mystical Theology*, is the one branch of theology that proved potent enough to achieve theology's end: union with the objective *finis beatificans* to the extent that it is possible in this life. Albert's view of the *corpus Dionysiacum*'s importance for the modality of his *scientia* of theology highlighted the value of Peter Lombard's *Sentences* and the *corpus Dionysiacum* as two different branches of theology, while also establishing an innovative relationship between them. The former branch follows modes of reasoning proper to humans; the latter transcends these in an intellectual movement that reaches out to its primary object, God, beyond creaturely ways of knowing.

Theology therefore now included two ways of knowing one and the same object under the aspect of its finality: *scientia affectiva* and *scientia mystica* or *scientia experimentalis*.[31] The two branches constituted two sides of the one coin of *scientia theologiae*, because they both pursued their subject matter teleologically. Biblical theology, as the third branch, entered the picture later on, when Albert began to write his commentary on the Gospel of Matthew in 1257–64. We will not discuss any of these intricate matters further here.

What all these considerations on Albert's *scientia* of theology tell us is that his intrinsic valorization of philosophy as its own *scientia* began in and with his thinking on theology as its own *scientia* — in comprehensively developing theological rationality before philosophical rationality. The moment of reflection that Albert integrated at the outset of his *Sentences* to explain what theology as a *scientia* is, how precisely it produces its knowledge, and what it aims for in producing knowledge — namely, the union of subject and object — became an immovable standard. None of Albert's subsequent works, whatever generic *scientia* they belonged to, could do without this moment of reflection.

As a consequence, when Albert applied similar reflections to philosophy for the first time — his first commentary on Aristotle's *Ethica Nicomachea*, composed in the Cologne classroom between 1250 and 1252 — there was no doubt that they spoke to the theological modality he had previously described. Albert's reflections demanded that philosophy be erected on its own, independent scientific norms, as its own, independent *scientia speculativa*. But how exactly did Albert express this independence in practice?

---

dupliciter potest accipi: scilicet quoad exponentem et quoad exposita; quoad exponentem sic: primo occurrit sensus ostendens historiam, et ideo historicus sensus est in intellectu, secundum quod refertur ad sensum. Circumstant autem adhuc tria intellectum: scilicet habitus illuminans, qui est fides, et sic in ipso est allegoricus sensus, qui aedificat fidem, sicut dicit Gregorius. Circumstat etiam ipsum intellectus practicus, et sic in ipso per reflexionem ad praxim sive opus est sensus moralis. Tertium quod circumstat ipsum est finis beatificans, et sic in ipso per conversionem ad ipsum est sensus anagogicus. Cum autem non plura cicumstant intellectum, non sunt plures sensus scripturae'.

31 See, for instance, Anzulewicz, 'The Systematic Theology of Albert the Great'.

## The Tools to Discipline Philosophy

Entire books could be written to answer that question; here, we raise three points that we consider central, and that allow us to appreciate the all-important relationship between the historical conditions under which the autonomy of philosophical *scientia* arose and the precise epistemic role the Arabic sources played in shaping that autonomy.

First, like his theology, Albert's philosophy was a *scientia* that was split into branches. Most generically, it was divided into a rational (*scientia rationalis*), a real (*scientia realis*), and a practical branch (*scientia moralis*). On the next level, real philosophy was divided into natural philosophy, metaphysics, and mathematics; in turn, natural philosophy was divided into altogether twenty-two different branches.[32]

In contrast to theology, where the divisions were made possible by different ways of knowing, philosophy was divided by subject matters of a particular genus of being, for instance, substances that move and change (as in the *Physics*). This is an approach that Aristotle had proposed in *Analytica Posteriora* I.28, and Albert followed his lead. At the beginning of his *Physica* and *Meteora* commentaries, especially, Albert goes into great detail as he divides the general scientific subject matter of natural philosophy — substances that move and change — into its varied branches.[33]

Although the natural scientific programme Albert thus composed was framed, at least in time, by two commentaries on authentic works of Aristotle, it added to these considerably. Albert penned no fewer than eight additional, completely autonomous works, encompassing matters that ranged from geographical location to the nature and origin of the soul: *De natura loci* (1251–54), *De nutrimento* (1256), *Liber de motibus animalium* (1256), *De spiritu et respiratione* (1256), *De aetate* (1256), *De morte et vita* (1256), *De intellectu et intelligibili* (1256), and *De natura et origine animae* (1263). He added commentaries on the pseudo-Aristotelian works *De causis proprietatum elementorum et planetorum* (1251–54), *Libri de mineralibus* (1254–57), and *De plantis* (1256).[34] Last but not least, he considerably expanded upon Aristotle's natural philosophical works, and in particular on *De animalibus*, on which he wrote a long commentary (the first commentary he wrote in Cologne before teaching the *De animalibus* in the form of *quaestiones* there between 1258 and 1263).

Albert certainly saw the chief value of these additional works and considerations as lying in their capacity to complement his own natural scientific programme,[35] which was the first of its kind in the Latin West and endured in

32 Krause and Anzulewicz, 'Albert the Great's *Interpretatio*', especially pp. 118–21.

33 See Albertus Magnus, *Physica*, I.1.4, ed. by Hossfeld, p. 6, v. 34–p. 8, v. 13.

34 See Krause and Anzulewicz, 'Albert the Great's *Interpretatio*', appendix 1.

35 Albertus Magnus, *De somno et vigilia*, I.1.1, ed. by Borgnet, p. 123a, and III.1.1, p. 178a: 'Et hoc ipsum quidem quod de divinatione dicit Aristoteles, breve quidem est et imperfectum, et habens

its uniquely comprehensive scope well into the sixteenth century. Indeed, what Albert took to hold together the generic branch of real philosophy as a whole, including its largest branch of natural philosophy, was its aim as a *scientia*: its aim of explaining reality as it is in itself.

This aim could be achieved only by correctly applying permissible instruments, and this is the second point we wish to highlight. For Albert — once again in reliance on Aristotle's logic — those instruments consisted in definition, which he believed could capture the essences of existing things, and syllogistic reasoning, which he believed could account for the proximate causes of those things. Albert saw possible further instruments for accomplishing the aim of accounting for reality as it is in itself: analogies and signs were useful descriptive tools, though they ranked less highly on the epistemic ladder of bringing forth true and certain knowledge.[36] Albert's panoply of philosophical instruments and their aims thus differed considerably from those of theology, where practices of exhortation and negation were built on the habit of faith (each, of course, in its own way) and ultimately aimed at union with God.

These instruments of reason — describing, defining, explaining — could only fulfil their aim of accounting for reality as such if they had some material to

plurimas dubitationes. Dico autem *breve*, quia carens probatione, sed simplex, et parum philosophiae habens videtur esse narratio, nec species somniorum neque probationem somnii aliquid significandi in se continens. *Imperfectum* autem est, quoniam licet sine magicis et astronomicis non possit ars interpretandi somnia adipisci, tamen solis physicis sufficienter scitur ex quibus et qualibus simulacris consistit somnium de quo debet esse divinatio: et hoc neque ab Aristotele, neque a Philosophis quidquam determinatum est. *Plurimas autem dubitationes habet*, quia in incerto relinquitur causa talium somniorum'.

36 See, for instance, Albertus Magnus, *Analytica posteriora*, I.5.1, ed. by Borgnet, p. 128a: 'Tertia autem ratio, quod potior est demonstratio, propter quam non errabitur de quo fiat demonstratio, hoc est, propter quam non dubitabitur de quo demonstretur, quam illa propter quam errabitur et dubitabitur de quo demonstretur. Est autem universalis demonstratio (secundum quod maxime universalis est) hujusmodi quae facit errare et dubitare de quo demonstretur: cujus probatio est, quia demonstratores procedentes ad magis universale, quod est analogum sive secundum analogiam commune multis, sicut proportionale quod commutabiliter est in numero, tempore, linea, solido, et plano, demonstrant de ipso secundum quod est aliquid in se praeter haec: unde neque in linea est secundum quod linea, neque in numero secundum quod numerus, neque est solidum secundum quod solidum, neque planum secundum quod planum, sed secundum quod est aliquid praeter haec. Si igitur haec demonstratio est magis universalis inter demonstrationes, et est de eo quod minus est quam particularis, et sic facit opinionem falsam, quod hoc scilicet quod minus est, magis sit quam id quod magis sit: indignior utique erit demonstratio universalis, quam particularis'; ibid., I.2.1, ed. by Borgnet, p. 21b–22a: 'Probatur tamen per signum, quod scire verum hujusmodi sit aliquid quale jam dictum est: et hoc est signum: quia quoties vere scientes et non vere scientes opinantur se scire (cum id quod omnes opinantur sit probabile) illud est signum, quod sit verum scire hoc modo qui dictus est, cum omnes etiam non vere scientes opinantur scire quando habent causam et sciunt quod illius causa est, et quod non contingit illud se aliter habere: ergo hoc est verum scire. Sed differentia est: quia scientes tunc arbitrantur se scire et vere sciunt: non scientes autem propter similitudinem quam habent cum scientibus opinantur se hoc modo vere scire, quamvis non vere sciant. Propter hoc omnis ejus cujus est scientia (quae simpliciter et per se est scientia) hoc est impossibile aliter se habere quam dictum est'.

work with and on. For Albert, that material consisted in the concrete scientific concepts, arguments, analogies, and signs he found in the source texts on the desk in front of him. These scientific objects were nothing other than cognitive abstractions from the things in the world. However, the abstractions were made not by Albert himself, or at least not by him alone, but by different thinkers in the long tradition of Aristotelian philosophy. Those thinkers' mental operations of imagination and thought together produced these abstractions — which now challenged Albert to order them anew, compare and contrast them, analyse and synthesize them, using his own reason in order to establish certain and comprehensive knowledge about them.

Just as he did with Aristotle's works, then, Albert viewed his Arabic sources as intellectual material that contained truths and errors, correct definitions and false ones, accurate demonstrations and flawed ones, proper inductions and improper ones, good analogies and bad analogies. He treated and used this material accordingly. Admittedly, he had clear preferences, as we saw in the passage quoted earlier in this introduction.[37] But especially in those areas of philosophy where advantages could not be made out on the basis of the expertise that his authorities possessed in either theology, medicine, or philosophy, Albert sifted through his sources with the single, lofty aim of establishing the truth about things in reality and reality as a whole in itself. The goal before Albert's eyes, in other words, was nothing less than to comprehensively institute the equivalence of the human mind with the world. That equivalence was mediated in part through his sources.

The criterion of comprehensiveness is evidenced by the way that Albert went about building his own philosophical programme as a whole; this is our third and final point. Alongside his concerns about the micro-level of particular themes in the body of knowledge he produced, Albert expanded the Aristotelian corpus considerably by composing eight books in natural philosophy alone, as discussed above. Importantly, some of these took their inspiration in content, layout, and even approach from independent Arabic texts in Latin translation. A good example is Albert's *De intellectu et intelligibili* (1256), which shows the impact of Alexander of Aphrodisias, Alfarabi, and Alkindi. Nevertheless, here too, Albert gave the works in his philosophical system their own identity and place, for example by counting his *De intellectu* among the works of natural philosophy rather than metaphysics.[38]

37 Albertus Magnus, *Commentarii in II Sententiarum*, d. 13C, a. 2, ed. by Borgnet, p. 247a, as quoted in note 3.

38 Albertus Magnus, *Physica*, I.1.4, ed. by Hossfeld, p. 7, v. 8–64: 'Sed scientia de animatis habet duas partes. Cum enim anima sit principium animatorum et principium oporteat cognoscere ante principiatum, oportet haberi scientiam de anima, antequam habeatur scientia de corporibus animatis. Scientia autem de anima duas necessario habet partitiones, quoniam aut est de ipsa anima et potentiis sive partibus eius aut scientia de operibus animae, quaecumque habet in corpore, et de passionibus eius, quas patitur in corpore, et scientia quidem de anima secundum se et potentias eius habet tradi in libris de anima dictis. Opera autem eius duplicia sunt, quia aut sunt animae in corpus, ita quod non per potentias, sed per se operatur anima, aut operator secundum potentias.

Perhaps the most distinctive mark that Albert left on philosophy in general was a superstructure he inherited from, but also created out of, his Arabic sources: the notion that the *scientia* of philosophy, which primarily tries to get to grips with the objects of science as they are in reality, also pursues another, secondary goal. This secondary goal is not to be mistaken for a goal intrinsic to the *scientia* of philosophy. Rather, it is extrinsic to the *scientia* of philosophy, but nonetheless useful in affecting the scientist, in the form of the realization of the scientist's human nature through his own pursuit of philosophy, the scientist's perfection as *homo inquantum homo solus intellectus* once all of that knowledge has been incorporated.

None of Albert's unique applications and arrangements of his tools to build an independent programme of philosophy is reducible to any of his predecessors, even if individual elements can be traced back to them. His achievement was to endow the ideas, views, and elements of knowledge he found in his sources with a new form of intellectual being — a new *forma totius* — that was fully present in his mind and has remained partially accessible to us in the extant philosophical works to this day.

The twelve chapters of this volume trace, in detail and mostly at the micro- and meso-levels of specific doctrines in single texts or groups of books, how Albert accomplished the momentous autonomy and comprehensiveness of his own philosophical programme. They examine in even greater detail the mechanisms that he employed to appropriate the many Arabic insights about reality into his philosophical *scientia*. Through the book's chronological structure, the chapters

---

Et opus animae quidem per substantiam animae factum in corpore est vita, cui mors opponitur. Et hoc opus determinatur in libro de causa vitae et mortis et causis longioris vitae. Opera vero animae alia sunt multiplicata secundum potentias vegetabilis, sensibilis et intellectualis animae partis, et opera quidem vegetabilis sunt nutrire et augere et generare. Sed duo illorum sufficienter determinantur in libro de generatione, scilicet generatio et augmentum. Tertium autem in genere habet determinari in libro de nutrimento. Opera vero sensibilis duo sunt in genere, scilicet sentire et movere secundum locum. Opus autem sensibilis secundum sentire tripliciter variatur; aut enim accipitur secundum comparationem sensus ad animal, scilicet secundum quod sensus egreditur vel ingreditur in animal, vel secundum comparationem sensus ad sensibile aut secundum reditum ex specie sensibili servata apud animam in rem prius acceptam in sensu. Et primum horum trium quidem in libris de somno et vigilia traditur, secundum autem in libro de sensu et sensato, tertium autem in libro de memoria et reminiscentia. Secundum autem quod motiva est anima sensibilis, dupliciter movet, scilicet secundum locum, aut mutando locum aut dilatando et constringendo corpus in eodem loco, et utrumque horum traditur in libro de motibus animalium. Hic autem motus est generalis omnibus animalibus sub disiunctione acceptis, quoniam omne animal aut movetur motu processivo aut dilatationis et constrictionis motu aut utroque. Facit autem specialem motum in habentibus pulmonem, qui est ad refrigerium pectoris per spiritum attractum, quem movet trahendo et retinendo et emittendo, et huius scientia traditur in libro de respiratione et inspiratione. Et ad adminiculum eius est liber Costa-ben-Lucae, quem composuit de differentia spiritus et animae. Opus autem animae secundum partem intellectualem tractatur in scientia subtili de intellectu et intelligibili. Quibus habitis sufficit addere scientiam de corpore animato vegetabili et sensibili, cuius differentiae quoad vegetabilia traduntur in libris de vegetabilibus, et quoad differentias animalium traditur scientia sufficiens in libris de animalibus. Et ille liber est finis scientiae naturalis'.

mirror the current of Albert's intellectual life and the history of his reliance on his Arabic sources when composing his philosophical programme.

Our book is framed by two contributions — by Jorge Uscatescu Barrón and David Twetten — that bracket the whole range of Albert's scholarly activity from the nascent stage of his *scientia,* in *De natura boni,* to the matured and fully independent development of the two *scientiae* of philosophy and theology, in *De causis* and *Summa theologiae* respectively. The papers between these two, with few exceptions, concentrate on a single work by Albert and select themes in whose development the Arabic sources played the leading part.

## Between Inheritance and Emergence

There is no doubt that many notions recur through Albert the Great's corpus from its very beginning to its very end. One is that of the good, to which **Jorge Uscatescu Barrón** devotes his attention in a study diachronically covering the two disciplinary realms of philosophical ethics and theology. Beginning with Albert's earliest work, *De natura boni,* and ending with one of his last, the *Summa theologiae,* Uscatescu Barrón examines the different meanings that Albert ascribes to the good against the background of selected philosophical and theological sources. Initially, Albert relies on an ontologically motivated definition of the good, which he himself traces back to Avicenna but which in fact, as Uscatescu Barrón points out, he borrows from Philip the Chancellor's *Summa de bono* — a source that profoundly influenced his early writings. Yet Albert did not shy away from adapting his loans in accordance with his own teachings. Those teachings, argues Uscatescu Barrón, became increasingly oriented on Aristotle's teleological understanding of the good and were the reason that Albert ultimately disfavoured Avicenna's definition of the good. Nonetheless, for Albert disfavouring one particular definition (as derived from one source) compared to another (as derived from another source) did not imply its rejection. As Uscatescu Barrón shows, Albert's interpretive resourcefulness was seemingly limitless: restricting the application of the Avicennian notion of the good to its teleological goal ultimately led Albert to rescue it in his *Summa theologiae* and to combine it once again with new meanings, which he now derived from an even greater pool of sources including Boethius, Algazel, and the *Liber de causis.*

In his chapter, Jorge Uscatescu Barrón reveals the historical depth of Albert's appropriations in their dependence on a variety of internal textual factors. The sources that Albert chooses for specifying the meanings of the good at any given moment change substantially with the disciplinary, and even the particular argumentative, contexts in which they are embedded. In the discipline of philosophical ethics, theological sources and their meanings of the good barely play a role, whereas in theology, philosophical sources may (but do not necessarily) take the lead. These are some of the reasons why Albert's selections of Arabic sources — Avicenna and Algazel in this case — seem to elude generalization as regards the mechanisms of his appropriations. In fact, the context-dependence of Albert's appropriation practices points to their emergent character, which itself will come to the fore elsewhere in the volume.

In the next chapter, **Richard C. Taylor** re-examines the complex history of Latin Averroism from a fresh perspective and with stimulating new results, stemming from his analysis of Albert's early appropriations of the long commentaries on Aristotle's *De anima* and *Metaphysica* by Averroes (Ibn Rushd, 1126–98). The conventional view of Latin Averroism distinguished between a First Averroism, characterized by seriously flawed and thus infertile misreadings among the Latin appropriations of Averroes's teachings, and a Second Averroism, marked by correct and highly productive, albeit controversial, interpretations. As Taylor shows, Albert's *De homine* contains the very misreading that characterized First Averroism, yet that misreading was far from resulting in blind alleys. On the contrary, Albert himself and later his student Thomas Aquinas developed their own teachings of the individual human intellect against the background of First Averroism's famous misreading, as well as against the background of other Peripatetic philosophers, such as Avicenna. Furthermore, Taylor corrects some conventional readings of Ibn Rushd's own doctrine in the literature, themselves tainted by Latin First Averroism, and alleviates the continued perplexity about these matters among contemporary scholars.

Richard C. Taylor's contribution allows us to understand in greater complexity the decisive role that Albert's *De homine* played in the building of his later scientific system. Located at the philosophical heart of the treatise, Albert's initial appropriation of Averroes's teachings on the nature of the human intellect was, despite its deviation from the original Ibn Rushd, bound to endure in his own teachings (and in those of his student Thomas Aquinas). And this is despite unequivocal corrections in his later interpretation of Averroes's genuine teachings. As well, Albert's own novel interpretive construction and attribution to Averroes of a monopsychism with the post mortem existence in a single unitary soul — a doctrine not found in Ibn Rushd — prompted him to refute that view and to assert his own teaching on the personal immortality of individual human beings.

What mattered most for Albert in his early appropriations of the Arabic material was not a correct reading of a single Arabic source. Rather, it was the doctrinal fit between different sources and the doctrinal fit with Latin convictions about the human soul, the afterlife, and happiness. Taylor's paper, like other recent historiography and, if in different ways, also like the chapters by Tracey, Müller, Anzulewicz, and Krause in this volume, suggests the need to attend more carefully to the appropriators' practices and contexts than has been commonly done so far in the history of philosophy.[39]

---

39 The history of science pays far more attention to these questions, not least because of the various 'turns' in its approaches. Many valuable studies could be listed here, but the three following examples are particularly suited to building methodological bridges: Sabra, 'The Appropriation and

Albert's overarching division between the two *scientiae* of philosophy and theology, and accordingly the different epistemic functions that he assigns to his sources, is very pronounced in his *Super Ethica*. In this first commentary on the *corpus Aristotelicum*, Albert grapples with the dilemma of subject matters overlapping between philosophical ethics and theology (such as ethical conduct and the virtues) and the different solutions proposed in each of the *scientiae*.

The stubborn question of the double truth of these subject matters is analysed by **Martin Tracey** in a chapter that reads Albert's solution in a novel way with attention to striking textual passages. The conventional reading, as propounded by René-Antoine Gauthier, is that, for Albert, philosophical *scientia* had to concur with theological *scientia* in its conclusions. At the time, this implied that both *scientiae* had to Christianize Aristotle's *Ethica Nicomachea*. Tracey shows that this is an insufficiently nuanced reading of Albert's commentary, since he insisted on different standards of rational reasoning rather than concurrence of conclusions. Albert allowed for the possibility of different solutions — for instance, in matters of fear of death, shame, justice and blameworthiness, usury and allowances (*permissiones*) — whenever supported by proper rational justification.

As Tracey suggests, this insistence on the correct way of reasoning is also perfectly consistent with Albert's discussions of his sources, such as the ancient Scholiast, on whose grounds he rejects Averroes's view of the human soul and ultimate happiness as a distortion of the truth of the matter. But what exactly is the epistemic role that Albert's rejection of Averroes plays in his *Super Ethica*? Martin Tracey's chapter provides detailed evidence that Albert's motivation to pursue the truth of a matter by means of the philosophical approach strongly influenced his appropriation practices. Invoking a particular source, Averroes in this case, to show that it violates a well-reasoned truth helped Albert to sustain his own coherence as what we wish to call a 'restrictive confirmation tool'. By putting restrictions on other possible reasonings and conclusions given beforehand, the source confirms or corroborates his own reasoning and conclusion. This particular epistemic function of source appropriation does not aim at carrying over any particular content from Averroes. Rather, it aims to borrow only formal elements, in which a particular teaching is presented, detaching these elements from its original context.[40] This is a factor in Albert's use of his Arabic sources that has hitherto escaped closer analysis, probably because (as will become clear in the papers by Müller and López-Farjeat) his appropriation of Averroes in his *De anima* followed different epistemic functions and does not show the same formal separation.

---

Subsequent Naturalization of Greek Science in Medieval Islam'; Ragep, Ragep, and Livesey, *Tradition, Transmission, Transformation*; Krause, Auxent, and Weil, 'Making Sense of Nature in the Premodern World'.

40 For an extensive discussion on this type of detachment, see Krause, 'Transforming Aristotelian Philosophy'.

At approximately the same time that Albert the Great penned his first commentary on Aristotle's *Ethica Nicomachea*, he also began his commentary project on what he called Aristotle's *philosophia realis*, including, as mentioned earlier in this introduction, the disciplinary realms of natural philosophy, metaphysics, and mathematics.[41] Contemporary scholars have repeatedly identified Albert's commentary on Aristotle's *Physica*, the commentary that initiated this project, as a turn away from theology, but others have disputed that view, given that his *Super Ethica* was written, or at least begun, somewhat earlier at Heilig Kreuz in Cologne.[42] There is certainly no doubt that Albert entered a new scholarly phase with his commentary on the *Physica*, and he admits as much openly at the very beginning of the book, but the full implications of this new phase may be much more visible with hindsight than they were from his own perspective.

Albert's more proximate intention was — if we are to believe the opening words of his *Physica*, the most general of the books on natural science or physics — to accede to the urging of his Dominican friars to teach them about physics, and to do so in a way that would supply them with a comprehensive *scientia* of it.[43] For Albert, this criterion of comprehensiveness implied rejecting Aristotle's opinion wherever it was false, but also, and more importantly, supplementing it wherever it was incomplete or unclear.[44] What was to be added, and derived from whom? This question can only be answered by focusing in on Albert's *digressiones* (his favoured way of commenting) and attending, as Tracey does in his contribution, to the exact nuances of his definitions and demonstrations.

**Josep Puig Montada**'s contribution does just that with regard to the eternity of motion, which Albert discusses in Book VIII of his *Physica*. Puig Montada takes us on a detailed tour of Albert's argument and the use of his sources, showing that Albert meticulously combed Aristotle, Boethius, Avicenna, and Averroes for what

---

41 Albertus Magnus, *Physica*, I.1.1, ed. by Hossfeld, p. 1, vv. 43–49: 'Cum autem tres sint partes essentiales philosophiae realis, quae, inquam, philosophia non causatur in nobis ab opere nostro, sicut causatur scientia moralis, sed potius ipsa causatur ab opere naturae in nobis, quae partes sunt naturalis sive physica et metaphysica et mathematica, nostra intentio est omnes dictas partes facere Latinis intelligibiles'.

42 Weisheipl, 'Classification of the Sciences in Medieval Thought'.

43 Albertus Magnus, *Physica*, I.1.1, ed. by Hossfeld, p. 1, vv. 9–22, as quoted in note 18 above.

44 Ibid., I.1.1, ed. by Hossfeld, p. 1, vv. 23–41: 'Erit autem modus noster in hoc opere Aristotelis ordinem et sententiam sequi et dicere ad explanationem eius et ad probationem eius, quaecumque necessaria esse videbuntur, ita tamen, quod textus eius nulla fiat mentio. Et praeter hoc digressiones faciemus declarantes dubia suborientia et supplentes, quaecumque minus dicta in sententia Philosophi obscuritatem quibusdam attulerunt. Distinguemus autem totum hoc opus per titulos capitulorum, et ubi titulus simpliciter ostendit materiam capituli, significatur hoc capitulum esse de serie librorum Aristotelis, ubicumque autem in titulo praesignificatur, quod digressio fit, ibi additum est ex nobis ad suppletionem vel probationem inductum. Taliter autem procedendo libros perficiemus eodem numero et nominibus, quibus fecit libros suos Aristoteles. Et addemus etiam alicubi partes librorum imperfectas et alicubi libros intermissos vel omissos, quos vel Aristoteles non fecit vel forte si fecit, ad nos non pervenerunt'.

he considered to be the truthful insights in their work. He discusses how Albert carefully weighed the right combination of his sources in order to reach his own solution: that motion is eternal once created, but that the world itself is created outside of time. This solution, as Puig Montada notes, built on different elements inherited from the different sources. Albert's understanding of eternity draws loosely on Boethius, his view on God's causal priority vis-à-vis the world echoes Averroes, and his position on the creation of movement by atemporal emanation resonates with Avicenna. Yet for Albert, none of this, particularly the creation of the world, amounted to demonstrative certainty. *Pace* Aristotle and following Maimonides, Albert maintained that the natural science of physics could yield only probable knowledge in these matters.

In our view, the attitude to Aristotle's fallibility expressed by Albert here is indicative of a value judgment that subordinates authority to the epistemic criteria of truth and comprehensiveness in *scientia*.[45] To a high degree and right from its start, Albert's commentary on Aristotle's *Physica* displays these two epistemic values, which accompany his commentary project as a whole. The sources — be they Greek, Arabic, or Latin — were above all instruments put to the service of these lofty ideals. As we read in Puig Montada's chapter, Albert chose them carefully and integrated them with deliberation. The reconciliation of different elements in Albert's natural philosophical argument for the eternity of motion thus results in a complex doctrine featuring a range of ordered characteristics that exceeds each of its parts and remains irreducible to them. But how exactly did Albert reconcile these characteristics in general terms?

Both Tracey's and Puig Montada's papers suggest, for different cases, how difficult it is to generalize the mechanisms of Albert's appropriations and practices of coordination around the epistemic values of truth, certainty, and comprehensiveness. What the two chapters do begin to reveal, however, is the extent to which those epistemic values and practices recur in Albert's works.

Another epistemic practice dispersed across Albert's works is the resolving of apparent divergences between teachings from varied sources, perhaps most visibly between Platonic and Aristotelian ones. But throughout, Albert subordinates his resolutions of Plato and Aristotle to the two higher scientific values we have already encountered. He insists that understanding and knowledge of the philosophical teachings in their precise relation to one another is foundational to achieving comprehensive and perfect *scientia* — that is, truth known with certainty — in both approach and content.[46]

If not as fundamentally as in his attempts to coordinate Plato with Aristotle, Albert applied the same approach to his Arabic sources, as **Irven M. Resnick**

45 Ibid., VIII.1.14, ed. by Hossfeld, p. 578, vv. 23–27: 'Et ad illum nos dicimus, quod qui credit Aristotelem fuisse deum, ille debet credere, quod numquam erravit. Si autem credit ipsum esse hominem, tunc procul dubio errare potuit sicut et nos'.

46 See, for instance, Anzulewicz, 'Albertus Magnus als Vermittler zwischen Aristoteles und Platon'.

suggests in his contribution on Albert's *De causis proprietatum elementorum* (1251–54), one of the four natural philosophical books to follow the *Physica* commentary. Resnick submits that Albert resolves a discrepancy between Avicenna and Averroes concerning the universal flood and the potential for the regeneration of species afterwards. He does so by selecting certain teachings in order to craft from them his own concerted solution. This is not unlike the commentary practices discussed by Uscatescu Barrón and Puig Montada. Resnick, however, notes that Albert's resolution of the discrepancy and his subsequent solution can only be appreciated if read against the background of his Latin tradition and theological commitments. Thus, when Albert endorses Avicenna's teachings concerning the possibility of a universal flood and the possibility of the regeneration from matter of most species based on celestial causality, and when he sides with Averroes concerning the possibility of the regeneration of perfect animals and humans on the basis of additional requirements such as coition, he does not lose sight of his own theological commitments — most importantly that God, as the first mover, utilizes different natural causes to produce effects in the world.

Albert's coordination of his divergent Arabic sources with his own commitments found acclaim from the Renaissance scholar Pomponazzi, who praised it as part of a shared medieval practice of appropriation. Resnick finds that praise a little misleading, since coordination like Albert's was not as widely shared a phenomenon among medievals as Pomponazzi thought.

Regardless of this historical appraisal, Albert's reconciliation of seemingly divergent teachings in his Arabic sources constitutes a synthesis of inheritance and emergence. As Resnick shows, Albert aspires to his highest epistemic criteria of truth and comprehensiveness of *scientia* by ordering, in the right manner, selected truths that he finds scattered throughout his different sources. In fact, Albert's practice of giving order to these different truths is one of the most widely shared practices among the Scholastics, and was also reflected upon in the numerous classifications of the sciences and the debates about the different intellectual operations required to reason correctly, which explicitly included that of ordering (*ordinare*).[47] Albert's practice of ordering truths contained in

47 See, for instance, Albertus Magnus, *Super Porphyrium De V universalibus*, Tr. de antecedentibus ad logicam, c. 7, ed. by Santos Noya, p. 15, vv. 20–37: 'Quae omnia fiunt actu rationis, qui est ratiocinatio, qui actus discursus rationis est ex uno in aliud. Et ideo tales scientiae a Dionysio vocantur "discursae disciplinae". Non autem potest sic ex uno in aliud discurrere ratio, nisi prius accipiat unum in alio esse per se vel per accidens vel unum ab alio esse divisum per se vel per accidens. Et hoc iterum esse non potest, nisi accipiat unum ordinatum ad aliud per se vel per accidens. Ordo autem est prioris et posterioris secundum naturam et esse, et sic accipit universale et particulare per se vel per accidens. Et sic invenit modum praedicandi unum de altero vel negandi. Et quoad ordinem inventa est Universalium scientia et scientia Praedicamentorum, et quoad modum unum educendi de alio inventa est scientia Divisionum. Rationis enim opus est ordinare, componere et colligere et resolvere ea quae collecta sunt. Quo opere utitur quasi instrumento in accipiendo scientiam, quando procedit a noto ad ignotum'. It should be noted that Albert composed his *Super Porphyrium De V universalibus*

his sources according to criteria of rank, subject matter, method, and suitability, among others, is thus woven into his view of the inmost texture of the human mind by way of its acquired tools of logic. This practice of ordering constitutes yet another emergent factor in Albert's oeuvre, as it points to the overall significance that he assigns to *scientia*, the acquisition of ultimate natural happiness.[48]

Divergent views on particular themes, not just in the Arabic sources — as discussed in Irven Resnick's paper — but already in Aristotle himself, have always presented a challenge to commentators on the Stagirite's works. This is partly because philosophers in the Aristotelian tradition aimed to give coherent interpretations of Aristotle, including the production of unified explanations. Commentators across the centuries embraced the ideals of coherence and unity, which went hand in hand with an overarching scientific programme that likewise had to be free from contradictions, and applied them to those accounts in Aristotle's work where it proved difficult to make out any coherence and unity.

Albert the Great's commenting practices were no exception to that when it came to Aristotle's seemingly contradictory accounts on the composition of material substances, the topic of **Adam Takahashi**'s contribution. Takahashi begins by showing that in his *Physica*, Aristotle endorses a hylomorphic explanation, whereas in *De caelo*, *De generatione et corruptione*, and *Meteora*, he draws on a materialist explanation in which the different mixtures of the four elements of fire, air, water, and earth account for the composition of material substances. Albert's reconciliation of these two seemingly contradictory accounts does not consider Aristotle's template alone. It also critiques an earlier, and in Albert's eyes false reconciliation of Aristotle's different accounts: that offered by Averroes. Takahashi shows that the disagreement between Albert and Averroes over the right kind of reconciliation turns on the question of explanatory reducibility or irreducibility. Can Aristotle's hylomorphism be reduced to the primary qualities of the elements or not? Averroes favoured a reductionist reconciliation, identifying the form of the elements with their primary qualities. For Albert, such reductionism could not hold in light of the truth. His own reconciliation gave precedence to Aristotle's hylomorphic account instead, and it did so by introducing into the debate the new concept of *inchoatio formae*, a quasi-active principle in matter that attracts different

around the same time that he wrote *De causis proprietatum elementorum*. This suggests that he might have thought about the ways of ordering his sources here along the lines outlined in this quotation.

48 Albertus Magnus, *Super Porphyrium De V universalibus*, Tr. de antecedentibus ad logicam, c. 3, ed. by Santos Noya, p. 6, vv. 16–29: 'Est autem non tantum necessaria, sed etiam utilis haec scientia. Si enim bonum et felicitas hominis est secundum optimae partis animae hominis perfectissimum actum, hoc est secundum intellectum contemplativum, nec contemplari poterit intellectus, nisi noverit contemplationis principia et sciat invenire quod quaerit contemplari, et diiudicare id ipsum quod iam contemplatur inventum, patet quod prae omnibus utilis est ad felicitatem haec scientia, sine qua non attingitur felicitatis actus et per quam ipse felix actum non impeditae recipit operationis. Haec enim scientia a phantasiis, quae videntur et non sunt, liberat, errores damnat, et ostendit falsitates et lumen dat rectae in omnibus contemplationis. Prae omnibus igitur desideranda est haec scientia'.

forms to the elements. These reductionist versus anti-reductionist tendencies, Takahashi shows, were not singular events in Averroes's and Albert's strategies of speaking the truth. On the closely related theme of the spontaneous generation of living beings, Averroes again proffered a reductionist explanation, granting cosmic heat the role of formative action, whereas Albert appealed to the concept of *virtus formativa* in analogy to the quasi-active power found in human semen.

Unlike the inheritance-oriented emergence of ordering we identified in Irven Resnick's paper, this emergence of two new explanatory concepts might be called a process of innovation bound to rejection. Key to Albert's reconciliation of Aristotle's templates were his very own concepts of *incohatio formae* and *virtus formativa,* which enabled him to unequivocally attribute the explanatory force to hylomorphism rather than to mixture of the elements. But without Averroes's prior, reductive reconciliation of Aristotle's templates, Albert might not have felt the need to propose the concept at all, let alone to mobilize its explanatory scope — in its modified form as *virtus formativa* — for the spontaneous generation of living composites. Albert's indebtedness to Aristotle, then, is but one component of the different historical layers to his explanation. His rejection of Averroes is closely tied to his invention of a new scientific concept. By throwing the concept of *incohatio formae* into the mix, Albert becomes able to rearrange the components of both of Aristotle's accounts in such a way that they constitute a new scientific explanation, one that adheres to the epistemic ideals of coherence and unity.

Albert's intellectual activities of ordering divergent views on a given theme and judging the truth value of those views also play a dominant role in **Luis Xavier López-Farjeat**'s contribution. Unlike Resnick, however, López-Farjeat shows that Albert's ordering activity in his *De anima* (1254–57) is already at a matured second stage and takes up a highly sophisticated perspective. In *De homine,* Albert had already devoted much attention to the question of how humans come to know and what the limits of their knowing are. Now, in *De anima,* he returns to these questions and engages in philosophically robust ways with the answers given by his Arabic predecessors. As López-Farjeat shows in detail, Albert now clusters the different views around the key elements of ontology and functionality, separating the wheat from the chaff. First, Albert identifies Alfarabi's and Avempace's views on the material nature of the possible intellect as mistaken; he next recognizes Avicenna's view of the separate agent intellect as 'giver of forms' as equally flawed, then reveals that Averroes's view on the functionality of the agent and possible intellects is true and must thus be adopted into his own teaching. As López-Farjeat observes, Albert made this move despite having previously rejected Averroes's view on the ontologies of both intellects.

The habit of revisiting key philosophical teachings and re-evaluating their truth value should be viewed as a hallmark of Albert's appropriation practices. The most obvious examples are his two *Ethics* commentaries, written with roughly ten years between them, but all teachings anthropological — including human cognition and intellection, the topic of López-Farjeat's paper — were included in this enterprise of continued re-evaluation. That does not make Albert less of a systematic or coherent thinker. On the contrary, his practice of returning to anthropological themes was inspired by his desire to formulate with ever-increasing precision and clarity what the human being is, does, and can do in its place in the cosmos. Clearly, such repetition also made his *scientia* of the human being more robust, giving it more coherence and unity. At the height of his intellectual activity, Albert synthesized and integrated a breadth and depth of elements in his anthropology that no other Latin thinker before or after him could match. López-Farjeat's paper is the first witness to this highly innovative feature of Albert's intellectual activities; further facets are revealed in Müller's, Anzulewicz's, and Krause's chapters.

Formal coherence and unity, the goal Albert had in mind when considering the composition of material substances, as we saw in Takahashi's paper, was also at work when he dealt with the teaching of intellectual memory. **Jörn Müller** shows that Albert's engagement with Avicenna's denial of intellectual memory was not an open rejection, but instead weighed philosophical against theological outlooks. His pupil Thomas Aquinas strongly held to the role of intellectual memory so as to account for the soul's natural knowledge in the afterlife,[49] but Albert did not worry about memory for the soul's natural knowledge in the afterlife. On the contrary, in his mature doctrine of intellectual memory in the *De anima*, Albert adopts Avicenna's conception of the *intellectus adeptus* and, as Müller shows, adds that intellectual memory matters from a formal rather than a material point of view. Acquired intelligibles matter because of their intellectual light (formal point of view) rather than because of their content (material point of view), and the complete acquisition of their intellectual light equals the acquisition of the *intellectus adeptus*. Albert here embraces an ascending and this-worldly teleology of intellectual development — one that he finds in his Arabic sources — while formulating his own account as regards the cause of such development.

Müller's chapter reveals Albert's systematic adoptions and rearrangements of the explanatory constituents known to him from his sources. Combining them with his own insights — this time, the insight of form in identity with the intellectual light of the intelligibles — Albert builds his very own overarching scientific explanations. The conditions under which Albert could do so were not the Aristotelian template and the Arabic sources per se, but the Aristotelian template and Arabic sources as they were in use among his Latin contemporaries. Müller's paper beautifully demonstrates both the importance of this additional factor of living debates and the difficulty of pinpointing its actual impact on the Latin solutions at hand. For us as historians of medieval philosophy, the living debates of which Albert was part are impossible to reconstruct comprehensively,

49 See, for instance, Cory, 'Embodied vs. Non-Embodied Modes of Knowing in Aquinas'.

not least because they belonged to a largely oral culture and have left few traces. All the more elusive is the question of what memory in all its implications truly meant for Albert — but his doctrine of a formal, light-bound, content-less memory may at least give us a glimpse.

In Albert's eyes, the human suitability for memory-building and intellectual perfection is not a given for every individual. It depends on certain prior conditions expressed by people's individual natures, conditions that Albert locates in the soul's faculties, the human body, and the environment that humans inhabit. A careful examination of Albert's teaching on these psychological, physiological, geographical, and climatic preconditions for an individual's ability to realize their true rational nature and bring it to intellectual perfection and happiness is offered by **Henryk Anzulewicz**, who takes us on a captivating tour through Albert's oeuvre. Paying special attention to two of Albert's works, *De natura loci* (1251–54) and *De animalibus* (1258), Anzulewicz finds that Albert is completely at ease with borrowing Aristotle's doctrine of the mean for his account of geographical and climatic factors, Galen's and Avicenna's medical teaching of complexion for his explanation of physiological factors, and a mixture of Greek, Arabic, and Latin sources for his account of psychological factors. As a whole, these borrowings offer an unusual insight into Albert's independent intellectual practice of ordering (a practice already described in a different context by Resnick), this time by way of typologies.

In Anzulewicz's view, Albert holds different types of geography and climate causally responsible for different types of human physical constitutions, which are expressed by different complexions of the body; these complexions, in turn, prove responsible for the different types of aptitudes that Albert finds in different humans for realizing their inborn rational nature and attaining perfection of their intellect. It is in this mature view of Albert's on the psychological, physiological, and geographical or climatic preconditions for an individual's aptitude for *scientia* that Anzulewicz finds the natural scientific reasons explaining why certain individuals fulfil their natural desire for knowledge while others fail to do so. Albert advanced a compound typology of explanatory factors in the suitability or unsuitability of certain humans for attaining their natural perfection, and these factors defy reductionism.

More than anything, the complex cooperative and inhibitory interactions of causality on the three different layers of explanation described by Anzulewicz reveal Albert's intellectual practice of ordering his sources by way of epistemic typologies. Once again, it is not the authoritative but the epistemic value of his sources that stands at the heart of Albert's appropriation practices. However, this time, the sources of epistemic significance lie outside the confines of Albert's *scientia naturalis*. Some of his insights are derived from Aristotle's ethics, a *scientia practica*, others from Galenic-Avicennian medicine, an *ars mechanica*.

This reveals two things about Albert's appropriation practices. First, they show that for his *scientia naturalis*, he relies on certain insights that derive from outside the canonical texts of Aristotle's *scientia naturalis*. Second, they reveal how Albert transforms the epistemic value of these external insights, raising medical insights, especially, to the more highly ranked *scientia naturalis* of the *De animalibus*. This factor of emergence in Albert's practices is, we contend, truly unique.

Albert's integrations of anatomical and physiological teachings from *medicina theorica* considerably enlarged the explanatory scope of his *scientia naturalis*. No longer did this *scientia* reign over the teachings of Aristotle and those of his followers as propounded in their books on sentient living beings (*animalia*) alone; it now also acquired a presiding authority over medical teachings that covered the overlapping subject matter of the body of the rational animal.[50] Albert thus transformed and expanded the Aristotelian *scientia de animalibus* on the level of content. But it is possible to detect repercussions of these and similar expansions in scientific scope on the level of *approach* as well. This is particularly the case with regard to Albert's integration of experiential insights and teachings from his Arabic sources.

In her contribution, **Katja Krause** discusses how Albert transformed experience (*experientia*) and doctrine (*doctrina*) from his Arabic sources. Attending first to Albert's *De animalibus*, she examines his integration there of the Galenic-Avicennian medical teaching that 'no bone, apart from teeth, has sensation'. Krause shows that Albert moves away from Avicenna's emphasis on experience as a transmitted piece of knowledge, bequeathed to Avicenna by Galen, to experience as an evidentiary piece of knowledge, warranting the truth of the matter. In this way, Albert turned transmitted experience into empirical verification.

Next, Krause studies Albert's attempt in *De anima* to demarcate taste from touch. Worried by the difficulty of reducing taste to touch (and the danger of ending up with four rather than five external senses), Albert builds his mature doctrine of taste on a formal component derived from Averroes's *Long Commentary on the De anima*: the form-matter relation of flavour to liquid. By identifying flavour as the formal cause and liquid as the material cause of taste, Albert devises the best possible explanation for the experience of tasting saltiness. Both examples of a formal transformation of his sources — the trans-historization of experience and the establishment of a particular teaching as the best explanation of a shared experience — suggest that Albert's appropriations of his sources went far deeper than issues of content alone. His concerns included not only truthfulness to the original, the threat of the double truth, and debates with his Latin interlocutors, but more importantly the epistemic value and function of any given piece of knowledge that he appropriated from his sources.

---

50 See, for instance, Cadden, 'Albertus Magnus' Universal Physiology'; Siraisi, 'The Medical Learning of Albertus Magnus'; Park, 'Albert's Influence on Late Medieval Psychology'; Jacquart, 'Die Medizin als Wissenschaftsdisziplin'; Krause, 'Grenzen der Philosophie'.

In fact, Albert's appropriations were marked as much by concerns of approach as by concerns of content — the former of which have largely gone unnoticed. When Albert turned transmitted experience into evidentiary experience, and when he identified a given doctrine as the best explanation for a shared experience, he determined the formal links between experience as evidentiary and doctrine as explanatory. Experience became proof for the truth of a teaching, and doctrine became the best explanation for a given sensory experience. In both cases, this is a precise epistemic determination that is nowhere found in his sources. What still remains to be investigated is whether these two connections between experience and doctrine themselves followed a new and wider epistemic programme of reconceiving the relationship between experience and explanation: how they were subject to Albert's scientific practices of ordering and rearranging, defining and demonstrating, and how they connected to the scientific goals he sought, truth, certainty, and comprehensiveness of knowledge.

**Amos Bertolacci**'s chapter brings us back to questions of content. Just like Puig Montada, Resnick, Takahashi, and Müller in their chapters, Bertolacci zooms in on how Albert reconciled his Arabic sources with each other, this time regarding the metaphysical doctrine of universals. Looking at a pre-existing disagreement among the Arabic sources, Bertolacci's study begins with a close reading of Averroes's portrayal of Avicenna's teaching on the nature of, and relationship between, the two universals of being and oneness. His aim is to show the extent to which Averroes already reinterpreted Avicenna's position. Bertolacci then discusses Albert's reading of both of these Arabic sources and unpacks the precise nature of Albert's doctrinal reconciliation between Avicenna's own position (as read by Albert in the Latin translation of Avicenna's *Metaphysics*), the reinterpreted Avicenna in Averroes's long commentary on Aristotle's *Metaphysica*, and Averroes's own position. The reconciliation, Bertolacci tells us, demanded an array of hermeneutical measures on Albert's part. These ranged from deliberately leaving out Averroes's name in the listed accusations against Avicenna's position, to sweeping Averroes's harshest accusations under the carpet and rephrasing other attacks, to slightly reinterpreting Avicenna's teachings. Albert applied all these to accomplish an entirely new product: a unified and coherent perspective on the nature and relation of the two universals of being and oneness, according to which Avicenna and Averroes are in almost perfect agreement with one another. The purpose of Albert's reconciliation, Bertolacci's discussion suggests, is to fully naturalize Arabic philosophy in general, and Arabic metaphysics in particular, within the world of Latin education.

Albert must have been among the first of the Latin thinkers to succeed in naturalizing the Arabic sources. The generations to come, as is well known, read Aristotle's *Metaphysica* not in isolation but in tandem with his Arabic commentators, whose voices were thus reinterpreted in ever-new polyphonies by Latin scholars until the early modern period. More than anything, Bertolacci's paper traces meticulously how selections, reinterpretations, and informational rearrangements needed to be made on the micro-level of a single doctrine. It also shows the hermeneutical precision that Albert required for his decisive advance of a doctrinal alignment that helped to catalyse the long-lived success of Aristotle's *Metaphysica* in the Latin West.

Albert's hermeneutical precision is also a key component in **David Twetten**'s expansively detailed examination of Albert cosmology. Twetten investigates Albert's treatment of emanation, particularly his mature teaching in *De causis* (1267), presenting the intricate play of inspiration of Greek (Proclus, Pseudo-Dionysius, John the Damascene), Arabic (*Liber de causis*, Isaac Israeli, Avicenna, Moses Maimonides), and Latin (Augustine, Boethius, Anselm of Canterbury, William of Auxerre, Philip the Chancellor, Robert Kilwardby, Bonaventure) provenance in Albert's mature solution.

The motivation of Twetten's chapter, however, lies elsewhere. The much-discussed concern among scholars about double truth merits Twetten's rigorous discussion: Is Albert's view reconcilable with his era's theological truth about creation or not? There is no need to worry about double truth, Twetten responds, because Albert's emanation scheme endorses neither mediate creation nor divine determinism, despite also aiming for the best philosophical explanation of the origin of all there is.

Tackling the question of divine determinism first, Twetten discusses Albert's doctrine of the divine will. At its heart stands the notion of God as *causa sui*, a notion that resolves all fears of determinism. Twetten then turns to Albert's articulation of a law-like scheme of the derivation of creatures from God that distinguishes true creation from information. The scheme's articulation follows the inspiration of Avicenna, writes Twetten, but it equally reveals Albert's own take: since *esse* alone is immediately caused by God *ex nihilo*, all forms other than *esse* require a different cause and a different type of causality. This other cause amounts to what Albert identifies as the Intelligence, and its type of causation is *informatio*, a bringing-about of diversity and composition by specifying *esse* through form. As a consequence, Twetten concludes, Albert is able to affirm equal omnipresence of God in all creatures, thus reducing all worries about mediate creation or divine determinism even further.

Perhaps more sumptuously than any other study in the volume, Twetten shows that although the sources Albert used supplied him with material and inspirational ideas, his own combination of that material was truly unique. In its approach, this composition complied with the high epistemic standards of the philosophical approach; in its aim, it aspired to the ultimate value of truth. For Albert, that implied there was no room for double truth. But contrary to his near-contemporaries — Thomas Aquinas and Duns Scotus are two that Twetten invokes explicitly — Albert did not simply fit new philosophical ideas and images into passed-down theological ones. He thought hard about his reconciliation on an abstract, highly sophisticated level of reason that, to this very day, has demanded extraordinary care and attention to detail from his readers in deciphering its true meaning and its pioneering take on reality.

Taken together, this volume's twelve chapters show exactly how Albert made use of his Arabic sources with the ends of truth, certainty, comprehensiveness, and human perfection in mind. They show him gleaning from his sources — Arabic, Greek, and Latin — with a consistent approach: he read and collected all the teachings available on a particular *quaestio*, determined the truth in light of certain principled criteria that he considered most foundational, and gave precedence to those teachings in his authorities that accounted for the world in its multiplicity and variety in the best possible way known to him. These authorities happened to be, in many instances, the Arabic sources that followed, explicated in greater detail, and completed Aristotle's philosophy as an enterprise sharing the very epistemic values that Albert would make his own.

However, Albert's use of these Arabic sources cannot be reduced only to their own teachings, nor is it modelled on the same epistemic values in all cases. This diversity in Albert's deployments applies particularly to those Arabic thinkers who make the most frequent appearances in his writings, Avicenna and Averroes. Their voices, like the voices of the other Arabic authorities in Albert, were never intended as endpoints to be reached, but rather as auspicious points of departure. Respected for their value of truth, for Albert the Arabs were stewards of philosophical teachings that he, with his own erudition, brought into the Latin world and shared with his peers and students alike.

His ultimate aim in so doing — knowledge for the sake of the perfection of the scientist — guided his use of these Arabic sources in a way that allowed new teachings and epistemic values to emerge. The epistemic processes of naturalization and corpus formation that the contributions to this book describe thus reveal how Albert created his own comprehensive medieval philosophical *scientia*, oscillating between his inheritance of the Arabic sources and the emergence of his own thinking.

## Works Cited

### *Primary Sources*

Albertus Magnus, *Analytica posteriora*, ed. by Auguste Borgnet, Editio Parisiensis, 2 (Paris: Vivès, 1890)

———, *Commentarii in II Sententiarum*, ed. by Auguste Borgnet, Editio Parisiensis, 27 (Paris: Vivès, 1893)

———, *De resurrectione*, ed. by Wilhelm Kübel, Editio Coloniensis, 26 (Münster: Aschendorff, 1958)

———, *Liber de natura et origine animae*, ed. by Bernhard Geyer, Editio Coloniensis, 12 (Münster: Aschendorff, 1955)

———, *Metaphysica, libri VI–XIII*, ed. by Bernhard Geyer, Editio Coloniensis, 16/2 (Münster: Aschendorff, 1964)

———, *Physica, libri I–IV*, ed. by Paul Hossfeld, Editio Coloniensis, 4/1 (Münster: Aschendorff, 1987)

———, *Physica, libri V–VIII*, ed. by Paul Hossfeld, Editio Coloniensis, 4/2 (Münster: Aschendorff, 1993)

———, *Quaestiones super libris de animalibus*, ed. by Ephrem Filthaut, Editio Coloniensis, 12 (Münster: Aschendorff, 1955)

———, *Summa de creaturis*, ed. by Auguste Borgnet, Editio Parisiensis, 34 (Paris: Vivès, 1895)

———, *Super I librum Sententiarum*, ed. by August Borgnet, Editio Parisiensis, 25 (Paris: Vivès, 1893)

———, *Super Porphyrium De V universalibus*, ed. by Manuel Santos Noya, Editio Coloniensis, 1/1A (Münster: Aschendorff, 2005)

Aristoteles Latinus, *Metaphysica, lib. I–X, XII–XIV, translatio anonyma sive 'media'*, ed. by Gudrun Vuillemin-Diem (Leiden: Brill, 1976)

Avicenna Latinus, *Liber de philosophia prima*, vol. 1 (I–IV), ed. by Simone Van Riet (Leuven: Peeters, 1977)

Dominicus Gundissalinus, *De divisione philosophiae: Über die Einteilung der Philosophie*, ed. and trans. by Alexander Fidora and Dorothée Werner (Freiburg im Breisgau: Herder, 2007)

### *Secondary Works*

Anzulewicz, Henryk, 'Albertus Magnus als Vermittler zwischen Aristoteles und Platon', *Acta Medievalia*, 18 (2005), 63–87

———, 'The Systematic Theology of Albert the Great', in *A Companion to Albert the Great: Theology, Philosophy, and the Sciences*, ed. by Irven M. Resnick (Leiden: Brill, 2013), pp. 15–67

Blankenhorn, Bernhard-Thomas, *The Mystery of Union with God: Dionysian Mysticism in Albert the Great and Thomas Aquinas* (Washington, DC: The Catholic University of America Press, 2015)

Bonin, Marie-Thérèse, *Creation as Emanation: The Origin of Diversity in Albert the Great's 'On the Causes and the Procession of the Universe'* (Notre Dame, IN: University of Notre Dame Press, 2001)

Brungs, Alexander, Vilem Mudroch, and Peter Schulthess, 'Institutionelle Voraussetzungen', in *Die Philosophie des Mittelalters 4: 13. Jahrhundert*, ed. by Alexander Brungs, Vilem Mudroch, and Peter Schulthess (Basel: Schwabe, 2017), part 1, pp. 41–91

Cadden, Joan, 'Albertus Magnus' Universal Physiology: The Example of Nutrition', in *Albertus Magnus and the Sciences: Commemorative Essays 1980*, ed. by James A. Weisheipl (Toronto: Pontifical Institute of Mediaeval Studies, 1980), pp. 321–39

Canguilhem, Georges, *Ideology and Rationality in the History of the Life Sciences*, trans. by Arthur Goldhammer (Cambridge, MA: MIT Press, 1988)

Cassirer, Ernst, *The Philosophy of the Enlightenment*, trans. by Fritz C. A. Koelln and James P. Pettegrove (Princeton, NJ: Princeton University Press, 1979)

Cory, Therese Scarpelli, 'Embodied vs. Non-Embodied Modes of Knowing in Aquinas: Different Universals, Different Intelligible Species, Different Intellects', *Faith and Philosophy*, 35 (2018), 417–46

Cunningham, Stanley, *Reclaiming Moral Agency: The Moral Philosophy of Albert the Great* (Washington, DC: The Catholic University of America Press, 2008)

Elkana, Yehuda, *Leben in Kontexten / Life in Contexts: Yehuda Elkana*, ed. by Reinhart Meyer-Kalkus (Berlin: Wissenschaftskolleg zu Berlin, 2015)

———, 'A Programmatic Attempt at an Anthropology of Knowledge', *Sociology of the Sciences Yearbook*, 5 (1981), 1–76

Goethe, Johann Wolfgang von, *Faust*, trans. by Bayard Taylor (New York: Arden, 1900)

———, *Faust: Ein Fragment* (Leipzig: Göschen, 1790)

Honnefelder, Ludger, 'Einleitung', in *Albertus Magnus und der Ursprung der Universitätsidee: Die Begegnung der Wissenschaftskulturen im 13. Jahrhundert und die Entdeckung des Konzepts der Bildung durch Wissenschaft*, ed. by Ludger Honnefelder (Berlin: Berlin University Press, 2011), pp. 9–26

Jacquart, Danielle, 'Die Medizin als Wissenschaftsdisziplin und ihre Themen', in *Die Philosophie des Mittelalters 4: 13. Jahrhundert*, ed. by Alexander Brungs, Vilem Mudroch, and Peter Schulthess (Basel: Schwabe, 2017), part 1, pp. 1595–1612

———, 'La place d'Isaac Israeli dans la médecine médiévale', *Vesalius*, special number (1998), 19–27

Janssens, Jules, 'Le *De Divisione philosophiae* de Gundissalinus: Quelques remarques préliminaires à une édition critique', in *De l'Antiquité tardive au Moyen Âge: Études de logique aristotélicienne et de philosophie grecque, syriaque, arabe et latine offertes à Henri Hugonnard-Roche*, ed. by Elisa Coda and Cecilia Martini Bonadeo (Paris: Vrin, 2014), pp. 559–70

Krause, Katja, 'Albertus Magnus zur Philosophie und Theologie: Die Rolle des Erkennenden', presentation at the Angelicum, Thomistic Institute, 30 March 2022, https://www.youtube.com/watch?v=hhNCJRGCTJo

———, 'Grenzen der Philosophie: Alberts des Großen Kommentar zu *De animalibus* und die Medizin', *Documenti e studi sulla tradizione filosofica medievale*, 30 (2019), 265–93

———, 'Transforming Aristotelian Philosophy: Alexander of Aphrodisias in Aquinas' Early Anthropology and Eschatology', *Przegląd Tomistyczny*, 21 (2015), 175–217

Krause, Katja, and Henryk Anzulewicz, 'Albert the Great's *Interpretatio*: Converting Libraries into a Scientific System', in *Premodern Translation: Comparative Approaches to Cross-Cultural Transformations*, ed. by Sonja Brentjes and Alexander Fidora (Turnhout: Brepols, 2021), pp. 89–132

———, with Maria Auxent and Dror Weil, 'Introduction: Making Sense of Nature in the Premodern World', in *Premodern Experience of the Natural World in Translation*, ed. by Katja Krause, Maria Auxent, and Dror Weil (New York: Routledge, 2022), pp. 7–19

Krause, Katja, and Tracy Wietecha, 'Albert the Great on Negative-Mystical Theology as the Epitome of Science', in *The Oxford Handbook of Apophatic Theology*, ed. by John R. Betz and Rik Van Nieuwenhove (Oxford: Oxford University Press, forthcoming)

Krois, John Michael, 'Cassirer's Revision of the Enlightenment Project', in *Philosophie der Kultur – Kultur des Philosophierens: Ernst Cassirer im 20. und 21. Jahrhundert*, ed. by Birgit Recki (Hamburg: Meiner, 2012), pp. 89–106

Miteva, Evelina, 'The Reception of Ibn Gabirol's *Fons vitae* in Albertus Magnus', in *Ibn Gabirol (Avicebron): Latin and Hebrew Philosophical Traditions*, ed. by Nicola Polloni, Marienza Benedetto, and Federico Dal Bo (Turnhout: Brepols, 2023), pp. 197–216

O'Meara, Thomas, ed. and trans., *Albert the Great: Theologian and Scientist* (Chicago: New Priory Press, 2013)

Park, Katharine, 'Albert's Influence on Late Medieval Psychology', in *Albertus Magnus and the Sciences: Commemorative Essays 1980*, ed. by James A. Weisheipl (Toronto: Pontifical Institute of Mediaeval Studies, 1980), pp. 501–36

Putallaz, François-Xavier, 'Die ersten Franziskaner', in *Die Philosophie des Mittelalters 4: 13. Jahrhundert*, ed. by Alexander Brungs, Vilem Mudroch, and Peter Schulthess (Basel: Schwabe, 2017), part 1, pp. 287–300

Ragep, F. Jamil, Sally P. Ragep, and Steven Livesey, eds, *Tradition, Transmission, Transformation: Proceedings of Two Conferences on Pre-modern Science Held at the University of Oklahoma* (Leiden: Brill, 1996)

Resnick, Irven, ed., *A Companion to Albert the Great: Theology, Philosophy and the Sciences* (Leiden: Brill, 2013)

———, and Kenneth F. Kitchell Jr., *Albertus Magnus and the World of Nature* (Chicago: Chicago University Press, 2022)

Rheinberger, Hans-Jörg, *On Historicizing Epistemology: An Essay*, trans. by David Fernbach (Stanford, CA: Stanford University Press, 2010)

Sabra, Abdelhamid I., 'The Appropriation and Subsequent Naturalization of Greek Science in Medieval Islam: A Preliminary Statement', *History of Science*, 25 (1987), 223–43

Siraisi, Nancy, 'The Medical Learning of Albertus Magnus', in *Albertus Magnus and the Sciences: Commemorative Essays 1980*, ed. by James A. Weisheipl (Toronto: Pontifical Institute of Mediaeval Studies, 1980), pp. 379–404

Strobino, Riccardo, *Avicenna's Theory of Science: Logic, Metaphysics, Epistemology* (Oakland: University of California Press, 2021)

Sturlese, Loris, 'Albert der Große', in *Die Philosophie des Mittelalters 4: 13. Jahrhundert*, ed. by Alexander Brungs, Vilem Mudroch, and Peter Schulthess (Basel: Schwabe, 2017), part 2, pp. 861–86

———, *Die deutsche Philosophie im Mittelalter: Von Bonifatius bis zu Albert dem Großen, 748–1280* (Munich: C. H. Beck, 1993)

Suarez-Nani, Tiziana, 'Die theologische Fakultät Paris in der ersten Hälfte des Jahrhunderts', in *Die Philosophie des Mittelalters 4: 13. Jahrhundert*, ed. by Alexander Brungs, Vilem Mudroch, and Peter Schulthess (Basel: Schwabe, 2017), part 1, pp. 269–86

Vost, Kevin, *St Albert the Great: Champion of Faith and Reason* (Charlotte, NC: TAN Books, 2011)

Wallace, William A., ed., 'Albertus Magnus', special issue, *American Catholic Philosophical Quarterly*, 70.1 (1996)

Weisheipl, James, ed., *Albertus Magnus and the Sciences: Commemorative Essays* (Toronto: Pontifical Institute of Mediaeval Studies, 1980)

———, 'Classification of the Sciences in Medieval Thought', *Mediaeval Studies*, 27 (1965), 54–90

Wood, Christopher, 'Source and Trace', *RES: Anthropology and Aesthetics*, 63/64 (2013), 5–19

JORGE USCATESCU BARRÓN

# Chapter 2. Albert the Great's Definition of the Good

## *Its Arabic Origin and its Latin Transformations*

With the arrival and increased use of Latin translations of the *corpus aristotelicum* and Arabic sources related to the *corpus* at the beginning of the thirteenth century, the subjects of philosophy and theology were profoundly reshaped. A fine example of these developments can be found in Albert the Great's definitions of the good in his ethical, metaphysical, and theological writings. During Albert's lifetime, Aristotle's major extant works on philosophy were disseminated widely among the Scholastic elites, giving Albert access to the Stagirite's *Metaphysics, Physics,* and *Nicomachean Ethics* as he rethought his expressions of Christian ethical teachings.

From the very start of his career, Albert grappled with the question of the good and ethics in general.[1] In his early works, he built his insights on doctrines of his Parisian teachers and on Aristotle's *Ethics* as read in three incomplete translations, the *Ethica vetus* (Books I and III), the *Ethica nova* (Book I), and

1 Studies on the concept of the good in Albert the Great are Schneider, *Das Gute*, pp. 39–100, especially pp. 39–55 and pp. 79–91; Ribes Montané, *Verdad y bien* (a general account, but in all a reliable exposition of Albert's metaphysics); Anzulewicz, '*Bonum* als Schlüsselbegriff' (on the central role of the concept of the good in Albert the Great, which runs through his entire oeuvre); Müller, 'Der Begriff des Guten im zweiten Ethikkommentar', especially pp. 318–44; Cunningham, *Reclaiming Moral Agency*, especially pp. 93–111 (on the metaphysics of the good, but only on some works). Only Schneider's survey focuses to some extent on the definitions of the good across all Albert's works. For a new and detailed reappraisal of Albert's concept of the good, see Uscatescu Barrón, *Der Begriff des Guten*, pp. 316–68. The present paper is not a resumé of the latter, but a reelaborated new version of some passages along with additions (e.g., on Algazel).

**Jorge Uscatescu Barrón** (jorge.uscatescu@t-online.de) teaches at the University of Freiburg and researches in ontology, ethics, aesthetics (antiquity, Middle Ages, Second Scholasticism, and phenomenology), and Indian philosophy (Upaniṣads).

*Albert the Great and his Arabic Sources*, ed. by Katja Krause and Richard C. Taylor, Philosophy in the Abrahamic Traditions of the Middle Ages, 5 (Turnhout: Brepols, 2024), pp. 43–68
BREPOLS PUBLISHERS 10.1484/M.PATMA-EB.5.136482

the *Ethica Borghesiana* (excerpts of Book VII).[2] Not until 1249 did Albert have a complete translation of *Nicomachean Ethics* at his disposal, made by Robert Grosseteste between 1240 and 1249.[3] This complete translation, together with Robert Kilwardby's *Expositio super libros Ethicorum* (1246–47), deeply affected Albert's teachings on the good and ethics,[4] as can be observed in his first commentary on the *Nicomachean Ethics, Super Ethica* (1249–52).[5] Albert's different ethical treatises thus reflect the changing availability of sources.

The first treatise Albert wrote was a relatively short work entitled *De natura boni*. It dates from before his time in Paris, when he was still in Germany (1230–40). For this treatise, Albert predominantly used a Latin source, Philip the Chancellor's (1160–1236) *Summa de bono* (1225–28). Due to the moral-theological approach that Albert favoured in this short treatise, he did not yet undertake a survey of the concept of the good.[6] Some years later, as a teacher at the University of Paris, Albert wrote a second treatise on the same subject, entitled *De bono* (1245–46), and now changed his approach to a metaphysical-ethical one. In this treatise, still influenced by the incomplete translations of Aristotle's *Nicomachean Ethics*, he ventured into the new realm of moral teachings strictly related to Aristotelian ethics, and conceived his draft of the 'metaphysics of the good' that he continued to endorse for more than three decades.

In this paper, I additionally draw on Albert's commentary on the *Sentences* (1246–49) to illuminate how, in his metaphysical approach, he elaborated his conception of the good by appropriating some Arabic ideas in order to form key definitions in his early works (I leave aside other aspects of his metaphysics of the good). Moreover, I show how Albert introduces into the discussion the Neoplatonic idea of the good as *diffusivum sui et sui esse*, in his commentary on Pseudo-Dionysius's *De divinis nominibus* (1250),[7] and how this idea helps him

2 Wilhelm Kübel in Albertus Magnus, *Super Ethica, libri I–V*, ed. by Kübel, p. xii. See in general Cordonier, De Leemans, and Steel, 'Die Zusammenstellung des "corpus aristotelicum"', especially pp. 153–54.

3 Callus, 'Date of Grosseteste's Translation', especially pp. 202–08, on the reception of Aristotle after the interdiction on reading his works in 1210 and 1215 and the recommendation of his ethical treatises by Cardinal Robert de Courçon in 1250; Gauthier, 'Le cours sur l'*Ethica nova*'; Gauthier, 'Notes sur les débuts'; Lafleur and Carrier, *Le 'Guide de l'étudiant'*; Buffon, 'Philosophers and Theologians on Happiness'; Zavattero, 'Éthique et politique'.

4 See the list in Thomas Aquinas, *Sententia libri Ethicorum, libri I–III*, I.2, praefatio, ed. by Gauthier, pp. *236–*238.

5 Albertus Magnus, *Super Ethica, libri I–V*, ed. by Kübel, prolegomena, p. vi a.

6 See Tarabochia Canavero, 'A proposito del tratatto'; Tarabochia Canavero, 'I *sancti* e la dottrina'.

7 Albertus Magnus, *Super Dionysium De divinis nominibus*, ed. by Simon. See also Albertus Magnus, *Summa theologiae I*, tr. 6, q. 26, cap. 2, a. 3, ed. by Siedler, p. 195, v. 40–p. 198, v. 84. It is well known that Albert composed two commentaries on Aristotle's *Nicomachean Ethics*: the *Super Ethica* around 1249–52, on the basis of the first complete Latin translation by Robert Grosseteste and with the aid of Averroes's *Middle Commentary* and Eustratius's commentary (*Super Ethica*, ed. by Kübel, prolegomena, pp. vi a and vii b); and the *Ethica*, around 1262. *Ethica* was edited by Auguste Borgnet in vol. 7 of the *Opera omnia*. There is now a critical edition (with extensive introduction) of Book I,

to achieve an increasingly solidified conception of his conception of the good. Finally, I discuss why Albert's *Summa theologiae* (1268–74) shows no essential change in the concept of the good, even though it certainly represents a great advance in the systematization of Albert's theological ethics.[8]

## Appropriating Avicenna's Definition of the Good in *De bono*

From the beginning of his *De bono*, Albert displays a marked awareness of the great variety of definitions of the good, which, as he remarks, cannot be reduced to one alone. He sees his immediate contribution to the question of the good as being to present briefly these traditional definitions along with their rational foundations, as well as replies to the objections levelled against each definition.[9] In this respect, Albert can be said to limit himself to relating the traditional definitions of the good to the arguments put forward for each one, without granting priority to any one of these definitions, each of which is considered equally valid in its own right. It should also be noted that he emphasizes the particular character of the definitions of first concepts.[10] In this section, I focus on the two definitions of the good that derive from Arabic sources, especially Avicenna's (980–1037) definition and its reception and development in the works of Albert.

Albert's list of three major definitions of the good begins with the one given by Aristotle in his *Nicomachean Ethics* (I.1, 1094a2–3): *Optime annuntiant bonum, quod omnia appetunt* ('They most adequately declare the good [as that] which all strive for').[11] The formulation of this passage preferred by Albert reveals a particular version of Aristotelian thought, and the Latin formulation giving a general definition of the good coincides exactly with the original.

In order to defend his own interpretation, which restricts the universal range of the Aristotelian definition, Albert provides two syllogistic arguments. In his

---

tr. 2, chapters 1–7 by Jörn Müller: 'Der Begriff des Guten im zweiten Ethikkommentar', pp. 345–70, though this particular passage is not relevant to the present paper. For the concept of the good in *Ethica*, see also Uscatescu Barrón, *Der Begriff des Guten*, pp. 335–42.

8 See Dionysius Siedler in Albertus Magnus, *Summa theologiae I*, ed. by Siedler, prolegomena, p. xvi b.

9 The following doctrinal opinions about the good are here omitted: *bonum diffusivum sui*, a formula otherwise well known to Albert (Albertus Magnus, *De bono*, ed. by Kühle et al. [hereafter '*De bono*'], I.1.1.21, p. 12, vv. 30–33); the good as goal (*De bono*, I.1.1.21, p. 12, vv. 34–35); and the good is something that there is better than not (Anselm, *Proslogion*, 5, ed. by Schmitt, vol. 1, p. 104, vv. 15). See Albertus Magnus, *Summa theologiae I*, tr. 6, q. 26, cap. 1, a. 2, II, ed. by Siedler, p. 175, vv. 50–51.

10 *De bono*, I.1.1.10, p. 6, vv. 63–66: 'Ad id autem quod obicitur, quod bonum non sit diffinibile, dicendum, quod non habet diffinitionem completissimam, sed tamen potest habere assignationes'.

11 Ibid., I.1.1.1, p. 1, v. 12. Albert's translation is taken from the *Ethica nova* (see editor's note, ibid.): 'Ideoque optime enunciant bonum, quod omnia optant' (*Ethica Nicomachea*, ed. Gauthier, p. 65, vv. 5–6). Moerbeke's later translation does not differ much from the older version: 'Ideo bene enunciaverunt, bonum quod omnia appetunt'. Quoted in Thomas Aquinas, *In decem libros Ethicorum*, ed. by Spiazzi, p. 3.

first argument, he reasons that all beings except God, who is naturally conjoined to the good (*cum bono coniuncto*), strive for an end. Since the good and end are the same, being that is ordered to an end (*ordinatio ad finem*) amounts to striving for the good. Therefore, all created beings strive for the good, *quod omnia appetunt*. This rests on two assumptions: that nothing is desired in vain (*vanum*),[12] and that desiderability and end are identical. The first of those assumptions derives from the more general insight expressed by Aristotle in several passages, according to which nature does not make anything in vain and operates teleologically. As the natural desire is a genuine expression of nature itself, it cannot be directed vaguely to something in general; the desired must be something that could be attained. The second assumption is borrowed from Aristotle's *Nicomachean Ethics* I.1, a very well-known passage in which what is final cause and the good are defined as what is desired, and so being desired and desirability characterize both final cause and the good.

The second argument takes into account the ontological structuring of a creature. Before existing, a creature is only a possible being, but its potentiality is directed to its actuality. Hence, a possible being can be defined as a being directed to its actuality as the end of its potentiality. In this argument, Albert observes an essential conjunction between actuality and being good, and infers from it that all beings desire the good. This point, in particular, anticipates the second definition of the good that Albert attributes to Avicenna: *bonum est indivisio actus a potentia* ('the good means the lack of difference between actuality and potentiality').[13] In fact, when we read the Persian philosopher's *Liber de Philosophia prima*, it soon emerges that he never gives a definition of the good of this kind.[14]

12 As is stated by Aristotle, *Pol.*, I.6, 1253a9; *On the Heavens*, I.4, 271a33. The thought is expressly assumed in Albertus Magnus, *Metaphysica*, II.8, ed. by Geyer, p. 100, vv. 38–45: 'nihil enim optat hoc naturali appetitu quod numquam contingit consequi, quia esset vanum et inutile naturale desiderium [...]. Infinitum igitur faciens in causa finali aufert rationem boni entium'.

13 *De bono*, I.1.1.1, p. 1, vv. 13–14 and p. 2, vv. 10–35. On this definition, see also Albertus Magnus, *Super Dionysium De divinis nominibus*, I, n. 60, ed. by Simon, p. 37, vv. 46–47. On this definition somehow expressed in Avicenna and its fortunes from Philip the Chancellor to Albert the Great, see Uscatescu Barrón, 'Zur Bestimmung des Guten'. In the present article, I develop particular aspects of Albert's appropriation of Avicenna's formula that are not discussed there or in *Der Begriff des Guten*, pp. 316–59, as well as Albert's consideration of an Algazelian definition of good, which was omitted in both my earlier studies.

14 For a discussion of Avicenna's concept of the good, see Uscatescu Barrón, *Der Begriff des Guten*, pp. 220–38. The process of appropriation of Avicennian philosophy is by no means limited to the concept of the good. It also extends to the more general concept of being. Being (*ens*) as the first known is the *prima intentio intellectus* to which all other concepts ultimately refer. See Albertus Magnus, *Super Dionysium De divinis nominibus*, IV, n. 5, ed. by Simon, p. 116, vv. 55–57. Being is the first concept that is conceived *prius et simplex* by the intellect (see Albertus Magnus, *Commentarii in I Sententiarum*, dist. 46 N, a. 14, ed. by Borgnet, p. 450a). On the other hand, *esse in actu* or *existentia* are prior to *essentia*. Although Albert has overcome the hylomorphism widespread in the twelfth and thirteenth centuries with the help of the established concepts *esse quod* and *esse quo*, which correspond to *existentia* or *esse in actu* and to *essentia* respectively, he still assigns *esse in actu* or *existentia* to form and *esse in potentia* to matter. See Ducharme, '*Esse* chez saint Albert le Grand',

Instead, in Avicenna's *Liber de Philosophia prima*, we find dispersed passages that treat of the good in closely related but much more complex ways.[15] In a passage explaining the relation between actuality and potentiality (*Liber de Philosophia prima*, IV.2),[16] Avicenna concisely expounds the various meanings of potentiality. In his defence of the ontological, epistemological, and definitional priority of actuality over potentiality, he speaks of the priority of actuality due to its perfection. Contrasting potentiality as 'not-yet-being', a privation, to actuality as being, a perfect state, Avicenna thus defines the good as the actuality of potentiality: 'Likewise, act [*effectus*] is prior to potency in its end and perfection. For potency is imperfection and act is perfection, and the good in all things [*in omni re*] is nothing but that which [*ipsam*] is in act'.[17]

The Latin medieval translation of the Arabic original in the passage above highlights the identity between the good and actuality, since it links *ipsam* to

---

especially pp. 223–24 with some quoted passages. For a general assessment of the concept of being in Albert, see Vargas, 'Albert on Being and Beings'.

15 The text was well known at the University of Paris before Albert; see Bertolacci, 'On the Latin Reception'. Unfortunately, Bertolacci does not mention Philip the Chancellor, who made use of Avicenna's *Metaphysics* perhaps more widely than Wicki's sources index suggests.

16 Latin citations are from the medieval translation edited by Simone Van Riet (Avicenna, *Liber de Philosophia prima*, ed. by Van Riet). The Latin text is very important for the transmission of the Arabic text, because the manuscripts of the Latin translation are much older than the oldest remaining Arabic manuscript. Moreover, the medieval reception of Avicenna rests exclusively on the translation into Latin and not on the Arabic text, which may have been accessible only to the translator and philosopher Gundissalinus. On Avicenna's metaphysics, see Cruz Hernández, *La metafísica de Avicena*; Verbecke's introduction in Avicenna, *Liber de Philosophia prima*, ed. by Van Riet; Verbecke, *Avicenna*; Gutas, *Avicenna and the Aristotelian Tradition*; Wisnovsky, *Avicenna's Metaphysics in Context*; see also the general introduction, the introductions to the books, and the notes by Olga Lizzini in Avicenna, *Metafisica*; also Bertolacci, *Reception of Aristotle's 'Metaphysics'*. In the last ten years other monographs on Avicennian metaphysics have been published, but his concept of the good has not been considered in either the older or the more recent publications.

17 Avicenna, *Liber de Philosophia prima*, IV.2, ed. by Van Riet, p. 212, vv. 42–44: 'Item effectus prior est potentia perfectione et fine. Potentia enim est imperfectio et effectus est perfectio, et bonum in omni re non est nisi ipsam esse in effectu'. This passage has seen a number of translations into modern European languages. The Italian translation by Olga Lizzini, for instance, runs as follows: 'L'atto è inoltre prima della potenza nella perfezione e nel fine; la potenza infatti è mancanza, mentre l'atto è perfezzione e il bene, in ogni cosa, si ha solo con l'esse in atto' (Avicenna, *Metafisica*, ed. by Lizzini and Porro, p. 409). One hundred years ago, before the critical text appeared, Max Horten offered a similar translation in German: 'Die Aktualität ist eine Vollkommenheit und das Gute, das sich in jedem Dinge befindet in Verbindung mit dem Aktuellsein. Überall aber, wo sich das Böse findet, ist auch etwas, was in gewisser Weise in Potenz ist' (Avicenna, *Die Metaphysik Avicennas*, ed. by Horten, pp. 273–74). The German translation shows that the Latin translation itself suggests the concept of connection or conjunction, which later on was rendered as *indivisio*. Georges Anawati translates similarly into French: 'La puissance est une imperfection et l'acte est une perfection. Et le bien dans toutes choses n'est que dans le fait d'être en acte; là où il y a le mal il y a quelque chose, qui d'une certaine manière est en puissance' (Avicenna, *La Métaphysique*, ed. by Anawati, vol. 1, p. 225). Finally, the English translation by Michael Marmura runs: 'Also, act is prior to potency in perfection and purpose. For potentiality is a deficiency, while actuality is a perfection. The good in all things is in conjunction with being actual' (Avicenna, *Metaphysics of 'The Healing'*, ed. by Marmura, p. 142).

*res,* and then implicitly refers to this in the subsequent passage: 'It is therefore manifest that that which is in act is a good inasmuch as it is in this way. However, that which is in potency, is bad, or the bad comes from it'.[18] Thus, while Avicenna's wording does not provide the precise formulation *indivisio actus a potentia,* the Latin translation of his *Liber de Philosophia prima* suggests that interpretation.

Familiar as he certainly was with Avicenna's passage and Aristotle's parallel discussions in his *Metaphysics,*[19] Albert might thus be considered the first Scholastic to have attributed the formula *indivisio actus a potentia* to Avicenna, with whose metaphysics he had been very well acquainted since his time in Paris.[20] In fact, however, Philip the Chancellor had already explicitly referred to the translated passage *bonum in omni re non est nisi ipsam esse in effectu,*[21] In the secondary literature, Louis-Bertrand Gillon was the first to trace this precise formula for the good to Philip the Chancellor, but he assumed wrongly that Avicenna was its true originator.[22]

It is true that in his *Summa de bono,* Philip discussed several definitions of the good. The third of these is introduced as follows: 'Likewise, another [definition] is derived from Aristotelian and other philosophers: "the good is to have the indivisibility of act from potency simply-speaking or in some way"'.[23] This passage

---

18 Avicenna, *Liber de Philosophia prima,* IV.2, ed. by Van Riet, p. 213, vv. 56–58: 'Manifestum est igitur quia id quod est in effectu bonum est inquantum sic est; quod vero est in potentia, est malum vel ab ipso est malum'. The Arabic expression *bil-fʿl* was translated as *in effectu,* although sometimes *in actu* is read. This expression also appears in the other translations, as is noted by Goichon, *Lexique de la langue philosophique d'Ibn Sīnā,* p. 277.

19 See Aristotle's *Metaphysics,* IX.9, 1051a13 and especially 19–21, where he writes — according to the old Latin translation, which Albert could have used (*Metaphysica,* ed. by Vuillemin-Diem, p. 181, vv. 4–7): 'Palam ergo quia non aliquid malum preter res ipsas; posterius enim natura malum potentia. Non ergo nec eis que sunt a principio sempiternis nichil inest nec malum nec peccatum nec corruptum (etenim differentia est malorum)'. In the English version in Aristotle, *Complete Works of Aristotle,* ed. by Barnes, p. 1660, the text runs: 'And therefore we may also say that in the things which are from the beginning, i.e. in eternal things, there is nothing bad, nothing defective, nothing perverted (for perversion is something bad)', suggesting the identity of perfection and eternity, and secondly that 'the actuality, then, is better' (a13).

20 *De bono,* I.1.1.1, p. 1, vv. 13–14. There has still been no serious attempt to appraise Albert's use of Avicenna's *Metaphysics,* which is surely embedded in the larger context of its reception from the early thirteenth century at the University of Paris. See Bertolacci, 'Albert's Use of Avicenna' (restricted to a few points in Albert's *Metaphysica*); also Bertolacci, 'Albert the Great and the Preface'.

21 Avicenna, *Liber de Philosophia prima,* IV.2, ed. by Van Riet, p. 212, vv. 42–44; Philip the Chancellor, *Summa de bono,* ed. Wicki, p. 7, vv. 34–41: 'bonum est habens indivisionem actus a potentia simpliciter vel quodammodo. "Simpliciter" dico ut in Primo; in divina enim essentia idem est potentia cum actu [...]. Alia secundum quid habent huiusmodi indivisionem, sed non simpliciter, cum quid sit ibi de potentia et ita de incompletione'; also ibid., p. 27, vv. 37–40: 'respondeo quod duplex est actus; est actus primus ipsius rei perfectio et secundum hoc bonum dicit actum, et est actus secundus quid de re egreditur, et sic dicit iustum, bonum autem non. Et sic primo modo idem est bonum dicere actum et substantiam'.

22 Gillon, *La théorie des oppositions,* p. 49.

23 Philip the Chancellor, *Summa de bono,* ed. by Wicki, p. 6, vv. 21–22: 'Item, alia extrahitur ab Aristotele et aliis philosophis: "Bonum est habens indivisionem actus a potentia simpliciter vel quodam modo"'.

reveals that Philip was well aware of the formula's Aristotelian origin and its Avicennian development (even though he does not explicitly allude to it here).[24] Philip took the idea expressed in this formula to be correct in two ways: simply speaking when applied to God, in whom actuality and potentiality coincide, 'and "in some way" when applied to creatures in whom actuality and potentiality coincide with one another only relatively [*secundum quid*]'.[25]

Moreover, in order to establish a homogeneous theory of the transcendentals based on the principle of negation, Philip had to adapt the definitions to his new conception. One had been already been defined by Aristotle negatively as *indivisio*,[26] but the other ones still had a positive definition. He therefore applied the concept of indivision to the concept of good by taking perfection as a key for understanding the good, an idea he had found in Avicenna (Aristotle), and defining it as indivision of act and potentiality. This new formula fits very well with the theory of transcendentals that Philip conceived for the first time in the history of philosophy,[27] and allows us to credit him with being the true originator of this 'Avicennian' formula.

Rather than Avicenna, therefore, Philip is the verbatim source of Albert's second definition of the good in his *De bono*, and also of one of his arguments to justify its use. Albert's attribution of the formula to Avicenna may be due to a Scholastic custom of ascribing views or opinions to authorities and not to contemporary masters, thus lending more weight to the opinion at stake. But Albert may also have been encouraged by the fact that Algazel, who was viewed by the Latins as a very close follower of Avicenna, had put forward a definition of the good — the third one under discussion in *De bono* — that could easily reflect another aspect of Avicenna's metaphysical idea. In this way, Albert could filiate two definitions of the good to two leading authorities in the Arabic world as part of Peripatetism, in which he includes himself.

Philip's authorship can be traced in Albert's second argument in favour of this formula, where he reasons negatively to the fundamental identity of actuality and the good and shows that being (*esse*) is due only to form or actuality, whereas potentiality is due to lack of form, namely to lack of matter. Another argument for

---

24 See text in note 23.

25 Philip the Chancellor, *Summa de bono*, ed. by Wicki, p. 7, vv. 35–45: '"Simpliciter" dico ut in Primo; in divina enim essentia idem est potentia cum actu. Et dicitur actus sine potentia secundum quod potentia sonat in quandam incompletionem. Et secundum hoc dictum est: "Nemo bonus nisi solus Deus", scilicet simpliciter indivisionem habens actus a potentia, sicut Primo cui totum est actus. Alia secundum quid habent huiusmodi indivisionem, sed non simpliciter, cum quid sit ibi de potentia et ita de incompletione [...]. In Primo ergo absoluta bonitas, in aliis secundum quid. Et secundum assimiliationem ampliorem cum actu qui est indifferens a potentia, id est cum Primo qui est pure actus sine potentia, dicetur magis bonum et secundum minorem assimilationem minus bonum'.

26 Ibid., p. 7, vv. 31–32: 'sicut nec unum cum dicitur unum est ens indivisum; "indivisum" enim point ens et privat ab ente divisionem'.

27 Ibid., p. 19, vv. 97–98. See note 38 below. For more details on Philip's theory of the transcendentals, see Uscatescu Barrón, *Der Begriff des Guten*, pp. 273–86.

the indivisibility of actuality and potentiality directly follows this argument, and suggests that if 'being actual' amounts to identity of actuality and potentiality, the identity of the good and 'being actual' follows. Just as in Philip's definition, Albert concludes here that this is the reason why the good can be understood as *indivisio actus a potentia*.[28] Especially in these two arguments, Albert thus shows a close overlap with his Latin predecessor in language and argumentation.

In addition to Philip's simplified formula, however, in his first argument Albert also adopted Avicenna's more complex explanation of the identity of actuality and potentiality — a formulation based on the idea of perfection as 'that which lacks nothing inasmuch as it has attained its end'.[29] In his argument for the 'Avicennian formula', this wording manifestly implies the identity of the good and what has attained its end, namely the perfect. Albert reveals its importance in a lengthy discussion that raises four different objections, only of course to refute them subsequently.[30] In his initial, general response to most of these contrary arguments, Albert stresses the naturality of striving and its commonality to all living beings, as opposed to the *appetitus perfectus*, which only pertains to sentient beings.[31]

In the first particular objection, Albert raises the additional difficulty that the term *indivisio* may sound like the term *privatio*.[32] His reply hints at the two Aristotelian passages mentioned above, with which Avicenna may have been familiar, and reasons that 'on the basis of these two authorities, some [people] have wished to derive the definition that the good amounts to an indivisibility between actuality and potency'.[33] Potentiality, however, is indifferent to goodness and badness, and only through its relation to actuality it is good, whereas in relation to privation it is bad. Here Albert borrows Avicenna's interpretation on evil as privation, as explained in the *Liber de Philosophia prima*, IX.6.[34] In reliance on Aristotle's *Physics* I.9, 192a16ff., Albert then defines actuality as a good, but

28 *De bono*, I.1.1.2, p. 2, vv. 25–35.

29 Ibid., vv. 10–24, especially vv. 10–13: 'Omne perfectum secundum quod huiusmodi est completum per finem; omne bonum est bonum secundum rationem finis; ergo omne bonum secundum quod huiusmodi est perfectum'.

30 Ibid., I.1.1.4, p. 3, vv. 25–49; I.1.1.7, p. 4, v. 48–p. 5, v. 23. See Uscatescu Barrón, *Der Begriff des Guten*, pp. 328–29.

31 On the various forms of striving-after, see Albertus Magnus, *Commentarii in III Sententiarum*, dist. 27 B, a. 4, ed. by Borgnet, p. 520b. On striving and will in Albert the Great, see Schneider, *Die Psychologie des Alberts des Großen*, especially vol. 1 / 1, pp. 255–92. Schneider argues that Albert does not care much about the appetitive faculty, and adds that 'his investigations on the will are relatively sparse' (pp. 261–62). On the striving after the highest good immanent to all beings, see pp. 278–82, where Schneider stresses Albert's Neoplatonic treatment of the issue.

32 *De bono*, I.1.1.4, p. 3, vv. 26–28.

33 Ibid., I.1.1.8, p. 5, vv. 33–35: 'Et ex istis duobus locis volunt quidam trahere hanc diffinitionem, quod bonum est indivisio actus a potentia'.

34 On Avicenna's conception of evil as privation, see Steel, 'Avicenna and Thomas Aquinas'.

unlike Avicenna, he refrains from identifying potentiality with badness.[35] Albert suggests that the Avicennian formula leaves room for the misunderstanding that indivisibility is a mixture of the good and the bad. In order to avoid that misconception, he suggests that actuality completes the good, which is necessarily perfect.[36] Indeed, indivisibility should be understood as actuality alone, which accrues to the perfect good especially.

In the second objection that Albert raises,[37] then refutes, the relation between actuality and potentiality is expanded upon: their indivisibility is taken to also entail a composition of actuality and potentiality. To refute this ontological assumption, Albert suggests in his response that the indivision of actuality and potentiality is not given in beings in which potentiality does not precede actuality: they are simple beings without composition. The other beings, however, possess a composition of actuality and potentiality, because the latter precedes the former. In this case, there is only a conjunction of potentiality and actuality, which implies a third element that unites them (*coniungens*).[38] Albert tries to distinguish sharply between indivisibility or identity and unifying conjunction, which necessarily supposes a third element, the unifier.

In the third objection, Albert presents an objector maintaining that neither potentiality nor actuality can lay claim to goodness on its own, but only together.[39] In response to this objection, Albert argues that an *indivisio actus a potentia* is also given in every being composed of actuality and potentiality, if we understand this *indivisio* instead as directedness of potentiality to actuality. This directedness should therefore be interpreted as *indivisio* not *simpliciter*, but *quodam modo*, following Philip the Chancellor.[40] The refutation of the third objection is ultimately based on the Avicennian argument that presents the absolute simple being as *ens per se sive ens necessarium*, without any cause in respect to which it could be possible.[41]

---

35 *De bono*, I.1.1.8, p. 5, vv. 24–47. Albert says only in lines 36–37: 'Potentia enim per se nec bona nec mala est, sed per privationem est mala et per actum est bona'. This phrase is only understandable in relation to Dionysian thought, as already noted.

36 Ibid., vv. 44–47: 'Ex hoc enim patet, quod actus est simpliciter bonum, potentia autem secundum quid; perfectum vero bonum est completum per actum. Et propter hoc illa diffinitio data est a bono perfecto'.

37 Ibid., I.1.1.4, p. 3, vv. 29–37.

38 Ibid., I.1.1.8, p. 5, vv. 48–66. Albert follows Philip, who also rejected the identification of *indivisio* with *coniunctio* (Philip the Chancellor, *Summa de bono*, ed. Wicki, p. 19, vv. 97–99): 'Item in diffinitione veri et boni sumitur "indivisio", non "coniunctio" vel "participatio" vel aliquid positivum, quia universalius dictum est per privationem'. For Philip, the *indivisio*, a negative determination, is more universal than a positive and therefore fits much better with the nature of the transcendental properties of being, which accordingly shows a negative structure. Albert rejects the equivalence of *indivisio* and *coniunctio* on another ground, namely that *coniunctio* presupposes a third term, which unifies both members of a composition. That is not the point made by Philip.

39 *De bono*, I.1.1.4, p. 3, vv. 38–41; I.1.1.3, p. 2, vv. 59–65.

40 See note 23.

41 Avicenna, *Liber de Philosophia prima*, I.7, ed. by Van Riet, p. 54, vv. 44–55.

In the fourth objection, finally, a contradiction between the first and second definition of the good is raised.[42] Albert rejects this criticism by arguing that striving after both the good and actuality is not different from the good itself, inasmuch as the striving is directed towards the good. Hence, directedness towards the good implies an indivision, and therefore entails goodness.[43]

Read in conjunction, the arguments against the four objections made by Albert establish an essential connection between the first three chapters of the fourth book of Avicenna's *Metaphysics*, in which the concept of potentiality is analysed and connected with the actuality and ultimately with perfection.[44] They enrich the complex formula that Philip the Chancellor took from Avicenna, by bringing together the indivisibility of actuality and potentiality with the idea of perfection as a finalistic process that leads to the good as acquired and then possessed perfection.[45] At the end of his discussion in *De bono*, Albert even speaks of an indivisibility of the end (*indivisio finis*) as the essence of the good (*ratio boni*). This connection is once again reminiscent of Philip's discussion, which assigned the indivisibility of actuality from potentiality (*indivisio actus a potentia*) to the concept of end, based on the idea of the good as perfect, since the achieved end renders equally perfect the being who strives after it as good.[46] This

---

42 *De bono*, I.1.1.4, p. 3, vv. 42–49.

43 Ibid., I.1.1.9, p. 6, vv. 30–32.

44 See Uscatescu Barrón, *Der Begriff des Guten*, pp. 225–31.

45 Avicenna himself had worked out this point, distinguishing between *finis* and *terminus*. The end is always in respect to an agent which intends it and makes it perfect on attaining the end. The movement is not made perfect, however, but is destroyed as soon as the mobile comes to the end (*terminus*). The good is that which is made perfect by going from potency to actuality: the acquisition of perfection. The good is what is attained in perfection, as far as it is in actuality. See Avicenna, *Liber de Philosophia prima*, VI.5, ed. by Van Riet, p. 340, v. 63–p. 341, v. 86, especially p. 341, vv. 66–71: 'Consideratus vero respectu recipientis quod perficitur per illum, cum ipsum est in potentia, est bonum quod adaptat ipsum, malum vero privatio est suae perfectonis, sed bonum quod est ei oppositum est esse et acquisitio in effectu'.

46 See Albertus Magnus, *Commentarii in I Sententiarum*, dist. 46 D, a. 2, ad 3, ed. by Borgnet, p. 427b: 'sed bonum dicit perfectionem quae est ex fine absolute'. Philip the Chancellor considers the definition of the good as *indivisio actus a potentia* the first and principal reason (Philip, *Summa de bono*, ed. by Wicki, p. 7, vv. 33–34: 'Dicimus autem rationem illam rationem primam et principalem: bonum est habens indivisionem actus a potentia simpliciter vel quodammodo'), whereas the other two definitions — good as 'diffusivum aut multiplicativum esse' and as final cause — are consequences of the primary definition (p. 7, v. 46–p. 8, v. 56, especially vv. 46–47: 'Altera ratio ipsius boni secundum quod dicitur: bonum est diffusivum aut multiplicativum esse, data est per posterius, per proprietatem consequentem bonum'). The third definition concerns the final causality as essence of the good. This definition presupposes a natural striving in all beings to a variable degree. Aware of the *scala naturae*, Philip remarks regarding the generality of this definition (ibid., p. 19, vv. 83–88): 'hec diffinitio: bonum est quod desideratur ab omnibus absolute dicta est summi boni, sed secundum mensuram bonitatis et unoquoque contrahitur, ut sic intelligatur: bonum in unoquoque genere est quod desideratur ab omni illius generis, sicut sanitas est bonum desiderativum a sanabili, unicuique enim actui diverso a potentia respondet potentia, et est accipere omne in unoquoque genere'. The good is essentially connected with the act and the perfection (ibid., p. 27, vv. 37–38: 'respondeo quod duplex est actus; est actus primus ipsius rei perfectio et secundum hoc bonum dicit actum'.

overlap also seems to suggest that even though Albert accepts all three definitions mentioned in his *De bono*, he prefers the definition of the good as end or as being directed towards an end, without explaining the relationship between final cause and directedness to a final cause.[47]

This preference is corroborated when Albert discusses the third definition of the good, which he derives from his second Arabic source, Algazel's *Metaphysics*. Albert's exposition corresponds to his view of Algazel as a follower of Avicenna (*insecutor Avicennae*,[48] or *abbreviator Avicennae*[49]), which goes so far as to suggest that Algazel's *Metaphysics*, the second book of *Maqāsid*, is no more than a brief summary of Avicenna's *Liber de prima philosophia*.[50]

In the Latin translation of this work, *Logica et philosophia*, Algazel (1058–1111) defines the good as actuality whose apprehension is followed by pleasure (*delectatio*).[51] But the text also defines the good as perfection. Whereas evil consists in privation or destruction, good inversely consists in perfection. In this

He understands the relation between act and final cause such that act is fundamentally referenced on final cause (ibid., p. 19, vv. 78–79): 'Respondeo ad primum quod "actus" non dicitur actio sed complementum per modum finis, Unde non est obiectio de forma nisi secundum quod finis'. Philip had seen the conception between perfection and final causality as it is stated in the Aristotelian philosophy, and Albert follows this tradition.

47 *De bono*, I.1.6.21, p. 11, vv. 82–86: 'Intentio enim entis est intentio simplicissimi, quod non est resolvere ad aliquid, quod sit ante ipsum secundum rationem. Bonum autem resolvere est in ens relatum ad finem'.

48 Albertus Magnus, *Metaphysica*, V.1, cap. 4, ed. by Geyer, p. 217, v. 28.

49 On Albert's relationship with Algazel, see Cortabarría Beitia, 'Literatura algazeliana' (on Albert's characterization of Algazel as Avicenna's follower, pp. 260–61).

50 Albertus Magnus, *Secunda pars Summae de creaturis*, tr. 1, q. 55, a. 3, et videtur quod, n. 7, ed. by Borgnet, p. 462a: 'Idem omnino dicit Algazel in sua *metaphysica*: quia dicta Algazelis non sunt nisi abbreviatio dictorum Avicennae'. In this question, Albert notes Avicenna's opinion that the *intellectus agens* is separate intelligence which belongs to the tenth order of secondary intelligences (n. 6, p. 462a), as seen in Avicenna's *Liber de Philosophia prima*, XI. Albert finds the same idea in Algazel's *Metaphysics* concerning this question, but he generalizes Algazel's doctrinal indebtedness to Avicenna. See Cortabarría Beitia, 'Literatura algazeliana', pp. 267–68, where the reference to Algazel in Albert's *De bono* is not mentioned. See note 51 for Albert on the Algazelian text.

51 Algazel, *Logica et philosophia*, ed. by Liechtenstein, fol. 44 ra: 'Bonum vero est perfectio, cuius apprehensio est delectatio' (= *Algazelis Metaphysica*, pars 1, tr. 5, in Algazel, *Metaphysics*, ed. by Muckle, p. 129, vv. 4–5, with the same wording). Albert says, instead: 'bonum est actus, cuius apprehensio est cum delectatione' (*De bono*, I.1.1.1, p. 1, vv. 16–17). The critical edition does not corroborate Albert's wording, but Albert's choice of *actus* instead of *perfectio* captures the Algazelian thought borrowed from Avicenna. The *Logica et philosophia* of Algazel (al-Ghazālī) is a Latin translation of *Maqāṣid al-falāsifa* going back to the end of the twelfth century in Toledo. In this work, dated 1091/92–94 (see Hourani, 'A Revised Chronology', esp. p. 292), Algazel largely draws on Avicenna's *Dānesh-nāmeh*. The work is divided into two parts. The Latin version of the *Logic* was edited by Charles Lohr: *Logica Algazelis*, ed. by Lohr. The Latin version of *Metaphysics* was critically edited by J. J. Muckle: *Algazel's Metaphysics*, ed. by Muckle (see the review by Alonso Alonso, 'Los Maqqsāṣid de Algazel'). The prologue contained in only one manuscript of the Latin version is missing in the oldest Arabic manuscripts. See Janssens, 'al-Ghazālī's *Maqāṣid al-Falāsifa*'. For the influence of Algazel in the Latin West in general, especially in Raimundus Lullus and Ramón Martí, see d'Alverny, 'Algazel dans l'Occident latin'.

formula, Algazel expresses the same thought as Avicenna, because perfection is opposed to potentiality as a form of privation.[52] The second part of the definition constitutes the third opinion that Albert adopts in his explanations. In his justification of this last definition of the good, Albert first maintains that nature and intelligence (*intelligentia*) are equal to one another inasmuch as they direct their respective actions to an end. The middle term of Albert's argument expresses the idea that *delectatio* arises from the confluence of two things which are mutually suitable (in the literal sense of the Latin): if the suitable is identified with the end, its apprehension is pleasant in itself. Later, Albert emphasizes the naturality of both *appetitus* and *delectatio*,[53] and consequently their universality. The natural striving for pleasure should be understood correspondingly as inscribed in all beings independently of whether they have sensation or not. This *appetitus naturalis* is flanked by a *delectatio naturalis* in a very broad sense, too, which Albert explicitly distinguishes from the *delectatio* as *passio in anima sensibili* addressed by Aristotle in his *Ethics*, and as related to a potency more universal than the sensible one in sentient beings. Only in this way can the Algazelian definition of the good attain a universal character.

Once again, we learn how Albert thinks of these relations through the objections he raises in order to respond to them.[54] The first objection that Albert presents links the suitable only to perceptible nature. But Albert warns that if this limitation were true, the third definition would not reach the general level required for a definition. In his response, he thus insists that appetite (*appetitus*), the notion at stake here, is of the most general kind. Hence, the required universality of the formula is assured.

The second objection is made on the basis of the concept of actuality, once again in its relation to the good and the bad. Actuality, the objector suggests, seems to be proper to the good and to the bad, because there are good acts and bad acts. If this is the case, however, actuality cannot pertain to being. In reply, Albert draws an important distinction between two meanings of actuality: in the first, actuality is an activity (*operatio*), and correspondingly there are good and bad acts; in the second, actuality is the fulfilment (*complementum*) of something else. Fulfilment, reasons Albert, can be understood either in a formal or a teleological sense. Whereas fulfilment by form (*complementum a forma*) differs from the essence of the perfect (*ratio perfecti*), fulfilment by the end (*complementum a fine*)

52 See Algazel, *Metaphysics*, I.5, ed. by Muckle, pp. 118–29.

53 *De bono*, I.1.1.10, p. 6, v. 33–p. 7, v. 12. On the formula *Delectatio est ex coniunctione convenientis cum convenienti*, see Bonaventura, *In primum librum Sententiarum*, dist. 1, a. 3, q. 1, concl., n. 1, ed. Quaracchi, p. 38b (see also editor's comment on the formula and its sources, p. 38a, n. 4). On the Avicennian origin of the expression, see Avicenna, *Liber de philosophia prima*, VIII.7, ed. by Van Riet, p. 431, vv. 50–433, esp. v. 102.

54 *De bono*, I.1.1.5–6, p. 3, v. 50–p. 4, v. 47; I.1.1.10, p. 6, v. 33–p. 7, v. 12.

corresponds to it, because the end is the ultimate being, beyond which there is nothing. It therefore necessarily implies the idea of perfection.[55]

The third definition of the good, which Albert attributes to Algazel, is contained in the second part of the sentence, and principally concerns the effect of the good (*tertia vero datur de proprio effectu consequente bonum*).[56] In the subsequent development of Albert's ethics, and in his more elaborate discussion of the different definitions of the good in his *Summa theologiae* and his commentary on Pseudo-Dionysius's *De divinis nominibus*, this definition no longer plays any important role, since it expresses the effect of being as actuality or as perfection on the natural appetite, but not the good itself. Indeed, from this point on Albert no longer relied on Algazel as a source for defining the good in his subsequent works, despite frequently invoking him in other metaphysical contexts.[57] What comes to the fore in these later discussions of the good, instead, is Albert's reworking of the Avicennian formula of the good in terms of the idea of the good as final cause.

From the beginning of his *De bono*, Albert intends a metaphysics of the good to be a prolegomena to his ethics, as developed in the pages that follow. The good is a transcendental concept in which the final cause of the appetite of beings and perfection are combined in the general context of a teleological understanding of nature in general and human action in particular. On the other hand, he takes as his point of departure the Boethian distinction between *primum bonum* and *bonum secundum*, but tries to attain a metaphysical speculative level in order to elaborate a general concept of the good.

## Avicenna's Aristotelian Formula Interpreted Neoplatonically

Although in the years after *De bono*, Albert does not explicitly mention the Avicennian formula, the idea implied in it is still recognizable in his definitions or descriptions of the good. In the account of the three transcendental concepts in his late *Ethics* commentary, *Ethica*, the good of a being consists in perfection in being, capability (*posse*), and acting (*agere*).[58] The *indivisio actus a potentia* can still

55 Ibid., I.1.1.10, p. 6, vv. 48–62. This thought can be also found in Philip the Chancellor's *Summa de bono*, ed. by Wicki, p. 19, vv. 78–79.

56 *De bono*, I.1.1.10, p. 7, vv. 11–12. Algazel, *Metaphysica*, ed. by Muckle, p. 129, vv. 1–5: 'Sed timor destruccionis essencie cum percipitur, forcior est timore destruccionis proprietatum, igitur malum est privacio, sed apprehensio privacionis est dolor; bonum vero est perfeccio cuius apprehensio est delectatio'.

57 For other philosophical and theological contexts in which Albert quotes and explains Algazel's thoughts, see Cortabarría Beitia, 'Literatura algazeliana', who does not mention any of the texts discussed here except the first referred to in note 50.

58 Albertus Magnus, *Ethica*, I.2.6, ed. by Borgnet, pp. 17–28, here p. 26a (cf. ed. by Müller, p. 366: 'Bonitas autem in ipso sicut ad perfectum agere suum perfectio').

be easily seen in this short definition. Good will be later defined as self-operation according to something's own nature if it is not impeded.[59]

Some years later, in 1270–74, Albert revisited Philip the Chancellor's formula *bonum est indivisio actus a potentia* in his final great work, the *Summa theologiae*. In this last approach to the concept of the good, Albert quotes the Aristotelian, Avicennian, and Anselmian definitions.[60] The objections now directed against the Avicennian formula differ considerably from those raised in *De bono* nearly thirty years earlier.

In the first objection in the *Summa theologiae*, the formula is dismissed as a determination of the true rather than of the good, because indivisibility concerns the form and not the end, which stands for the good.[61] The objector sees here a definition of *esse perfectum* that is actually correct.[62] In his response, Albert repeats what he said before, pointing out that *indivisio actus a potentia* should not be understood according to the last natural and self-possessed virtue or capacity of any good whatsoever if it is not impeded in perfect operation. The final cause guides the operation produced by the efficient cause. In this way, Albert defines perfection as a process guided by a final cause.[63]

In a second objection, Albert adduces Boethius's statement that good proceeds from good, and suggests the determination of good as the effect of an efficient cause.[64] With the help of the *Liber de causis*,[65] the objector places the problem in the context of the theology of creation, and characterizes being as the first product of creation. In his response, Albert interprets the Boethian dictum *bonum a bono* (in Albert's own wording) by subordinating the efficient cause to the final cause, because good is always interpreted according to finality. The good is accordingly determined by a final cause, which moves an efficient cause,[66] whereas being is determined by an efficient cause and is therefore simply

---

59 Ibid., p. 26b (ed. by Müller, p. 362).

60 Albertus Magnus, *Summa theologiae I*, tr. 6, q. 26, cap. 1, a. 2, II: 'Quae sit determinatio boni secundum intentionem communem', ed. by Siedler, p. 175, v. 39–p. 177, v. 53. The following arguments add to what I wrote in Uscatescu Barrón, *Der Begriff des Guten*, pp. 326–27.

61 Ibid., p. 176, vv. 12–14: 'Hoc enim est esse perfectum ex forma et non ex fine, et haec est potius determinatio veri quam boni'.

62 Interestingly, Albert builds this objection with the help of his own argument, as set out in his commentary on the *Sentences*, for an essential connection between form and truth.

63 Albertus Magnus, *Summa theologiae I*, tr. 6, q. 26, cap. 1, a. 2, II, ed. by Siedler, p. 177, vv. 20–27: 'Ad id quod obicitur de hac determinatione, quod "bonum est indivisio actus a potentia", dicendum, quod bona est, sed intelligitur de indivisione actus, qui est secundum virtutem ultimam boni uniuscuiusque naturalem et propriam non impeditam in operatione perfecta, et sic redit ad primam et sic intelligendo non determinatur per formam, sed per finem ultimum'.

64 Ibid., tr. 6, q. 26, cap. 1, a. 2, II, p. 176, vv. 15–25. See Boethius, *De hebdomadibus*, ed. by Moreschini, p. 191, vv. 107–18.

65 *Liber de causis*, IV.37–38, ed. by Pattin, p. 142: 'Prima rerum creatarum est esse et non est ante ipsum creatum aliud'.

66 Earlier, he had expressed this more exhaustively. See Albertus Magnus, *Summa theologiae I*, tr. 6, q. 26, cap. 1, a. 2, I, ed. by Siedler, p. 174, vv. 38–41: 'Et sic dicitur bonum in efficiente, quia disponitur a fine. Finis enim movet efficientem et est movens immobile, efficiens autem movens motum'.

created.[67] Albert thus emphasizes the indivisibility of the good with regard to the final cause. The Avicennian formula, originally taken as a general definition of good, is now used as a definition of the last good according to the last natural and ultimate potency (*virtus ultima*) if it is not impeded in its activity, and so it is determined by the last final cause. This is a more restricted version of the Avicennian formula and implies a more limited conception of perfection, which means here only the last perfection of a being. Nevertheless, some pages earlier, Albert has presented the identity of actuality and potentiality as equal to the identity of actuality and form, that he interprets also as an end and as what is related to end. With this explanation, Albert endows the Avicennian formula with the universality required for a transcendental concept.[68]

Besides these two well-known definitions, which were already treated in his early *De bono*, Albert offers a third definition of the good, one that suggests a reliance on Anselm of Canterbury instead of Algazel. This new definition holds that 'the good per se is that whose existence is better than its non-existence' (*per se bonum est, quod omnibus melius est esse quam non esse*).[69] Albert accepts this formula without restriction to the *bonum per se*, but when it is applied to the *bonum secundum quid*, he remarks that its validity depends on the thing for which something is good and on the particular occasion.[70] In this final passage on the good (*communis intentio boni*), Albert does not favour any single formula, but suggests the validity of all the three definitions in the senses he has already explained in response to each objection. Avicenna's formula has to be adapted to the finalistic conception of the good in a Neoplatonic way.[71]

This Neoplatonic touch in Albert's definition of the good can be better appreciated in two additional works that Albert wrote between *De bono* and the *Summa theologiae*. In his commentaries on Dionysius's *De divinis nominibus* (1250) and on the anonymous *Liber de causis* (after 1268),[72] Albert appropriates a wide range

67 Ibid., tr. 6, q. 26, cap. 1, a. 2, II, p. 177, vv. 28–45; p. 177, vv. 40–45: 'Ex quo accipitur, quod etiam bonum in efficiente determinatur ex fine. Et sic patet, quod non est eadem determinatio boni et entis. Ens enim determinatur per causam efficientem simplicem, bonum autem per causam finalem, quae movet efficientem'.

68 Ibid., tr. 6, q. 26, cap. 1, a. 2, I, p. 174, v. 85–p. 175, v. 4: 'Ad aliud dicendum, quod bonum determinatur per formam non secundum se, sed secundum quod forma est finis et ad finem, qui est ut vigilia et non ut somnus; sic enim inducit bonum. Inductivum autem boni bonum est, ut dicit Proclus. Finis autem ultimus est actus ut vigilia'.

69 Ibid., tr. 6, q. 26, cap. 1, a. 2, II, p. 175, vv. 50–51. See Anselm, *Proslogion*, V, ed. by Schmitt, vol. 1, p. 104, v. 16.

70 Albertus Magnus, *Summa theologiae I*, tr. 6, q. 26, cap. 1, a. 2, II, ed. by Siedler, p. 177, vv. 46–53.

71 More generally on God as the first goodness and Albert's Neoplatonic interpretation, see Schneider, *Das Gute*, pp. 55–85.

72 In the latter commentary, Albert repeats in many ways the definition of the good as *bonum communicativum* and defines *bonitas* accordingly, as *dispositio ad emittendum*. See Albertus Magnus, *De causis et processu universitatis a prima causa*, I.3, cap. 5, ed. by Fauser, p. 40, vv. 46–58: he quotes Avicenna, Alfarabi, and Algazel as authorities. Avicenna, especially, interprets Aristotle in a Neoplatonic way. His references to emanation from the first being are analogous to those of the

of Neoplatonic ideas, many of which do not seem to be immediately compatible with the heritage of Aristotle's corpus. For instance, in his *De divinis nominibus*, Albert modifies the traditional definition of the good as desirable in the following way. By connecting it to the well-known Boethian idea that everything strives for what is similar to itself,[73] Albert affirms the general tendency to the good that all beings naturally have in order to show that all beings strive after something that is similar to themselves. This similarity, Albert reasons, concerns the first good (rather than the proper good for each one), and the tendency towards it is found in all grades of being by way of their proportionality (*proportio*) to the final end.[74] As a consequence, Albert no longer seems to conceive of the good only as something desirable in horizontal terms (as he did mostly, but not exclusively, in *De bono*[75]), but rather in vertical terms alone, as something related to an ontologically higher transcendental end rather than to an ontologically or operationally immanent end (the proper good for each being).

This impression appears to be confirmed in the more intensive discussion of Aristotle that we also find in Albert's *De divinis nominibus*. It leads him to a manifest reappraisal of the Avicennian formula and a subsequent transformation of the Neoplatonic definition of the good as *bonum diffusivum sui* into a final cause.[76] Albert repeatedly describes this Neoplatonic definition as something that shares itself without limitation,[77] thus seeming at first sight to reinterpret the original definition of the good more in terms of efficient causality. Yet Albert forestalls such an interpretation by pointing out the priority of the final cause over the three other causes. If, Albert reasons in *De divinis nominibus*, an efficient cause induces the formal cause in matter (material cause) and is superior to form and matter,[78] and if the efficient cause is directed to and by a final cause by

---

author of the *Liber de causis* and other Neoplatonists. See Avicenna, *Liber de Philosophia prima*, IV.3, ed. by Van Riet, p. 216, vv. 23–25: 'Plus quam perfectum autem est id cui est esse quod debet habere et ab eo exuberat esse ad caeteras res'. On this point, see Lizzini, *Fluxus (fayḍ)*.

73 Albertus Magnus, *Super Dionysium De divinis nominibus*, cap. 4, n. 158, ed. by Simon, p. 243, vv. 49–51; Boethius, *De hebdomadibus*, ed. by Moreschini, p. 188, vv. 49–52. See Uscatescu Barrón, *Der Begriff des Guten*, pp. 176–77.

74 Albertus Magnus, *Super Dionysium De divinis nominibus*, cap. 2, n. 6, ed. by Simon, p. 47, vv. 38–64; ibid., cap. 4, n. 58, p. 165, vv. 38–67.

75 But in *De bono*, Albert is aware of the transcendental good all things strive for in different degree (*De bono*, I.1.6.17, p. 10, vv. 26–34). It could not be otherwise, because he is interpreting Boethius's *De hebdomadibus* rigorously.

76 Albertus Magnus, *Super Dionysium De divinis nominibus*, cap. 4, n. 5, ed. by Simon, p. 116, vv. 39–61: 'Est enim bonum semper secundum respectum ad finem secundum aliquem modum, vel prout est in efficiente; et sic dicitur bonum, quod convertitur cum ente, quia est a bono [...] et bonitas subiecti relatione ad primum efficiens finale'.

77 Ibid., cap. 1, n. 18, p. 9, vv. 51–52: 'boni enim est communicare se, sicut possibile est'; ibid., cap. 2, n. 47, p. 75, vv. 22–23: 'quia boni est communicare seipsum et vocare res in esse'.

78 Ibid., cap. 4, nn. 55–57, p. 163, v. 21–p. 165, v. 33. See Schneider, *Das Gute*, p. 67, n. 123 against the interpretation of the formula *bonum diffusivum sui* as efficient cause, and Uscatescu Barrón, *Der Begriff des Guten*, pp. 349 ff.

inducing form in matter,[79] then the final cause is superior to both the efficient and the formal cause. Indeed, because of this superiority, it fits the highest rank of essential communicability of the good, which is best ascribed to it in the Neoplatonic formula *bonum diffusivum sui*. This priority of the good as final cause in the Neoplatonic sense may provoke the worry that it diffuses itself substantially in all things, thus losing its special ontological status. Yet according to Albert's reading, it does not challenge the Scholastic theory of the transcendentals, which remains untouched inasmuch as the good as final cause has to be added to being, as we shall see below.

Albert confirms this priority of being over the other transcendentals (the one [*unum*], the true [*verum*], and the good [*bonum*]) in a number of his works, including the *De bono*, his commentary on the *Sentences*,[80] and his *De divinis nominibus*.[81] Albert's Neoplatonic reinterpretation of the Avicennian formula is simultaneous with his development of the theory of transcendental concepts from *De bono* onwards. In noticeable contrast to Philip the Chancellor, who defended a theory of the transcendentals based on the notion of *indivisio* in the opening pages of his *Summa de bono*, Albert already gives up negation as a main criterion for transcendentality in his *De bono*. In place of Philip's reduction of the transcendentals to *indivisio* as a conspicuous form of negation, he defines the 'one' by the attribute of *indivisio*, and speaks of the other two transcendentals of the true and the good as attributes that add further modes of speaking (*modi significandi*) or intentions to being (*secundum intentionem*). The one, Albert argues, adds an indivisibility in itself (*indivisio in se*) to being and a divisibility from others (*divisio ab aliis*).[82] The true constitutes either a relation to form (*relatio ad formam*) or a relation to that by which a thing is formally (*relatio ad id quo est res formaliter*), and the good adds a relation to the end (*relatio ad finem*), also referred to as an inseparability from the end (*indivisio a fine*).[83] The Avicennian formula of the good is not suited to describing a transcendental, but the idea of indivision contained in it is still valuable in Albert's analysis of the transcendentals.

In his commentary on the *Sentences*, composed between *De bono* and *De divinis nominibus*, Albert develops an even clearer understanding of the

79 Albertus Magnus, *Summa theologiae I*, tr. 6, q. 26, cap. 1, a. 2, II, ed. by Siedler, p. 177, vv. 40–45. The text is quoted in note 90 below.

80 Albertus Magnus, *Commentarii in I Sententiarum*, dist. 46 N, a. 13, sol., ed. by Borgnet, pp. 447b–449b.

81 *De bono*, I.1.10.38, p. 21, vv. 11–16. A similar theory of the transcendentals can be found in Albertus Magnus, *Super Dionysium De divinis nominibus*, cap. 4, n. 5, ed. by Simon, p. 116, vv. 1–64. For a general overview on Albert's theory of the transcendentals, see Uscatescu Barrón, *Der Begriff des Guten*, pp. 342–55.

82 *De bono*, I.2.2.46, p. 26, vv. 49–50.

83 Ibid., I.1.10.38, p. 20, vv. 41–47: 'Dicendum, quod verum, secundum quod est declarativum esse vel id quod est, sicut supra est explanatum, cum ente convertitur secundum supposita, secundum autem intentiones nominum ens est prius ad verum. Verum enim super ens addit relationem ad formam sive ad id quo est res formaliter, sicut et bonum super ens addit relationem ad finem'.

transcendentals. Here, the one is again defined as indivisibility (*indivisio*), but at same time as privation (*privatio*). The true is now defined as a relation to a form down to the least exemplar (*relatio ad formam ad minus exemplarem*) and identified with manifestation (*ratio manifestationis*). Finally, the good is conceived as a relation to an extrinsic end (*relatio ad finem extra*).[84] As before, the true and the good add relational aspects to being: the true adds a relation to an idea (*respectus ad ideam*), and the good a relation to an end (*respectus ad finem*). Albert further divides this relation to an end into two different kinds.[85] If the good is the terminus or end of efficient causality, it is interchangeable with being inasmuch as all things come from the first good (*a bono*).[86] But the good can also be understood as that which comes to being through physical motion brought about by a final cause.[87] If it is regarded in the first way, namely as given in an efficient cause, it has to be thought as a created good issuing from a good (*bonum a bono*).[88] If, however, it is a final cause, all beings have to be thought as related to a

84 Albertus Magnus, *Commentarii in I Sententiarum*, dist. 46 N, a. 14, ed. by Borgnet, p. 450a. Later in the *Sentences* commentary, he adopts a less differentiated view on the final cause: 'Ad ultimum dicendum, quod bonum per essentiam, est idem bonum quod convertitur cum ente in quantum est, et non addit super ens nisi relationem ad finem' (Albertus Magnus, *Commentarii in II Sententiarum*, dist. 36 K, a. 7, ad ultimum, ed. by Borgnet, p. 593b).

85 Albertus Magnus, *Commentarii in I Sententiarum*, dist. 1 G, a. 20, sol., ed. by Borgnet, p. 46a: 'Addit et bonum super ens relationem ad finem: tamen duplex est relatio ad finem, scilicet secundum quod finis est terminus motus causae efficientis, vel secundum quod finis est per intentionem in efficiente: et primo modo dicuntur bona quae sunt a bono: secundo modo quae sunt ad bonum quod intendit efficiens, id est, quod movet eum ad operandum'.

86 This Boethian formula (*De hebdomadibus*, ed. by Moreschini, p. 191, vv. 108–18) should be interpreted not only as the definition of efficient cause (as Schneider, *Das Gute*, p. 83, does), but also as an end. See Albertus Magnus, *Summa theologiae I*, tr. 6, q. 26, cap. 1, a. 2, I, ed. by Siedler, p. 174, vv. 36–41; Albertus Magnus, *Super Dionysium De divinis nominibus*, cap. 4, n. 5, ed. by Simon, p. 116, vv. 41–42: 'et sic dicitur bonum, quod convertitur cum ente, quia est a bono'. See Uscatescu Barrón, *Der Begriff des Guten*, pp. 183–90.

87 Albertus Magnus, *Super Dionysium De divinis nominibus*, cap. 4, n. 5, ed. by Simon, p. 116, vv. 60–61: 'et bonitas subiecti relatione ad primum efficiens finale'. Schneider, *Das Gute*, p. 87, n. 213, reads 'efficiens formale'. At issue, however, is a final cause, namely of the last goal, which is also the first final cause, i.e., of the last final cause, which is both the first and the last efficient cause.

88 See Albertus Magnus, *Super Dionysium De divinis nominibus*, cap. 4, n. 5, ed. by Simon, p. 116, vv. 39–44: 'Est enim bonum semper secundum respectum ad finem secundum aliquem modum, vel prout est in efficiente; et sic dicitur bonum, quod convertitur cum ente, quia est a bono, vel secundum quod est acquisitum per motum physicum, et sic est bonum in rebus mobilibus'. In this passage, Albert clearly identifies *esse* with *esse creatum* and *bonum transcendens* with *bonum a bono* or *bonum creatum* as given in the First Efficient Cause and produced by It. Moreover, the second kind of relation to final cause seems to be considered in a different way than it was in the passage from *In Sent.* (see note 85), because here he restricts the good to what can be acquired in movement, by emphasizing only the kinetic dimension of the good as *terminus motus*. In this way, Albert restricts the good to the realm of things in movement, excluding immobility and therefore numbers from the good. In another passage, however, he applies goodness to mathematical entities. See in *Commentarii in I Sententiarum*, dist. 1 G, a. 20, sol., ed. by Borgnet, p. 46b: 'Ad aliud dicendum, quod mathematica non habent finem qui acquiritur per motum physicum, quia separata sunt secundum rationem diffinitivam et a motu et a materia: sed non carent intentione boni quod est a bono et ad bonum primum'. This is an opinion

practical end. Indeed, Albert considers the principal relation implied by the good to be the relation to final cause intended by efficient cause.[89] This explanation of the twofold conception of end can also be understood universally. On the one hand, Albert implicitly distinguishes between end (*terminus*) and final cause, so that the final cause could be the end of an agent inasmuch as it is its natural potency. On the other, *terminus* can be interpreted as a relation to an extrinsic end that is intended by an efficient cause that acts toward that end. Albert here refers to Boethius's *De hebdomadibus*, in which the relation between the *primum bonum* and the *secundum bonum* (*ens creatum*) is discussed, but he criticizes Boethius for having worked out the relation between the two kinds of the good as a relation between efficient cause and effect. Whereas *esse* is *esse creatum* insofar as it is caused by an efficient agent, good is defined as related to a final cause, which moves the efficient one.[90]

With regard to the transcendental concepts, in his commentary on the *Sentences*, Albert holds that a privation (oneness), and two rational relations (truth and goodness) are added to the principal transcendental of being.

In his final words on the matter, his *Summa theologiae*, Albert turns to the question of the convertibility of the transcendentals. Instead of using the terms of relations (*relationes*) or modes of signification (*modi significandi*), he now prefers to speak of modes of being (*modi existendi*) and confirms the conceptual posteriority of the good (*intellectus boni*) to being.[91] For these relations as characterizations of the transcendentals are, strictly speaking, *relationes rationis* and are therefore interchangeable with the *modi significandi*, which are implied by the *relationes rationis*.[92] In these changes of terminology, there is no rectification or retraction of a philosophical position held before, but perhaps efforts to capture a more adequate expression.

---

opposed to that of Aristotle, who separated the mathematical and geometric entities from movement and finally excluded them from goodness. See *Met.*, II.2, 996a22–32.

89 *Commentarii in I Sententiarum*, dist. 1 G, a. 20, sol., ed. by Borgnet, p. 46a: 'Si autem considerentur bonum et ens secundum supposita, sic convertuntur: quia licet bonum sit ad efficiens ut est bonum, et ens ad efficiens ut est ens, [...] ideo comitatur bonum semper ipsum ens, et non separatur ab ipso secundum suppositum, licet separetur secundum intentionem'. See also ibid., dist. 3 F, a. 15, ad ultimum, ed. by Borgnet, p. 108b.

90 Albertus Magnus, *Summa theologiae I*, tr. 6, q. 26, cap. 1, a. 2, II, ed. by Siedler, p. 177, vv. 28–45: 'Ad id quod obicitur de Boethio, dicendum, quod bonum non est a bono per causam efficientem solum, sed per causam efficientem fine secundum intentionem dispositam [...]. Ex quo accipitur, quod etiam bonum in efficiente determinatur ex fine. Et sic patet, quod non est eadem determinatio boni et entis. Ens enim determinatur per causam efficientem simplicem, bonum autem per causam finalem, quae movet efficientem'.

91 Ibid., tr. 6, q. 28, *De comparatione istorum trium ad invicem et ad ens utrum scilicet ens unum verum et bonum convertantur*, p. 213, v. 1–p. 214, v. 80.

92 Albertus Magnus, *Super Dionysium De divinis nominibus*, cap. 4, n. 5, edited by Simon, p. 116, vv. 28–36: 'sunt autem quaedam relationes quae non innascuntur ex mutatione rei, sed potius ex mutatione alterius, sicut in columna inmobili causatur dextrum ex motu hominis; et tales relationes sunt potius rationes quam res [...] et modum fundatum super tales relationes addit verum et bonum supra ens'. See also Uscatescu Barrón, *Der Begriff des Guten*, pp. 350–51.

## Conclusion

In his long elaboration of the concept of the good from his first work, *De natura boni* (1230–40), until his last, *Summa theologiae* (1280), a conceptual development can be traced in Albert's thoughts on the good — beginning with the Avicennian formula coined by Philip the Chancellor and closing with the final reinterpretation of form in the sense of the good as end.

Albert praises the Avicennian formula of the good, but ultimately prefers the Aristotelian, finalistic conception of good as directedness to the end. The reason for this choice may lie in the fact that the suggested identity of actuality and potentiality implied in the formula means rather the perfection of being, which is obviously not a property of being. But for Albert, there is no difference between goodness and perfection, because he endorses both Avicenna's and Algazel's definitions, which equate good with perfection and actuality. From *De bono* onwards, this formula is completely absent, but Albert does not offer any reason for having abandoned it. It is likely that the finalistic conception of the good endorsed by Aristotle became overwhelmingly evident, especially after the *Nicomachean Ethics* was made available in a full translation. Moreover, in his *Metaphysics*, Albert had already given up negation or privation as base for the three transcendental concepts and favoured a relational conception of true and good. This explains also why he no longer identifies *bonum* with *finis* or final cause. Instead he stresses his relational conception of the good by defining it as relatedness to an end or a final cause. This metaphysical choice was no longer compatible with the Avicennian formula as a definition of the good, coined and developed by Philip the Chancellor, since that expresses a decidedly 'absolutist', non-relational conception of the good. Nevertheless, some resonances of this formula can still be detected in the works after *De bono*.

Finally, Albert's finalistic conception is reshaped in his commentaries on Dionysius Areopagita's *De divinis nominibus* and the *Liber de causis*. In both works, Albert immensely promoted the interpretation of the good as *bonum diffusivum sui*, mainly in the sense of final cause conjoined to an efficient cause. This diminished the role of the Avicennian formula in the explanation of the good.

In this twist of thinking, Albert has finally come to successfully reinterpret the Avicennian formula connecting the form, which concerns actuality, with the efficient cause. He does this in order to pave the way for unifying the efficient cause with the final, as far as it guides the operation of the efficient cause. In his *Summa theologiae*, Albert rescues the Avicennian formula and defends it from objections by reinterpreting it in terms of final and formal causality in this Neoplatonic way.

## Works Cited

### *Primary Sources*

Albertus Magnus, *Commentarii in I Sententiarum*, ed. by Auguste Borgnet, Editio Parisiensis, 26 (Paris: Vivès, 1893)

———, *Commentarii in II Sententiarum*, ed. by Auguste Borgnet, Editio Parisiensis, 27 (Paris: Vivès, 1893)

———, *Commentarii in III Sententiarum*, ed. by Auguste Borgnet, Editio Parisiensis, 28 (Paris: Vivès, 1894)

———, *De bono*, ed. by Heinrich Kühle, Karl Feckes, Bernhard Geyer, and Wilhelm Kübel, Editio Coloniensis, 28 (Münster: Aschendorff, 1951)

———, *De causis et processu universitatis a prima causa*, ed. by Winfried Fauser, Editio Coloniensis, 17/2 (Münster: Aschendorff, 1993)

———, *Ethica*, ed. by Auguste Borgnet, Editio Parisiensis, 7 (Paris: Vivès, 1891)

———, *Metaphysica, libri I–V*, ed. by Bernhard Geyer, Editio Coloniensis, 16/1 (Münster: Aschendorff, 1960)

———, *Physica, libri I–IV*, ed. by Paul Hossfeld, Editio Coloniensis, 4/1 (Münster: Aschendorff, 1987)

———, *Secunda pars Summae de creaturis*, ed. by Auguste Borgnet, Editio Parisiensis, 35 (Paris: Vivès, 1896)

———, *Summa theologiae sive De mirabili scientia Dei, libri I pars I, Quaestiones 1–50 A*, ed. by Dionysius Siedler, Editio Coloniensis, 34/1 (Münster: Aschendorff, 1978)

———, *Super Dionysium De divinis nominibus*, ed. by Paul Simon, Editio Coloniensis, 37/1 (Münster: Aschendorff, 1972)

———, *Super Ethica, libri I–V*, ed. by Wilhelm Kübel, Editio Coloniensis, 14/1 (Münster: Aschendorff, 1968–72)

Algazel, *Logica*, in Charles H. Lohr, '*Logica Algazelis*: Introduction and Critical Text', *Traditio*, 21 (1965), 223–90

———, *Logica et philosophia*, ed. by Petrus Liechtenstein (Venice, 1506; repr. Frankfurt am Main: Minerva, 1969)

———, *Metaphysics: A Mediaeval Translation*, ed. by Joseph T. Muckle, St Michael's Mediaeval Studies (Toronto: The Institute of Mediaeval Studies, 1933)

Anselmus Cantuariensis, *S. Anselmi Cantuariensis archiepiscopi opera omnia*, ed. by Franz S. Schmitt, 6 vols (Stuttgart-Bad Cannstatt: frommann-holzboog, 1968)

Aristotle, *The Complete Works of Aristotle: The Revised Oxford Translation*, ed. by Jonathan Barnes (Princeton, NJ: Princeton University Press, 1984)

———, *Ethica Nicomachea*. Translatio antiquissima lib. II–III sive 'Ethica Vetus' et translatio antiquioris quae supersunt sive 'Ethica Nova', 'Haferiana', 'Borghensiana', ed. by Renatus Antonius Gauthier, Aristoteles latinus 26/1–3, fasc. 1–5 (Leiden: Brill, 1972–74)

———, *Metaphysica, lib. I–X, XII–XIV, translatio anonyma sive 'media'*, ed. by Gudrun Vuillemin-Diem, Aristoteles Latinus, 25/2 (Leiden: Brill, 1976)

Avicenna, *Metafisica: La scienza delle cose divine (Al-Ilāhiyyāt) dal Libro della Guarigione (Kitāb al-Šifā'): Testo arabo a fronte, testo latino in nota*, trans., introduction, and notes by Olga Lizzini; preface and revision of the Latin text by Pasquale Porro (Milan: Bompiani, 2002)

———, *Die Metaphysik Avicennas*, ed. by Max Horten (Halle an der Saale: R. Haupt, 1907; repr. Frankfurt am Main: Minerva, 1960)

———, *La Métaphysique du Shifā'*, ed. by Georges C. Anawati, 2 vols, Études musulmanes, 21 (Paris: Vrin, 1978–85)

———, *The Metaphysics of 'The Healing': A Parallel English-Arabic Text*, ed. by Michael E. Marmura (Provo, UT: Brigham Young University Press, 2005)

Avicenna Latinus, *Liber de Philosophia prima sive Scientia divina: Édition critique de la traduction latine médiévale*, ed. by Simone Van Riet, introduced by Gérard Verbeke, 2 vols (Leiden: Brill, 1977–80)

Boethius, *De hebdomadibus*, in Boethius, *De consolatione philosophiae. Opuscula theologica*, ed. by Claudio Moreschini, Bibliotheca scriptorum Graecorum et Romanorum Teubneriana (Munich: Saur, 2005)

Bonaventura, *Commentaria in quatuor libros Sententiarum, tomus I, In primum librum Sententiarum*, ed. Quaracchi, Doctoris seraphici s. Bonaventurae S.r.e. episcopi cardinalis Opera Omnia, 1 (Ad Claras Acquas: Ex typographia Collegii s. Bonaventurae, 1883)

*Liber de causis*, in Adriaan Pattin, 'Le *Liber de causis*: Édition établie à l'aide de 90 manuscrits avec introduction et notes', *Tijdschrift voor filosofie*, 28 (1966), 90–203

Müller, Jörn, 'Der Begriff des Guten im zweiten Ethikkommentar des Albertus Magnus. Untersuchung und Edition von *Ethica*, Buch 1, Traktakt 2', *Recherches de théologie et philosophie médiévales*, 69 (2002), 318–70

Philippus Cancellarius, *Philippi Cancellarii Parisiensis Summa de bono*, ed. by Nikolaus Wicki, 2 vols (Bern: Francke, 1985)

Thomas Aquinas, *In decem libros Ethicorum Aristotelis ad Nicomachum expositio*, ed. by Raymundus M. Spiazzi, 3rd ed. (Taurini: Marietti, 1977)

———, *Sententia libri Ethicorum, lib. I–III*, ed. by René A. Gauthier, Sancti Thomae de Aquino opera omnia, 47/1 (Rome: Ad Sanctae Sabinae, 1969)

### Secondary Works

Alonso Alonso, Manuel, 'Los "Maquasid" de Algazel: Algunas deficiencias de la edición canadiense', *Al-Andalus*, 25 (1960), 445–54

Anzulewicz, Henryk, '*Bonum* als Schlüsselbegriff bei Albertus Magnus', in *Albertus Magnus: Zum Gedenken nach 800 Jahren. Neue Zugänge, Aspekte und Perspektiven*, ed. by Walter Senner (Berlin: Akademie-Verlag, 2001), pp. 112–40

Bertolacci, Amos, 'Albert's Use of Avicenna and Islamic Philosophy', in *A Companion to Albert the Great: Theology, Philosophy, and the Sciences*, ed. by Irven M. Resnick (Leiden: Brill, 2013), pp. 601–11

———, 'Albert the Great and the Preface of Avicenna's Kitāb- al-Śifā'', in *Avicenna and His Heritage: Acts of the International Colloquium Leuven-Louvain La Neuve, septembre 8–11, 1999*, ed. by Jules Janssens and Daniel de Smet (Leuven: Leuven University Press, 2002), pp. 131–52

———, 'On the Latin Reception of Avicenna's Metaphysics before Albertus Magnus', in *The Arabic, Hebrew and Latin Reception of Avicenna's Metaphysics*, ed. by Dag Nikolaus Hasse and Amos Bertolacci (Berlin: De Gruyter, 2012), pp. 197–223

———, *The Reception of Aristotle's 'Metaphysics' in Avicenna's 'Kitāb al-Šifā'': A Milestone of Western Metaphysical Thought* (Leiden: Brill, 2006)

Buffon, Valeria, 'Philosophers and Theologians on Happiness: An Analysis of Early Latin Commentaries on the Nicomachean Ethics', *Laval théologique et philosophique*, 60 (2004), 449–76

Callus, Daniel A., 'The Date of Grosseteste's Translations and Commentaries on Pseudo-Dionysius and the Nicomachean Ethics', *Recherches de théologie ancienne et médiévale*, 14 (1947), 186–210

Cordonier, Valérie, Pieter De Leemans, and Carlos Steel, 'Die Zusammenstellung des "corpus aristotelicum" und die Kommentartradition', in *Die Philosophie des Mittelalters 4: 13. Jahrhundert, Erster Halbband*, ed. by Alexander Brungs, Vilem Mudroch, and Peter Schulthess (Basel: Schwabe, 2017), pp. 149–61

Cortabarría Beitia, Ángel, 'Literatura algazeliana en los escritos de San Alberto Magno', *Estudios filosóficos*, 11 (1962), 255–76

Cruz Hernández, Miguel, *La metafísica de Avicena* (Granada: Universidad de Granada, 1949)

Cunningham, Stanley, *Reclaiming Moral Agency: The Moral Philosophy of Albert the Great* (Washington, DC: The Catholic University of America Press, 2008)

d'Alverny, Marie-Thérèse, 'Algazel dans l'Occident latin', in d'Alverny, *La transmission des textes philosophiques et scientifiques au Moyen Âge*, ed. by Charles Burnett (Aldershot: Variorum, 1994), pp. 3–24

Ducharme, Léonard, '*Esse* chez saint Albert le Grand', *Revue de l'Université d'Ottawa*, section spéciale (1957), 209–52

Gauthier, René A., 'Le cours sur l'*Ethica nova* d'un maître de Paris (1235–1240)', *Archives d'histoire doctrinale et littéraire du Moyen Âge*, 42 (1975), 71–141

———, 'Notes sur les débuts (1225–1240) du premier "averroïsme"', *Revue des sciences philosophiques et théologiques*, 66 (1982), 321–74

Gillon, Louis-Bertrand, *La théorie des oppositions et la théologie du péché au XIIIe siécle* (Paris: Vrin, 1937)

Goichon, Amélie-M., *Lexique de la langue philosophique d'Ibn Sīnā (Avicenne)* (Paris: Desclée de Brouwer, 1938)

Gutas, Dimitri, *Avicenna and the Aristotelian Tradition: Introduction to Reading Avicenna's Philosophical Works* (Leiden: Brill, 1988)

Hourani, George F., 'A Revised Chronology of Ghazālī's Writings', *Journal of the American Oriental Society*, 104 (1984), 289–302

Janssens, Jules, 'al-Ghazālī's *Maqāṣid al-Falāsifa*, Latin Translation of', in *Encyclopedia of Medieval Philosophy: Philosophy between 500 and 1500*, ed. by Henrik Lagerlund (Dordrecht: Springer, 2011), vol. 1, pp. 387–90

Lafleur, Claude, and Joanne Carrier, *Le 'Guide de l'étudiant' d'un maître anonyme de la Faculté des Art de Paris au XIII^e^ siècle: Édition critique provisoire du ms. de Barcelona. Arxiu de la Corona d'Aragó, Ripoll 109, fol. 134^ra^–158^va^* (Québec: Faculté de philosophie, 1992)

Lizzini, Olga, *Fluxus (fayḍ): Indagine sui fondamenti della metafisica e della fisica di Avicenna* (Bari: Edizioni di pagina, 2011)

Müller, Jörn, *Natürliche Moral und philosophische Ethik bei Albertus Magnus* (Münster: Aschendorff, 2001)

Ribes Montané, Pedro, *Verdad y bien en el filosofar de Alberto Magno* (Barcelona: Balmes, 1974)

Schneider, Artur, *Die Psychologie des Alberts des Großen nach den Quellen dargestellt*, 2 vols (Münster: Aschendorff, 1903–06)

Schneider, Johannes, *Das Gute und die Liebe nach der Lehre Alberts des Großen* (Paderborn: Schöningh, 1967)

Steel, Carlos, 'Avicenna and Thomas Aquinas on Evil', in *Avicenna and His Heritage: Acts of the International Colloquium Leuven-Louvain La Neuve, septembre 8–11, 1999*, ed. by Jules Janssens and Daniel de Smet (Leuven: Leuven University Press, 2002), pp. 171–96

Tarabochia Canavero, Alessandra, 'A proposito del tratatto *De bono naturae* nel *Tractatus de natura boni* di Alberto Magno', *Rivista di filosofia neo-scolastica*, 76 (1984), 353–73

———, 'I *sancti* e la dottrina dei trascendentali nel Commento alle Sentenze di Alberto Magno', in *Was ist Philosophie im Mittelalter? Akten des X. Internationalen Kongresses für Mittelalterliche Philosophie der Société Internationale pour l'Étude de la Philosophie Médiévale, 25. bis 30. August 1997 in Erfurt*, ed. by Jan A. Aertsen and Andreas Speer (Berlin: De Gruyter, 1998), pp. 515–21

Uscatescu Barrón, Jorge, *Der Begriff des Guten: Eine historisch-systematische Untersuchung. Band 2: Mittelalterliche Philosophie von Augustinus bis Suárez* (Freiburg: Karl Alber, 2020)

———, 'Zur Bestimmung des Guten als Ungeteiltheit von Akt und Potenz bei Avicenna und ihrer Rezeption in der christlichen Scholastik des Hochmittelalters', *Salzburger Jahrbuch für Philosophie*, 51 (2006), 29–62

Vargas, Rosa M., 'Albert on Being and Beings: The Doctrine of *Esse*', in *A Companion to Albert the Great: Theology, Philosophy, and the Sciences*, ed. by Irven M. Resnick (Leiden: Brill, 2013), pp. 627–48

Verbeke, Gérard, *Avicenna: Grundlegung einer neuen Metaphysik* (Opladen: Westdeutscher Verlag, 1983)

Wisnovsky, Robert, *Avicenna's Metaphysics in Context* (Ithaca, NY: Cornell University Press, 2003)

Zavattero, Irene, 'Éthique et politique à la Faculté des arts de Paris dans la première moitié du XIIIe siècle', in *Les débuts de l'enseignement universitaire à Paris (1200–1245 environ)*, ed. by Jacques Vergier and Olga Weijers (Turnhout: Brepols, 2013), pp. 205–45

RICHARD C. TAYLOR

# Chapter 3. Albert the Great and Two Momentous Interpretive Accounts of Averroes*

Several of the Arabic philosophical accounts of the human soul crafted by Ibn Sīnā, Ibn Rushd and others became available to thinkers of the European Christian Latin tradition via translations made in the second half of the twelfth century and in the thirteenth century. The German Dominican scholar Albert the Great displayed throughout his life an interest in and knowledge of the Arabic tradition in translation perhaps more intensive and comprehensive than that of any of his peers in the thirteenth century. Albert's student Thomas Aquinas shared his eagerness to benefit from the philosophical works of the two major Muslim philosophers Ibn Sīnā and Ibn Rushd, the few available works of al-Fārābī,[1] the short but monumentally important metaphysical work widely known among the Latins as the *Liber de causis*, and other available texts.[2] As a young Italian

* In composing this paper, I benefited greatly from advice and criticism by Henryk Anzulewicz, Fouad Ben Ahmed, Josep Puig Montada, Jules Janssens, and David Twetten, to whom I extend my sincere thanks.

1 On the importance of al-Fārābī's writings in the European Christian context, see de Libera, 'Existe-t-il une noétique "averroïste"?', and de Libera's discussion of acquired intellect (العقل المستفاد *al-ʿaql al-mustafād, intellectus adeptus*) in his 'Averroïsme éthique et philosophie mystique' and *Métaphysique et noétique*, pp. 265–328.

2 See Burnett, 'Arabic Philosophical Works Translated into Latin'. For al-Fārābī, see pp. 816–17. The Arabic *Kalām fī maḥḍ al-khair* (Discourse on the Pure Good) in a version apparently attributed to Aristotle (e.g., *Liber Aristotelis de expositione bonitatis purae* in Aosta, Seminario Maggiore 3-B-38) was translated into Latin by Gerard of Cremona before 1187 and was not unsuitably dubbed the *Book of Causes* for its account of the First Cause (God) and other transcendent entities. For a Latin text, see Pattin, 'Le *Liber de causis*'. For some suggested revisions, see Taylor, 'Remarks on the Latin Text'.

**Richard C. Taylor** (richard.taylor@marquette.edu) is professor of philosophy at Marquette University, annual visiting professor at the KU Leuven, and director of the 'Aquinas and "the Arabs" International Working Group'.

*Albert the Great and his Arabic Sources*, ed. by Katja Krause and Richard C. Taylor, Philosophy in the Abrahamic Traditions of the Middle Ages, 5 (Turnhout: Brepols, 2024), pp. 69–116
BREPOLS PUBLISHERS 10.1484/M.PATMA-EB.5.136483

student in Paris, where he arrived *c.* 1242, this aspiring theologian met Albert and later worked with him in 1248–52 at Cologne, where Albert had opened a new Dominican *studium*.[3] There, Albert made the unusual decision to begin by commenting on the works of Pseudo-Dionysius. In his commentary to *On the Divine Names*, he demonstrated particularly well for Thomas and his other students the value of using the *Metaphysics* of Avicenna and the *Liber de causis* from the Arabic tradition, as well as writings by Boethius, Anselm, and others from the Latin tradition, to explain the metaphysics both of creation and of being as a divine name in the writings of Dionysius.[4] In his commentary on Dionysius's *On the Divine Names*, Albert also discussed issues in philosophical psychology and even set out a brief account of monopsychism that was based largely on the philosopher 'Averroes'[5] — albeit without explicitly mentioning the name of his source.[6]

Among the very early works of Albert is the section of his *Summa de creaturis* named *De homine* (*c.* 1242), a work especially notable for its presentation of (i) a teaching on natural human knowing that Albert believed to be in full accord with the thinking of Averroes.[7] In his *Super Ethica* (1250–52), however, Albert set aside his earlier interpretation of Averroes regarding human soul and intellectual understanding.[8] In its place, he presented another (ii) interpretation destined to

---

Recently three volumes of conference papers on Proclus and the *Liber de causis* were published by Dragos Calma: Calma, *Reading Proclus and the Book of Causes*.

3 On this and relevant contexts, see Mulchahey, 'The Studium at Cologne'. For dating of Albert's works, I follow generally the chronology provided on the website of the Albertus-Magnus-Institut: https://institutionen.erzbistum-koeln.de/albertus-magnus-institut/albertus_magnus/leben/, but see Rigo, 'Zur Redaktionsfrage der Frühschriften des Albertus Magnus'. For dating the career and works of Thomas, I follow Porro, *Aquinas*, and Torrell, *Saint Thomas Aquinas*, pp. 327–29.

4 See, for example, Albertus Magnus, *Super Dionysium De divinis nominibus*, ed. by Simon, cap. 5.

5 In this chapter, I try to maintain a distinction between the original Arabic teachings of Ibn Rushd and those attributed to Averroes in the Latin tradition. One should perhaps consider the same sort of distinction for the Arabic texts of Ibn Sīnā and the Latin texts of Avicenna and likewise for those of al-Fārābī and those of Alfarabius, but I will not do so here.

6 See note 8 below.

7 It is not clear whether he had fully realized the need to correct this attribution before or after crafting his commentaries on the works of Dionysius (begun *c.* 1248). Still, his mention of an Averroistic monopsychism suggests he had. See note 8. A dated account of Albert and Averroes is Miller, 'An Aspect of Averroes' Influence on St Albert'. Miller's understanding of the complex teaching of Ibn Rushd on the separate intellects is flawed — particularly regarding the Material Intellect — and leads to an unsound analysis of Albert's account of Averroes. Also see Hayoun and de Libera, *Averroès et Averoïsme*, p. 80. I discuss relevant texts, and also the early natural epistemology of Aquinas, which follows the account of Albert, in Taylor, 'Remarks on the Importance of Albert the Great's Analyses'.

8 Note, however, that in his question-commentary on Dionysius's *On Divine Names* (*c.* 1248–50), Albert provides a short account of monopsychism that attributes it to *plures philosophorum* without mention of Averroes. The editors of the critical edition note key passages of Averroes's *Long Commentary on the De anima* that appear to contribute at least in part to the view Albert sets out. See Albertus Magnus, *Super Dionysium De divinis nominibus*, ed. by Simon, p. 136, v. 57–p. 137, v. 11: 'remotis autem individuantibus non remaneret nisi id quod commune est, et ita ex omnibus animabus non remaneret nisi una anima, ita scilicet quod remanerent animae rationales solum in

be influential, premised on his own conception of the philosophical doctrine of Averroes on the post mortem existence of human soul (a point requisite for his explicit attribution to Averroes of a doctrine of monopsychism). Both of these interpretations played important roles in Albert's own understanding of the issues at stake and also in the development of philosophy in the Latin tradition of Christian Europe during the thirteenth century and beyond. The former interpretation (i) was momentous as a model for the foundational account of the natural epistemology of Thomas Aquinas in his *Commentary on the Sentences*. The latter interpretation (ii) was highly influential for its novel account of a doctrine of monopsychism, which Albert attributed to Averroes in his learned works dealing with this issue.

Although both interpretations by Albert were crucial in the development of European Christian thought, neither one of them is an authentic position of the Muslim philosopher Ibn Rushd. The two works by Ibn Rushd of chief importance for Albert were the *Long Commentary on the De anima of Aristotle*[9] and the *Long Commentary on the Metaphysics of Aristotle*,[10] which offered challengingly complex philosophical teachings new even in the Arabic tradition, especially Ibn Rushd's doctrine on the separate eternal Material Intellect shared by human knowers and his doctrine on human fulfilment and happiness.

Albert and other Christian theologians and philosophical thinkers of the thirteenth century later received Michael Scot's Latin translations of these texts into the various contexts of their own religious and philosophical commitments and evolving understandings of the Aristotelian or Graeco-Arabic Peripatetic tradition in its diverse forms. Attributing their accounts of the reasoning of Averroes to the Cordoban philosopher himself, they conceived themselves to be engaging with the genuine teachings of Ibn Rushd, and not merely with what they were

---

intelligentia influente huiusmodi actum corpori; et secundum hoc dicunt plures philosophorum intellectum separari et ponunt exemplum de candelis multis, quae illuminantur ex uno igne, quod extinctis candelis non remanet nisi ignis communis.' For the doctrine, see below at note 101. Presumably this is the text to which Hayoun and de Libera, *Averroès et Averoïsme*, refer at p. 87. Cf. Albertus Magnus, *Super Ethica*, ed. by Kübel, p. 72, vv.12–19, p. 79, vv. 76–80, p. 453, v. 40 ff.

9 Averroes, *Commentarium magnum in Aristotelis De anima libros*, ed. by Crawford. Hereafter *Commentarium magnum De anima* for specific references. For an English translation with introduction, see Averroes, *Long Commentary on the 'De anima' of Aristotle*, ed. and trans. Taylor. Hereafter *Long Commentary on the De anima, English* for specific references; for general references to this I will simply use *Long Commentary on the De anima*. Surviving Arabic fragments of this work are included in the notes to the English translation. Regarding this work and more on the Arabic fragments, see Averroes, *L'original arabe du Grand Commentaire d'Averroès au 'De anima' d'Aristote*, ed. by Sirat and Geoffroy, and Averroes, *De la faculté rationnelle*, ed. by Sirat and Geoffroy. The complete project on the fragments remains to be realized.

10 Averroès, *Tafsīr mā baʿd aṭ-Ṭabīʿat*, ed. by Bouyges, hereafter *Tafsīr mā baʿd aṭ-Ṭabīʿat*. The Latin is available in *Aristotelis opera cum Averrois commentariis*, Venice 1562–74, hereafter cited as *Long Commentary on the Metaphysics, Latin*. An English translation by Charles Genequand, *Ibn Rushd's Metaphysics*, will be cited as *Long Commentary on the Metaphysics, Arabic-English*. A critical edition of the Latin is in preparation by Dag Nikolaus Hasse and his team at Würzburg.

willing to consider rough and possibly inaccurate interpretations. In view of these considerations, four things are necessary: (i) clear accounts of the genuine teachings of Ibn Rushd on the issues, together with (ii) an explication of teachings attributed to Averroes, (iii) an account of Albert's changing views of the meaning of Averroic texts,[11] and (iv) an explanation of how these Latin interpretations came to be attributed to Averroes by Albert despite being incongruous or even simply incompatible with the teachings of the Cordoban. These points require us to delve deeply into the Commentaries of Ibn Rushd on Aristotle's *De anima* and *Metaphysics*, in order to understand the actual teachings of the Cordoban philosopher, before proceeding to the Latin texts of Averroes as read and interpreted by Albert.

In what follows, I first explicate the evolving teachings of Ibn Rushd on human soul and intellectual understanding, along with his neglect or even outright denial of post mortem existence of the human soul in his Short, Middle, and Long Commentaries on the *De anima* of Aristotle. I then turn to human happiness and related teachings in Ibn Rushd's *Long Commentary on the Metaphysics*. Second, I recount a 'Tale of Two Averroisms' on the confusing and confused understandings of what have been called 'First Averroism' and 'Second Averroism'. Third, I explain in detail some of Albert's key interpretive misconstruals of Ibn Rushd's teachings in the Latin translations of the *Long Commentary on the De anima* and the *Long Commentary on the Metaphysics* and the impact these had on the formation of his understanding of doctrines of Averroes. This will make it clear that on the two issues discussed here, Albert was not properly in dialogue

11 In his introduction to *Renaissance Averroism*, Giglioni is careful to distinguish the descriptive terms 'Averroan', 'Averroist', and 'Averroistic': 'In this volume, the name "Ibn Rushd" denotes the actual historical figure, whereas his literary incarnation in translations and philosophical treatises in the Latin West will be referred to as "Averroes"'. Giglioni goes on to explain that '"Averroan" refers to any philosophical view that belongs directly to Ibn Rushd and is synonymous with "Rushdian". "Averroist" refers to opinions held by any follower of Ibn Rushd in the Latin West during the late Middle Ages, the Renaissance and — though less and less frequently — during the seventeenth and eighteenth centuries. Finally, "Averroistic" refers to the generic cultural label denoting a pronounced rationalistic attitude, of a vaguely Aristotelian ilk, towards questions of philosophical psychology (in particular, the nature of the human mind and its survival after death of the body), natural determinism and, above all, the relationships between philosophical freedom and dogmatic truths, often of a religious kind. Averroistic thinkers looked (and still look) at Averroes as the philosopher who denied the personal identity of human beings, of course, but also as an incarnation of Machiavellian dissimulation in politics and religion, as one of the heroes of the *libertinage érudit*, as a precursor of seventeenth-century materialism, as a pantheist and even an atheist'. Giglioni, *Renaissance Averroism*, pp. 1–2. These proposed distinctions, however, do not capture the present case, in which Albert claims that his own understanding of key texts of the Latin Long Commentaries on the *De anima* and *Metaphysics* provides a sound interpretation of the thought of Ibn Rushd/ Averroes. Hence, I follow the suggestion of my colleague Josep Puig Montada and use the term 'Averroic' to denote texts of the medieval Latin translations of Ibn Rushd's works and decline to use 'Averroan'.

with Ibn Rushd.[12] I supplement this section with brief remarks on the profound influence of Albert's interpretations on the teachings of his student, Thomas Aquinas. I conclude with summary remarks on the issues discussed in this chapter and their importance for understanding the influence of philosophy from the Arabic tradition in Latin translation. I also add a short appendix on the concept of 'acquired intellect'.

## Ibn Rushd on Human Intellect

Ibn Rushd's efforts to know the nature of human intellectual understanding were dynamic. At least three distinct stages can be discerned in his commentaries on the *De anima*, though one feature of his teaching on human soul remained constant throughout: the lack of an account of post mortem existence of human soul. The *Long Commentary on the Metaphysics*, a later work that refers to the teachings of the *Long Commentary on the De anima* and seems to be largely in accord with them, also maintains that same feature.[13]

In his early *Short Commentary on the De anima*, Ibn Rushd follows in part the thought of Alexander of Aphrodisias and in part that of Ibn Bājja,[14] by holding that the term *material intellect* denotes not a substance but a disposition or power of the human soul for understanding and receiving intelligibles. Such understanding and receiving takes place subsequent to the abstraction of those intelligibles from forms or intentions apprehended by the external and internal senses. These forms or intentions depend upon our activities to become what Ibn Rushd calls 'material intelligibles', that is, intelligibles formed through intellect on the basis of sensory experience of the world. Ibn Rushd specifically follows Ibn Bājja in describing this process as a disposition of the forms of the imagination. By that, he may have meant that there is a kind of certification of the forms in the imagination through an enhanced modality when the separate Agent Intellect is formally (albeit not ontologically) in the soul to bring about the abstraction or separation of intelligibles and the reception of the intelligibles in the soul's

---

12 At the beginning of Chapter 6 of his invaluable account of Albert's philosophical psychology and intellectual understanding, *Métaphysique et noétique*, Alain de Libera writes: 'La noétique d'Albert est le cœur vivant de sa pensée, le foyer de son système, le principal terrain de son engagement philosophique. À la fois réception du péripatétisme, dans le commentaire sur le *De anima*, et profession de foi péripatéticienne, dans le *De intellectu et intelligibili*, la psychologie albertinienne naît aussi d'un dialogue de pensée avec Averroès' (p. 265). The importance Albert's engagement or philosophical dialogue with Averroes has long been evident to scholars of medieval European philosophy. Yet for a sound knowledge of the history of philosophy, the question of his engagement with Ibn Rushd's actual teachings is even more important.

13 I address this issue directly in two articles: 'Personal Immortality in Averroes' Mature Philosophical Psychology' and 'Averroes on the Ontology of the Human Soul'.

14 See my discussions of these sources for Ibn Rushd in my introduction to *Long Commentary on the De anima, English*, pp. lxxxi–lxxxiii and lxxxix–xciii. On Ibn Bājja, see Genequand, 'Introduction'. Also see Wirmer, *Vom Denken der Natur zur Natur des Denkens*.

'material intellect'. Regarding this presence of the Agent Intellect in the soul as something that 'can belong to us', Ibn Rushd writes:

> For this reason one can consider that its understanding[15] can be ours *ultimately* [*bi-ākhiratin*].[16] I mean insofar as it is *form for us* [*ṣūra la-nā*] and it is such that it has generated for us as necessary an eternal intelligible, since it is itself an intellect whether or not we have intellectual understanding of it without its existence as intellect being from our activity as is the case for the material intelligibles. This state is what is known as uniting or conjoining.[17]

The notion that the Agent Intellect becomes 'form for us' (*ṣūra la-nā*) should be understood not in the sense that essences or quiddities in the Agent Intellect come to be in us simply by its efficient or agent causality, but rather in the sense that, in some fashion, in our understanding of intelligibles the Agent Intellect is an external actualizing form that has a presence in us by way of the abstraction and understanding of an intelligible in our material intellect. This is the notion of an acquired intellect (العقل المستفاد, *intellectus adeptus*) belonging to the human soul, found in various forms in Alexander of Aphrodisias, Themistius, and al-Fārābī, elaborated and used by Ibn Sīnā and Ibn Rushd, and later developed further by Albert. (More on the acquired intellect can be found in the appendix to this chapter).

Ibn Bājja held abstraction to be an exercise of intellect in virtue of which an individual human being may somehow, through abstractions, rise to the level of the Agent Intellect, and perhaps beyond, in a transcendent unity and a post mortem existence. Ibn Rushd did not assert any immortality for the individual human soul or intellect. He only provided an epistemological account that he deemed sufficient to answer the question of the nature of the human material

---

15 عقله *aqla-hu*.

16 Note that the phrase here, بآخرة *bi-ākhiratin*, 'afterwards', 'later', in manuscripts may be written identically with بأخرة *bi-akharatin*. Still, the two are here synonymous and distinct from بالآخرة *bi-l-ākhirati* 'in the afterlife', which will be discussed below.

17 ولذلك يظن ان عقله ممكن لنا بآخرة . أعني من حيث هو صورة لنا. ويكون قد حصل لنا ضرورة معقول ازلي . إذ كان في نفسه عقلا سواء عقلناه نحن أو لم نعقله . لا أن وجوده عقلا من فعلنا كالحال في المعقولات الهيولانية . وهذه الحال هي التي تعرف بالاتحاد والاتصال. Averroes, *Short Commentary: Talkhīs Kitāb al-Nafs*, ed. by El-Ahwani, p. 89, vv. 3–7; *Epitome de Anima*, ed. by Gómez Nogales, p. 127, vv. 7–11 (reading بآخرة *bi-akhiratin*, not بآخرة *bi-ākhiratin*; *La Psicología de Averroes*, ed. by Gómez Nogales. My translation and emphases. The new edition by David Wirmer with French translation by J.-B. Brenet, in *L'intellect*, (p. 157, vv. 6–13) has: ولذلك < أمكن أن يظن > أن عقله ممكن لنا بآخرة . أعني من حيث هو صورة لنا. ويكون قد حصل لنا ضرورة معقول أزلي . إذ كان في نفسه عقلا سواء عقلناه نحن أو لم نعقله . لا أن وجوده عقلا من فعلنا كالحال في المعقولات الهيولانية . وهذه الحال هي التي تعرف بالاتحاد والاتصال ('C'est pourquoi l'on peut estimer que son intellection nous est possible à la fin, je veux dire en tant que forme pour nous, et il nous sera <alors> nécessairement <comme> un intelligible éternel, puisqu'il est en lui-même intellect, que nous l'ayons intelligé ou pas, <et qu'il n'est donc> pas <tel> que son existence comme intellect <découle> de notre acte, comme c'est le cas <en revance> des intelligibles matériels. Cela, c'est l'état connu comme étant "l'union" or "la jonction"'). Regarding بآخرة 'ultimately', see Brenet's note 289, pp. 255–56, and his Introduction.

intellect,[18] an account which has the imagination as that through which a human being can rise to a level of intellectual understanding of universals above the apprehension of particulars. It should be noted that in this account, insofar as the imagination is for Ibn Rushd a power belonging to an individual human being composed of body and soul, the imagination, and the material intellect as a disposition of the imagination, those all remain perishable with the perishable nature of their subject, the human soul.[19]

In composing the *Middle Commentary*, Ibn Rushd appears to have become aware of a problem posed by his earlier teaching. If the imagination is essentially a power of a particular human being composed of body and soul, even if it is in some fashion not wholly identical with body, still the imagination is a bodily power and the power called material intellect dependent upon it is individuated by its relationship to the human being to which it belongs. Given that, the imagination cannot be the proper subject for intelligibles in act, that is, it cannot receive universal notions without drawing them into its own materiality and consequent particularity of subject, contrary to the very notion of the universal. As Marc Geoffroy has shown,[20] here Ibn Rushd draws on a celestial model according to which the permanent heavenly bodies are eternally moved by associated souls that are not composed hylomorphically with the heavenly bodies in the fashion of the natural hylomorphic composition of natural sublunar beings. In the *Middle Commentary*, Ibn Rushd now holds that the material intellect must be true intellect as an immaterial power associated with the individual human rational soul. This is so because human knowledge of universals requires an immaterial reception in order to avoid the problems of matter and particularity. In this way, the material intellect is now conceived as a separate but associated power, the existence of which depends upon the individual human being to whom it belongs. This satisfies the need for an immaterial subject capable of receiving abstracted intelligibles without particularizing them, an abstraction that again comes about thanks to the presence of the Agent Intellect in the soul:

> It is clear that, in one respect, this intellect is an agent and, in another, it is *a form for us* [*ṣūra la-nā*], since the generation of intelligibles is a product

18 Note that in his early *Epitome of Metaphysics*, Ibn Rushd indicated that the Agent Intellect emerges as 'the last [mover] in the order of these moving [causes], which should be determined as the mover of the sphere of the moon'. Averroes, *On Aristotle's 'Metaphysics'*, ed. by Arnzen, p. 170. In Ibn Rushd's commentaries on *De anima* and *Long Commentary on the Metaphysics*, the Agent Intellect and later the Material Intellect are epistemological posits in the formation of accounts of human knowing. Certainly, in the *Middle Commentary on the De anima* and the *Long Commentary on the De anima* they have no cosmological role in the ontological constitution of sublunar or celestial worlds. Nevertheless, in the *Commentarium magnum De anima*, at p. 442 he calls the Material Intellect the lowest of separate substances.

19 See my introduction in *Long Commentary on the De anima*, English, pp. xxii–xxviii.

20 Averroes, *La béatitude de l'âme*, pp. 71 ff. Cf. Taylor, 'The Agent Intellect as "Form for Us"'; and *Long Commentary on the De anima, English*, introduction, pp. lxix and lxxi. Also see Taylor, 'Abstraction and Intellection'.

> of our will. When we want to think something, we do so, our thinking it being nothing other than, first, bringing the intelligible forth and, second, receiving it. The individual intentions in the imaginative faculty are they that stand in relation to the intellect as potential colors do to light. That is, this intellect renders them actual intelligibles after their having been intelligible in potentiality. It is clear, from the nature of this intellect — which, in one respect, is *a form for us* [*ṣūra la-nā*] and, in another, is the agent for the intelligibles — that it is separable and neither generable nor corruptible, for that which acts is always superior to that which is acted upon, and the principle is superior to the matter. The intelligent and intelligible aspects of this intellect are essentially the same thing, since it does not think anything external to its essence. There must be an Agent Intellect here, since that which actualizes the intellect has to be an intellect, the agent endowing only that which resembles what is in its substance.[21]

Here Ibn Rushd clearly states again the two functions of the Agent Intellect: it is that which brings about the abstraction generating the intelligible known, and it is also 'form for us' (in a non-quidditative sense) as an actualizing power from outside but now present and available to us by our willing. Yet this conception of the material intellect as an associated power still entails the perishability of the material intellect, since the material intellect exists as a power of the perishable human soul. What is particularly noteworthy in the *Middle Commentary* is that in the course of sketching his new view, Ibn Rushd distinctly and very explicitly rejects the possibility that the material intellect could be a substance in its own right.[22] As it happens, that is precisely the teaching which will become his in the *Long Commentary*, as I will discuss below.

---

21 وبين أن هذا هو من جهة فاعل ومن جهة صورة لنا إذ كان توليد المعقولات إلى مشيئتنا ، وذلك أنه متى شئنا أن نعقل شيئا ما عقلناه ، وليس عقلنا أياه شيئا غير تخليق المعقول أولا وقبوله ثانيا. والشئ الذى يتنزل من العقل منزلة الألوان التى بالقوة من الضوء هى المعانى الشخصية التى فى القوة الخيالية ، أعنى أن هذا العقل يصيرها بالفعل معقولات بعد أن كانت بالقوة . وهذا العقل الذى هو صورة لنا من جهة وفعال للهعقولات من جهة بين من أمره أنه مفارق وأنه غير كائن ولا فاسد وذلك أن الفاعل يجب أبدا أن يكون أشرف من المفعول والمبدأ أشرف من الهيولى . وهذا العقل هو الذى العقل والمعقول منه شئ واحد بذاته إذ كان لا يعقل شيئا خارجا عر ذاته . وإنما كان واجبا أن يكون هاهنا عقل فعال لأن الفاعل للعقل يجب أن يكون عقلا إذ كان لا يعطى الفاعل إلا شبيه ما فى جوهره
Averroës, *Middle Commentary on Aristotle's De anima*, ed. and trans. by Ivry, p. 116, vv. 10–21 (translation very slightly modified; emphasis added), hereafter cited as Averroes, *Middle Commentary on Aristotle's De anima*.

22 Averroes, *Middle Commentary on Aristotle's De anima*, p. 112, vv. 8–13 (translation very slightly modified):
وذلك أن بهذا الموضع الذى قلناه نتخلص من أن نضع شيئا مفارقا فى خوهره استعداد ما. لوضعنا الاستعداد موجودا له لا من طبيعته بل من قبل اتصاله بالجوهر الذى فيه هذا الاستعداد بالذات وهو الإنسان. وبوضعنا أن هاهنا شيئا يلحقه هذا الاستعداد بنوع من العرض نتخلص من أن يكون العقل الذى بالقوة استعداد فقط.
'For, by our position as stated, we are saved from positing something separate in its substance as a certain disposition, positing [instead] that the disposition found in it is not due to its [own] nature but due to its conjunction with a substance which has this disposition essentially — namely, man — while, in positing that something here is associated incidentally with this disposition, we are saved from [considering] the intellect in potentiality as a disposition only'.

Although the focus of this section is on Ibn Rushd's commentaries on *De anima*, it is worthy of mention that in *Epistle 1 On Conjunction* — a work posterior to the *Middle Commentary* but prior to the *Long Commentary on De anima* — Ibn Rushd returns to the issue of the material intellect and, as noted by Geoffroy and Steel,[23] for the first time raises the question of what could prevent the conception of the material intellect as a separate substance (the view he definitively rejected in the *Middle Commentary*, as indicated above). Rather than providing a complete account of that issue, Ibn Rushd chooses merely to say that it is something requiring further study.

The teachings of the Greek and Arabic traditions on the intellect receive detailed study by Ibn Rushd in the *Long Commentary on the De anima* — this was the profound study referred to in *Epistle 1 On Conjunction*.[24] Careful consideration of his analysis and attention to his sources reveals what had previously prevented his acceptance of the conception of the material intellect as a separate substance shared by all human beings. Simply put, in the earlier commentaries Ibn Rushd had not accounted for the plurality of diverse human material intellects and the way in which science, knowledge, and discourse depend upon a common set of intelligibles. This consideration became evident to him in his third reading of the *Paraphrase* of Themistius,[25] which — in the sole extant Arabic version rendering the Greek — has the following:

> وليس ينبغى أن يعجب من أن نكون كلنا معشر المركبين من الذى بالقوّة والذى بالفعل وكل واحد منّا إنما وجوده من قبل ذلك الواحد نرجع إلى واحد هو العقل الفعّال فإنّه لولا ذلك من أين كانت تكون لنا العلوم المتعارفة مشتركةً ومن أين كان يكون الفهم للحدود الأول وللقضايا الأول متماثلا بلا تعلّم فإنّه خليق أن يكون لو لم يكن لنا عقل واحد نشترك فيه كلّنا لم نكن أيضا[26]

---

23 Averroes, *La béatitude de l'âme*, pp. 210 and 261.

24 See *Long Commentary on the De anima, English*, introduction, pp. xlii–xlix.

25 The *Paraphrase of the De anima* by Themistius is mentioned in each of the *De anima* commentaries, but its greatest influence on Ibn Rushd was in his reasoning in the *Long Commentary on the De anima*. Regarding the intellects, Themistius seems to write of four intellects: the common intellect, the imperishable potential intellect, the imperishable active intellect, and the Productive (Arabic العقل الفعال *al-ʿaql al-faʿʿāl*, Active Intellect). For Themistius, the potential intellect and the active intellect in the human being are the imperfect and perfect phases of the individual human intellect, while the Productive Intellect is the unique transcendent intellect that penetrates each human potential intellect and assists it in its transition to actual intellect. Since the Productive Intellect is wholly actual and thinks all the forms eternally, it too can be called an actual or active intellect though it is not identified with the individual human intellect. The perishable common intellect is passive and bodily, so it is not properly intellect. The four intellects mentioned by Themistius reduce to three, the human intellect as potential and as active and the Productive Intellect. See Themistius, *Paraphrasis in libros Aristotelis De anima*, ed. by Heinze (hereafter *De anima paraphrasis, Greek*), pp. 98–107; *An Arabic Translation of Themistius' Commentary on Aristoteles De anima*, ed. by Lyons (hereafter *Commentary on Aristotle's De anima, Arabic*), pp. 169–96; *On Aristotle's On the Soul*, trans. by Todd (hereafter *On Aristotle's On the Soul, English*), pp. 122–34.

26 Themistius, *Commentary on Aristotle's De anima, Arabic*, p. 188, v. 17–p. 189, v. 4.

> There need be no wonder that we all are, as a group, composites of what is in potency and of what is in act. All of us whose existence is by virtue of this one are referred back to a one which is the Agent Intellect. For if not this, then whence is it that we possess known sciences in a shared way? And whence is it that the understanding of the primary definitions and primary propositions is alike [for us all] without learning? For it is right that, if we do not have one intellect in which we all share, then we also do not have understanding of one another.[27]

εἰ δὲ εἰς ἕνα ποιητικὸν νοῦν ἅπαντες ἀναγόμεθα οἱ συγκείμενοι ἐκ τοῦ δυνάμει καὶ ἐνεργείᾳ, καὶ ἑκάστῳ ἡμῶν τὸ εἶναι παρὰ τοῦ ἑνὸς ἐκείνου ἐστίν, οὐ χρὴ θαυμάζειν. πόθεν γὰρ αἱ κοιναὶ ἔννοιαι; πόθεν δὲ ἡ ἀδίδακτος καὶ ὁμοία τῶν πρώτων ὅρων σύνεσις καὶ τῶν πρώτων ἀξιωμάτων; μήποτε γὰρ οὐδὲ τὸ συνιέναι ἀλλήλων ὑπῆρχεν ἄν, εἰ μή τις ἦν εἰς νοῦς, οὗ πάντες ἐκοινωνοῦμεν.[28]

> There is no need to be puzzled if we who are combined from the potential and the actual [intellects] are referred back to one productive intellect, and that what it is to be each of us is derived from that single [intellect]. Where otherwise do the notions that are shared [*koinoi ennoiai*] come from? Where is the untaught and identical understanding of the primary definitions and primary axioms derived from? For we would not understand one another unless there were a single intellect that we all shared.[29]

Ibn Rushd accepts this principle as an account of the unity of abstracted intelligibles in act in the Material Intellect.[30] But how can the Material Intellect receive those intelligibles without particularizing them? His response to this question is to assert that the Material Intellect is an immaterial entity unique in its species, not a determinate particular (المشار أليه *al-mushār ilai-hi, hoc aliquid*) which contracts what it receives to its own particularity.[31] For Ibn Rushd, the particularity

27 My translation from Taylor, 'Themistius and the Development of Averroes' Noetics', pp. 15–16, n. 44. Whether Ibn Rushd precisely follows the meaning of Themistius himself is a complex issue that cannot be pursued here. It is clear, however, that he studied the account of Themistius carefully but rejected it, albeit making use of some insights for his own different view.

28 Themistius, *De anima paraphrasis, Greek*, p. 103, v. 36–p. 104, v. 3.

29 Themistius, *On Aristotle's On the Soul, English*, p. 129.

30 See Taylor, 'Intelligibles in Act in Averroes'.

31 Ibn Rushd was well aware of the novel explanation he was crafting. See *Commentarium magnum De anima*, III.5, 656–77, pp. 409–10: 'Opinandum est enim quod iste est quartum genus esse. Quemadmodum enim sensibile esse dividitur in formam et materiam, sic intelligibile esse oportet dividi in consimilia hiis duobus, scilicet in aliquod simile forme et in aliquod simile materie. Et hoc necesse est in omni intelligentia abstracta que intelligit aliud; et si non, non esset multitudo in formis abstractis. Et iam declaratum est in Prima Philosophia quod nulla est forma liberata a potentia simpliciter, nisi prima forma, que nichil intelligit extra se sed essentia eius est quiditas eius; alie autem forme diversantur in quiditate et essentia quoquo modo' ('One should hold that it [i.e., the material intellect] is a fourth kind of being. For just as sensible being is divided into form and matter, so too intelligible being must be divided into things similar to these two, namely, into something similar to

engendered by materiality is what bars intelligibility. Since the Material Intellect is both a unique entity in its species and immaterial, it can be understood as receiving intelligibles in act without particularizing them. Those intelligibles in act come to be in the Material Intellect through the abstractive power of the Agent Intellect and through individual human efforts to abstract or transfer the form intelligible in potency apprehended by exterior and interior senses into a new mode of being as immaterial intelligible in act. This is done by will and individual effort on the part of particular human beings who employ sensed images received into the imagination and denude them of accidental features as much as possible by the cogitative power in order to reveal the form or intention of a thing — though what results still remains a particular.

In the *Long Commentary on De anima*, Ibn Rushd is quite explicit about the presence of the Agent Intellect 'in the soul'. Regarding the text of Aristotle, he writes:

> Now he gives the way on the basis of which it was necessary to assert *the agent intelligence to be in the soul*. For we cannot say that the relation of *the agent intellect in the soul* to the generated intelligible is just as the relation of the artistry to the art's product in every way. For art imposes the form on the whole matter without it being the case that there was something of the intention of the form existing in the matter before the artistry made it. It is not so in the case of the intellect, for if it were so in the case of the intellect, then a human being would not need sense or imagination for apprehending intelligibles. Rather, the intelligibles would enter into the material intellect from the agent intellect, without the material intellect needing to behold sensible forms. And neither can we even say that the imagined intentions are solely what move the material intellect and draw it out from potency into act. For if it were so, then there would be no difference between the universal and the individual, and then the intellect would be of the genus of the imaginative power.[32]

---

form and into something similar to matter. This is [something] necessarily present in every separate intelligence which understands something else. And if not, then there would be no multiplicity {410} in separate forms. It was already explained in First Philosophy that there is no form free of potency without qualification except the First Form which understands nothing outside itself. Its being is its quiddity. Other forms, however, are in some way different in quiddity and being', *Long Commentary on the De anima, English*, pp. 326–27).

32 *Long Commentary on the De anima, English*, pp. 350–51 (emphasis added); *Commentarium magnum De anima*, III.18, 34–51, p. 438: 'Modo dat modum ex quo oportuit ponere in anima intelligentiam agentem. Non enim possumus dicere quod proportio intellectus agentis in anima ad intellectum generatum est sicut proportio artificii ad artificiatum omnibus modis. Ars enim imponit formam in tota materia absque eo quod in materia sit aliquid existens de intentione forme antequam artificium fecerit earn. Et non est ita in intellectu; quoniam, si ita esset in intellectu, tunc homo non indigeret, in comprehendendo intelligibilia, sensu neque ymaginatione; immo intellecta pervenirent in intellectum materialem ab intellectu agenti, absque eo quod intellectus materialis indigeret aspicere formas sensibiles. Neque etiam possumus dicere quod intentiones ymaginate sunt sole

Regarding the Material Intellect, the Agent Intellect and the theoretical intellect or intellect in a positive disposition (*intellectus in habitu*, العقل بالملكة, *al-ʿaql bi-l-malakati*), Ibn Rushd writes:

> [T]here are three parts of the intellect in the soul, one is the receptive intellect, the second is that which makes [things], and the third is the product [of these]. Two of these three are eternal, namely, the agent and the recipient; the third is generable and corruptible in one way, eternal in another way.[33]

Thanks to the presence of the separate abstracting Agent Intellect and separate receptive Material Intellect 'in the soul', stated clearly by Ibn Rushd, the individual human being can by will achieve knowledge, that is, realize the theoretical intellect as a result of the transfer of the form or intention from the being of an intelligible in potency to that of an intelligible in act. However, the individual human soul is not eternal or immaterial or per se an intellect, but rather perishable with the perishing of the body.

Since the abstractive activity of the Agent Intellect brings about an intelligible in act in the receptive Material Intellect as the subject of intelligibles in act, there is a sense in which these two intellects are one in this activity while two in description and being. It is this which Ibn Rushd describes when he writes the following:

> Generally, when someone will consider the material intellect with the agent intellect, {451} they will appear to be two in a way and one in another way. For they are two in virtue of the diversity of their activity, for the activity of the agent intellect is to generate while that of the former is to be informed. They are one, however, because the material intellect is actualized through the agent [intellect] and understands it. In this way we say that two powers appear in the intellect conjoined [*continuatus*] with us, of which one is active and the other is of the genus of passive powers.[34]

---

moventes intellectum materialem et extrahentes eum de potentia in actum; quoniam, si ita esset, tunc nulla differentia esset inter universale et individuum, et tunc intellectus esset de genere virtutis ymaginative'. Note: ibid., III.36, 34–51, pp. 499–500: 'Quoniam, quia illud per quod agit aliquid suam propriam actionem est forma, nos autem agimus per intellectum agentem nostram actionem propriam, necesse est ut intellectus agens sit forma in nobis' ('For, because that in virtue of which something carries out its proper activity is the form, while we carry out {500} our proper activity in virtue of the agent intellect, it is necessary that the agent intellect be form in us', *Long Commentary on the De anima, English*, p. 399).

33 *Long Commentary on the De anima, English*, pp. 321–22; *Commentarium magnum De anima*, 36.5, 570–74, p. 406: '[I]n anima sunt tres partes intellectus, quarum una est intellectus recipiens, secunda autem est efficiens, tertia autem factum. Et due istarum trium sunt eterne, scilicet agens et recipiens; tertia autem est generabilis et corruptibilis uno modo, eterna alio modo'.

34 *Long Commentary on the De anima, English*, p. 360; *Commentarium magnum De anima*, 3.20, 213–22, pp. 450–51: 'Et universaliter, quando quis intuebitur intellectum materialem cum intellectu agenti, apparebunt esse duo uno modo et unum alio modo. Sunt enim duo per diversitatem actionis eorum; actio enim intellectus agentis est generare, istius autem informari. Sunt autem unum quia intellectus materialis perficitur per agentem et intelligit ipsum. Et ex hoc modo dicimus quod intellectus

Nevertheless, the basis for distinguishing two distinct intellects is clearly stated:

> [I]n view of our having asserted that the relation of the imagined intentions {439} to the material intellect is just as the relation of the sensibles to the senses (as Aristotle will say later), it is necessary to suppose that there is another mover which makes [the intentions] move the material intellect in act (and this is nothing but to make [the intentions] intelligible in act by separating them from matter).[35]

This reasoning, he continues, 'forces the assertion of an agent intellect different from the material intellect and different from the forms of things which the material intellect apprehends'.[36] The Agent Intellect and the Material Intellect, then, are two distinct substances, with the Agent Intellect not knowing the world directly since it has no receptivity and the Material Intellect knowing only abstracted forms derived from the world thanks to its relationship with the Agent Intellect.[37]

In Book 3, Comment 36, Ibn Rushd explains further that the Agent Intellect is present in us such that we are ourselves knowers acting by will: 'For because that in virtue of which something carries out its proper activity is the form while we carry out our proper activity in virtue of the Agent Intellect, it is necessary that the agent intellect be form in us'.[38] In this sense, then, the Agent Intellect is acting with and in us as we form the disposed intellect (العقل بالملكة *al-ʿaql bi-l-malakati, intellectus in habitu*) as the theoretical intellect (العقل النظري *al-ʿaql al-naẓarī, intellectus speculativus*) in which we acquire intelligibles formed by abstraction in the Material Intellect. This is as much in us as the acquired intellect (العقل المستفاد *al-ʿaql al-mustafād, intellectus adeptus*) insofar as we are connected via the theoretical intelligibles present both in the human theoretical intellect and in the Material Intellect by the activity of the Agent Intellect:[39]

---

continuatus nobiscum, apparent in eo due virtutes, quarum una est activa et alia de genere virtutum passivarum'.

35 *Long Commentary on the De anima, English*, p. 451; *Commentarium magnum De anima*, 3.18, 51–57, pp. 438–39: 'Unde necesse est, cum hoc quod posuimus quod proportio intentionum ymaginatarum ad intellectum materialem est sicut proportio sensibilium ad sensus (ut Aristoteles post dicet), imponere alium motorem esse, qui facit eas movere in actu intellectum materialem (et hoc nichil est aliud quam facere eas intellectas in actu, abstrahendo eas a materia)'.

36 *Long Commentary on the De anima, English*, p. 451; *Commentarium magnum De anima*, 3.18, 58–60, p. 439: '[H]ec intentio cogens ad ponendum intellectum agentem alium a materiali et a formis rerum quas intellectus materialis comprehendit'.

37 'Intelligentia enim agens nichil intelligit ex eis que sunt hic'. *Commentarium magnum De anima*, 3.19, 15–16, p. 441 (*Long Commentary on the De anima, English*, p. 353: 'The agent intelligence understands nothing of the things which are here').

38 *Long Commentary on the De anima, English*, p. 399; *Commentarium magnum De anima*, 3.36, 586–90, pp. 499–500: 'Quoniam, quia illud per quod agit aliquid suam propriam actionem est forma, nos autem agimus per intellectum agentem nostram actionem propriam, necesse est ut intellectus agens sit forma in nobis'.

39 To be clear, this does not mean humans have a personal immaterial intellect. Rather, it represents the function of individual human awareness of the intelligibles in act realized in the Material Intellect. Ibn Rushd seeks to characterize this as the Agent Intellect as 'form for us' as acquired intellect. The route

> It is evident that when that motion [i.e., conjoining with the Agent Intellect through the theoretical intelligibles] will be complete, immediately that intellect will be conjoined with us in all ways. Then it is evident that its relation to us in that disposition is as the relation of the intellect which is in a positive disposition [i.e., العقل بالملكة *al-ʿaql bi-l-malakati, intellectus in habitu*] in relation to us. Since it is so, it is necessary that a human being understand all the intelligible beings through the intellect proper to him and that he carry out the activity proper to him in regard to all beings, just as he understands by his proper intellection all the beings through the intellect which is in a positive disposition [i.e., العقل بالملكة *al-ʿaql bi-l-malakati, intellectus in habitu*] when it has been conjoined [*continuatus*] with forms of the imagination.[40]

This is the way in which knowledge of intelligibles comes to be in, and to be proper to, an individual human being thanks to the assistance of the Agent and Material Intellects, which also have presence *in the soul*. Marvelling at this, Ibn Rushd recalls the view of Themistius:

> In this way, therefore, human beings, as Themistius says, are made like unto God in that he is all beings in a way and one who knows these in a way, for beings are nothing but his knowledge and the cause of beings is nothing but his knowledge. How marvelous is that order and how mysterious is that mode of being![41]

That is, through this process humans, unlike other mortal animals, are able to become intellectual knowers by way of coming to have intelligibles of all things in

---

for this attribution is circuitous: the human individual provides particular images which the Agent Intellect abstracts and actualizes in another mode of being in the Material Intellect (as light actualizes the opaque medium of sight to be transparent and to receive colours); the Material Intellect (as subject of being) thereby has the theoretical intelligibles (unique intellectual forms) which it then makes available to the individual who provided the original images (the subject of truth); given that this process was made possible only thanks to the involvement of the Agent Intellect, an external source, the process can be traced back to the Agent Intellect and in the individual knower it can be called the acquired intellect (*intellectus adeptus*, العقل المستفاد *al-ʿaql al-mustafād*).

40 *Long Commentary on the De anima, English*, p. 399; *Commentarium magnum De anima*, 3.36, 607–16, p. 500: 'Et manifestum est quod, cum iste motus complebitur, quod statim iste intellectus copulabitur nobiscum omnibus modis. Et tunc manifestum est quod proportio eius ad nos in illa dispositione est sicut proportio intellectus qui est in habitu ad nos. Et cum ita sit, necesse est ut homo intelligat per intellectum sibi proprium omnia entia, et ut agat actionem sibi propriam in omnibus entibus, sicut intelligit per intellectum qui est in habitu, quando fuerit continuatus cum formis ymaginabilibus, eis omnia entia intellectione propria'.

41 *Long Commentary on the De anima, English*, p. 399; *Commentarium magnum De anima*, 3.36, 617–22, p. 501: 'Homo igitur secundum hunc modum, ut dicit Themistius, assimilatur Deo in hoc quod est omnia entia quoquo modo, et sciens ea quoquo modo; entia enim nichil aliud sunt nisi scientia eius, neque causa entium est aliud nisi scientia eius. Et quam mirabilis est iste ordo, et quam extranaeus est iste modus essendi!'

them, though for humans knowledge is posterior to the things known, while for God knowledge is causally prior to the things.[42]

For the sake of what follows, five conclusions from our study of the commentaries on *De anima* are important to note here. First, in the *Long Commentary* Ibn Rushd held the existence of the Agent Intellect and the Material Intellect to be two distinct substances, not one substance under two different descriptions. Second, he held these to have existences of their own as separately existing immaterial substances and not to be essential, ontologically intrinsic powers belonging to each human soul. Third, in all three of his commentaries on *De anima*, Ibn Rushd holds that the Agent Intellect should be described as 'form for us', صورة لنا *ṣura lanā*, *forma nobis*. In the *Long Commentary* he holds that these intellects — now understood as separately existing substances — can come to be present 'in the soul' with the Agent Intellect providing the power of intellectual abstraction and with the Material Intellect shared by all human knowers being the locus of the unique set of immaterial intelligibles in act common to all human knowers. Fourth, in none of the three commentaries is there provision for a life after death for the individual human soul. That is, there is no provision for the immortality of the human soul in Ibn Rushd's commentaries on *De anima*. Fifth, though we are conjoined via the Material Intellect to the Agent Intellect in the process of attaining and possessing abstractive knowledge — something marvellous and, as it were, divine — there is no ontological or substantial unity with the Agent Intellect and no clearly established, reasoned grounds in support of an ascent to it via another different kind of knowing of it and other transcendent intellectual substances.[43]

---

42 See Kogan, *Averroes and the Metaphysics of Creation*, pp. 229–48.

43 This is in accord with the view of Herbert A. Davidson in *Alfarabi, Avicenna, and Averroes, on Intellect*: 'Conjunction with the active intellect occurs, in Averroes' several accounts, during the life of the body and not in the hereafter. None of the accounts envisions anything ecstatic or properly mystical in the conjunction of the human intellect with the active intellect' (p. 330). Furthermore, in additional to bodily faculties, '"Practical intelligible thoughts" likewise do not survive; they are tied to the "imaginative faculty" and perish together with it. Human theoretical thoughts that grow out of images presented by the imaginative faculty suffer an identical fate [...] individual human consciousness of theoretical thoughts perishes together with the faculties on which consciousness of such thoughts depends [...]. The Long Commentary, from which the last quotations are taken, has a unique conception of the material intellect and its relation to the human soul. But compositions belonging to other stages of Averroes' career make equally plain that "theoretical intelligible thoughts", that is to say, human scientific knowledge at the "mathematical", "physical", and even the "metaphysical" levels, all "perish" together with the human imaginative faculty. Metaphysical knowledge, no less than physical knowledge, is rooted in images furnished by the imaginative faculty, since it consists in abstractions made from propositions presented by the science of physics'. Ibid., pp. 337–38. Nevertheless, Ibn Rushd does assert a uniting with the Agent Intellect when all potential intelligibles have been realized and the Material Intellect ceases to exist. This is discussed in more detail below.

## Ibn Rushd on the Afterlife and Human Fulfilment in his *Long Commentary on the Metaphysics* of Aristotle

In his Long Commentary on Aristotle's *Metaphysics*, Book Lam, Ibn Rushd read in the Arabic text 17 of Aristotle,

واما إن كان شئ ما يبقى بأخرة فقد ينبغى ان نبحث عن ذلك و ذلك أن نبحث الأشياء فلا مانع يمنع مثال ذلك إن كانت النفس حالها هذه الحال لا كلها لا كن العقل وذلك أن كلها لعله الا يمكن

> As for whether something remains *afterwards* [*bi-akharatin*], it may be necessary that we investigate it, for regarding some things there may not be anything preventing it, for example, the situation for the soul though not the whole of it but rather for the intellect, since for the whole it is perhaps not possible.[44]

My translation of this text is quite similar to that of Charles Genequand, with the exception of the prepositional phrase 'afterwards' (بأخرة *bi-akharatin*), which he renders as 'in an afterlife'.[45] Here the Arabic renders the Greek *ὕστερόν*. The context for Aristotle is the coexistence of formal causes with their effects, with an example of the shape as formal cause coming into being with a bronze sphere. Does the form or shape necessarily remain existing with the bronze taken away? Aristotle writes:

> Whether any form remains also afterwards [*ὕστερόν*] is another question. In some cases there is nothing to prevent this, e.g. the soul may be of this nature (not all of it, but the intelligent part; for presumably all of it cannot be).[46]

To paraphrase Aristotle, the formal cause of something must remain with the effect so long as the effect is what it is as caused by the formal cause. The shape of the bronze sphere is there as essential to the thing as bronze sphere. The implication is that after the separation of the formal cause (the spherical shape) from the bronze, the *sphere* no longer exists. To expand, if soul is form of the body and the individual is the composite of the two, the composite human will not remain as such after the death of the body and neither will the soul remain, unless there is perhaps some other special consideration concerning 'the intelligent part'

---

44 *Tafsīr mā ba'd aṭ-Ṭabī'at*, p. 1486, v. 13–p. 1487, v. 2. My translation and emphasis.

45 Genequand's translation is: 'We must inquire whether anything can last *in an afterlife*, for in certain things there is nothing to prevent it; for instance, if the soul is in that situation, it is not the whole soul, but the intellect only; for the whole it is perhaps impossible'. *Long Commentary on the Metaphysics, Arabic-English*, p. 103. My emphasis.

46 Aristotle, *Metaphysics*, 1070a24–26, ed. by Ross, English in *Aristotle's Metaphysics*, trans. by Tredennick, p. 131: (εἰ δὲ καὶ ὕστερόν τι ὑπομένει, σκεπτέον· ἐπ᾽ ἐνίων γὰρ οὐδὲν κωλύει, οἷοω εἰ ἡ ψυχὴ τοιοῦτον, μὴ πᾶσα, ἀλλ᾽ ὁ νοῦς· πᾶσαν γὰρ ἀδύνατον ἴσως). Lindsay Judson's recent translation (*Metaphysics Book Λ*, trans. by Judson, p. 25) renders this passage as: '(Whether something remains afterwards too has to be considered, since in some cases nothing prevents it; for example whether the soul is of such a sort — not all soul but intellect; for perhaps it is impossible for all soul to remain)'.

of soul. Such might be considered possible for the soul, though not all of it but only the intellect. But there is no assertion of an afterlife of individuals here. True, the context of Aristotle's discussion, which includes a reference to Plato and the Forms and mention of the soul, may lure one — in the present context, it is Albert in accord with his Christian commitments and a general understanding of the afterlife in the Abrahamic tradition of Islam — into reflective consideration of the notion of an afterlife and prompt that notion. Still, بأخرة *bi-akharatin* is not the proper Arabic phrase to indicate the afterlife; the correct phrase would be بالآخرة *bi-l-ākhirati* 'in the afterlife', that is, الدار الآخرة 'the ultimate abode'. Here the phrase بأخرة *bi-akharatin* is merely an expression with the more common meaning of 'afterwards', 'later', 'eventually', or 'ultimately'. The phrase occurs two more times in Ibn Rushd's Comment 17 on this text and is incorrectly rendered as 'in an afterlife' in the English translation by Genequand.[47]

The same phrase بأخرة *bi-akharatin* occurs repeatedly in the *Long Commentary on the Metaphysics* and is rendered by Michael Scot in his Latin translation — both there and apparently in his translation of the *Long Commentary on the De anima* — as *in postremo*, with the meanings of 'afterwards', 'later', 'eventually' or 'ultimately'. The phrase بالآخرة *bi-l-ākhirati* 'in the afterlife' is not found in the *Long Commentary on the Metaphysics*.

In Ibn Rushd's Comment 17 on this text of Aristotle's, he speaks in his own voice and declares that the Agent Intellect is like form in the Material Intellect bringing about (يفعل *yafʿalu*) intelligibles in act received into the Material Intellect. This latter is not truly matter but rather has the role *as the place* (كالمكان *ka-l-makān*) into which abstracted intelligibles are received. These Intellects are two distinct substances, both eternal without generation or corruption. In the case of the Material Intellect, which in the *Long Commentary on the De anima* Ibn Rushd (in a novel philosophical teaching) calls a unique fourth kind of being,[48] it is something eternal but can also apprehend things of the natural world of generation and corruption. Envisioning the ultimate achievement of all knowledge, towards the end of this Comment he imagines a time when human perfection is achieved with all knowledge attained, and the potentiality as well as the need for intellectual abstraction by way of the two separate Intellects is no longer present. Then, when all potentiality of intellect has been eliminated by being actualized

---

47 *Tafsīr mā baʿd aṭ-Ṭabīʿat*, pp. 1487–88; *Long Commentary on the Metaphysics, Arabic-English*, p. 103. The occurrence of 'after-life' in Genequand's translation may be a typographical error for 'afterlife'. In Genequand's case, the morphological differences in the two Arabic phrases provide a distinction that needed to be recognized but was not. As we shall see below, such was not the case for Albert, who took Michael Scot's sound Latin rendering of the Arabic as *in postremo* to mean 'in the afterlife' rather than 'afterwards' or 'later'. Further, note that Ibn Rushd's Comment 17 begins with a quotation from Alexander, who raises the question of the individual human soul or intellect persisting after death only to dismiss it altogether. After explaining this and mentioning that most commentators think an individual receptive (*scil.*, 'material') intellect survives (as in al-Fārābī and Ibn Sīnā), Ibn Rushd sets out his own teaching.

48 *Commentarium magnum De anima*, p. 409; *Long Commentary on the De anima, English*, p. 326.

and intellect has been fully realized, this will be highest happiness for us through an intellection in which act and substance are one and the same:

> Since [the Agent Intellect] conjoins [اتصل *ittaṣala*] with the Material Intellect, its act, insofar as [the Agent Intellect] unites with [the Material Intellect], is not its substance and what it acts on is a substance which belongs to another, not to itself. Owing to this it is possible that an eternal thing intellectually understands [يعقل *yaʿqilu*] what is generated and corruptible. So if this intellect [i.e., the Material Intellect] is divested of potentiality when it reaches human perfection, then it is necessary that this act which is other than it also cease. Hence, in this state either we are not at all intellectually understanding by this intellect, or we are intellectually understanding insofar as its act is its substance and it is impossible that we should at any time not be intellectually understanding by it. Thus, it remains that when this intellect is free from potentiality, we are intellectually understanding by it insofar as its act is its substance, and this is ultimate happiness [السعادة القصوى *al-sa ʿāda al-quṣwā*].[49]

Here Ibn Rushd imagines a future when all intellectual abstraction by way of sense powers comes to perfection and completion in the realization of all intelligibles in act. Before then, abstraction and the attainment of intelligibles in act require sense powers and memory, the Agent Intellect's power of abstraction, and the Material Intellect's receptivity as the place (not matter) into which the abstracted intelligibles are received. When all intelligibles in potency have been garnered, there is no longer any use or need for the Material Intellect and its potentiality; as a consequence, it will cease to exist. Then either there is no human intellectual understanding, or we will be eternally understanding by the Agent Intellect alone. Such an imagined future is the moment of the attainment of ultimate happiness.

This hyperbolic imagined scenario is replete with complex issues and consequences and prompts a grand array of questions worthy of pursuit at another

---

49 I have made some significant changes to the translation of Genequand, *Long Commentary on the Metaphysics, Arabic-English*, pp. 104–05; *Tafsīr mā baʿd aṭ-Ṭabīʿat*, p. 1490, vv. 2–10:
لما اتصل بالعقل الهيولانى كان فعله من جهة ما يتصل به غير جوهره وكان ما يفعله هو جوهر هو لغيره لا لذاته ولذلك
امكن ان يكون شيء ازلى يعقل ما هو كا ئن فاسد فان كان هذا العقل يتعرى عند بلوغ الكمال الانسانى عن القوة فقد يجب
ان يبطل منه هذا الفعل الذى هو غيره فاما ان نكون فى تلك الحال غير عاقلين اصلا بهذا العقل او نكون به عاقلين من حيث
فعله جوهره ومحال ان نكون فى وقت من الاوقات غير عاقلين به فقد بقى ان نكون اذا برئ هذا العقل من القوة عاقاين به
من حيث فعله جوخره وهى السعادة القصوى.
The corresponding Latin translation has: 'Sed cum fuerit copulatus cum intellectu materiali erit actio eius secundum quod copulatur cum eo actio alia a substantia eius. et fuit aliud [corr: illud] quod agit substantia et est alii non sibi. et ideo possibile est ut aliquod aeternum intelligat aliquod generabile et corruptibile. Si igitur iste intellectus denudetur apud perfectionem humanam a potentia, necesse est ut destruatur ab eo hec actio que est alia ab eo. et tunc aut non intelligimus omnino per hunc intellectum: aut intelligemus secundum quod actio eius est substantia eius. et impossibile est ut in aliqua hora non intelligamus per ipsum. Relinquetur igitur cum iste intellectus fuerit denudatus a potentia ut intelligamus per ipsum secundum quod actio est substantia eius et est ultima prosperitas' (*Long Commentary on the Metaphysics, Latin*, 303B–D). It is because of the use of 'prosperitas' here for السعادة *al-saʿād* that I render *prosperitas* in the texts of Albert as 'happiness'.

opportunity. For present purposes, it should be noted that this conjoining with the Agent Intellect is by no means supposed to be a description of the experience of individual human beings before that final moment. For mortal human individuals, the highest happiness that can be achieved is in the attainment of intellectual knowledge in the present life through willed scientific study. (What precisely a human being would then be and what knowing would be is by no means clear in the event of perfection that Ibn Rushd fantastically imagines).

The phrase بأخرة *bi-akharatin* also occurs in Comment 38 of Book Lam. The context is a discussion of the cause of the eternal movement of the heavens, where Aristotle says in the Arabic:

> Heaven and nature, then, are in agreement with such a principle; heaven and nature, then, depend [on it]. Its sojourn [*ḥulūl*] is in accordance with that which is most excellent, which belongs to us for a short time, but for it is eternally so.[50]

Citing a different translation — 'it is on such a principle then, that the heaven and nature depend; we enjoy something like a happy state for a short time' — Ibn Rushd explains,

> He means: it is evident that the heavens and nature are conjoined [*ittaṣalat*] with a principle which is an intellect in the highest state of pleasure, happiness and bliss, similar to our own state of conjoining for a short time with the intellect which is our principle.[51]

Ibn Rushd understands this second version to indicate that our perishable souls can have intellectual understanding and fulfilled happiness for but a short time, whereas the heavens as incorruptible have unending intellectual happiness ultimately through the Unmoved Mover. In his teaching in the *Long Commentary on the De anima*, intellectual fulfilment for human knowers comes about through the external and internal sense powers and the abstractive power of the Agent Intellect, which moves intelligibles in potency to a new mode of being as intelligibles in act and 'places' them into the receptive Material Intellect. For us, these separate intellects together constitute our principle and are movers and ends. But our fulfilment and happiness comes about by the active principle bringing about intellectual abstraction and intelligibles in act in the receptive principle. Ibn Rushd writes:

---

50 *Long Commentary on the Metaphysics, Arabic-English*, p. 155; *Tafsīr mā baʿd aṭ-Ṭabīʿat*, p. 1608, v. 8–p. 1609, v. 2:
فادا السماء و الطبيعة متعلقتان و الحلول على ما هو فاضل جدا الذى يكون لنا زمانا يسيرا هكذا لذلك دائما.

51 *Long Commentary on the Metaphysics, Arabic-English*, p. 156 (I revise Genequand's 'in contact with' to 'conjoined with' and 'contact' to 'conjoining'); *Tafsīr mā baʿd aṭ-Ṭabīʿat*, pp. 1611, v. 15–1612, v. 3:
يريد انه قد تبين ان السماء والطبيعة قد اتصلت بمبدا هو عقل فى غاية اللذة والسرور والغبطة كحالنا نحن فى الاتصال بالعقل الذى هو مبدانا زمانا يسيرا.

> It clearly appears from that that Aristotle thinks that happiness for men *qua* men consists in their conjoining with the intellect which has been shown in the *De anima* to be principle, mover, and agent for us. This is because the separate intellects insofar as they are separate must be principle in both of two ways: insofar as they are mover and insofar as they are ends. Hence, the Agent Intellect, insofar as it is separate and principle for us, must move us in the way the lover moves the beloved [يحرك العاشق المعشوق *yuḥarriku al-ʿāshiqu al-maʿshūqa*]. And, if every motion must be conjoined with the thing which moves it in the manner of the end, then we must be conjoined ultimately [بأخرة *bi-akharatin*] with this separate intellect,[52] so that we depend on the likes of this principle, on which the heavens depend, as Aristotle says, although this happens for us but for a short time.[53]

Here Ibn Rushd explains that, whereas finality is key in the cases of intellects, celestial souls, and ensouled celestial bodies, the Agent Intellect (lover) moves the receptive Material Intellect and us (the beloved) to the fulfilling realization of knowledge through conjoining.[54]

---

52 Note that the intellect referred to here is the Agent Intellect, not the Unmoved Mover.

53 *Long Commentary on the Metaphysics, Arabic-English*, p. 157 (Arabic added); *Tafsīr mā baʿd aṭ-Ṭabīʿat*, p. 1612, v. 8–p. 1613, v. 4: ومن هنا يظهر كل الظهور ان ارسطاطاليس يرى ان السعادة للناس بما هم ناس انما هو اتصالهم بالعقل الذى تبين فى كتاب النفس انه مبدا محرك وفاعل لنا و ذلك ان العقول المفارقة بما هى مفارقة يجب ان تكون مبدا لما هى له مبدا بالنحوين جميعا اعنى من جهة ما هى محرك ومن جهة ما هى غاية فالعقل الفعال من جهة ما هو مفارق ومبدا لنا قد يجب ان يحركنا على جهة ما يحرك العاشق المعشوق وان كانت كل حركة فقد يجب ان تتصل بالشىء الذى يحركها على جهة الغاية فواجب ان نتصل باخرة بهذا العقل المفارق حتى نكون قد علقنا بمثل هذا المبدا الذى علقت به السماء كما يقول ارسطو وان كان ذلك لنا زمانا يسي *Long Commentary on the Metaphysics, Latin*, 321F–G: 'Et ex hoc quidem apparet bene quod Aristoteles opinatur quod forma hominum in eo quod sunt homines non est nisi per continuationem eorum cum intellectu, qui declaratur in libro De anima esse principium agens et movens nos. Intelligentiae enim abstractae in eo quod sunt abstractae debent esse principia eorum quorum sunt principia duobus modis: secundum quod sunt moventes et secundum quod sunt finis. Intelligentia enim agens inquantum est abstracta et est principium nobis necesse est ut moveat nos secundum quod amatum amans. Et si omnis motus necesse est ut continuetur cum eo a quo fit secundum finem, necesse est ut in postremo continuetur cum hoc intellectu abstracto, ita quod erimus dependentes a tali principio a quo coelum dependet, quamvis hoc sit in nobis modico tempore, sicut dixit Aristoteles'. I have substantially revised Genequand's translation. Note in particular that I correct his 'the beloved moves the lover' into 'the lover moves the beloved'. Also note that in the Latin *secundum quod amatum amans, amans* might tempt a reader to understand the phrase differently and out of context as 'the beloved [moves] the lover'. See also note 54 below.

54 In 1984, the same year as the publication of Genequand's translation, a French translation by Aubert Martin appeared, in Averroès, *Grand Commentaire de la 'Métaphysique d'Aristote'* (*Tafsīr mā baʿd aṭ-Ṭabīʿat*). Both translations were based on the edition of Bouyges, and both worked to convey valuable translations of the work, but their methodological foci were different. While Genequand was largely concerned with the philosophical reasoning, Martin explicitly focused on lexical and philological considerations and somewhat less on the philosophical. For the passage considered here, Martin (pp. 233–34) presents a translation in accord with the one I have provided. In support of his reading, Martin (p. 234, n. 10) references a parallel passage in Ibn Rushd's *Middle Commentary on the Metaphysics*, available in the 1947 printing of texts in *Rasāʾil Ibn Rushd*. There, Ibn Rushd explains that the existence of the motions of the heavens is due to separate immaterial intellects insofar as celestial

Three considerations should be noted here. First, the doctrine that Ibn Rushd provides in the *Long Commentary on the Metaphysics* is in accord with that of the *Long Commentary on the De anima*. Human happiness is found in the achievement of intellectual understanding through a conjoining with the separately existing substances, Agent Intellect and Material Intellect. For individual human knowers, however, such happiness takes place only during short earthly lives, since there is no post mortem existence for them in an afterlife. Second, the function of the Material Intellect is to be the shared immaterial locus of abstracted intelligibles attained through the apprehension of things of the world, which it receives thanks to the abstractive light of the Agent Intellect. In the *Long Commentary on the De anima*, Ibn Rushd also explains that the Material Intellect, as intellect, understands not only the abstracted intelligibles separated from matter but also the Agent Intellect itself, since the Material Intellect is intellect with entailed powers. He adds that its nature as intellect does not undermine its ability to understand other separate forms, namely, intellects. Note that he does not say that we, too, through our conjoining with the Material Intellect, come to know separate intellectual substances.[55] Third, Ibn Rushd multiple times uses the phrase بأخرة *bi-akharatin*, which in one case is correctly translated by Genequand as 'ultimately', though his other translations as 'in an afterlife' are incorrect. This is important since, in the discussion of Albertus Magnus below, translation, interpretation, and meaning will again be a major focus, simply because the phrase بأخرة *bi-akharatin* is rendered into Latin by Michael Scot as *in postremo*. In Latin, unlike in Arabic, this phrase more easily allows for two very different meanings: one as 'later', 'ultimately', and the like, and another as 'in the afterlife'.

## A Tale of Two Averroisms

In his 1982 paper 'Notes sur les débuts (1225–1240) du premier "averroïsme"', René Antoine Gauthier provides a valuable multifaceted study of the initial entry of the translated works of Averroes into Latin Europe.[56] This includes an account

---

bodies conceive them (*'alà jihati t-taṣawwuri bi-l-'aqli*) so as to cause desire in the celestial bodies, 'just as the form of the lover moves the beloved' (*kamā yuḥarriku ṣūratu al-'āshiqi al-ma'shūqa*). Ibn Rushd, *Middle Commentary on the Metaphysics, Rasā'il Ibn Rushd* ed., p. 141. Note that Genequand provides a translation of the entirety of Ibn Rushd's commentary on Book Lam / Lambda, whereas Martin does not translate Texts and Comments 42–50. A critical edition by Maroun Aouad of the Middle Commentary with French translation and English introduction appeared in late 2023, after the present article went to print, as volume 11 of the Brill series Islamicate Intellectual History. See https://brill.com/edcollbook/title/62324?language=en.

55 See *Commentarium magnum De anima*, 3.5.679–83, p. 410; *Long Commentary on the De anima, English*, pp. 327–28. Some years ago, Steven Menn valuably brought to my attention that my translation of *intellectibus simplicibus* at *Commentarium magnum De anima*, 3.20, 277, p. 453 as 'with the simple intelligibles' (*Long Commentary on the De anima, English*, p. 362) should be emended to 'with the simple intellects'.

56 Parts of this section draw on my introduction to the *Long Commentary on the De anima*, pp. xcix–cv.

of what he holds to be a common reading of those works regarding the nature of the human intellect. With extraordinary attention to detail and a critical eye to sources and their interpretation by other scholars, Gauthier examines the letter of King Manfred of Sicily to the scholars of Paris and determines that the works Manfred mentions sending to Paris could not have been those of Averroes, since, as Gauthier demonstrates, the letter turns out to have been written for Parisian scholars around 1263. The major works of Averroes were widely available well before that date. This may seem inconsequential, but it is clear that only when misconceptions and the misreading of documents are pointed out in detail can we correctly understand important issues such as the one at stake here: the dating of the entry of the works of Averroes into the scholarly world of the thirteenth-century theologians.[57]

Having dealt with the challenging issue of the letter of Manfred, Gauthier proceeds to examine in detail R. de Vaux's account of Roland of Cremona regarding the first entry of the works of Averroes.[58] Again, Gauthier applies meticulous care and a wide and deep knowledge of secondary literature to the question of whether Roland's *Summa*, presumed to be written around 1230, indicates by absence of reference to Averroes that the works of Averroes were not yet available at that date. I will not rehearse here all the details of Gauthier's analyses, but just indicate that he traces the scholarship that gradually moved the date of this work from 1230 to 1233 and then to 1236, then finally to 1244. It is certain that the works of Averroes were widely circulating by that date. Hence, the *Summa* of Roland has no value with regard to the dating of the entry of the translated works of Averroes — another misconception that came to be set aside through the accumulated work of scholars such as Lottin, Doucet, and Cremascoli and the critical analyses of Gauthier.[59]

Next, with an account too brief, Gauthier turns to a valuable consideration of the biography and work of Michael Scot, the presumed translator of all Averroes's Long Commentaries,[60] concluding that Michael was in service to Frederick II from September 1220 up to his death in 1235. Gauthier goes on to argue that Michael's first translations of Averroes were made between 1220 and 1224: those of the *Long Commentary on the De anima* and the *Long Commentary on the Metaphysics*, which came to be known from 1225.[61]

Before the availability of Averroes's works, the dominant account of the soul in the Arabic tradition available to Latin readers was that found in the translated works of Avicenna — in his *De anima* and relevant passages of his *Metaphysics* — as well as the abbreviated account of his teachings in Algazel's *Summa theoreticae*

57 Gauthier, 'Notes sur les débuts', pp. 322–30.

58 De Vaux, 'La première entrée d'Averroès chez les latins'.

59 Gauthier, 'Notes sur les débuts', pp. 330–31.

60 For a recent discussion of the translations of Michael Scot and others, see Hasse, *Latin Averroes Translations*.

61 Gauthier, 'Notes sur les débuts', pp. 331–34.

*philosophiae*.[62] In those works, Ibn Sīnā/Avicenna taught that the rational soul makes instrumental use of the body, yet is not properly form of the body but transcends the body, since the rational soul is an immaterial substance in its own right.[63] The intellectual development of the soul in the natural world takes place thanks to its exercise of powers resident in the physical brain in relation to abstracted or separated images obtained through perception of the world and things in it. However, the activity that brings about intelligibles in act in the rational soul, that is, the final level of abstraction, is not that of a power in the brain or even in the soul itself alone, but rather an activity of the rational soul in relation to the Agent Intellect. For at that stage of the rational soul's preparation, the Agent Intellect connects with the soul and emanates upon the soul a flow of intelligibles which the soul retains so long as it is in contact with the Agent Intellect.[64] This doctrine of Avicenna seems very clear in the Latin translations and proved to be widely influential in Europe, with some theologians venturing to hold that the Agent Intellect is God. Other works were available and studied, such as the *De intellectu* of al-Fārābī and the *De intellectu* of Alexander of Aphrodisias,[65] along with works authored by the translator Domingo Gundisalvi.[66] Various interpretations of Avicenna's accounts were available and a dominant explanation of the Peripatetic thought that he presented generated controversy, even condemnation.[67] For the Latins, the arrival of translations of Ibn Rushd's *Long Commentary on the De anima* and *Long Commentary on the Metaphysics* — with complete texts of Aristotle and Averroes's detailed commentary with critical explanations — was welcomed as a challenging alternative to what Avicenna presented.

Also in 1982, Gauthier published a treatise written by a master of arts around 1225, entitled *De anima et potenciis eius*.[68] The author of this work cites Averroes's *Long Commentary on the Metaphysics* as well as the *Long Commentary on the De*

62 This is al-Ghazālī's *Maqāṣid al-Falāsifa*, sometimes rendered as *The Intentions of the Philosophers*, translated by Gundissalinus and John of Spain. For further information on this work in its Latin translation, see Minnema, 'Algazel Latinus'. For a brief account of the text, its source and its use, see Janssens, 'Al-Ghazālī's *Maqāṣid al-falāsifa*', cited by Minnema, p. 154, n. 5.

63 See Janssens, 'Ibn Sīnā's Ideas of Ultimate Realities', p. 255.

64 In the Latin tradition, the interpretation of Avicenna largely involves this emanative account. Among modern scholars of the Arabic writings of Ibn Sīnā there has been a lively controversy about how to reconcile his various accounts, including discussion of a naturalistic understanding in which the Agent Intellect is absent. I review some of the literature on intellectual understanding in Ibn Sīnā and provide my own account in Taylor, 'Avicenna and the Issue of the Intellectual Abstraction of Intelligibles'. For a substantial recent treatment, see Alpina, *Subject, Definition, Activity*; for a succinct account of Ibn Sīnā's metaphysics of human soul, see Gutas, 'Ibn Sina [Avicenna]'.

65 On the importance of Alexander and al-Fārābī in Latin translation for the European Christian tradition, see de Libera, *Métaphysique et noétique*, pp. 264–328.

66 See Polloni, *The Twelfth-Century Renewal of Latin Metaphysics*. See Burnett, 'Arabic Philosophical Works Translated into Latin', for a list of translations and translators with dates.

67 Perhaps the most valuable account of Avicenna on soul and intellect in Latin remains Hasse, *Avicenna's 'De anima' in the Latin West*.

68 Gauthier, 'Le traité *De anima et de potenciis eius*'.

*anima* and presents Averroes as holding, in contrast to Avicenna, that the agent intellect is a power of the human soul — the doctrine that Gauthier calls 'First Averroism'.[69] To be specific, he adds, for this author there are two intellects, the agent intellect and the possible or receptive material intellect, but they are not really different; rather, they are substantially identical but distinguishable by reason as joined in the single substance of the human rational soul. In this case, consideration of the intellect as united to the body indicates that the possible intellect is perishable in its content yet immortal in its substance, whereas the agent intellect just in itself, as subsistent, is immortal. Such, Gauthier remarks, 'is the doctrine which reigned without challenge at the faculty of arts from 1225 to 1250, and was maintained even beyond that date'.[70] Gauthier concludes his impressive study with the determination that the first entry of the works of Averroes can now be placed as early as 1225. Yet, as we will see below, evidence of the so-called 'First Averroism' doctrine in fact antedates the translations of Averroes.

As Gauthier notes, the issue of 'First Averroism' and 'Second Averroism' was addressed by Dominique Salman in an article published in 1937. According to Salman, initially Averroes was welcomed as a corrective to Avicennian thought, which required a separate Agent Intellect; Averroes was read as having held the agent intellect and material intellect to be powers of the individual soul. A second, very different, understanding of Averroes's teaching on intellect later came to the fore among theologians of Europe, one that in Salman's analysis in fact reflects the genuine teaching of Ibn Rushd:

> For Averroes, the possible intellect was unique and separated just like the agent intellect. These two immaterial substances, which are the last in their order, can indeed unite and know in the world of spirits; however, they bring about an act of human intellection if they come into contact in the same phantasm provided for them by the cogitative power of the animal, still mortal, which is the human individual.[71]

69 Gauthier 'Notes sur les débuts', p. 335. His article proceeds with a detailed study of texts, devoting sections to Averroes at Oxford *c.* 1230–32, Averroes at Paris *c.* 1225–40, the *Commentary on the Sentences* of Hugh of Saint-Cher (1231–32), the *Summa* of Philip the Chancellor (*c.* 1232), the question *On Divine Knowledge* in works attributed to Alexander of Hales (*c.* 1236?), the *De intelligentiis* of Adam of Puteorum Villa (*c.* 1240), the *De virtutibus* of William of Auvergne (1228–31), and William's citation of Averroes in his *De universo* (*c.* 1233–35).

70 Gauthier 'Notes sur les débuts', p. 335.

71 Salman, 'Note sur la première influence d'Averroès', p. 204. Below I present B. Carlos Bazán's account of the teachings in the Latin texts of Averroes that support this account by Salman. Bazán ('Was There Ever a "First Averroism"?') holds that the Latin expresses clearly enough the real view of Ibn Rushd on the separate and eternal Agent and Material Intellects. He also recognizes the early thirteenth-century Latin view that the agent and possible (material) intellects are powers of the individual human soul, at least partially in line with the views mentioned just above by Gauthier. As will be made clear below, Albert in *De homine* (1242) likewise holds those two intellects to be powers

That is, Salman recognizes the two forms taken by readings of Averroes, and rightly notes that Averroes's (and Ibn Rushd's) true teaching is that the Agent Intellect and the Material Intellect exist as separate immaterial and eternal substances.

The distinction between 'First Averroism' and 'Second Averroism' also features in Gauthier's 1984 introduction to the critical Leonine edition of Thomas Aquinas's *Sententia De anima*. There, Gauthier explicitly (and incorrectly) contends that the real teaching of Ibn Rushd, as reflected in the *Long Commentary on the De anima*, is precisely that of 'First Averroism', the account that the agent intellect and the possible (material) intellect are powers of individual human souls. As for 'Second Averroism', the doctrine that the Agent Intellect and the Material Intellect are immaterial substances existing separately and ontologically distinct from the human soul, this (he incorrectly maintains) was a false creation on the part of the thirteenth-century Christian theologians and not the genuine teaching of Averroes (that is, of Ibn Rushd). In this, Gauthier asserts his own understanding of Ibn Rushd and moves away from Salman's correct account of Ibn Rushd, which affirmed that the Cordoban himself had taught the existence of two separate intellectual substances, the Agent Intellect and the Material Intellect. This assertion by Gauthier is in contradiction to the account of the thought of Ibn Rushd on intellect that I provided above.

For the question of the nature and cogency of the account of 'First Averroism' and 'Second Averroism' that is our concern here, we must return to some additional remarks by Gauthier in his 1982 article, immediately following his important account of the discovery of the citation of Averroes on the intellect in the *De anima et potenciis eius*. Gauthier tells us that this citation is quite surprising, since modern studies have more commonly cited the 'Second Averroism' that argues for the separation of the Agent Intellect and the possible or receptive Material Intellect. Gauthier supports his (incorrect) view now by citing a presumed authority on the Arabic texts of Averroes: 'But do we have reason to be surprised? S. Gómez Nogales has written recently that, on the problem of the intellect, one thing is for sure: "Averroes is not an 'Averroist', in the sense of the term Second Averroism"'.[72] Were the view of Gómez Nogales correct, certainly it would create a serious problem in light of the analysis of the thought of Ibn Rushd I provided above: there, I explained in detail that the doctrine of intellect in the *Long Commentary on the De anima* is precisely the doctrine of the separate Agent Intellect and separate Material Intellect. But the account of Gómez Nogales — whom Gauthier considers an authority in this matter — is incorrect.

---

of the human soul and asserts this to be a genuine teaching of Averroes. That is, Albert shares in what Gauthier describes as a common understanding among Christian thinkers in 1225–50.

72 Gauthier, 'Notes sur les débuts', p. 335: 'Mais avons-nous raison de nous étonner? S. Gomez Nogales a écrit récemment que, dans ce problème de l'intellect, une seule chose est sûre: "Averroès n'est pas averroïste", au sens où le second averroïsme entend ce mot'.

In his 1976 article 'Saint Thomas, Averroès et l'averroïsme', Salvador Gómez Nogales, editor and translator of Averroes's *Short Commentary on the De anima*, selected as a concrete example of the paradox that Averroes is not an Averroist the problem of the unicity of the human intellect — that is, whether or not Averroes taught the existence of two separate substances, the Agent Intellect and the Material Intellect, in his doctrine of human intellectual understanding. Gómez Nogales spells out his methodology with remarks worth considering at length:

> The adversaries of Averroes, among them St Thomas Aquinas, all the Averroists, and even, among the moderns, some Arabists who ordinarily tend to defend Arabic thought, such as Asín Palacios, all agree that Averroes defended the unity of the human intellect. The issue, however, is not clear. There are some expressions in Averroes which show clearly that he admits the unity of the human intellect, yet if one accepts this point of view, one encounters in Averroes a manifest contradiction. I have reached a conclusion which has been affirmed *a posteriori* by three different procedures.[73]

The first of these procedures consists of noting that some (unnamed) Arab researchers have studied the issue without having read the articles of Gómez Nogales but, employing the same sources, have reached the same conclusions: 'Averroes, they say, did not defend the unity of the intellect as has been thought in the West'.[74] I will not recount the rest of his arguments here; suffice it to note with some irony that Gómez Nogales criticizes at length the analyses of several prominent scholars who support the view that there are two separately existing intellects, the Agent Intellect and the Material Intellect, and proceeds to read into the text of the *Long Commentary on the De anima* his own interpretation, which is more in accord with the earlier account of the *Short Commentary on the De anima* of Ibn Rushd, though even the understanding of that work by Gómez Nogales is sorely unsound. Among his conclusions in the article, Gómez Nogales writes the following: 'Averroes was not an Averroist. If it is true that there have been Averroists who have admitted the unity of the human intellect, this is not the case for Averroes, who admits the individual immortality of the human soul even in the case of the material intellect'.[75] Yet, as I have shown above, there is no provision for individual immortality in any of Ibn Rushd's commentaries on *De anima*. Gómez Nogales is also unaware of Averroes's teaching on the Agent Intellect as 'form for us' and of the characterization of the presence of the two separate intellects 'in the soul' discussed above. He directly contradicts

73 Gómez Nogales, 'Saint Thomas, Averroès et l'averroïsme', p. 166.

74 Ibid., p. 167. The Arabic texts of the Short and Middle Commentaries on the *De anima* are extant and in each Ibn Rushd holds that there are two intellects, the particular human material intellect and the transcendent separate Agent Intellect (the latter of which is common to the Arabic tradition). Aside from fragments, the full text of the *Long Commentary on the De anima* survives only in the Latin translation of Michael Scot. To access this, significant skill in Latin is required.

75 Gómez Nogales, 'Saint Thomas, Averroès et l'averroïsme', p. 177.

the teaching of the *Long Commentary on the De anima* that the two intellects are eternal substances ontologically distinct from one another and ontologically independent of human soul.

Gauthier's account of 'First Averroism' and 'Second Averroism' is an inaccurate understanding of Ibn Rushd and the teachings in the Latin translation of the *Long Commentary on the De anima*. As I have shown, that view was based on a seriously misleading article by Gómez Nogales, who presented an unsound account of the theory of intellect in the mature Averroes. Key to capturing the real meaning of the text of Ibn Rushd's *Long Commentary* is an understanding of his engagement with reasoning in the *Paraphrase of the De anima* by Themistius and of the meaning of the teaching that the Agent Intellect must be 'form for us' and intrinsically present 'in the soul', an understanding achieved by few readers of the *Long Commentary* in either medieval or modern times.[76] These problematic readings are in large measure due to the complexity of the philosophical issues involved and the novelty of Averroes's unprecedented teaching on the Material Intellect in his *Long Commentary on the De anima*.

Gauthier's inaccurate account, which rejected the traditional understanding of Ibn Rushd/Averroes as asserting the existence of the separate Agent Intellect and separate Material Intellect, contributed to confusion among scholars. Yet this does not necessarily undermine his thesis that until 1250, some Latin thinkers held that the agent intellect and the possible (material) intellect are powers of the human soul. The existence of this particular teaching has been convincingly affirmed by B. Carlos Bazán.[77]

Bazán argues persuasively in an article published in 2000 that in the pre-1250 period, the doctrine of Averroes on both the Agent Intellect and Material (Possible) Intellect as separate substances was clear enough for Latin Christian readers of the *Long Commentary on the De anima*. The notion of the Agent Intellect as a unique separate substance was commonplace in the Arabic tradition. It was found in translated writings of al-Fārābī, was even a hallmark of the very clear teaching of Avicenna, and was evident in Averroes. As for the unique separate and shared Material Intellect in the historically novel teaching of Ibn Rushd, this too was known and is witnessed by Richard Rufus in his study of Averroes. In Richard's *Contra Averroem*, the question *An intelligentiae separatae sint res individuae* is followed by a detailed discussion, closely based on the texts of Averroes, that displays a clear understanding of the teaching of Ibn Rushd on the separate Material Intellect. For example, Richard writes, 'this seems to have moved Averroes himself even reasonably to assert the possible intellect in us to be one in all'.[78] For Bazán, what

76 See Taylor, 'Intellect as Intrinsic Formal Cause in the Soul'; Taylor, 'Themistius and the Development of Averroes' Noetics'.

77 Bazán, 'Was There Ever a "First Averroism"?', p. 37.

78 *Contra Averroem* 1.2 (Dictum 1, tractatus 2 in Erfordia, Bibl. univ., Amploniana, Quarto 312, fol. 81vb): 'Ergo hoc videtur movisse ipsum Averroem et rationabiliter in ponendo intellectum possibilem in nobis omnibus unum et etiam universale, sicut prius dictum est'. This work is

Gauthier called 'First Averroism' is in fact an original creation by Christian Latin theologians who, for the first time, set out the doctrine that the agent intellect is a power of the individual human soul:

> The doctrine of the agent intellect as a faculty of the soul is an original contribution of the Latin Masters to the reading of Aristotle's *De anima* III, 4–5. The importance of this contribution should not be minimized by a label such as 'First Averroism', which risks obscuring its originality.[79]

That claim, however, is accurate only if limited to the medieval Arabic and Christian philosophical traditions, since both Philoponus and Themistius in different ways held there to be an agent intellect in the human soul.[80]

Albert the Great himself must be considered a member of the group of Latins to which Gauthier refers, and to which Bazán points, at least since his *De homine* (*c.* 1242). Albert attributes to Averroes the very doctrine that Gauthier mentions as common in 1225–50 and Bazán ascribes to the creativity of early thirteenth-century Christian theologians. As I shall now show, Albert in the *De homine* unequivocally states that according to Averroes, the agent intellect and the possible (that is, material) intellect are powers of the individual human soul. He also goes further, explicitly asserting that the view that these are separate substances is distinctively false. Albert himself embraced these understandings for his own teaching in accord with the views of his times.

## Explaining Albert's Two Momentous Interpretive Misconstruals

By 1242, if not before, Albert already held in several works that it is the view of Averroes that ultimate happiness is attained by the rational human soul after death through a conjoining with the First Mover. In his *De resurrectione*, Albert cites

sometimes also called *De ideis*. My thanks to Rega Wood for permission to quote this passage from her unpublished edition. Thanks also to Timothy Noone for sharing his transcription of the text and for discussion of the work of Richard Rufus. There is some disagreement on the dating of this work. For present purposes it is sufficient to say that it may have been composed in the early 1230s or as late as 1240 — that is, either a few years prior to Albert's *De homine* (*c.* 1242) or perhaps even as late as to be contemporaneous with parts or most of it. As will become clear below, Albert was aware of this sound interpretation of the teaching of Ibn Rushd on the two separate Intellects, but dismissed it as an incorrect understanding of Averroes.

79 Bazán, 'Was There Ever a "First Averroism"?', p. 37. He continues: 'The doctrine appeared during the first three decades of the thirteenth century, even before Averroes' writings were known or had a decisive influence. Latin Masters of Arts, such as John Blund, and Theologians, such as Philip the Chancellor (whose "Summa de bono" was written between 1228 and 1236, and who quotes Averroes only once), held that the agent intellect is a power of the individual soul'.

80 Richard Sorabji remarks that 'Themistius and Philoponus give a role in concept formation to active intellect, but regard it as human'. Sorabji, *The Philosophy of the Commentators*, p. 104. See his selected translations of Themistius and Philoponus at pp. 107 and 117 and the references there.

Averroes's *Long Commentary on the Metaphysics*, Book XI (XII), Comment 51 on Aristotle 1073a3 ff., writing:

> Further, the Philosopher says in *Metaphysics* I that divine science is the goddess of the sciences. And the Commentator on book XI [XII] says that the question of intellect and of the knowledge that God himself has 'is what is desired by all'.[81]

Ibn Rushd himself connects Aristotle's discussion of the ultimate object of God's intellection (namely, God himself) with the notion from *Metaphysics* I.1 that all humans by nature desire to know.[82] Albert follows him in this and in the discussion of the most noble object of intellection at *Long Commentary on the Metaphysics* XI (XII), Comment 51. Regarding this latter point, Albert concludes for Averroes and holds for himself that the First Mover, which is God as First Intellect, is the end to be sought and to be known by human beings in the contemplation that is ultimate happiness. At another location, Albert apparently refers to Averroes again: 'The tenth [apparition] certifies the transition of the risen blessed to beatification, because, as a certain philosopher says, that conjunction with the Prime Mover, that is, with God, is the end of happiness [*prosperitatis*]'.[83] This notion is also found in *De quattuor coaequaevis*, where, after discussing Plato, Albert writes: 'Hence also the philosophical position is that the end of the soul's happiness is to be conjoined with the First Mover through contemplation'.[84] In the *De homine*, Albert writes in his own behalf that 'the potency of the possible intellect after death will be perfected [*complebitur*] by the agent intellect and by forms which are in the separate intellects [...]. For the philosophers say that the soul after death returns [*convertitur*] to the First Mover, and this is its end of

81 Albertus Magnus, *De resurrectione*, ed. by Kübel, p. 328, vv. 33–36: 'Praeterea, PHILOSOPHUS dicit in I METAPHYSICAE, quod scientia divina dea scientiarum est. Et SUPER XI dicit COMMENTATOR, quod quaestio de intellectu et scientia dei per se "est desiderata ab omnibus"'. See Aristotle, *Metaph.* I.2, 983a6.

82 *Tafsīr mā baʿd at-Ṭabīʿat*, p. 1693, vv. 10–11: ان هذا المطلب لما كان هو اشرف المطالب المطالب التي في الله وهو ان يعلم ماذا يعقل وكان كل انسان يتشوقه بالطلع *Long Commentary on the Metaphysics, Arabic-English*, p. 191: '[T]his object of research is the noblest of the objects of research dealing with God and consists in knowing what is His object of intellection which every man desires by nature'; *Long Commentary on the Metaphysics, Latin*, 335D: '[I]sta quaestio est nobilissima omnium quae sunt de deo, scilicet scire quid intelligit, et est desiderata ab omnibus naturaliter'.

83 Albertus Magnus, *De resurrectione*, ed. by Kübel, p. 284, vv. 76–80: 'Decima certificat transitum beatorum resurgentium ad beatificationem, quia, sicut dicit quidam philosophus, coniunctio cum primo motore, idest cum deo, finis est prosperitatis'. Regarding my translation of 'prosperitas' in Albert as 'happiness', see note 49.

84 *De quattuor coaequaevis*, ed. by Borgnet, p. 312b: 'Unde etiam positio philosophica est quod finis prosperitatis animae post mortem, est quod continuetur primo motori per contemplationem'.

happiness'.[85] Elsewhere in the *De homine*, we find Albert explicitly attributing this doctrine to Averroes: 'Averroes, in his commentary on *Metaphysics* XI, says that the rational soul remains after death and it will have its end of happiness, if it conjoins with the First Mover. And he calls the First Mover the Principle of the Universe, which is God'.[86] Later in his career, Albert uses his understanding of the acquired intellect (*intellectus adeptus*, العقل المستفاد *al-ʿaql al-mustafād*), taken from al-Fārābī, to explain the grounds for his own doctrine on how this return and conjoining is attained.[87] Two teachings are implicit in this understanding of the intellectual apprehension of God as the ultimate felicitous end of human beings: first, an affirmation of a post mortem existence of human beings and, second, based on that, a conception of human intrinsic intellectual powers. In his *Super Ethica*, Albert changed his understanding of the second aspect after realizing his misconception of the teaching of Averroes, and instead attributed to Averroes the post mortem perdurance only of one Soul alone, contained in the tenth Intelligence, a form of monopsychism not found in Ibn Rushd.

### *Albert's Interpretive Misconstrual of the Natural Epistemology of Averroes in the 'De homine'*

Albert provides a detailed account in the *De homine* of how he crafted a theory explaining that the agent intellect and the receptive intellect are powers of the individual human soul in accord with what Gauthier considered commonplace in 1225–50. This he does by quoting with precision and at length sections of the Latin text of Avicenna's *De anima* and of Averroes's *Long Commentary on*

85 *De homine*, ed. by Anzulewicz and Söder, p. 429, vv. 15–20: '[P]otentia intellectus possibilis post mortem complebitur ab intellectu agente et a formis, quae sunt in intelligentiis separatis, et ideo non erit supervacua. Dicunt enim philosophi quod anima post mortem convertitur ad motorem primum, et hoc est finis prosperitatis eius'. The editors indicate that this view is also found in several other places in *De homine*: p. 465, vv 53–58; p. 466, vv 8–23; p. 473, v. 18. See also *Ethica*, I.7.17, ed. by Borgnet, p. 133b: 'Propter quod dicit Averroes super XI philosophiae primae, quod finis prosperitatis animae post mortem est, si coniungatur ad motorem primum'.

86 *De homine*, ed. by Anzulewicz and Söder, p. 465, vv. 53–56: 'Averroes super XI Metaphysicae dicit quod anima rationalis manet post mortem et finis prosperitatis eius erit, si coniungetur primo motori. Et appellat primum motorem universitatis principium, quod est deus'. Note that earlier in the discussion, at vv. 27–24, Albert understands the phrase *in postremo* to mean after the death of the body.

87 See, for example, his commentary on the *Metaphysics* (*c.* 1262), Albertus Magnus, *Metaphysica*, ed. by Geyer, p. 527, vv. 46–59: 'Et quia nos iam ALIBI docuimus qualiter homo adipiscitur intellectum suum, etiam iam ex isto potest sciri, qualiter adepto intellectu proprio adipisci potest intellectum substantiarum divinarum et qualiter ista adeptio stat in intellectu substantiae primae, quae est lux omnium intellectuum et intelligibilium per seipsam. Et cum omnes homines natura scire desiderent et illud desiderium naturale stet in fine et ratione et causa ommum intellectuum et intelligibilium, pro certo stabit desiderium in scientia intellectus substantiae primae et propter adipiscendum desiderat scire alia, et quando pervenitur ad Ipsum, stat et habet finem felicitatis contemplativae'. Also see his discussion on pp. 472–73. For an analysis and discussion of this teaching, see de Libera, *Métaphysique et noétique*, chap. 6.

*the De anima*. From these, Albert wove his first substantial account of natural human knowing. Avicenna held that the human rational soul is an entity that is intellectual, receptive, immaterial, eternal *a parte post*, and distinct from the body and the senses. The human rational soul uses the body and its senses in the fashion of a tool or instrument regarding the perceptual world and then connects with the Agent Intellect, a unique eternal separate substance containing all the forms, to bring to perfection or completion human intellectual understanding in an individual rational soul. These teachings, which Albert found in the Latin texts of Avicenna, he rejected in favour of what he understood to be the account of Averroes.[88] Ibn Rushd taught that the human soul is form of the body and depends for human scientific knowledge on sensation, but also on some sort of a natural relationship with the separate Agent Intellect and the separate receptive Material Intellect and their abstractive powers, as I have already indicated. Albert dismissed Avicenna's conception of the human rational soul as requiring a connection to the Agent Intellect that is separate in substance and replete with forms from which the world derives.[89] Regarding Averroes, however, Albert's account is more complex.

As explained above, in Albert's time there were two competing interpretations of intellect in the thought of Averroes in his *Long Commentary on the De anima*. One largely accorded with the genuine teaching of Ibn Rushd that the Agent Intellect and the Material Intellect are separately existing immaterial substances through which human beings have scientific knowledge of universals (confirmed by Richard Rufus and discussed by Salman and Bazán). The other contended that agent intellect and receptive material intellect are two immaterial (that is, unextended and incorporeal) powers of each individual human soul.[90] In his *De homine*, Albert is well aware of these two interpretations and rejects the first, instead asserting that the proper understanding of Averroes is that the agent intellect and the material intellect are powers of the individual human soul.[91] That is, Albert reads the text of Averroes in accord with the common view of the

88 'Albert clearly rejects the views of the philosophers who say that the Agent Intellect is separate and efficient cause of human knowing. He writes against "others" (i.e., Avicenna) that he rejects the connection between the intellect as the tenth in the emanative hierarchy of the heavens and the function of the Agent Intellect. The notion that "the human possible intellect moves a human being to be connected to the agent intellect of the tenth order" (*intellectus humanus possibilis movet hominem ad hoc quod conformetur intelligentiae agenti decimi ordinis*) and that "in this way the goodnesses flow from the agent intellect into the possible intellect" (*et hoc modo fluunt bonitates ab intelligentia agente in intellectum possibilem*) is something Albert will have none of (*nos nihil horum*)'. Taylor, 'Remarks on the Importance', pp. 140–41.

89 Albertus Magnus, *De homine*, ed. by Anzulewicz and Söder, p. 408, v. 68 and p. 412, vv. 5–68.

90 This is the issue of 'First Averroism' and 'Second Averroism'. In my introduction to the *Long Commentary on the De anima, English*, pp. xcix–civ, I explain the series of errors that led several important twentieth-century scholars to make very bewildering statements about the interpretations of Averroes by the Latins. See also Hayoun and de Libera, *Averroès et Averoïsme*, pp. 78–82.

91 See Taylor, 'Remarks on the Importance', p. 143, where I indicate that Albert himself bears witness to two interpretations in the *De homine*: at 411.52–53, 'uterque istorum intellectuum erit in nobis

soul and its powers on the part of his own predecessors and peers. According to Albert's account, for which he credits Averroes, all natural knowledge comes through the senses, and intelligible species are abstracted from the content of the external and internal senses by the immaterial power of an intrinsic agent intellect and received as intelligibles in act in the immaterial power of receptive possible intellect.[92]

On the basis of a theory of knowledge *not* found in the *Long Commentary on the De anima* of Ibn Rushd, Albert attributed this theory to Averroes and adopted it for himself. His account of natural epistemology in *De homine* was later largely followed by his student Thomas Aquinas in the latter's *Commentary on the Sentences of Peter Lombard*, written in Paris 1252–56, though without the doctrinal misattribution to Averroes.[93] In 1248–52, Albert and Thomas worked together in Cologne, where the young Dominican was assigned to assist Albert in his work of teaching and research.[94] It is impossible to think that they did not discuss in detail philosophical teachings in translated Arabic works as well as what Albert had written in his *De homine*. Many texts from the Arabic tradition are cited and used in the commentaries on Dionysius that Albert wrote with Thomas present. In 1250–52, Albert completed the first Latin commentary on the *Nicomachean Ethics* with Thomas again present wholly or for the most part. By the time of that work, Albert had realized his earlier mistake in *De homine* regarding the incorrect attribution to Averroes of the teaching that the active and possible intellects are intrinsic powers of the individual soul and set out a very different account in its place.[95] The correction is reflected in the work of Thomas, though without mention of Albert (following the custom in his day). Irrespective of the fact that Albert had misunderstood this point, however, the German Dominican's work proved to be an invaluable and lasting foundational starting point for Thomas's thought on the nature of human knowing and the powers of the soul, as will become clear later in this chapter.

---

existens et non separata substantia' ('both of those Intellects will be existent in us and not separate substance').

92 See ibid., pp. 143–45. In the opening lines of the *Posterior Analytics*, I.1, 71a1–2, Aristotle states that all reasoned teaching and learning arises from prior knowledge. In the final chapter of *Posterior Analytics*, II.19, 100a3 ff., he identifies this as what is apprehended through sense perception.

93 I provide a short account of the natural epistemology of Aquinas in the opening pages of Taylor, 'Remarks on the Importance'. For a detailed study of Aquinas's first substantial engagement with the Arabic tradition on this with translation of the key text, see Taylor, 'Aquinas and "the Arabs"'.

94 See Mulchahey, 'The *Studium* at Cologne'.

95 Albertus Magnus, *Super Ethica*, ed. by Kübel, pp. 451, v. 3–453, v. 89. In his *Super Ethica*, in a *solutio* at p. 71, vv. 73–85, Albert explains that philosophy is not sufficiently able to know whether the souls of the dead continue to exist after death. In the response to the third objection at p. 72, vv. 12–19, he notes that Averroes holds that all souls exist as one after death and adds that this is contrary to the faith. Note that the discussion on pp. 451–53, which mentions the 'Commentator', often refers to Eustratius of Nicaea, as indicated by Wilhelm Kübel, editor of Albert's *Super Ethica* (with the exception of p. 452, vv. 80–83 and p. 453, vv. 63–70, where it is a reference to Averroes).

### *Albert and the Monopsychism of Averroes*

In his early *De homine*, Albert considers, in a supporting *sed contra*, the view of Averroes in Book 3 of the *Long Commentary on the De anima* and then, in a passage I have already mentioned, remarks:

> [T]he potential of the possible intellect after death will be completed by the agent intellect and by forms which are in the separate intellects, and for this reason it will not be superfluous. For the philosophers say that the soul after death returns to the First Mover and this is its end of happiness.[96]

Also in *De homine*, in the course of a discussion titled 'Whether or not the rational soul is corrupted with the corruption of the body',[97] Albert indicates he will set out first the views of philosophical authors, then probable arguments, then demonstrative and necessary arguments, and next discuss them and provide his own solution to the issue. Among the texts he cites is Aristotle, *Metaphysics* XI (XII) 3, 1070a 24–26, the first of the two key texts from *Metaphysics* XI (XII), chapter 3 which I discussed in my account of the teaching of Ibn Rushd above. Here in his *De homine*, Albert quotes part of the text from the Arabic translation into Latin, *Si autem remanet in postremo* ('If, however, it remains *afterwards*') and goes on to paraphrase the rest of the text with his own understanding of *in postremo*: 'There should be investigation regarding this. For in certain cases it is not impossible, for example, if the soul is of such a disposition, nevertheless not the whole, but the intellect'.[98] Albert then explains that the issue is whether after death the whole soul remains in existence, including the sensible and vegetative powers of the soul, or just the intellectual part. A few lines later, Albert cites the commentary of Averroes on *Metaphysics* XI (XII), this time referencing Comment 38, as already mentioned:

> Averroes, in his commentary on *Metaphysics* XI, says that the rational soul remains after death and its end [of happiness] will come to belong to it, if it conjoins with the First Mover. And he calls the First Mover the Principle of the Universe, which is God.[99]

Here it is clear that Albert attributed to Averroes the view that human soul has an afterlife in virtue of its intellectual power. What of Albert's *Super Ethica*, written about a decade later in 1250–52?

In the *Super Ethica*, as noted earlier, Albert interestingly states that the issue of the soul post mortem and its ultimate happiness is properly speaking not a matter for philosophers. It is beyond the ken of philosophers and is, rather, an issue that

---

96 See note 85 above.

97 Ibid., p. 464: 'Utrum corrumpatur anima rationalis curruptione corporis, an non'.

98 *Long Commentary on the Metaphysics, Latin*, 302-I: '[Q]uaerendum est de hoc. In quibusdam enim non est impossibile, verbi gratia, si autem anima talis est dispositionis, non tota tamen, set intellectus'.

99 See note 85.

belongs to theology and faith. In his *solutio* to an article on whether philosophy can know the state of the soul post mortem, Albert explains:

> It should be said that the notion that the souls of the dead remain [in existence] after death cannot be sufficiently known through philosophy. On the supposition that they remain [in existence], nothing at all can be known through philosophy regarding their state and how they are related to the things which come to pass concerning us. Rather, these things are known by a higher infused non-natural light, which is the habit of faith.[100]

In response to objection 3 in the same article, he asserts that something superior can do whatever something inferior can do, but in a more eminent way. Hence, human intellect's understanding of things known by sensing is through a mode superior to that of sense. He continues:

> Similarly, a separated soul has a more noble operation which cannot be known by us through philosophy [...] and if intellect is not a particular form,[101] it cannot be demonstrated that many souls remain distinct [in existence] but rather for all there will be one soul, as the Commentator asserts in his *Commentary on the De anima*. In this way he expounds the authority of Aristotle that is introduced, although it is contrary to faith.[102]

---

100 Albertus Magnus, *Super Ethica*, ed. by Kübel, p. 71, vv. 73–79: 'Dicendum, quod hoc quod animae defunctorum remaneant post mortem, non potest per philosophiam sufficienter sciri. Et supposito, quod remaneant, de statu earum et qualiter se habeant ad ea quae circa nos flunt, omnino nihil sciri per philosophiam potest, sed haec cognoscuntur altiori lumine infuso non naturali, quod est habitus fidei'. Also see ibid., p. 72b: 'Solutio: Dicendum, quod, sicut dictum est, philosophus nihil habet considerare de statu animae separatae, quia non potest accipi per sua principia. Unde qualiter se habeat anima separata ad ea quae fiunt hic, et qualiter iuvatur per ea, nihil pertinet ad philosophum, sed at theologum'. Later in his *De natura et origine animae* (1258), Albert explains the use of the notion of light employed by Averroes and Ibn Bājja (Abubacher) in asserting that humans share in one intellect and refutes it in detail. See Albert, *De natura et origine animae*, II, cap. 4 and 9; also I, cap. 5 and 6.

101 That is, a determinate particular form. Regarding the sense of *situalis* here, see Albertus Magnus, *De homine*, ed. by Anzulewicz and Söder, p. 154, vv. 1 ff.

102 Albertus Magnus, *Super Ethica*, ed. by Kübel, p. 72, vv. 10–17: 'Similiter anima separata nobiliorem habet operationem, quae nobis per philosophiam non potest esse nota [...] et si intellectus non sit forma situalis, non potest demonstrari, quod remaneant plures animae distinctae, sed omnibus una, sicut ponit Commentator in libro De Anima, et hoc modo exponit auctoritatem Aristotelis inductam, licet sit contra fidem'. The editor of Albert's text identifies Albert's reference to Averroes as referring to what is written in three places in the *Long Commentary on the De anima, Latin*, III.5: p. 401, vv. 424 ff.; p. 403, vv. 73–76; p. 407, vv. 593–96. This last reference should probably be corrected to pp. 406–07, vv. 575–83. In each of these passages, Ibn Rushd's discussion is clearly about one common shared intellect, not soul. In the same order: *Long Commentary on the De anima, English*, p. 317: 'The second question, how the material intellect is one in number in all individual human beings, neither generable nor corruptible, and the intelligibles [are] existing in it in act (this is the theoretical intellect), [yet it is also] enumerated in virtue of the numbering of individual human beings, generable {402} and corruptible through generation and corruption of individuals, this question is very difficult and has the greatest ambiguity'; ibid., p. 318: 'For this reason one should

In the *solutio* of the next article, he adds:

> It should be said that, as was said, philosophy has no business considering the state of the separated soul, because [that state] cannot be accepted through its principles. Hence, how the separated soul is related to things which take place here and how it may be aided by them does not pertain to the philosopher but to the theologian.[103]

This did not, however, stop him from considering the teachings of the philosophers, in particular those of Averroes, on the afterlife and ultimate human fulfilment and happiness.

Later in the *Super Ethica*, Albert repeats his earlier view of the philosophers that after death human intellect is linked to separate intellects.[104] In the second objection, Albert cites Averroes (as 'the Commentator') in Comment 38 on Book XI (XII) of the *Metaphysics* as saying that 'this is our ultimate happiness, that our soul is conjoined with the intelligences acting on our souls'.[105] In his response to this objection, Albert writes the following, which is quite in accord with what is found in his *De quattuor coaequaevis* and *De homine*:

> Averroes says many heretical things. If we nevertheless wished to support him in this issue, it should be said that our happiness will be in the conjoining to the intelligence not with respect to being but with respect to object, when the soul after death will contemplate the simple quiddities such as the intelligence.[106]

---

hold the opinion that if there are some living things whose first actuality is a substance separate from its subjects, as is thought concerning the celestial bodies, it is impossible that there be found more than one individual from one species of these'; ibid., p. 322: 'On the basis of this account we have held the opinion that the material intellect is one for all human beings and also {407} on the basis of this we have held the opinion that the human species is eternal, as was explained in other places. The material intellect must not be devoid of the natural principles common to the whole human species, namely, the primary propositions and singular conceptions common to all [human beings]. For these intelligibles are unique according to the recipient and many according to the intention received'. Also see *Super Ethica*, ed. by Kübel, p. 453, vv. 44–47: 'Non manet nisi una anima, quia cum individuatio animae non sit nisi ex corpore substracto, hoc per quod efficiebatur proprium, remanebit unum commune'. Albert sides there, instead, with Avicenna: coming to be individuated depends on the body, but once individuated, the human soul is a substance having its own *esse*. For the context, see note 106.

103 Albertus Magnus, *Super Ethica*, ed. by Kübel, p. 72, vv. 57–62: 'Dicendum, quod, sicut dictum est, philosophus nihil habet considerare de statu animae separatae, quia non potest accipi per sua principia. Unde qualiter se habeat anima separata ad ea quae fiunt hic, et qualiter iuvatur per ea, nihil pertinet ad philosophum, sed ad theologum'.

104 Ibid., p. 452, vv. 69–70: 'Sexto videtur, quod sit ponere continuationem intellectus ad intelligentias post mortem'.

105 Ibid., p. 452, vv. 80–82: 'Commentator in XI Metaphysicae dicit, quod haec est ultima prosperitas, quod anima nostra continuatur ad intelligentias agentes in animas nostras'.

106 Ibid., p. 453, vv. 63–70: 'Averroes multas haereses dicit; unde non oportet, quod sustineatur. Si tamen in hoc volumus eum sustinere, dicendum, quod prosperitas nostra erit in continuatione

Why Albert would say this becomes clear when we consider his *solutio*:

> It should be said that the error of some of the Arabs was that our intellect does not remain after death according to being, but only according to essence. In this way they said that it remains in the intelligence of the tenth [procession] from which it flows, and in this way there **remains only one soul**. [This is] because, since the individuation of the soul is only in virtue of body, when that through which it was made proper has been removed, then there will remain one common thing. But this is heresy.[107]

As indicated earlier regarding Comment 38 of the *Long Commentary on the Metaphysics*, the Arabic بأخرة (*bi-akharatin*), a prepositional phrase that translated Aristotle's ὕστερον, is suitably rendered in an adverbial sense as 'afterwards', 'hereafter', or even perhaps 'eventually' and 'ultimately' in its appearances in the *Long Commentary on the Metaphysics*.[108] In the Latin translations by Michael Scot, this Arabic phrase is rendered *in postremo* in its many instances, each of which can well and suitably be rendered with the same meanings as the Arabic. To put it simply, the Latin translation is certainly correct. Were the Arabic بالآخرة (*bi-l-ākhirati*), the sense would easily be understood rather as 'in the afterlife', for *al-dār al-ākhira*, 'the ultimate abode', and could also correctly be translated as *in postremo*. But that construction is not found in the Arabic texts. Yet in Latin each occurrence of بأخرة (*bi-akharatin*) is soundly rendered as *in postremo*. Hence, while the Latin translation is not wrong here, the translation *in postremo* is liable to the possibility of misconstrual and misinterpretation. This is, in fact, precisely what we find in the *Super Ethica* and the earlier works of Albert discussed in this chapter. Albert could have understood the Latin phrase *in postremo* in the adverbial sense as found in Arabic, but instead chose to read it as meaning 'in the afterlife' or 'in the hereafter'. With this misinterpretation, Albert affirms for Averroes precisely what was denied in the philosophical teachings of Ibn Rushd, namely, the afterlife of human soul.

---

ad intelligentiam non secundum esse, sed secundum obiectum, quando anima post mortem contemplabitur simplices quiditates sicut intelligentia'.

107 Ibid., p. 453, vv. 40–47: 'Dicendum, quod quorundam Arabum error fuit, quod intellectus noster non manet post mortem secundum esse, sed secundum essentiam tantum. Et sic dicebant, quod manet in intelligentia decimi, ex quo fluit, et sic non manet nisi una anima, quia cum individuatio animae non sit nisi ex corpore subtracto, hoc per quod efficiebatur proprium, remanebit unum commune. Sed haec est haeresis'. This teaching is in fact a construction by Albert based on a mixture of the teachings of Avicenna, Algazel, and Averroes. Albert's understanding of Averroes is spelled out clearly in his responses to the first two objections. In the first response, he explains that Averroes in context does not mean a separation of the human individual passible intellect, which is a bodily power, 'because after death the very essence of soul remains [in existence]' (*quia in anima post mortem manet ipsa essentia animae*), ibid., p. 453, vv. 57–58.

108 My thanks to Dag Hasse, who is currently preparing a critical edition of the Latin text of Averroes's *Long Commentary on the Metaphysics*, for helpful discussion of this Latin phrase in email correspondence in August 2020. This phrase *in postremo* is also found in the *Long Commentary on the De anima* without reference to the afterlife.

In the context of the Arabic discussions of Ibn Rushd, who does not hold post mortem existence for human soul, such a reading makes no sense. As we have seen, what Ibn Rushd held was that the Agent Intellect and the Material Intellect are each unique, separately existing, eternal substances available to mortal human knowers. In Albert's Latin religious context, however, 'in the afterlife' was an expected and obvious choice, one rather understandable since Ibn Rushd was unique among the major Arabic-writing philosophers in Latin translation in quite clearly denying the post mortem existence of the individual human soul.[109]

With this interpretation of Averroes as affirming a post mortem existence of human soul, Albert was able to complete his own understanding of the teachings of Averroes in the form of a doctrine of Latin Averroism that is not found in the writings of Ibn Rushd. This is the teaching of monopsychism, of one essential soul into which all individual human souls are resolved in a unity at the death of the body. This second misconstrual by Albert of texts in the Latin translations of work of Ibn Rushd — one essential soul to which all individual human souls return at death of the body — contributed to the formation of a form of monopsychism which later became foundational to further developments of Latin Averroism.

This doctrine was set out in Albert's *Super Ethica*, composed in 1250–52 while Thomas was still his assistant in Cologne. Hence, it is not surprising to find young Thomas using what he had learned from his German teacher for his own reasoning in his *Commentary on the Sentences*, written in Paris immediately following his time in Cologne. Albert's interpretation is reflected in the title and content of Thomas's first account of natural epistemology in the context of translations from the Arabic tradition in his *Commentary on the Sentences* II, d. 17, q. 2, a. 1: 'Whether there is **one soul** or intellect for all human beings'.[110] In addition, Albert's account of Averroes's monopsychism is clearly reflected by Aquinas at *Commentary on the Sentences* II, d. 19, q. 1, a. 1. There, in the context of his consideration 'Whether the human soul is corrupted with the corruption of the body', Aquinas provides an account surely based on Albert's conception of monopsychism:

> The third position is that of those who say that the intellective soul is partly corruptible and partly incorruptible, because that part of the soul which is proper to this body is corrupted when the body has been corrupted; moreover, that part which is common to all [i.e., soul itself] is incorruptible. For they assert the intellect to be one in substance for all — some the agent [intellect], others the possible [intellect], as was said above [d. 17]. And [they say] this is an incorruptible substance, and that in us there are only phantasms illuminated by the light of the agent intellect, which move

---

109 Regarding the possibility of a similar view in one of al-Fārābī's lost works, see Neria, 'Al-Fārābī's Lost Commentary on the Ethics'.

110 Thomas Aquinas, *Scriptum super libros Sententiarum Magistri Petri Lombardi Episcopi Parisensis*, ed. by Mandonnet, pp. 420–30. See the analysis and translation of this article in Taylor, 'Aquinas and "the Arabs"'. My emphasis.

the possible intellect, in virtue of which we are intelligent insofar as we are conjoined to separate intellect through them. From this if it follows that, if that which is proper is destroyed with only what is common remaining, then only one substance from all the human souls would remain when bodies have dissolved. The reasons supporting this position and how it can be disproved, [are treated] above in distinction 17.[111]

## Conclusions

This chapter has focused on two misconstruals in the interpretation of the philosophical thought of Ibn Rushd by Albertus Magnus, misconstruals that had momentous influence. Both concern the nature of human intellectual understanding as discussed in the Long Commentaries of Ibn Rushd on the *De anima* and the *Metaphysics*, and neither involves mistranslation of the Arabic into Latin on the part of Michael Scot. Rather, each misinterpretation was likely motivated, at least in part, by deep-seated religious and cultural beliefs.

First, Albert's misconstrual of the texts of Averroes in his *De homine* is momentous for its influence on his student Thomas Aquinas and surely others as well. In that work, Albert critically examined the teachings of Avicenna, rejecting the Avicennian notion of an external transcendent Agent Intellect emanating forms to complete the process of knowing on the part of the individual rational soul. He went on to set out an account he attributed to Averroes, which held that the agent intellect and the possible (material) intellect are immaterial powers of the individual human soul. He explicitly rejected those interpretations of Averroes holding that the intellects are separate eternal substances which play key roles in the formation of human intellectual knowledge, as is precisely the teaching of Ibn Rushd. This misconstrual permitted Albert to form an account of human intellectual understanding through individuals' experience of the world by way of external and internal sense powers, powers of the brain, and abstraction of intelligibles in potency in things by the individual, intrinsic agent intellect to form

111 Thomas Aquinas, *Scriptum super libros Sententiarum*, vol. 2, pp. 482–83: '*Tertia positio est eorum qui dicunt, animam intellectivam secundum quid corruptibilem esse, et secundum quid* incorruptibilem; quia secundum hoc quod de anima est huic corpori proprium, corrumpitur corrupto corpore; secundum autem id quod omnibus est commune, incorruptibilis est. Ponunt enim intellectum esse unum in substantia omnium; quidam agentem, et quidam possibilem, ut supra dictum est, [dist. 17]: et hunc esse substantiam incorruptibilem, et in nobis non esse nisi phantasmata illustrata lumine intellectus agentis, et moventia intellectum possibilem, quibus intelligentes sumus, secundum quod per ea continuamur intellectui separato. Ex quo sequitur quod si id quod est proprium, destruitur, tantum communi remanente, ex omnibus animabus humanis una tantum substantia remaneat, dissolutis corporibus. Haec autem positio quibus rationibus innitatur, et quomodo improbari possit, supra dictum est, [17 dist.]'. This quotation from Aquinas is largely in accord with the account of Albert's monopsychism interpretation of Averroes in the *Super Ethica*. Notice especially the use of *proprium, commune, remanet*, and *corrupto corpore* or *dissolutis corporibus*. The reference Aquinas mentions is to *Commentary on the Sentences*, II, d. 17, q. 2, a. 1.

intelligibles in act in the individual intrinsic possible (material) intellect. With this, Albert's *De homine* (1242) offered a view of human intellectual apprehension — largely in accord with his times — that was adopted as foundational by his student Thomas in his *Commentary on the Sentences* (1252) and other works.[112]

B. Carlos Bazán has shown that the Latin texts of the *Long Commentary on the De anima* by Averroes could be read clearly enough for Latins to understand the actual views of Ibn Rushd/Averroes that the Agent Intellect and the Material Intellect are separately existing immaterial and eternal substances.[113] Bazán also confirmed Gauthier's view that there developed among Christian theologians in the early thirteenth century a doctrine novel for the Arabic and Latin medieval traditions, asserting that the agent intellect is a power of the individual human soul.[114] It was in this period that debates flourished on the nature of the human soul and its relation to the body. Is the human soul a *hoc aliquid* or determinate particular substance in its own right, such that it lives on after the death of the body? The Christian doctrine of the resurrection of body requires a reuniting of body with soul and an eternal post mortem existence for each human being. Given that the human being is a created composite of body and soul, how are the two related? Is it sufficient to propose that the soul has a certain *unibilitas* in relation to body and to secure the unity of the human being in that way?[115]

Early thirteenth-century Latin theologians' rejection of the Avicennian notion of the unique separate Agent Intellect shared by all human individual rational souls (material or possible intellects) and the assertion that the agent intellect and possible (material) intellect are powers of the individual soul were important positive steps towards a resolution of the lingering Augustinian problem of soul-body dualism. What remained to be addressed in detail was just how body is necessary for the human soul. Avicenna had provided an account of the rational soul's use of the body with its powers of physical senses and brain as a tool for perfecting the soul. That perfection of soul, however, involved both the influence of the separate Agent Intellect and the denial of an essential unity of body and soul in the human being. Albert adopted this account of the powers of the soul found in his predecessors and read it into the texts of Averroes as a genuine doctrine of the Cordoban. He also explicitly rejected the Avicennian separate Agent Intellect. In doing so, what Albert gained from his reading of Averroes was an account of how human knowing is grounded in the sensory apprehension of things experienced in the world, beginning with external senses, then the common sense's formation of a particular image, next the cogitative power's denuding of the extraneous from the particular image, then the deposit of the particular image in the brain power of memory. Memory then supplies the image to the power of the (human individual's) agent intellect for abstraction and the formation of the intelligible in

---

112 See Taylor, 'Abstraction and Intellection in Averroes and the Arabic Tradition'.

113 See Bazán, 'Was There Ever a "First Averroism"?', pp. 32–33.

114 See ibid., pp. 33 ff.

115 See Bieniak, *The Soul-Body Problem*.

the (human individual's) possible intellect. In this way, the necessity of the body — with its sensory and brain powers — for the attainment and perfection of the human soul is clearly established.

In sum, Averroes was understood as providing a teleological response to the question of why the soul requires the body. It is this account of human knowing (*sans* Albert's misunderstanding of the real doctrine of Ibn Rushd/Averroes on separate Agent Intellect and Material Intellect) that became the teaching of Thomas Aquinas a decade after Albert completed his *De homine*.[116] It is precisely this sophisticated account in Averroes of the relation of phantasm and cogitative power behind abstraction that supplies Aquinas with materials (not found in Avicenna) for his naturalized epistemology.

Second, Albert's attribution to Averroes of monopsychism and an afterlife of human soul is momentous in its contribution to the development of Latin Averroism. The formation of this school of thought required first ascribing to Averroes the notion of the afterlife of human soul, not found in the commentaries on *De anima* and *Metaphysics* by Ibn Rushd despite being common to philosophical thinkers of the religious traditions of Islam, Christianity, and Judaism. Ibn Rushd is an outlier on this in his philosophical teachings. Nevertheless, Albert read that doctrine into the texts by Averroes where he found the Latin phrase *in postremo* and interpreted it as 'in the afterlife'. To this incorrect attribution of a doctrine of the afterlife to the texts of Averroes, Albert seems to have wedded a reading that for Averroes (and Ibn Rushd albeit in a different conception), the Unmoved Mover of Aristotle's *Physics* and *Metaphysics* is the ultimate object of human knowing and happiness (*prosperitas*). This interpretation is evidenced in Albert's early *De resurrectione* and also in *De quattuor coaequaevis*, where he asserts that 'the philosophical position is that the end of the soul's happiness is to be conjoined with the First Mover through contemplation'. In his *De homine*, Albert explains that 'the philosophers say that the soul after death returns [*convertitur*] to the First Mover, and this is its end of happiness', later adding the point I have quoted above: 'Averroes, in his commentary on *Metaphysics* XI, says that the rational soul remains after death and it will have its end of happiness if it conjoins with the First Mover. And he calls the First Mover the Principle of the Universe, which is God'.[117] Yet for Ibn Rushd, while the First Principle is the formal and final cause for all things — and in this way is the ultimate cause drawing all things into being and perfection — there is no doctrine of a personal post mortem contemplative return to God.

Albert asserts in the *Super Ethica* that philosophy has nothing to tell us about the rational soul and its end after death; rather, this is a matter of faith infused by a higher non-natural light. Nevertheless, in each case he proceeds to explain that, while Averroes says many heretical things, the Cordoban does hold that after

---

116 Regarding the synthetic and critical use of the teachings of Avicenna and Averroes by Aquinas for the doctrine of the soul in Thomas Aquinas, see Blackerby, 'Contextualizing Aquinas's Ontology of Soul'.

117 See note 86.

death we as intellect will contemplate simple quiddities and achieve happiness in that way. In this context, Albert tells us in the solution that some of the Arabic philosophers held that in the tenth intellect from which the form of soul flows, there is just one soul to which individual souls return, not in individual being after the death of the body but only 'according to essence' due to the end of bodily individuation. There he calls this heresy and goes on to explain briefly Avicenna's doctrine on the post mortem existence of the human soul.[118] Then, in the response to the second objection, he explains how one might be able to sustain the view of Averroes.[119] Here, Albert himself crafts and attributes to Averroes a doctrine of monopsychism that is not found in Ibn Rushd.

Albert's misconstrual of Ibn Rushd's separate intellects as powers of the human soul in the *De homine* led to the momentously valuable account of individual human intellectual understanding on the part of his student Thomas Aquinas. His misconstrual of Ibn Rushd's teaching on the afterlife led to the momentous consequence of the development of a novel doctrine of monopsychism which he attributed to Averroes. The positive value of the latter came to be found in the Latin thinkers' responses to this Averroism and the development of sophisticated accounts of individual personal immortality and ultimate happiness in the afterlife in the context of Christian teachings.

118 Albertus Magnus, *Super Ethica*, ed. by Kübel, p. 453, vv. 40–54.
119 The reader of the Latin text should take care regarding Albert's referent in his use of the word *Commentator*. See note 95.

## Appendix: Some Remarks on the Acquired Intellect

In Alexander of Aphrodisias, the conception of the acquired intellect involved sense powers and the external, eternal Agent Intellect coming to have a transitory presence of a sort in the individual perishable human soul for the apprehension of immaterial intelligibles. Isḥāq's Arabic translation seems, however, to have offered opportunities for new issues, solutions, and understandings. Marc Geoffroy provides an intriguing account of Alexander and the translation of Isḥāq, proposing that al-Fārābī's notion of the acquired intellect, العقل المستفاد *al-ʿaql al-mustafād, intellectus adeptus,* was formed in connection with his study of the *Theology of Aristotle* edited by al-Kindī from the *Plotiniana Arabica.*[120] In al-Fārābī's رسالة في العقل *Risala fī-l-ʿaql, De intellectu,* which was available in a twelfth-century Latin translation,[121] the acquired intellect involves the intellectual transformation or completive perfection of a particular human from being a perishable entity into being an immaterial imperishable substance, eternal *a parte post,* when it has reached the point of no longer needing the body in the consideration of intelligible forms (cf. Aristotle, *De anima* III.4, 429b5–10).

For Themistius (whose *Paraphrase of the De anima* was also translated by Isḥāq and was known by al-Fārābī, Ibn Sīnā, and Ibn Rushd), human understanding also involves sense powers and the unique transcendent Agent Intellect's necessary presence assisting the imperishable, immaterial, and incorporeal individual human intellect with its agent and receptive intellects to form a proper understanding of intelligibles.[122] In Ibn Sīnā, the acquired intellect (العقل المستفاد *al-ʿaql al-mustafād, intellectus adeptus*) denotes the actual moment of the active apprehension of an intelligible by the human soul in conjoining with the Agent Intellect.[123] For Ibn Rushd in his *Long Commentary on the De anima,* acquired intellect denotes the apprehended intelligibles in the perishable soul and can be identified with the habitual intellect (العقل بالملكة *al-ʿaql bi-l-malakati, intellectus in habitu*) and the

120 See Geoffroy, 'La tradition arabe du Περὶ νοῦ d'Alexandre d'Aphrodise': Geoffroy, 'Averroès sur l'intellect comme cause agent et cause formelle'. This is also discussed in Geoffroy's doctoral dissertation, a version of which has been published in two parts as Geoffroy, 'Sources et origines de la théorie de l'intellect d'Averroès'. See also Taylor, 'The Agent Intellect'; Taylor, 'Intellect as Intrinsic Formal Cause in the Soul'.

121 The edition of the Latin text is contained in Gilson, 'Les sources greco-arabes de l'Augustinisme avicennisant'.

122 See Themistius, *De anima paraphrasis,* Greek, p. 103, v. 20–p. 104, v. 13 and p. 98, v. 12–p. 99, v. 10; *Commentary on Aristotle's De anima, Arabic,* p. 187, v. 18–p. 189, v. 15 and p. 169, v. 4–p. 197, v. 9; *On Aristotle's On the Soul, English,* pp. 128–29 and 122–23.

123 هذا ضرب من العقل بالفعل. وهو القوة تحصل للنفس أن تعقل بها تشاء فإذا شاءت اتصلت وفاض فيها الصورة المعقولة. هى العقل المسفاد بالحقيقة وهذه القوة هى العقل بالفعل فينا من حيث لنا أن نعقل. وأما العقل المستفاد فهو العقل بالفعل من حيث هو كمال *Avicenna's De anima (Arabic Text),* ed. by Rahman, pp. 247–48; Avicenna Latinus, *Liber De anima seu Sextus de naturalibus,* ed. by Van Riet, p. 150, vv. 62–67: 'Hic enim modus intelligendi in potentia est virtus quae acquirit animae intelligere cum voluerit; quia, cum voluerit, coniungetur intelligentiae a qua emanat in eam forma intellecta. Quae forma est intellectus adeptus verissime et haec virtus est intellectus in effectu [...] secundum quod est perfectio'.

theoretical intellect (العقل النظري *al-ʿaql al-naẓarī, intellectus speculativus*).[124] Albert was aware of the use of this notion in Avicenna and Averroes early on, as we can see in his early works right up to *Super Ethica* (1250–52). Yet in those earlier works, he may have conflated the use of the term in Avicenna and Averroes, in a tendency that favoured the meaning in Averroes,[125] before he had access to the *De intellectu* of al-Fārābī. Albert does evidence knowledge of the acquired intellect with the meaning found in al-Fārābī's *De intellectu* later in his *De anima* and importantly in his *Ethica*.[126] Albert's later use of *intellectus adeptus* as a human power that apprehends separate substances, including God, eventually became the foundation of a so-called 'Averroistic mysticism' in the Latin tradition.[127]

## Works Cited

### *Primary Sources*

Albertus Magnus, *De homine*, ed. by Henryk Anzulewicz and Joachim R. Söder, Editio Coloniensis, 27/2 (Münster: Aschendorff, 2008)

———, *De quattuor coaequaevis* (*Summa de de creaturis*, pars I), ed. by Auguste Borgnet, Editio Parisiensis, 34 (Paris: Vivés, 1895)

———, *De resurrectione*, ed. by Wilhelm Kübel, Editio Coloniensis, 26 (Münster: Aschendorff, 1958)

———, *Ethicorum lib. X*, ed. by Auguste Borgnet, Editio Parisiensis, 7 (Paris: Vivès, 1891)

———, *Metaphysica, libros quinque priores*, ed. by Bernhard Geyer, Editio Coloniensis, 16/1 (Münster: Aschendorff, 1960)

———, *Super Dionysium De divinis nominibus*, ed. by Paul Simon, Editio Coloniensis, 37 (Münster: Aschendorff, 1972)

———, *Super Ethica*, ed. by Wilhelm Kübel, Editio Coloniensis, 14 (Münster: Aschendorff, 1968–87)

———, *Super IV libros Sententiarum*, ed. by Auguste Borgnet, Editio Parisiensis, 29–30 (Paris: Vivès, 1894)

Aristotle, *Aristotle's Metaphysics*, trans. by Hugh Tredennick, Loeb Classical Library, 18 (London: William Heinemann, 1935)

---

124 This is also discussed in my introduction to the *Long Commentary on the De anima, English*, pp. lix, lxv, lxvii–lxxv.

125 See Albertus Magnus, *De homine*, ed. by Anzulewicz and Söder, pp. 396–99 and 430; *Super IV libros Sententiarum* I, d. 3, a. 20, ed. by Borgnet, p. 118b–119a, and II, d. 3. a. 6, p. 70b; *Super Dionysium De divinis nominibus*, ed. by Simon, pp. 133–34, 199.

126 See Wietecha, 'Albert the Great's Ethical Commentaries'.

127 See de Libera, 'Existe-t-il une noétique "averroiste"?' and his discussion of acquired intellect (العقل المستفاد *al-ʿaql al-mustafād, intellectus adeptus*) in de Libera, 'Averroïsme éthique et philosophie mystique', and *Métaphysique et noétique*, pp. 265–328. Also see Flasch, *D'Averroès à Maître Eckhart*.

———, *Aristotle's Metaphysics: A Revised Text with Introduction and Commentary*, ed. by W. D. Ross (Oxford: Clarendon Press, 1924)

———, *Metaphysics Book Λ*, trans. with introduction and notes by Lindsay Judson (Oxford: Clarendon Press, 2019)

Averroes, *Averroès, Grand Commentaire de la 'Métaphysique' d'Aristote (Tafsīr mā baʿd aṭ-Ṭabīʿat). Livre Lam-Lambda*, trans. and annotated by Aubert Martin (Paris: Les Belles Lettres, 1984)

———, *Averroès, La béatitude de l'âme: Editions, traductions et études*, ed. and trans. by Marc Geoffroy (Paris: Vrin, 2001)

———, *Averrois Cordubensis Commentarium magnum in Aristotelis De anima libros*, ed. by F. Stuart Crawford (Cambridge, MA: The Mediaeval Academy of America, 1953)

———, *De la faculté rationnelle: L'original arabe du Grand Commentaire (Sharḥ) d'Averroes au 'De Anima' d'Aristote (III, 4–5, 429a10–432a14)*, ed. by C. Sirat and M. Geoffroy (Rome: Aracne, 2021)

———, *Epitome de Anima*, ed. by Salvador Gómez Nogales (Madrid: Instituto 'Miguel Asin', 1985)

———, *Ibn Rushd's Metaphysics: A Translation with Introduction of Ibn Rushd's Commentary on Aristotle's Metaphysics, Book Lambda*, trans. by Charles Genequand (Leiden: Brill, 1984)

———, *L'intellect: Commendium du livre De l'âme*, ed. by David Wirmer and trans. by Jean-Baptiste Brenet (Paris: Vrin, 2022)

———, *Long Commentary on the 'De Anima' of Aristotle*, trans. with an introduction and notes by Richard C. Taylor, with Thérèse-Anne Druart (New Haven, CT: Yale University Press, 2009)

———, *Middle Commentary on Aristotle's 'De anima'*, ed. and trans. by Alfred L. Ivry (Provo, UT: Brigham Young University Press, 2002)

———, *Middle Commentary on the Metaphysics* (كتاب ما بعد الطبيعة), in *Rasāʾil Ibn Rushd* (Hyderabad: Daʾirat al-Maʿārif al-ʿUthmāniyya, 1947)

———, *On Aristotle's 'Metaphysics': An Annotated Translation of the So-Called 'Epitome'*, ed. by Rüdiger Arnzen (Berlin: De Gruyter, 2010)

———, *L'original arabe du Grand Commentaire d'Averroès au 'De anima' d'Aristote*, ed. by C. Sirat and M. Geoffroy (Paris: Vrin, 2005)

———, *La psicología de Averroes: Comentario al libro sobre el alma de Aristóteles*, trans. by S. Gómez Nogales (Madrid: Universidad Nacional de Educación a Distancia, 1987)

———, *Short Commentary: Talkhīs Kitāb al-Nafs*, ed. by A. F. El-Ahwani (Cairo: Imprimerie Misr, 1950)

———, *Tafsīr mā baʿd aṭ-Ṭabīʿat*, ed. by Maurice Bouyges, Bibliotheca arabica scholasticorum, série arabe, 5/2 (Beirut: Dar El-Machreq, 1973)

Averroes Latinus, *Aristotelis Metaphysicorum libri XIIII: Cum Averrois Cordubensis in eosdem Commentariis* [Long Commentary on the *Metaphysics*], Aristotelis Opera cum Averrois commentariis, 8 (Venice, 1562; repr. Frankfurt am Main: Minerva, 1962)

Avicenna, *Avicenna's 'De anima' (Arabic Text) Being the Pyschological Part of Kitāb al-Shifā'*, ed. by F. Rahman (London: Oxford University Press, 1960)

———, *Avicenna's Psychology: An English Translation of 'Kitāb al-Najāt', Book II, Chapter VI with Historico-Philosophical Notes and Textual Improvements on the Cairo Edition*, trans. by Fazlur Rahman (London: Oxford University Press, 1952)

———, *Liber de anima seu Sextus de naturalibus: Édition critique de la traduction latine médiévale*, ed. by Simone Van Riet with an introduction by Gérard Verbeke, vol. 1 (books IV–V), vol. 2 (books I–III) (Leuven: Peeters, 1968–72)

Pattin, Adriaan, 'Le *Liber de causis*: Édition établie à l'aide de 90 manuscrits avec introduction et notes', *Tijdschrift voor Filosofie*, 28 (1966), 90–203

Themistius, *An Arabic Translation of Themistius Commentary on Aristoteles De anima*, ed. by M. C. Lyons (Oxford: Cassirer, 1973)

———, *Paraphrasis in libros Aristotelis De anima*, Commentaria in Aristotelem Graeca, 5/3, ed. by R. Heinze (Berlin: Georg Reimer, 1899)

———, *On Aristotle's 'On the Soul'*, trans. by Robert B. Todd (Ithaca, NY: Cornell University Press, 1996)

Thomas Aquinas, *Scriptum super libros Sententiarum Magistri Petri Lombardi Episcopi Parisensis*, ed. by P. Mandonnet, vol. 2 (Paris: P. Lethielleux, 1929)

### *Secondary Works*

Alpina, Tommaso, *Subject, Definition, Activity: Framing Avicenna's Science of the Soul* (Berlin: De Gruyter, 2021)

Bazán, B. Carlos, 'Was There Ever a "First Averroism"?', in *Geistesleben im 13. Jarhundert*, ed. by Jan A. Aertsen and Andreas Speer (Berlin: De Gruyter, 2000), pp. 31–53

Bieniak, Magdelena, *The Soul-Body Problem at Paris, ca. 1200–1250: Hugh of St-Cher and His Contemporaries* (Leuven: Leuven University Press, 2010)

Blackerby, Nathan, 'Contextualizing Aquinas's Ontology of Soul: An Analysis of His Arabic and Neoplatonic Sources' (doctoral thesis, Marquette University, 2017, available at https://epublications.marquette.edu/dissertations_mu/725/)

Burnett, Charles, 'Arabic Philosophical Works Translated into Latin', in *The Cambridge History of Medieval Philosophy*, ed. by Christina van Dyke and Robert Pasnau (Cambridge: Cambridge University Press, 2010), vol. II, pp. 814–22

Calma, Dragos, ed., *Reading Proclus and the 'Book of Causes'*, 3 vols (Leiden: Brill, 2019–22)

Davidson, Herbert A., *Alfarabi, Avicenna, and Averroes, on Intellect: Their Cosmologies, Theories of the Active Intellect, and Theories of Human Intellect* (Oxford: Oxford University Press, 1992)

de Libera, Alain, 'Averroïsme éthique et philosophie mystique: De la félicité intellectuelle à la vie bienheureuse', in *Filosofia e teologia nel Trencento: Studi in ricordo di Eugenio Randi*, ed. by Luca Bianchi (Turnhout: Brepols, 1994)

———, 'Existe-t-il une noétique "averroiste"?', in *Averroismus im Mittelalter und in der Renaissance*, ed. by Friedrich Niewöhner and Loris Sturlese (Zurich: Spur, 1994), pp. 51–80

———, *Métaphysique et noétique: Albert le Grand* (Paris: Vrin, 2005)

de Vaux, R., 'La première entrée d'Averroës chez les latins', *Revue des Sciences philosophiques et théologiques*, 22 (1933), 193–243

Flasch, Kurt, *D'Averroès à Maître Eckhart: La naissance de la 'mystique' allemande de l'ésprit de la philosophie arabe* (Paris: Vrin, 2008)

Gauthier, René Antoine, 'Notes sur les débuts (1225–1240) du premier "averroïsme"', *Revue des sciences philosophiques et théologiques*, 66 (1982), 321–74

———, ed., 'Le traité *De anima et de potenciis eius* d'un maître ès arts (vers 1225)', *Revue des sciences philosophiques et théologiques*, 66 (1982), 3–55

Genequand, Charles, 'Introduction', in *Ibn Bāǧǧa (Avempace): La Conduite de l'isolé et deux autre epîtres. Introduction, édition critique du texte arabe, traduction et commentaire*, ed. by Charles Genequand (Paris: Vrin, 2010), pp. 1–88

Geoffroy, Marc, 'Averroès sur l'intellect comme cause agente et cause formelle et la question de la "jonction"', in *Averroès et les averroïsmes juif et latin: Actes du colloque tenu à Paris, 16–18 juin 2005*, ed. by Jean-Baptiste Brenet (Turnhout: Brepols, 2007), pp. 77–110

———, 'Sources et origines de la théorie de l'intellect d'Averroès (I)', *Mélanges de l'Université Saint-Joseph*, 66 (2014–15), 181–302

———, 'Sources et origines de la théorie de l'intellect d'Averroès (II)', *Mélanges de l'Université Saint-Joseph*, 67 (2019–20), 139–231

———, 'La tradition arabe du Περὶ νοῦ d'Alexandre d'Aphrodise et les origines de la théorie farabienne des quatre degrés de l'intellect', in *Aristotele e Alessandro di Afrodisia nella tradizione araba*, ed. by Cristina D'Ancona and Giuseppe Serra (Padua: Il Poligrafo, 2002), pp. 191–231

Giglioni, Guido, 'Introduction', in *Renaissance Averroism: Arabic Philosophy in Early Modern Europe*, ed. by Anna Akasoy and Guido Giglioni (Dordrecht: Springer, 2013), pp. 1–34

Gilson, Étienne, *Les sources greco-arabes de l'Augustinisme avicennisant* (Paris: Vrin, [1929] 1981)

Gómez Nogales, Salvador, 'Saint Thomas, Averroès et l'averroïsme', in *Aquinas and Problems of His Time*, ed. by Gerard Verbeke and Daniel Verhelst (Leuven: Leuven University Press, 1976), pp. 161–77

Gutas, Dimitri, 'Ibn Sina [Avicenna]', *The Stanford Encyclopedia of Philosophy* (Fall 2016 Edition), ed. by Edward N. Zalta, https://plato.stanford.edu/archives/fall2016/entries/ibn-sina/

Hasse, Dag Nikolaus, *Avicenna's 'De anima' in the Latin West: The Formation of a Peripatetic Philosophy of the Soul 1160–1300* (London: The Warburg Institute, 2000)

———, *Latin Averroes Translations of the First Half of the Thirteenth Century* (Hildesheim: Olms, 2010)

Hayoun, Maurice-Ruben, and Alain de Libera, *Averroès et Averoïsme* (Paris: Presses Universitaires de France, 1991)

Janssens, Jules, 'Al-Ghazālī's *Maqāṣid al-falāsifa*, Latin Translation of', in *Encyclopedia of Medieval Philosophy*, ed. by H. Lagerlund (Dordrecht: Springer, 2011), pp. 387–90

———, 'Ibn Sīnā's Ideas of Ultimate Realities, Neoplatonism and the Qur'ān as Problem-Solving Paradigms in the Avicennian System', *Ultimate Meaning and Reality*, 10 (1987), 252–71

Kogan, Barry, *Averroes and the Metaphysics of Creation* (Albany: SUNY Press, 1985)

Miller, Robert, 'An Aspect of Averroes' Influence on St Albert', *Mediaeval Studies*, 16 (1954), 57–71

Minnema, Anthony H. 'Algazel Latinus: The Audience of the *Summa Theoricae Philosophiae*, 1150–1600', *Traditio*, 69 (2014), 153–215

Mulchahey, M. Michèle, 'The *Studium* at Cologne and Its Role within Early Dominican Education', *Listening: Journal of Religion and Culture*, 43 (2008), 118–47

Neria, Chaim Meir, 'Al-Fārābī's Lost Commentary on the *Ethics*: New Textual Evidence', *Arabic Sciences and Philosophy*, 23 (2013), 69–99

Polloni, Nicola, *The Twelfth-Century Renewal of Latin Metaphysics. Gundissalinus's Ontology of Matter and Form* (Toronto: Pontifical Institute of Mediaeval Studies, 2020)

Porro, Pasquale, *Aquinas: A Historical and Philosophical Profile*, trans. by Joseph Trabbic and Roger W. Nutt (Washington, DC: The Catholic University of America Press, 2015)

Rigo, Catarina, 'Zur Redaktionsfrage der Frühschriften des Albertus Magnus', in *Albertus Magnus und die Anfänge der Aristoteles-Rezeption im Lateinischen Mittelalter: Von Richardus Rufus bis zu Franciscus de Mayronis*, ed. by Ludger Honnefelder, Rega Wood, Mechthild Dreyer, and Marc-Aeilko Aris (Münster: Aschendorff, 2005), pp. 325–74

Salman, Dominique, 'Note sur la première influence d'Averroès', *Revue néoscolastique de philosophie*, 40 (1937), 203–12

Sorabji, Richard, *The Philosophy of the Commentators 200–600 AD: A Sourcebook*, vol. 1, *Psychology (with Ethics and Religion)* (London: Duckworth, 2004)

Taylor, Richard C., 'Abstraction and Intellection in Averroes and the Arabic Tradition: Remarks on Averroes, Long Commentary on the "De Anima Book 3", Comment 36', in *Sujet libre: Pour Alain de Libera*, ed. by Jean-Baptiste Brenet and Laurent Cesalli (Paris: Vrin, 2018), pp. 321–25

———, 'The Agent Intellect as "Form for Us" and Averroes's Critique of al-Fārābī' (2005), reprinted with revisions in *Universal Representation and the Ontology of Individuation*, ed. by Gyula Klima and Alexander W. Hall (Newcastle upon Tyne: Cambridge Scholars, 2011), pp. 25–44

———, 'Aquinas and "the Arabs": Aquinas's First Critical Encounter with the Doctrines of Avicenna and Averroes on the Intellect', *IN 2 SENT*. D. 17, Q. 2, A.1', in *Philosophical Psychology in Arabic Thought and the Latin Aristotelianism of the 13th Century*, ed. by Luis X. López-Farjeat and Jörg Tellkamp (Paris: Vrin, 2013), pp. 142–83

———, 'Averroes on the Ontology of the Human Soul', *Muslim World*, 102 (2012), 580–96

———, 'Avicenna and the Issue of the Intellectual Abstraction of Intelligibles', in *Philosophy of Mind in the Early and High Middle Ages*, ed. by Margaret Cameron (Abingdon: Routledge, 2019), pp. 56–82

———, 'Intellect as Intrinsic Formal Cause in the Soul according to Aquinas and Averroes', in *The Afterlife of the Platonic Soul*, ed. by Maha El-Kaisy Friemuth and John M. Dillon (Leiden: Brill, 2009), pp. 187–220

———, 'Intelligibles in Act in Averroes', in *Averroès et les averroïsmes juif et latin: Actes du colloque tenu à Paris, 16–18 juin 2005*, ed. by Jean-Baptiste Brenet (Turnhout: Brepols, 2007), pp. 111–40

———, 'Personal Immortality in Averroes' Mature Philosophical Psychology', *Documenti e studi sulla tradizione filosofica medievale*, 9 (1998), 87–110

———, 'Remarks on the Importance of Albert the Great's Analyses and Use of the Thought of Avicenna and Averroes in the *De homine* for the Development of the Early Natural Epistemology of Thomas Aquinas', in *Die Seele im Mittelalter: Von der Substanz zum funktionalen System*, ed. by Günther Mensching and Alia Mensching-Estakhr (Würzburg: Königshausen & Neumann, 2018), pp. 131–58

———, 'Remarks on the Latin Text and the Translator of the *Kalam fi mahd al-khair/Liber de causis*', *Bulletin de philosophie médiévale*, 31 (1989), 75–102

———, 'Themistius and the Development of Averroes' Noetics', in *Medieval Perspectives on Aristotle's 'De Anima'*, ed. by Russell L. Friedman and Jean-Michel Counet (Leuven: Peeters, 2013), pp. 1–38

Torrell, Jean-Pierre, *Saint Thomas Aquinas*, vol. 1: *The Person and His Work* (Washington, DC: Catholic University of America Press, 2005)

Wietecha, Tracy, 'Albert the Great's Ethical Commentaries and al-Farabi's *De Intellectu*', in *Homo, Natura, Mundus: Human Beings and Their Relationships*, ed. by Roberto Hofmeister Pich, Alfredo Carlos Storck, and Alfredo Santiago Culleton (Turnhout: Brepols, 2020), pp. 345–56

Wirmer, David, *Vom Denken der Natur zur Natur des Denkens: Ibn Bāǧǧas Theorie der Potenz als Grundlegung der Psychologie* (Berlin: De Gruyter, 2014)

MARTIN J. TRACEY

# Chapter 4. Albert's Invocations of Averroes in his Account in *Super Ethica* of the Relation between Philosophical and Theological Ethics

Religious believers undertaking to exposit Aristotle's writings and evaluate their claims face the question of the relation of Aristotle's teachings to those of their own faith traditions. Are Aristotle's various doctrines compatible, incompatible, or incommensurable with those of faith? For medieval authors, reflection on that question often opened another, more general one: What is the relation between philosophical discourse on a subject and theological discourse on it?

Like Averroes before him, Albert the Great confronts those particular and general questions.[1] One important place that he does so is in his *Super Ethica commentum et quaestiones* (hereafter *SE*), a work deriving from lectures on Aristotle's *Nicomachean Ethics* that Albert gave between 1250 and 1252 at the Dominican house of religious formation in Cologne.[2] In what follows, I examine Albert's account within that work of the relation between philosophical and theological discourses on morality, reflect on its significance, and illumine Albert's reasons for holding it. As we shall see, Albert believes that the two discourses give different

1 For a recent discussion of Averroes's efforts to account for the place of philosophy in relation to religious knowledge, with some remarks concerning the contribution of Averroes's *Middle Commentary* on Aristotle's *Nicomachean Ethics* to that endeavour, see Black, *History of Islamic Political Thought*, pp. 120–24.

2 On the basis of textual evidence different from my own, Alain de Libera reaches a complementary conclusion regarding Albert's understanding of the relation between faith and reason in the *Super Ethica*: 'il y a une manière de parler [...] "selon les opinions des philosophes", une autre, tout opposée "selon le théologien"'. De Libera notes that the account Albert defends on this matter in *SE* departs significantly from a more Augustinian one that he defends in his *Sentences* commentary. De Libera, *Raison et foi*, p. 266.

**Martin J. Tracey** (mtracey@ben.edu) is professor of philosophy at Benedictine University (Lisle, IL). He specializes in Latin moral thought of the thirteenth century.

*Albert the Great and his Arabic Sources*, ed. by Katja Krause and Richard C. Taylor, Philosophy in the Abrahamic Traditions of the Middle Ages, 5 (Turnhout: Brepols, 2024), pp. 117–144
BREPOLS PUBLISHERS 10.1484/M.PATMA-EB.5.136484

answers to certain fundamental moral questions, and do so because they proceed from different principles. Albert advocates that position in part because it enables him to defend the rightness of Aristotle's reasoning about morality even as he acknowledges that Aristotle defends moral doctrines that are contrary to those of faith. The path leading Albert to this position begins with his examination of certain of Averroes's interpretations of Aristotle — in particular, whether the doctrines Averroes ascribes to Aristotle are indeed Aristotle's own, and whether the doctrines themselves are true.

Albert's *SE* was the first continuous commentary by a medieval Latin author on the whole of Aristotle's *Nicomachean Ethics* (hereafter *EN*). Study of the *EN* made a profound impression on Albert's thought.[3] In turn, his interpretation of the work made a profound impression on Thomas Aquinas, who, as a student, prepared the *reportatio* of the lectures from which Albert's text derives and later a kind of index of the principal themes of the *EN* and of Albert's commentary on it.[4] Albert delivered his lectures at the Dominican *studium* in Cologne early in his career, at roughly the same time that he launched his great project of making Aristotle intelligible to the Latins by paraphrasing all his known works.[5] His decision to lecture on Aristotle's *EN* was bold; at the time he made it, many Christian leaders, including prominent ones within his own religious order, thought that studying the *EN* was dangerous for believers insofar as it seemed to lend authority to doctrines contrary to faith.[6]

The *SE* consists of a literal exposition of Aristotle's text together with over five hundred Scholastic questions examining the truth, adequacy, and sufficiency of diverse claims contained in it. The *EN* is the first Aristotelian work upon which Albert comments,[7] and his *SE* contains some of his earliest reflections on the

3 Stanley Cunningham discusses how and why 'Aristotle's *Nicomachean Ethics* exercised a lasting influence on the thought of Albert', noting that Albert 'devoted a great deal of time and effort both to the exposition of Aristotle's ethical thought and to the elaboration of his own moral doctrine'. Cunningham, *Reclaiming Moral Agency*, p. 26. For a study of the *SE* in the context of Albert's other moral writings, see Tracey, 'Albert's Moral Thought'.

4 For general notes on Thomas's *Tabula libri Ethicorum*, see Torrell, *Saint Thomas Aquinas*, vol. 1, pp. 229–30. For some brief recent remarks regarding the impact of Albert's interpretation of the *EN* upon Thomas Aquinas, see Perkhams, 'Einleitung', pp. 17–23.

5 The *Nicomachean Ethics* is among the works that Albert will later paraphrase; Albert is thus the author of two different commentaries on the text. See Dunbabin, 'Two Commentaries'. For the chronology of Albert's works, see Albertus-Magnus-Institut, *Albertus Magnus*, pp. 28–31.

6 For discussion of the boldness of Albert's decision, see Sturlese, *Die deutsche Philosophie*, pp. 337–38. See also Mulchahey, *'First the Bow is Bent in Study'*, pp. 256–57. For general discussion of the 'sympathetic' and 'antagonistic' reactions to the *EN* by thirteenth-century Latin readers, see Luscombe, 'Ethics in the Thirteenth Century', pp. 668–69. Luscombe discusses criticisms of particular Aristotelian moral teachings by several pre-1250 authors: Alexander Neckham, William of Auvergne, Philip the Chancellor, Richard Fishacre, and Roger Bacon.

7 Note that while the *EN* is indeed the first Aristotelian work upon which Albert comments, the *Super Ethica* is not the first text in which Albert engages Aristotle's writings in depth. His earlier work *De homine* is a compilation of Aristotelian psychological teachings that draws heavily on Aristotle's *De anima* and *Parva naturalia*.

relation of philosophical and theological *modi loquendi* — a topic to which he returns in later works.[8] The *EN* had already been a subject of commentary by several different masters in the Faculty of Arts at Paris in the second quarter of the thirteenth century. However, those authors did not comment on any but the work's first three books. Their commentaries show their struggle to understand Aristotle, and their propensity to read him as if he were a man of their own time with ideas and values fundamentally similar to their own.[9] The Arts masters' Christianized reading of the *EN* became much less tenable once Robert Grosseteste published his translation of Aristotle's complete work in 1247/48.[10] Grosseteste not only made the whole of the *EN* available in Latin, but also supplied translations of Greek and Byzantine commentaries on the text, together with his own text-critical comments and notes.[11] The compilation of Greek commentaries that Grosseteste translated included works by two Byzantine Christian authors: the commentaries on *EN* I and *EN* VI by Eustratius, the metropolitan of Nicaea *c.* 1200, and commentaries on *EN* V, IX, and X by Michael of Ephesus, professor at Constantinople in the first half of the eleventh century. It also included a collection of anonymous scholia on Books II, III, IV, and V, probably from the third century.[12]

It would take time for Latin authors to disentangle the teachings of the historical Aristotle from those ascribed to him by Christianizing interpreters. According to René-Antoine Gauthier, arguably the most influential twentieth-century scholar of the Latin reception of Aristotle's *EN*, Albert the Great played a large and not entirely laudable role in that process.[13] On the one hand, Gauthier celebrates Albert as one of the first Latin authors to have grasped how profoundly Aristotle's moral teaching differs from Christian moral teaching, and presents

8 For an argument in support of the view that Albert's account of the relation between theology and philosophy in *Super Ethica* is not one premised upon any simple harmony or complementarity of faith and reason, together with an argument on behalf of the methodological autonomy of philosophical ethics, see Müller, *Natürliche Moral*, pp. 48–58.

9 Wieland, *Ethica – scientia practica*, pp. 143–97. For further discussion of some of the characteristic misunderstandings of the early Arts masters, see Buffon, 'Structure of the Soul'; Zavattero, 'Le bonheur parfait', pp. 311–17.

10 Among the pre-1250 commentators, Robert Kilwardby stands out in various ways for having recognized at a relatively early date the errancy of some Christianizing misreadings. See Celano, 'Robert Kilwardby', pp. 149–62.

11 Although Grosseteste is not the first medieval author to have translated the whole of Aristotle's text into Latin — Burgundio of Pisa enjoys that distinction — his integral Latin translation appears to be the first to have reached a large number of readers. For recent discussion of Grosseteste's work and its significance, see Luscombe, 'Ethics in the Thirteenth Century', pp. 669 ff.

12 For more information about the compilation of Greek commentaries that Grosseteste translated, see Mercken, 'Introduction', pp. 3*–29*.

13 With Jean Yolif, René Gauthier translated the *Ethics* into French and wrote a seminal, multivolume commentary on the text. He prepared the critical edition of Thomas Aquinas's *Sententia libri Ethicorum* and *Tabula libri Ethicorum*, as well as the critical editions of thirteenth-century Latin translations of the *Ethics*. The claims about Albert's interpretation of the *Ethics* are found chiefly in his much-cited study of three Latin commentaries from the late thirteenth and early fourteenth century, Gauthier, 'Trois commentaires'.

Albert's *SE* as the best of the 'countless' medieval Latin commentaries on the text, not least because of its success in uncovering teachings of the historical Aristotle.[14] On the other hand, Gauthier faults Albert for having erected an exegesis of the *EN* that obscures the incompatibility of some of Aristotle's key moral ideas with Christian ones. He claims that Albert confected and disseminated his own Christianizing reading of the *EN* as part of a plan to thwart any misguided initiative by 'ignorant blasphemers of philosophy' to ban its study.[15] According to Gauthier, Albert reasons that philosophy's blasphemers will not act to ban the study of the *EN* if they are led to believe that it agrees with faith.

Gauthier's Albert labours to defend a 'concordizing' reading of the *EN* not least because he recognizes the threat posed by Grosseteste's integral translation of the *EN* and provision of tools for reading it rightly.[16] Left unchecked, Grosseteste's translation and tools threaten to unmask the authentic Aristotle hidden beneath the Christianizing interpretations of pre-1250 Arts commentators.[17] Albert's future liberty to indulge his enthusiasm for Aristotle is doomed unless he can succeed in explaining away, cleansing, hiding, or otherwise masking Aristotle's paganism.[18] His Christianizing exegesis of the *EN* in the *SE* aims to do just that, and it succeeds, albeit for a short time and at a high price. Albert's exegesis, though it only managed to retard the recognition of a more historical Aristotle for some twenty years, nevertheless inhibited a historically conscious study of his texts in Roman Catholic circles indefinitely thereafter.[19] That is not its darkest legacy, according to Gauthier: in pre-Reformation continental Europe and beyond, Albert's Christianizing exegesis made possible a misguided Christian moral theology — one that prizes fidelity to a pagan philosopher over fidelity to the Bible and the Fathers.

As sophisticated as Gauthier's account of Albert's purposes in the *SE* may be, it fails to explain satisfactorily some of the commentary's central features.[20] Below I foreground two such features, the ones that are hardest to explain: Albert's presentation in the *SE* of particular ways in which Aristotle's teaching diverges from that of faith, and his explanation of how it is possible for Aristotle to diverge from what is known by faith and nevertheless to reason rightly and speak truthfully.

Albert's *quaestiones* in the *SE* do not inquire after the Christianity of Aristotle's teaching as such. On the contrary, they inquire after the truth, appropriateness, or

14 Gauthier, *L'Éthique à Nicomaque*, p. 123.

15 Cf. Gauthier, 'Trois commentaires', pp. 269 et passim.

16 Ibid., pp. 244–45.

17 Ibid., pp. 293.

18 Ibid., pp. 253 and 293.

19 Ibid., p. 269. Cf. Gauthier's account of Albert's role, through Thomas and Thomists, in promulgating an interpretation of the *EN* that is 'la négation de l'enseignement exprès de l'Aristote historique'. Gauthier, *L'Éthique à Nicomaque*, p. 275.

20 For critical discussion of Gauthier's ascription of a 'Christianization project' to both Albert the Great and Thomas Aquinas, see McInerny, 'Aristotle and Thomas'.

sufficiency of particular Nicomachean utterances. Before we can understand what Albert means when he argues that some particular Nicomachean dictum is true, we need to know what criterion of truth he employs. Now, if we are to read the *SE* as a systematic attempt to Christianize the *EN*, we might suppose that coherence with faith is the relevant criterion in these *quaestiones*. So construed, what Albert might mean (at least in part) when he says that some Nicomachean teaching is true is that it coheres with or does not contradict the teaching of faith. But Albert's presentation of particular ways in which Aristotle's teaching contradicts faith shows that coherence with faith cannot be (or at any rate is not always) the criterion that Albert has in mind. In his eyes, to claim that a certain interpretation is *contra fidem* is not necessarily to claim either that its content is false or that it is a misinterpretation of Aristotle.

If Albert's aim in the *SE* were to Christianize the *EN*, we should hardly expect to find acknowledgements and explanations of this kind in it. To acknowledge places where Aristotle affirms what faith denies or denies what faith affirms would be to undermine, and not to advance, the thesis of their fundamental harmony. If paganizing or contra-Christianizing interpretations of the *EN* appeared at all in a Christianizing commentary, they would presumably appear as interpretations targeted for attack. A Christianizing Albert might well criticize some interpretation that would have Aristotle affirm what faith denies, but he would surely not advance such an interpretation in his own voice, much less explain how such an interpretation can be both true and *contra fidem*. Yet Albert does advance such interpretations and explanations in the *SE*, and no sound account of his commentarial purposes can leave them out.[21]

## Rational or *contra fidem*?

### *Averroes*

In *SE* I.13, Albert states that nothing can be demonstrated contrary to what is determined by faith. The reason this is so, he explains, is that faith is not contrary to reason. Indeed, it seems for Albert that faith cannot be contrary to reason,

21 See David Twetten's observation, discussing Albert's 'metaphysics of first causes', that 'we can no longer ignore the fact that Albert's philosophy leaves room for heterodox views'. Twetten, 'Albert the Great', p. 276. Twetten further observes that Albert's philosophical writings 'contain a non-Christian element that helped provoke the crisis between faith and reason in the middle ages' (p. 277). Twetten presents Albert as 'the infamous purported theorist of "two contrary truths" missing and wanted since 1277' (p. 277). For another presentation of Albert as 'la racine d'une manière de concevoir l'autonomie de la philosophie conduisant à la double vérité', see Bianchi, *Pour une histoire*, p. 41.

because what faith determines is true and what reason determines is true, and no truth clashes with another.[22]

A thinker who has this view about the harmony of faith and reason, and who wishes to affirm that Aristotle's determinations about human action in the *EN* cohere with what reason determines about human action, would be strongly inclined to read the *EN* concordistically. He would be loath to admit that Aristotle attempts to demonstrate anything *contra fidem*. For on the coherentist theory, to admit that Aristotle attempts as much is to admit that he attempts to do what cannot rationally be done. Indeed, on this view, it seems perfectly valid to reason from what one knows by faith to conclusions about what Aristotle, reasoning rightly, is likely to have demonstrated or taught. Seen from that perspective, any interpretation that would have Aristotle teaching something *contra fidem* appears as a probable misinterpretation — as a reading that can only be right if Aristotle reasons wrongly. If one assumes, as Albert does, that Aristotle very rarely reasons wrongly, then, on these assumptions about the harmony of faith and reason, the determinations of faith provide a negative check on Aristotle's teachings: any interpretation that would have Aristotle teaching something *contra fidem* would flag itself as a probable misinterpretation.[23]

Albert seems to employ faith in just this way in *SE* I.13. At the centre of that chapter is an *auctoritas* from Averroes, a thinker whose understanding of the relation of Aristotle's teachings to Islamic faith traditions plays a central role in Albert's account of Aristotle's teachings to Christian ones. At issue is whether Aristotle's statement in *De anima* III.5 that 'intellect is incorporeal' reveals anything concerning the Greek philosopher's beliefs about the state of the human soul after death.[24] Albert's Averroes believes that it does, and that what it reveals is that for Aristotle only one common human soul, rather than many distinct human souls, survives death. Although Albert does not address the adequacy of this interpretation explicitly, he notes with surprise that Averroes reads Aristotle as he does 'even though' (*licet*) that way of expositing the pertinent Aristotelian

22 Albertus Magnus, *Super Ethica commentum et quaestiones*, I.13, ed. by Kübel (hereafter *SE*), p. 71, vv. 79–84: 'Sed tamen contra ea quae fide determinata sunt, nihil potest demonstratio esse, eo quod fides non est contra rationem, quia nulla veritas alii discordat, sed est supra rationem, PS. (CXXXVIII, 6): "Mirabilis facta est scientia tua". Et ideo indiget lumine fidei'. For Averroes's discussion of related ideas — including the idea that 'truth is a unity' such that 'truth does not contradict truth but rather is consistent with it and testifies to it' — see Taylor, '"Truth Does Not Contradict Truth"'.

23 Although Albert holds that Aristotle very rarely errs, he does explicitly affirm in his *Physica* that Aristotle was not a god 'but a man', who 'could err just as we can too'. For discussion of that famous locus, see Synan, 'Introduction', pp. 11–12.

24 *SE*, I.13, p. 71, vv. 53–55. Albert's reading of Aristotle's teachings on this matter is shaped profoundly by the *Liber de causis*, which he regards to be a work of the Aristotelian tradition. For a rich discussion of the place of the *Liber de causis* within Albert's philosophical programme, see Krause and Anzulewicz, 'From Content to Method'.

authority is *contra fidem*.[25] By putting it that way, Albert implies that Averroes misinterprets Aristotle, and that one sign of his having done so may be that his reading would have Aristotle teaching something *contra fidem* — specifically, denying the immortality of individual human souls that faith affirms.[26] If right reason cannot be *contra fidem*, Aristotle cannot deny personal immortality and reason rightly.

Gauthier's theory about the role of faith in Albert's reading of Aristotle cannot, of course, be sustained on the basis of a single oblique argument. We will want to see other loci in the *SE* where Albert employs or, better still, endeavours to legitimize the putative rule for reading Aristotle's texts. The scarcity of such loci is a first sign that the theory is mistaken. Nevertheless, a second text that might be taken to reinforce it comes in *SE* VI.8.[27] There, too, Albert comments on the accuracy of one of Averroes's readings, and seems to afford faith a role in assessing its accuracy.

The *auctoritas* from Averroes arises within a *quaestio* concerning whether one ought to posit the union of the human intellect with the intelligences after death. One argument for believing that the soul is united with the intelligences after death is Averroes's statement, in his *Metaphysica*, that the final well-being of human souls for Aristotle lies in being unified with the intelligences acting on our souls.[28] Albert's reply begins by stating that Averroes 'utters many heresies' (*multas haereses dicit*) and that for that reason 'it is not appropriate' (*non oportet*) that the interpretation of Aristotle he advances in his *Metaphysics* commentary 'be sustained' (*sustineatur*).[29] Albert reasons that Averroes's *auctoritas* here may be ignored on the grounds that any *auctor* who utters many heresies *ipso facto* utters many falsities, that the interpretation of Aristotle in the *auctoritas* in question may well be one of them, and that interpreting Aristotle rightly is hindered by attending to the *auctoritates* of such *auctores*. This locus thus would appear

---

25 *SE*, I.13, p. 72, vv. 14–19: 'et si intellectus non sit forma situalis, non potest demonstrari, quod remaneant plures animae distinctae, sed omnibus una, sicut ponit COMMENTATOR in libro DE ANIMA et hoc modo exponit auctoritatem Aristotelis inductam, licet sit contra fidem'.

26 It is worth underscoring that Albert's account of the human relation to the separate substances is very complicated and seems to change over the course of his career and in accordance with contextual matters. His limited engagement with this subject in the *SE* is only one of many places where he discusses the matter. For some recent discussion of Albert's account of this matter, see Krause, 'Transforming Aristotelian Philosophy', pp. 184–89.

27 *SE*, VI.8, p. 452, vv. 69–83.

28 *SE*, VI.8, vv. 80–83. Note that Richard Taylor contests the view that Averroes himself has a philosophical doctrine regarding the *post mortem* existence of human beings. See Taylor, 'Personal Immortality'; Taylor, 'Averroes on the Ontology of the Human Soul'. Here and throughout, we are speaking of the Averroes of Albert's interpretation, who must not be conflated with the historical Averroes (Ibn Rushd), just as the Aristotle of Albert's interpretation must not be conflated with the historical Aristotle. See also Taylor in this volume.

29 *SE*, VI.8, p. 453, vv. 63–65: 'Ad secundum dicendum, quod Averroes multas haereses dicit; unde non oportet, quod sustineatur'. For a locus in which Albert ponders whether Plato's authority is 'to be sustained' or not, see *SE*, V.14, p. 376, vv. 17–27.

to confirm the legitimacy in Albert's eyes of assuming in one's interpretation of Aristotle that what the Philosopher teaches, however it may differ from the teachings of faith, is unlikely to clash with or contradict them.

The evidence of these two passages should not be neglected. *SE* I.13 and *SE* VI.8 might be taken to articulate the underlying theoretical justification for Albert's Christianization of Aristotle. Indeed, they may seem to lend Albert's putative 'entreprise de "christianisation"' a certain theoretical warrant, and hence to make the hypothesis of Christianization more persuasive.[30] Even so, on their apparent assumptions about faith and reason, Albert's reason for Christianizing Aristotle's *EN* would not necessarily be to hide its paganism and thereby to sustain the widespread ignorance of 'le véritable Aristote' that permits him to indulge his enthusiasm for Aristotle's writings.[31] Instead, it would be to reduce the likelihood of grossly misinterpreting Aristotle's *EN* — a text with a history of misinterpretation on the part of Latin authors, and one that had only just become widely available to them in an integral translation.

Yet the two passages should not be taken as supporting Gauthier's Christianization hypothesis. As we shall see, there is a reason why Albert does not himself say in *SE* I.13 that Averroes's interpretation is improbable *because* it is *contra fidem*, just as there is a reason why Albert proceeds in *SE* VI.8 to explain a way that Averroes's interpretation *may be sustained*. The reason is his emergent recognition that it is possible for Aristotle to speak truthfully about human action even when he affirms what faith denies or denies what faith affirms.

An important sign of that recognition is the effort Albert makes in *SE* I.15 to present Averroes's reading of the *De anima*'s view about the *post mortem* soul as a misinterpretation of Aristotle's texts. If inferences from faith were themselves sufficient to establish that Averroes's interpretation on this matter could only be right if Aristotle were to reason wrongly, Albert would have no pressing need to argue, as he does extensively in *SE* I.15, that Averroes's interpretations are clearly contrary to Aristotle's own principles, and presumably can be recognized as such without any assistance from faith.[32] In an effort to manifest the infidelity of the interpretation that Averroes gives to the texts of the Greek philosopher, Albert contends that the Muslim philosopher's view is that there is only one common soul, which remains when all human bodies are destroyed.[33] He then tries to show that this reading is an inference from two prior interpretations that he ascribes to Averroes regarding Aristotle's psychology. One of the theses — that the human soul is a non-composite or simple form — is, for Albert, a

30 Gauthier, 'Trois commentaires', p. 246.

31 Ibid., p. 293.

32 *SE*, I.15, p. 79, v. 59–p. 80, v. 22.

33 *SE*, I.15, p. 79, vv. 59–79. To be precise, Averroes himself holds neither for a single common soul nor for an afterlife. Rather, human beings share in the common separately existing material (namely, receptive) intellect during finite human worldly life. See Averroes, *Long Commentary on the De anima*, trans. by Taylor, pp. lxv–lxvi.

correct interpretation. However, the other — that without matter, there is nothing to individuate the human soul — is incorrect.[34] For Albert, it is the fact that Averroes's interpretation is premised upon a misreading of Aristotle, and not the fact the teaching of Averroes's Aristotle denies what faith affirms, that explains what makes Averroes's interpretation false. Albert's criticisms of Averroes in *SE* I.13 and *SE* VI.8 provide the strongest warrants in the *SE* for thinking that what Albert means when he affirms in the *SE* that Aristotle's teachings are true is that they cohere with and do not contradict the faith. Nevertheless, as we have seen, even in these passages where Albert would have been sorely tempted to invoke faith in order to support his argument that Averroes's readings are misreadings, he refuses to make that move.

### *The Ancient Scholiast on* EN *II–IV and Michael of Ephesus*

The criticisms that Albert makes of Averroes in the *SE* examined thus far do not concern the interpretation of Aristotle's *EN* per se, but rather his *De anima* and *Metaphysics*. We began by examining them because they seem to make the best case of any texts in the *SE* that Albert affords a role to faith in the assessment of both what Aristotle teaches and whether Aristotle's teaching is true. They show Albert entertaining the principle that it is impossible for an utterance to be true and to be *contra fidem*, but ultimately resisting the use of this principle, or at least any explicit or exclusive reliance on it, by arguing that particular interpretations of Aristotle's texts are inaccurate.

Albert's criticisms of certain readings by one of Aristotle's late antique commentators — the pagan author(s) known to specialists as the 'Ancient Scholiast' — show his reasons for resisting.[35] In at least four places, according to Albert, the Ancient Scholiast attributes to Aristotle the view that a person may licitly 'do evil that good may come'.[36] Albert vacillates as to whether that view is Aristotle's own. In at least two of those places, Albert appears to believe that the view is not Aristotle's own, and on account of a now familiar line of reasoning: namely, it is *contra fidem* to do evil that good may come, what is *contra fidem* is *contra rationem*, and Aristotle rarely subscribes to irrational views. However, in at least one of the remaining loci, Albert is inclined to believe that the un-Christian view is indeed

---

34 *SE*, I.15, p. 79, vv. 61–63: 'Unde Aristoteles in II De Anima probat, quod anima non est materia nec hoc aliquid sive compositum, sed forma'; ibid., vv. 74–75: 'Si enim esset tantum forma principium individuationis eius'.

35 On the identity of the 'Ancient Scholiast', see Mercken, 'Introduction', pp. 14*–22*. In referring to 'the Ancient Scholiast', I do not overlook Mercken's point that the 'anonymous "commentary" on Books II, III, IV, and V' is a 'collection of scholia' that 'may have accumulated over many years and be the work of several scholiasts' (p. 14*).

36 See *SE*, II.7, p. 124, v. 50–p. 125, v. 83; III.1, p. 142, vv. 11–62; IV.14, p. 288, v. 37–p. 289, v. 13; V.15, p. 380, v. 43–p. 381, v. 37.

Aristotle's own, and that Aristotle's defence of that view remains a rational one despite the fact that the view itself is *contra fidem*.

Let us begin by collecting the four loci of relevance. The Scholiast first attributes to Aristotle the view that one may licitly do evil that good may come as a correction to Aristotle's inclusion of adultery in a list of acts in *EN* II.7 that Albert calls *secundum se mala*. Intending to correct a potential misunderstanding of Aristotle's meaning, the Scholiast claims that for Aristotle adultery is not, as that text might lead one to conclude, always and everywhere base and blameworthy. On the contrary, the Scholiast's Aristotle teaches that adultery is sometimes permissible — that a person may licitly commit it, for example, if by doing so he may achieve something of 'great usefulness' (*magna utilitas*) for his city.[37] The second attribution is occasioned by Aristotle's remarks in *EN* III.1 concerning whether a person may blamelessly do evil in order to avoid greater evils or to pursue some good (*EN* 1105a3–6). For Aristotle, as the Scholiast explains, a person ought to lie and commit adultery 'at times' (*aliquando*) — for example, when by doing so he may free his fellow citizens from oppression.[38] The third attribution is occasioned by Aristotle's discussion of the moral virtue of truthfulness in *EN* IV.7. Albert's Scholiast there observes that, for Aristotle, it is permissible to lie to someone in certain situations — when, for example, one may through a lie avoid some 'great loss' (*magnum damnum*).[39] He makes his fourth and final attribution while expositing Aristotle's discussion of the virtue of decency or *epieikia* in *EN* V.10. It is not only licit, but an act of *epieikia* for a patriot to commit adultery with a tyrant's wife when by doing so he may acquire the access to the tyrant that he needs to assassinate him.[40]

There is little doubt that in Albert's eyes it is *contra fidem* to suggest that a person may ever licitly lie or commit adultery for the sake of some *magna utilitas*. The suggestion that a person may at times licitly lie shocks Albert's objector in *SE* III.1, who invokes the *sancti* in defence of the thesis that a person ought not

37 *SE*, II.7, p. 125, vv. 47–53. We speak here of the Aristotle of Albert's Scholiast, and not of the Scholiast's Aristotle per se.

38 *SE*, III.1, p. 142, vv. 27–31.

39 *SE*, IV.14, p. 288, vv. 43–46.

40 *SE*, V.15, p. 380, vv. 60–71. Albert later approves a fourth example of a decent act that he reads in Eustratius: It is decent for a leader (*dux*) to violate a law commanding the building of walls to protect a city if he perceives that the security afforded by walls softens the souls of its citizens and so makes them less able to defend it. See *SE*, VI.16, p. 488, vv. 74–83. Albert takes issue with one of the Commentator's three examples of 'decent acts', which is to say, acts that are contrary to civic laws but nevertheless just because they remedy inequities caused by the universality of those laws. According to him, it is decent to violate the law enjoining the return of borrowed items when the borrowed item is a sword and the lender demanding its return is in a murderous rage. Likewise, it is decent to violate a civic ordinance which forbids any climbing on the walls of the city if the climber does so as part of a plan to kill a tyrant who has captured the city. Albert finds no reason to doubt either of these examples. Only the third of the Commentator's three examples draws his criticism: his claim that it is an act of Aristotelian *epieikia* for a person to commit adultery with a tyrant's wife if this helps him to assassinate the tyrant.

to lie for any reason whatsoever, 'even if by doing so he may avoid death'.[41] In an argument *contra*, Albert presents the principle that a person may do evil that good may come as contrary to the teachings of St Paul in Romans and to St Augustine. He underscores the point by analogizing the principle to Augustine's doctrine that the person who sins venially in order to save another person from mortal sin acts neither nobly nor licitly but rather basely and illicitly.[42]

Our question is this: Does Albert's belief that it is *contra fidem* to do evil that good may come inform his understanding on whether or not Aristotle teaches that view in the *EN*? If we approach that question with Albert's criticisms of Averroes's contra-Christian interpretations in mind, we might suspect that it does — that the fact of its being *contra fidem* itself warrants, in Albert's mind, the expectation that this is not Aristotle's teaching in the *EN*. Indeed, his confidence in the validity of such reasoning could explain why Albert dismisses the Scholiast's first two attributions as he does: unconditionally and very briefly. Concerning the first attribution, Albert says that the view that a person may licitly commit adultery with the wife of a tyrant is not Aristotle's opinion, as the Scholiast claims, but that of unnamed others.[43] Concerning the second, Albert takes the same line: the view that a person may lie to liberate his city from an oppressor is neither true nor Aristotle's opinion.[44]

By contrast, in replying to the third attribution, Albert reads Aristotle to affirm in *EN* IV.7 that a person may licitly lie in certain circumstances — for example, when his doing so promotes the public good. More exactly, lying is licit when it helps 'society stand together' (*ad consistentiam civilitatis*) or 'avoid something by which the society is more jeopardized' (*ut evitetur aliquid per quod magis civilitas periclitetur*).[45] Albert's change of perspective here is surprising. The surprise is lessened somewhat if we recall that the fundamental question for Albert is what the letter of Aristotle's text affirms, not what it must be taken to affirm in order to cohere with prior determinations. It may well be the case that Aristotle denies in *EN* III.1 that a person may lie to liberate his city but affirms in *EN* IV.7 that a person may licitly lie *ad consistentiam civilitatis*. However that may be, we recognize that Albert attributes to Aristotle in *SE* IV.15 an opinion very similar to the one that he says is neither true nor Aristotle's own in *SE* III.1.[46]

---

41 *SE*, III.1, p. 142, vv. 30–31: 'quia SANCTI dicunt, quod pro nulla re est mentiendum, etiam pro vitanda morte'.

42 *SE*, V.15, p. 380, vv. 72–79: 'Sed contra hoc ultimum obicitur; APOSTOLUS enim dicit ROM. III (8), quod qui dicunt "faciamus mala, ut veniant bona, eorum damnatio iusta est"; ergo cum adulterium sit secundum se malum, pro nullo bono faciendum est. AUGUSTINUS etiam dicit, quod pro nullo bono faciendum est aliquod peccatum adeo, quod nec venialiter peccandum est, ut alius a peccato mortali liberetur'.

43 *SE*, II.7, p. 125, vv. 47–53.

44 *SE*, III.1, p. 142, vv. 59–61.

45 *SE*, IV.14, p. 288, vv. 64–85.

46 Although Albert does not seem concerned about tensions between his reading of *EN* IV.7 and *EN* III.1, he does seem concerned about tensions between his reading of *EN* IV.7 and *EN* VII.7.

More surprisingly still, Albert sketches a theory as to why it is the case that lying is at times licit philosophically speaking but never licit theologically speaking. As Albert expresses his theory here, it is in accordance with the perfection of 'civic' or earthly virtue at times to 'turn to a lie', be it of words or deeds — one may do so 'without detriment to virtue' (*sine detrimento virtutis*). By contrast, it is never in accordance with the perfection of theological virtue to lie 'in any way' (*nullo modo*).[47] Aristotle says what he does, Albert explains, because he is 'speaking philosophically'. Yet how can moral philosophy permit what moral theology forbids? Albert says that it can do so because the ultimate goal of human action, theologically speaking, is an infinite good, whereas the ultimate goal of human action philosophically speaking is a finite good. An infinite good is, as it were, so good that no person should ever turn from it to a lesser good 'out of consideration for any gain or loss' (*pro nullo damno vel lucro*).[48] Yet to lie is of necessity to turn from the infinite good to a lesser good, and so to lose the infinite good. To lose an infinite good is always to come out with less, since no matter how great the finite good to be gained by lying or to be lost by not lying, its gain cannot be more profitable than that of possessing an infinite good, nor its loss costlier than that of forfeiting the infinite good. Matters are different when the ultimate goal is a finite good — even a finite good as noble and divine as that of the common good. For to lie is not of necessity to turn from or lose that good; indeed, lying in certain circumstances may promote its attainment. Lying is licit when what is gained for the community outweighs (*praeponderare*) what is lost for it by doing so.[49]

Evidently, an action is good philosophically speaking because it promotes the common good, which is the ultimate end of human action philosophically speaking. By the same token, an action is good theologically speaking because it promotes the attainment of the divine good, which is the ultimate good of human action theologically speaking. It is because lying can promote the common good but can never promote the divine good that lying is sometimes good philosophically speaking but never so theologically speaking.

If lying can sometimes promote the common good for Albert's Aristotle, adultery never can. That is one of the reasons why Albert argues in *SE* V.15 that

---

Specifically, he seems to worry that the argument he makes on behalf of the licitness of lying for Aristotle in *EN* IV.7 will undermine his suggestion that lying counts among acts that Aristotle regards as wrong per se in *EN* VII.7. Albert tries to soften this tension by explaining that for Aristotle in *EN* IV.7, lying remains wrong per se even though it may sometimes be licitly performed. This is possible, he explains, because of the relevant sense of per se; lying is wrong *per se* or 'in itself' in *EN* IV.7 in the sense that is wrong when done for its own sake, which is to say, when not 'adjoined to something for the sake of which truth is sometimes to be set aside, civically speaking' ('sed "per se" dicitur malum, idest non adiuncto alio pro quo quandoque dimittenda sit civiliter veritas') (*SE*, IV.14, p. 288, vv. 82–84).

47 *SE*, IV.14, p. 288, vv. 64–85.

48 *SE*, IV.14, p. 288, vv. 69–74.

49 *SE*, IV.14, p. 288, vv. 74–78.

the Byzantine commentator Michael of Ephesus misinterprets Aristotle when he ascribes to him the view that it would be an act of the virtue of decency (*epieikia*) for a man to commit adultery with a tyrant's wife if by doing so he could acquire the intimacy necessary to assassinate him.[50] Albert notes here that one could with justification say that Michael speaks falsely in attributing that view to Aristotle. However, as we will see below, having noted that possibility, Albert retreats from it, proceeding to consider what 'can be said' (*potest dici*) if one wishes to sustain Michael's reading.[51]

Why is it possible for Albert's Aristotle to lie for the sake of the common good, but not to commit adultery for it? Albert suggests an explanation in *SE* II.7, when he endeavours to explain why, in Aristotle's eyes, it is not always evil to kill a human being, but is always evil to commit adultery. Killing a human being is licit when it promotes the common good because the benefit (*utilitas*) it brings belongs to or falls within the act of killing itself; there is no way to separate the good that comes from killing that human being from the killing of him. Adultery is never licit because the benefit it brings for the community lies outside (*extra*) the act itself.[52]

The metaphysics of action presupposed in this reply raises more questions than one can responsibly resolve on the basis of Albert's brief remarks. Albert refuses to admit that adultery can be adjoined to a good end in the way lying can. Nevertheless, even after Albert has argued in support of that view, he again proceeds to consider what can be said if one wishes to sustain Michael's reading of Aristotle. It is not clear whether the remarks that follow have the status of a thought experiment or of an alternative interpretation with as strong a claim to Aristotelicity as the first. I favour the latter construal, because the passage seems to me to reinforce the account that we saw Albert develop in *SE* IV.14 of the difference between the source of the goodness of good acts in philosophical discourse about human action and the source of their goodness in theological discourse.

The justification for adultery that Albert considers here begins by arguing that the sexual acts the assassin performs with the tyrant's wife do not necessarily constitute acts of adultery in Aristotle's eyes.[53] The way that Albert argues on behalf of this hypothetical redescription of the assassin's act is not important

50 For discussion of Michael of Ephesus, his eleventh-century Greek commentary on *EN* V, and its Latin reception, see Mercken, 'Introduction', pp. 22*–28*.

51 *SE*, V.15, p. 381, vv. 24–36.

52 *SE*, II.7, p. 125, vv. 47–53.

53 Albert uses a similar argument in *SE*, II.7, p. 125, vv. 32–46. In *SE*, V.15, he wishes to show that Aristotle can consistently affirm that it is always blameworthy *and* that the assassin's sexual acts with the tyrant's wife are praiseworthy. By contrast, in *SE*, II.7, he wishes to show that Aristotle can consistently affirm that while homicide is always blameworthy, the execution of a thief is praiseworthy. Aristotle may consistently regard the execution of a thief (*fur*) as an act of virtue and insist that every taking of a human life is vicious because the execution of a criminal is an act not of homi-cide but of furi-cide.

here. What is important is Albert's insistence, again, that what makes an action good, philosophically speaking, is its being ordered to or promoting the common good. The assassin's non-adulterous sleeping with a tyrant's wife is permissible philosophically speaking if the benefits it brings to the community outweigh the harms. It can bring such benefits as long as it is not tied by its very nature to an evil end (as a proper act of adultery would be, but the assassin's redefined action is not). The vital point here is that once again, albeit hypothetically, Albert suggests there is a fundamental difference between what makes an action good philosophically speaking and what makes it good theologically speaking. A certain calculus concerning the attainment of the common good can evidently justify actions in philosophy which, considered theologically, are never justified. When the ultimate end is an infinite one, as is the end of human action theologically speaking ('having God and eternal life'), neither lying nor adultery is ever justified.[54] When it is finite, lying may well be.

Albert's criticisms of Averroes in *SE* I, the Ancient Scholiast in *SE* II–V, and Michael of Ephesus in *SE* V seem at first glance to reinforce the notion that when Aristotle's teachings are true, they cohere with Christian faith and, indeed, they must so cohere. Yet on further inspection, as we have seen, Albert's criticisms actually undermine that notion by suggesting in different ways that contradicting faith is not a good indicator of the truth or falsity of Aristotle's teaching.

The shift from reinforcing to undermining is less pronounced in Albert's discussion of the *auctoritates* from Averroes in *SE* I than it is in his discussion of the Ancient Scholiast. With respect to Averroes, we see it when Albert ultimately refuses to argue directly from the fact that Averroes interprets Aristotle as teaching heresy (in his *Metaphysics* and *De anima*) to conclusions that Averroes misinterprets Aristotle. If Albert seems initially to entertain the legitimacy of such an inference, he clearly does not rely upon it to make his case that Averroes misreads Aristotle. Instead, as we saw above, he argues that Averroes's reading is mistaken because it presupposes an account of the individuation of human souls that is not Aristotle's own.

The shift as regards the Scholiast is much less subtle. If Albert seems inclined to believe that the Scholiast's first two attributions to Aristotle of a thesis that is *contra fidem* are probable misinterpretations *because* that thesis is *contra fidem*, he later himself asserts that what Aristotle teaches is indeed *contra fidem*, before proceeding to explain why it is perfectly rational for Aristotle to teach what he does. Convinced that *EN* IV.7 actually does teach that it is right to lie at times, Albert makes no effort, *pace* Gauthier, to hide or obscure the opposition of that teaching to the faith's teaching that one may under no circumstances licitly lie. On the contrary, he calls attention to the difference and tries to explain it. In a similar way, Albert makes no effort to explain away or disguise the reading of

54 Cf. Albert's two proposals in *SE*, VI.14, p. 482, vv. 37–51, following Eustratius, to explain where the falsity lies in the practical syllogism: 'libertas est expetenda; sed per adulterium possum acquirere libertatem' (ibid., p. 481, vv. 80–81).

Aristotle, defended by Michael of Ephesus, whereby an adulterous act, performed under the right circumstances, would be an act of virtue. Although Albert regards that reading to be both *contra fidem* and contrary to Aristotle's meaning, he nevertheless explains at some length why, philosophically speaking, the doctrine may be correct.

Albert's criticisms of Averroes, the Ancient Scholiast, and Michael of Ephesus offer what, I believe, are the best evidence in the *SE* for the (erroneous) view that 'coherence with faith' is a criterion Albert recommends for determining whether Aristotle's utterances are true. Yet as we have seen, even these texts, with their explicit references to reason, faith, and heresy, do not sustain that notion. My own strongest evidence for believing that 'coherence with faith' is not a criterion that Albert recommends is yet to come. So far, my argument has been largely negative and indirect. I have tried to show that Albert's criticisms of three 'paganizing' interpreters of Aristotle's metaphysical and ethical teaching do not support the logic underlying Gauthier's Christianization hypothesis — in particular, its presupposition that Albert will never foreground places where Aristotle's teaching contradicts faith, much less explain why it is possible to deny what faith affirms and nevertheless to reason rightly.

A negative argument concerning loci in Albert's text that fail to support Gauthier's logic does not suffice, however; I wish to show two features of Albert's text that actively undermine it. Albert's discussion of the Scholiast's interpretation in *SE* IV.14 and of Michael of Ephesus's interpretation in *SE* V.15 has anticipated both of these features: first, Albert's presentation of places where Aristotle's teaching contradicts that of faith; and second, his attendant explanations of how Aristotle speaks rationally even when he speaks *contra fidem*.

We will consider three such contradictions. Although Albert notes many particular differences between the two ways of speaking about human action, he regards only a select number of those differences as contradictions. He endeavours to explain and justify the possibility of a smaller number still of these contradictions. The four texts I discuss in what follows are thus precious ones. The sequence of my consideration is intended to reflect a gradation in the degree or extent of difference between the philosophical way of speaking about morality and the theological way: The first contradiction concerns the account each discourse gives of a particular virtue, bravery. The second concerns their respective lists of particular virtues and, specifically, the place of shame in each. The third concerns the standards relative to which each judges actions just or unjust. The fourth concerns both the specific determination each reaches about the morality of a particular practice, usury, and the manner in which each reaches that determination.

## Rational and *contra fidem*

### *Fear of Death, and the Ultimate End*

In *SE* III.1, Albert examines the truthfulness of a dictum at *EN* 1117b10–11: namely, that a 'brave man' (*fortis*) is more saddened by death than is a 'non-brave man' (*non fortis*), be he cowardly or rash. One argument the objector gives for doubting this claim is that the seven martyrs of Maccabees were brave men who seemed less saddened by 'crucifixion and death' than cowardly or rash men would have been.[55] Albert sees a genuine contradiction here between theological and philosophical ways of speaking about human action. Theologically speaking, virtuous human beings are indeed less saddened by death than vicious ones, whereas philosophically speaking, virtuous human beings are not less saddened by death than vicious ones but on the contrary are more saddened by it.

Albert explains the doctrinal difference on this matter by appeal to the different ultimate ends of human action within the two discourses. For the *theologus*, that end is the contemplation of God in heaven through a virtue that is not acquired but infused. Accordingly, when the *theologus* considers the feelings that virtuous Christians may have when facing death, he knows that the sadness or pain they may feel is attenuated by their belief that dying well will not thwart but rather will promote the ultimate end of their actions. For the *ethicus*, the ultimate end is the contemplation of God on earth through a virtue that is not infused but acquired. When the *ethicus* considers the feelings that brave men may have when facing death, he does so in the knowledge that death, as Aristotle says, is a boundary whose crossing terminates all further opportunities to act virtuously. As such, it is something that will rightly sadden the *fortis*, whose ultimate goal in life, like that of any virtuous man, is to act in accordance with perfect virtue (that is, to contemplate philosophically), and who orders all his actions in life so as to acquire the dispositions and resources that can enable him to do so. Aristotle's *fortis* will thus be more saddened about crossing this boundary than the cowardly or rash man will be, because these men have no such goal in life, and for this reason do not rightly understand the price of crossing the boundary.[56]

In advancing his solution, Albert not only protects the rationality of Aristotle's *contra fidem* utterance at *EN* 1117b10–11, but also reminds his auditors that Aristotle's discussion of the brave man's attitude toward death coheres with discussions by other moral philosophers who have spoken truthfully about it, even if what they say seems at first glance quite contrary. Albert makes that point in reply to an objection that several important moral philosophers taught that virtuous men do not fear death but instead despise it: 'Seneca and Cicero and other moral philosophers teach that death is to be despised' (SENECA *et* TULLIUS

---

55 *SE*, III.11, p. 196, vv. 5–6 and vv. 23–25. Cf. II Maccabees 7.

56 *SE*, III.11, p. 196, vv. 72–93.

*at alii morales philosophi docent vilipendere mortem*). While it may seem that those philosophers contradict Aristotle, Albert explains that their view is not that death is 'to be held as a matter of little concern' (*parvipendae*), but rather that it is to be feared less than virtuous action is to be loved.[57] He suggests that this is a point they share with Aristotle, and one that would not prevent them from affirming both Aristotle's thesis about the brave man's sadness before death and his teaching that, despite this sadness, the *fortis* will nevertheless choose to suffer death in some circumstances. In seeking to explain why it is that Aristotle's *fortis* will choose death in some circumstances, Albert notes in passing that it is easier to explain why pagans who believe in some reward after death will choose to die in those circumstances than it is to explain why a rationalist such as Aristotle would, who recognizes that reason can know nothing about the state of the soul after death. Nevertheless, the fact that some pagans die for God is a sign not that unaided human reason can indeed know of such rewards, but rather that philosophers who speak in this way either are not speaking rationally or have been influenced by religious traditions.[58] The brave man's sadness at death thus becomes an indication of the purity of Aristotle's 'rationalism' and its harmony with veridical philosophical authorities.

In recognizing that Aristotle's teaching on fear of death contradicts that of faith while trying to show the rationality of Aristotle's teaching, Albert clearly refuses to Christianize Aristotle in Gauthier's sense. He does not, for example, seek a way to deny that Aristotle's *fortis* fears death more than his non-*fortis*. Instead, he calls attention to the difference between theological and philosophical discourses on this matter, and presents it as a necessary consequence of a difference concerning the ultimate end of human action. If the contradiction here seems petty or arcane, it should not be forgotten that one of the 219 *articuli* condemned at Paris in 1277 concerns the fear of death (*articulus* 213/178: *Quod finis terribilium est mors*).[59] Thus, even if the point at issue here is less fundamental than some others, it is still not one that a reader aiming to persuade Latin readers of the Christianity of Aristotle's moral teaching would necessarily care to admit. That being said, Albert does concede more fundamental differences between the two discourses, to one of which we now turn.

### Shame and the Uprooting of Vice

Philosophers 'speaking ethically' about human action contradict theologians 'speaking theologically' in diverse ways. One way, we learn in *SE* IV.16, concerns the list of human excellences or virtues. Whereas philosophers say that shame (*verecundia*) is not a virtue, theologians say that it is, at least in a broad sense

57 *SE*, III.11, p. 196, vv. 21–22 and vv. 68–71.

58 For a similar attestation to the greater purity of Aristotle's rationalism, see *SE*, I.13, p. 71, vv. 86–90.

59 Cf. Hissette, *Enquête sur les 219 articles*, pp. 304–07.

(*largo modo*).[60] Albert's analysis of a general difference between philosophers and theologians vis-à-vis a particular Nicomachean dictum reminds us that when he refers to 'speaking philosophically' about human action, he has in mind speaking about it as Aristotle does in the *EN*. His Aristotle merits the honorific *ethicus* in addition to *philosophus* on account of the rigorously reasoned thinking about human action in that text. The *EN* becomes, for Albert, the central point of reference for what it means to 'speak ethically' or 'speak philosophically' about human action. By the same token, the Lombard's *Sentences* are the central point of reference for what it means to 'speak theologically' about it. Having completed his *Commentary on the Sentences* within two or three years of his Cologne lectures on the *EN*, Albert was aware of the ambiguities and tendentiousness of the *Sentences*: still, he does not associate what the faith says about human action with what the theologian (*theologus*) says about it nearly as closely as he associates what reason says about human action with what the *ethicus* says about it in the *EN*. Nevertheless, the *Sentences* provides Albert with some orientation as to what faith says about certain technical matters that he otherwise could not draw upon when engaging with Aristotle's many particular dicta in the *EN*.

One such matter, not nearly the most technical, is the list of human excellences.

The Lombard was not the only *auctor* upon whom Albert could draw in evaluating Aristotle's arguments in *EN* IV.9 that shame (*verecundia*) is not a virtue. Albert takes St Gregory the Great to affirm that it is, as surely as the Lombard does. Indeed, Gregory goes so far as to call fear (*timor*) one of the seven gifts of the Holy Spirit — a fact that Albert suggests should be recognized as a very strong affirmation that shame is a virtue, since Gregory calls 'fear' (*timor*) what Aristotle calls 'shame', and gifts are just especially efficacious virtues. The adequacy of the complex translations among authoritative vocabularies that Albert makes in this *quaestio* — among the *Moralia*, the *Sentences*, and a *Gloss* on Matthew (which associates *verecundia/timor* with *paenitentia*) — are not of interest at this point. What we wish to understand is this: How does Albert account for the fact that what Aristotle calls 'shame' and what Christian authorities call 'fear' and 'repentance' is not a virtue *ethice loquendo*, but is a virtue *theologice loquendo* (or at any rate, is a virtue *largo modo*)?

The reason Albert gives is that theological discourse *de moribus* considers grace, whereas philosophical discourse does not, and this difference entails a different explanation of the 'rooting out' (*deletio*) of what the *ethicus* calls 'vice' and the theologian calls 'sin'. Limited to a knowledge of those natural causes that can be known by unaided human reason, the *ethicus* knows only that human beings root out a vice by implanting its opposed virtue. By contrast, the theologian knows that God's grace roots out vice, and that it does so by perfecting the virtuous action that roots out sin or vice. Virtuous action alone is not sufficient to

60 *SE*, IV.16, p. 299, v. 71–p. 300, v. 64.

uproot sin in the perfect way that God's grace does. It must be accompanied by shame (or fear), satisfaction, and repentance (or sadness, *dolor*). Shame makes the sinner withdraw from the evil to which sin has disposed him. Satisfaction takes away the eternal punishment that his sinful acts merit. Repentance stops him from finding pleasure in the lesser good that he sought in his sin. The theologian speaks of shame, satisfaction, and repentance as virtues 'in a broad sense' because he distinguishes the role of virtuous actions in uprooting sin/vice from the role of the three semi-virtues that share in that *deletio*, making it more thorough and complete than it would be without them. On this account, shame, satisfaction, and repentance are not excellences that human beings acquire through their natural activities, but gifts infused by grace.[61]

If Albert's argument here raises many questions, its central argumentative end is clear: Aristotle reasons rightly when he denies that shame is a virtue, since his account of virtue considers only its natural causes to be knowable by unaided human reason, and shame is, even theologically speaking, only a semi-virtue and one whose cause is supernatural and therefore unknown to natural human reason. The difference in their ways of speaking here impugns neither the rationality of Aristotle's determinations nor the authority of Gregory, the Lombard, or the anonymous Glossator. Albert's explanation of the causes of this doctrinal difference provides a fuller sense of the discursive difference beginning to take shape: not only do the ends of the two discourses differ, but also the causes or principles by which those ends are attained.

### *Justice and Blameworthiness*

Albert offers additional information about the two *modi loquendi* in *SE* V.14. At issue is the truthfulness of Aristotle's suggestion in *EN* 1136b30–33 that a servant who performs an act of injustice at his lord's bidding does not necessarily act unjustly. Albert understands Aristotle to argue that a person who acts unjustly in this way does so in a fundamentally different sense than a person who performs an act of injustice from choice after deliberation (and hence in a way that expresses a settled disposition to act unjustly). Albert wonders about the truthfulness of Aristotle's formulation, asking: Does a servant who performs an act of injustice at the command of his lord act unjustly?

His solution offers both a civic or philosophical answer and a theological answer to the question. Philosophers, unsurprisingly, answer the question as Aristotle does: the servant who performs such an act is not necessarily unjust. In explaining why Aristotle speaks that way, Albert reminds his auditor that lords have the right to beat servants who disobey their orders. Civil law gives a lord 'full power' (*plenam potestatem*) over his servant, including the power to 'subdue'

61 *SE*, IV.16, p. 300, vv. 37–55.

(*opprimere*) the servant should he resist his commands.[62] Accordingly, if a servant performs the act of injustice for fear of lashes, he is not counted as unjust, civically or philosophically speaking. The reason is that such a servant acts from a feeling of fear rather than a habit of injustice, and actions performed on the basis of feelings — especially a feeling like the one in question, which Albert suggests would befall any reasonable person, and so even a brave 'man standing firm' (*constantem virum*) — do not determine whether a person is unjust. For to be unjust is to possess a habit of injustice, and habits are acquired by actions performed by choice, from deliberation, and repeatedly. A philosopher reasoning rightly about this matter will thus recognize that a servant who performs an act of injustice is not necessarily unjust, and so answer this question as Aristotle himself does in *EN* 1136b30–33.

To a theologian, matters appear quite different. For both the earthly lord and his servant are servants of the Lord God, and acts of injustice, whatever their causes, are acts against God.[63] Because a servant may never blamelessly do anything at the command of his human lord that is against the Lord God, he may never blamelessly perform any act of injustice that his earthly lord commands. On the theoretical level, this problem ought not arise for the *theologus* speaking theologically about injustice in the same way that it arises for the *ethicus* speaking ethically, since divine law forbids human masters to 'subdue' their servants. A Christian servant should never find himself moved to perform an act of injustice for fear of lashes, if his earthly master obeys Paul's teaching in Romans 6:20.[64]

The difference between the way that philosophers and theologians speak about this matter suggests that theological morality demands more from human beings than philosophical morality does. Philosophically speaking, a servant who acts for fear of lashes does not choose the action or perform it deliberately. The slave's action in such circumstances is neither morally good nor morally bad, but rather morally indifferent. Theologically speaking, if the action that the slave performs is *contra Deum*, then irrespective of whether he acts from fear of lashes, his action is most certainly morally bad.[65]

### *Usury and permissiones*

Albert articulates a very direct practical difference between theological and philosophical ways of speaking *de moribus* in *SE* V.16. He finds occasion to do so in a remark that his Latin Aristotle makes in *EN* 1138a6–7: what law commands

62 *SE*, V.14, p. 375, vv. 30–44.

63 *SE*, V.14, p. 375, v. 9.

64 *SE*, V.14, p. 375, vv. 30–37.

65 *SE*, VIII.4, p. 606, vv. 29–35: 'Ad tertium dicendum, quod in civilibus nihil prohibet aliquem esse nec bonum nec malum, qui scilicet nec habet habitum virtutis nec corruptionem civilitatis, sicut sunt rustici, ut dicit COMMENTATOR in X METAPHYSICAE, sed secundum THEOLOGUM nihil est indifferens, quia etiam vanum et otiosum computatur in pravum'.

(*iubet*) a citizen not to do, law forbids (*prohibet*) him to do. The remark provides a premise that is needed to complete the argument of the lines preceding it. It makes clear why it follows that *if* what is generally unjust is what the law *forbids* a citizen to do, and the law *commands* a citizen *not* to kill himself, then for a citizen to kill himself is generally unjust. To see how this follows, one needs to recognize a certain equivalence, namely, that 'what the law commands a citizen not to do' is 'what the law forbids him to do'. Albert's *quaestio* examines the universality of that equivalence. He asks: Does law forbid a citizen to do everything that it commands him not to do?

It seems that law does not forbid (*prohibeat*) a citizen to do everything that it commands (*iubet, praecipit*) him not to do. An objector needs only a single counterexample in order to topple Aristotle's dictum that the law forbids a citizen to do everything that it commands him not to do. Albert's example is usury. Usury, the objector notes, is not forbidden in civic law (as codified, at any rate, by Justinian); on the contrary, it is permitted. Yet it is a practice that civic law commands citizens not to do. (We return below to the problem of where and how it commands this). The implication is clear, even if hard to follow through all the negations: If usury is a practice that civic law commands (*praecipit*) citizens not to do but does not forbid (*prohibet*) them to do, then Aristotle's remark is false, because not as universally valid as he represents it.[66]

The way Albert untangles this knot is itself not of interest here; suffice it to say that he finds a way of saving Aristotle's utterance. Let us attend instead to the speculation about philosophical and theological discourses *de moribus* that the example of usury occasions in the *quaestio*'s replies to objections. For here the discursive modes part ways to such an extent that the *philosophus* permits a practice that the *theologus* forbids. The civil law permitting usury in Justinian's Code conforms with right natural reasoning about action, despite the fact that divine law forbids usury. How can this be? How can reason permit what faith forbids?

According to Albert, the *ethicus* and the *theologus* must reach different conclusions about the licitness of usury, because each reaches his conclusion based on his respective conception of the ultimate end of deliberately chosen actions. This explanation is familiar, although Albert here expresses somewhat differently what these ends are. To begin with, the moral scientist conceives the ultimate end as a good attainable on earth — a *bonum civile*. Laws that express scientific knowledge concerning action — what is to be done, what is to be avoided, and what is to be permitted — are developed with the aim of facilitating the attainment of the best human good for a great number (*multitudo*) of one's fellow citizens. By contrast, the *theologus* conceives the ultimate end as a good attained in heaven. Divine laws express God's knowledge concerning action; they show what human beings

66 *SE*, V.16, p. 384, vv. 29–32: 'Praeterea, quaedam sunt permissa in legibus civilibus, sicut usuram accipere, quae tamen non sunt praecepta, quia sunt praeter intentionem legislatoris; sed nulla permissa sunt prohibita'.

should do and avoid as God's plan for Creation is realized. According to that plan, every individual human being will achieve the degree of perfection that is possible for him to achieve.[67]

The crucial difference between these two ultimate ends is not that one is heavenly and the other earthly. Nor is it that one is 'local' and the other 'universal'. It is rather the way that the *ethicus* and *theologus* believe the ultimate end may be realized. For the *theologus*, God's plan for humankind is realized in a way that neither permits nor requires any bad action on the part of individuals. This is so because God's law is able to promote the good of the whole human race — indeed, to bring every individual to the state of perfection he was created to possess, without permitting any individual human being to act badly. This is the reason why divine law, in contrast to civil law, does not contain any 'allowances' (*permissiones*). The philosopher's ultimate end, because sought on earth and by much less efficacious means, is realized differently. In some situations, promoting this ultimate end, which is a good for the vast majority (*multitudo*), requires permitting certain individuals to act in bad ways.[68] Their bad actions, if not directly intended by the civic legislator, are foreseen and permitted *pro bono publico*. Usury is just such a bad way of acting, because although individual human beings who perform it act badly, 'many useful results' (*multas utilitates*) follow from it for a city, even though its performance by individuals is not for this or any reason altruistic or noble.[69]

Philosophically speaking, as Aristotle himself shows, usury remains a vicious practice that is opposed to the virtue of liberality. However, Albert takes Aristotle's larger moral theory to show why legislators would nevertheless be faithful to Aristotle's thinking if they permit an action that Aristotle himself regards as bad for individuals — in a sense, if they allow evil (individual evil actions) that good (the common good) may come. Albert's reasoning is clear, even if certain of its premises seem neither true nor transparently Nicomachean: good legislation seeks to promote the common good, and usury, although bad for individuals, can, on account of the 'crookedness of matter' (*obliquitas materiae*), be good for the multitude.[70] Although civic legislators permit usury, they do not and cannot command any individuals to practise it: on the contrary, they tacitly command all citizens not to practise it, by crafting laws whose manifest aim is to make all citizens good, thereby making it known that practising usury does not make a person good, even though it is permitted.

If Albert does not answer all of the questions that his remarks raise here, his main argumentative aims are once more clear. He wishes, first, to defend the rationality of the Nicomachean dictum that gives rise to the *quaestio*. At the same time, he evidently wishes to argue that it is rational for civil law to permit usury,

67 *SE*, V.16, p. 384, vv. 70–88.
68 *SE*, V.16, p. 384, vv. 70–72.
69 *SE*, V.16, p. 384, vv. 70–88.
70 *SE*, V.16, p. 384, vv. 70–75.

and hence that laws faithful to Aristotle's moral thinking would permit it. One challenge that attends this latter effort is how to explain why Aristotle himself describes usury as a vicious practice (and hence presumably as a practice that human beings should avoid), even though Aristotle's larger reasoning about the way the best earthly good is to be attained, as Albert interprets it, permits the practice.

## Conclusion

There are passages beyond the four we have examined in the *SE* in which Albert distinguishes the ways that faith speaks about human action from the way that Aristotle speaks about it.[71] The passages we have studied suffice, however, to illustrate both the fact and the extent of the differences between the two discourses for Albert. They show that if Albert believes there is a harmony between Christian and Nicomachean discourses *de moribus*, their harmony is not such that the two never contradict each other. He argues that Aristotle can indeed reason rightly and nevertheless deny what faith affirms or affirm what faith denies. Aristotle can do so because his ideas about human action turn on a certain conception of the ultimate end and the means by which that end is achieved. The relative imperfection of that end and of those means explains why the two discourses speak as differently about human action as they do — for example, why theology forbids without exception actions that philosophy permits under certain well-reasoned circumstances.

It is worth considering what Albert's discussion of such divergences discloses regarding his understanding of truth. When he maintains above, for example, that theology judges a particular lie to be morally wrong that philosophy judges to be morally right, what is he affirming about the morality of that lie? Is it neither morally right nor morally wrong? On the contrary, the evidence presented above shows that on Albert's view the lie is both morally right and morally wrong.[72] Evidence of that kind helps plausibilize a striking thesis that in recent years has won broad acceptance: namely, that Albert's account of the relation of philosophy and theology was a source of inspiration for the doctrine of the twofold truth (*duplex veritas*) associated with the 'Latin Averroists' of the thirteenth century.[73]

---

71 For example, *SE*, I.10, pp. 55–56; IV.16, pp. 299–300; X.16, pp. 774–75.

72 To some readers, it may seem unproblematic to affirm that a proposition is 'true philosophically' insofar as it coheres with the principles and goals of moral-philosophical discourse, and that it is 'true theologically' insofar as it coheres with principles and goals of moral-theological discourse. Nevertheless, those affirmations are ones that seem to have offended gravely Stephen Tempier, Bishop of Paris and author of the Condemnation of 1277. For detailed discussion of the meaning and impact of Tempier's famous denunciation of the doctrine of 'double truth' (*duplex veritas*) in his prologue to the Condemnation, see Bianchi, *Pour une histoire*, pp. 19–22.

73 De Libera, *Albert le Grand et la philosophie*, pp. 21–28.

A full examination of Albert's account of truth and its relation to those of his thirteenth-century intellectual successors is beyond the scope of this study.[74]

Albert's account in the *SE* of the relation between philosophical and theological ethics explains why on his view it is invalid for Christians to reason from their religious beliefs to conclusions about what Aristotle is likely to have taught. A thinker like Albert who does not believe in the validity of such inferences could not, I submit, in good conscience intentionally Christianize Aristotle's *EN*. However that may be, it is now clear that if Albert Christianizes the *EN* on some points, he also paganizes it — that is to say, he not only shows ways that Aristotle's teaching agrees with the faith, but also shows ways that it disagrees with it. For Albert, the theological and philosophical modes of speaking are properly neither complementary nor contradictory but rather incommensurable.[75] They are so because they proceed on the basis of different principles, which is to say, in respect of different causes. Moral-philosophical discourse diverges from moral-theological discourse because of the difference in their final and efficient causes — in their ultimate ends and the means by which they are realized.

74 Albert's way of accounting in the *SE* for differences between philosophical and theological accounts of moral subjects seems to have served as 'un modello metodologico' for late thirteenth-century Arts commentators on Aristotle's *EN*. See Bianchi, *Il vescovo e i filosofi*, p. 161. For an excellent recent discussion of those Arts commentaries on the *EN*, see Costa, 'L'*Éthique à Nicomaque*'.

75 Such incommensurability is one of the key findings of Alain de Libera's seminal work, *Albert le Grand et la philosophie* — a work whose stated purpose is to 'revise the conventions for reading Albert' so as to uncover the German Dominican's 'true intellectual visage' beneath layers of misreading. A key way that de Libera seeks to do that is by disclosing, among other things, the similarity of Albert's account of the relation between philosophy and theology to those of the Latin Averroists, and the difference of Albert's account from that of Thomas Aquinas. See de Libera, *Albert le Grand*, pp. 10 and 37–43. See also de Libera, 'Philosophie et théologie'. For some critical engagement with de Libera's incommensurability thesis, see Tracey, 'Albert's Reading of Aristotle's Moral-Philosophical Treatises', pp. 313–16.

## Works Cited

### Primary Sources


Albertus Magnus, *Super Ethica commentum et quaestiones*, ed. by Wilhelm Kübel, Editio Coloniensis, 14/1–2 (Münster: Aschendorff, 1968–87)

Averroes (Ibn Rushd) of Cordoba, *Long Commentary on the 'De Anima' of Aristotle*, trans. with an introduction and notes by Richard C. Taylor, with Thérèse-Anne Druart (New Haven, CT: Yale University Press, 2009)


### Secondary Works


Albertus-Magnus-Institut (ed.), *Albertus Magnus und sein System der Wissenschaften: Schlüsseltexte in Übersetzung, Lateinisch-Deutsch* (Münster: Aschendorff, 2011)

Bianchi, Luca, *Il vescovo e i filosofi: La condanna parigina del 1277 e l'evoluzione dell'aristotelismo scolastico* (Bergamo: Pierluigi Lubrina Editore, 1990)

———, *Pour une histoire de la 'double vérité'* (Paris: Vrin, 2008)

Black, Antony, *The History of Islamic Political Thought: From the Prophet to the Present* (Edinburgh: Edinburgh University Press, 2011)

Buffon, Valeria A., 'The Structure of the Soul, Intellectual Virtues, and the Ethical Ideal of Masters of Arts in Early Commentaries on Aristotle's *Nicomachean Ethics*', in *Virtue Ethics in the Middle Ages: Commentaries on Aristotle's 'Nicomachean Ethics', 1200–1500*, ed. by István Bejczy (Leiden: Brill, 2008), pp. 11–30

Celano, Anthony J., 'Robert Kilwardby on the Relation of Virtue to Happiness', *Medieval Philosophy and Theology*, 8 (1999), 149–62

Costa, Iacopo, '*L'Éthique à Nicomaque* dans la Faculté des Arts de Paris avant et après 1277', *Archives d'histoire doctrinale et littéraire du Moyen Âge*, 79 (2012), 71–114

Cunningham, Stanley B., *Reclaiming Moral Agency: The Moral Philosophy of Albert the Great* (Washington, DC: The Catholic University of America Press, 2008)

de Libera, Alain, *Albert le Grand et la philosophie* (Paris: Vrin, 1990)

———, 'Philosophie et théologie chez Albert le Grand et dans l'école dominicaine allemande', in *Die Kölner Universität im Mittelalter*, ed. by Albert Zimmermann (Berlin: De Gruyter, 1989), pp. 49–67

———, *Raison et foi: Archéologie d'une crise d'Albert le Grand à Jean-Paul II* (Paris: Éditions du Seuil, 2003)

Dunbabin, Jean, 'The Two Commentaries of Albertus Magnus on the Nicomachean Ethics', *Recherches de théologie ancienne et médiévale*, 30 (1963), 232–50

Gauthier, René-A., *L'Éthique à Nicomaque: Introduction, traduction, commentaire*, ed. by René-A. Gauthier and Jean Y. Jolif, 4 vols (Louvain: Édition Béatrice Nauwelaerts, 1970)

———, 'Trois commentaires "averroïstes" sur l'Éthique à Nicomaque', *Archives d'histoire doctrinale et littéraire du Moyen Âge*, 16 (1947–48), 188–336

Hissette, Roland, *Enquête sur les 219 articles condamnés à Paris le 7 mars 1277* (Louvain: Publications Universitaires, 1977)

Krause, Katja, 'Transforming Aristotelian Philosophy: Alexander of Aphrodisias in Aquinas' Early Anthropology and Eschatology', *Przegląd Tomistyczny*, 21 (2015), 175–217

———, and Henryk Anzulewicz, 'From Content to Method: The "Liber de causis" in Albert the Great', in *Reading Proclus and the 'Book of Causes'*, vol. 1: *Western Scholarly Networks and Debates*, ed. by Dragos Calma (Leiden: Brill, 2019), pp. 180–208

Luscombe, David, 'Ethics in the Thirteenth Century', in *Albertus Magnus und die Anfänge der Aristoteles-Rezeption im lateinischen Mittelalter*, ed. by Ludger Honnefelder, Rega Wood, Mechthild Dreyer, and Marc-Aeilko Aris (Münster: Aschendorff, 2005), pp. 657–84

McInerny, Ralph, 'Aristotle and Thomas: Père Gauthier', in *Aquinas on Human Action: A Theory of Practice* (Washington, DC: The Catholic University of America Press, 1992), pp. 161–77

Mercken, H. Paul F., 'Introduction', in *The Greek Commentaries on the Nicomachean Ethics of Aristotle in the Latin Translation of R. Grosseteste* (Turnhout: Brepols, 1973), vol. 1, pp. 1*–66*

Mulchahey, Marian Michèle, *'First the Bow is Bent in Study…': Dominican Education before 1350* (Toronto: The Pontifical Institute of Mediaeval Studies, 1998)

Müller, Jörn, *Natürliche Moral und philosophische Ethik bei Albertus Magnus* (Münster: Aschendorff, 2001)

Perkhams, Matthias, 'Einleitung', in *Thomas von Aquin: Sententia Libri Ethicorum I et X – Kommentar zur Nikomachischen Ethik, Buch I und X. Lateinisch / Deutsch*, ed. by Matthias Perkhams (Freiburg im Breisgau: Herder, 2014), pp. 11–58

Sturlese, Loris, *Die deutsche Philosophie im Mittelalter: Von Bonifatius bis zu Albert dem Großen, 748–1280* (Munich: C. H. Beck, 1993)

Synan, Edward, 'Introduction: Albertus Magnus and the Sciences', in *Albertus Magnus and the Sciences: Commemorative Essays 1980*, ed. by James A. Weisheipl (Toronto: The Pontifical Institute of Mediaeval Studies, 1980), pp. 1–12

Taylor, Richard C., 'Averroes on the Ontology of the Human Soul', *Muslim World*, 102 (2012), 580–96

———, 'Personal Immortality in Averroes' Mature Philosophical Psychology', *Documenti e studi sulla traduzione filosofica medievale*, 9 (1998), 87–110

———, '"Truth Does Not Contradict Truth": Averroes and the Unity of Truth', *Topoi*, 19 (2000), 3–16

Torrell, Jean-Pierre, *Saint Thomas Aquinas: The Person and His Work*, trans. by Robert Royal, rev. ed., 2 vols (Washington, DC: The Catholic University of America Press, 2005)

Tracey, Martin, 'Albert's Moral Thought', in *A Companion to Albert the Great: Theology, Philosophy, and the Sciences*, ed. by Irven M. Resnick (Leiden: Brill, 2013), pp. 347–79

———, 'Albert's Reading of Aristotle's Moral-Philosophical Treatises on Pleasure vis-à-vis Three Recent Perspectives on his Thought', in *Albertus Magnus. Zum Gedenken nach 800 Jahren: Neue Zugänge, Aspekte und Perspektiven*, ed. by Walter Senner OP (Berlin: Akademie Verlag, 2001), pp. 311–25

Twetten, David, 'Albert the Great, Double Truth, and Celestial Causality', *Documenti e studi sulla tradizione filosofica medievale*, 12 (2001), 275–358

Wieland, Georg, *Ethica – scientia practica: Die Anfänge der philosopischen Ethik im 13. Jahrhundert* (Münster: Aschendorff, 1981)

Zavattero, Irene, 'Le bonheur parfait dans le premiers commentaires latins de l'*Éthique à Nicomaque*', *Revue de théologie et philosophie*, 139 (2007), 311–27

JOSEP PUIG MONTADA

# Chapter 5. Albert and 'the Arabs'*

## *On the Eternity of Motion*

In a recent handbook on Albert the Great, the Latin medieval theologian and philosopher who lived from 1200–80, a chapter by David Twetten, Steven Baldner, and Steven C. Snyder is devoted to his physics, recognizing the importance of this philosophical discipline for Albert.[1] Baldner addresses a question — 'Is motion everlasting?' — that is inextricably linked to the question of the world's eternity, and the research cited by the authors corroborates its import for Albert's thought. My article proposes to make a modest contribution to this twofold issue, with particular consideration of the relationship between Albert the Great and the Andalusian philosopher Averroes (1126–98).[2]

Aristotle begins Book VIII of his *Physics* by asking whether movement has come into being and ceases to be, or whether it is eternal, 'belonging to all things as their deathless and never-failing property' (250b13–14). Since Aristotle believed in the eternity of the world, as did most ancient Greeks, the second answer seems more plausible to him, but he nonetheless needs a convincing explanation for it. To obtain this, Aristotle goes back to the definition of movement that he established in an earlier book of the *Physics*, in which movement is the actualization of potentiality as such (201a10). Now, however, he paraphrases the definition as the actualization of the mobile as mobile (251a10). According to this definition, he explains, the mobile must already have been in existence — but how would this mobile in existence have come about?

* I thank David Twetten for his careful reading and excellent suggestions and Kate Sturge for her help with the style editing.

1 Twetten, Baldner, and Snyder, 'Albert's Physics'.

2 I focus here on the *Physica*, not on the early and late theological works of Albert.

**Josep Puig Montada** (puigmont@ucm.es) is professor emeritus of Arabic and Islamic studies at the Universidad Complutense, Madrid. He works on medieval philosophy and contemporary Islamic thought and society.

*Albert the Great and his Arabic Sources*, ed. by Katja Krause and Richard C. Taylor, Philosophy in the Abrahamic Traditions of the Middle Ages, 5 (Turnhout: Brepols, 2024), pp. 145–166
BREPOLS PUBLISHERS 10.1484/M.PATMA-EB.5.136485

Aristotle contemplates two different possibilities: either it must have come into being at a given moment, or it must have been eternally. If the mobile came into being, a prior movement would have to have occurred to create it — and this would lead to an infinite regress. If, however, the mobile was in existence but not moving, a movement prior to it must have occurred to prompt its own movement — which, again, would lead to an infinite regress. On the basis of these arguments and by logical exclusion of their negation, Aristotle thought he had proved the eternity of movement. However, the argument was not as evident as it seemed, and commentators on Aristotle's works had much to say about its premises.

Among the commentators discussing Aristotle's argument on perpetual motion was Albert the Great, who had at his disposal Aristotle's *Physics*, Avicenna's *Physics*, and Averroes's Long Commentary on Aristotle's *Physics* in Latin translations, as well as other sources. In his critical edition of Albert's commentary on the *Physics*, Paul Hossfeld included the *translatio vetus* of Aristotle, basing his edition on five thirteenth-century manuscripts; some years later, Jozef Brams published a larger critical edition.[3] The translation Albert would have read dates from the time of James of Venice (d. 1147), along with the translation included in Averroes's Long Commentary, made from the Arabic by Michael Scot (1175–1235).

Avicenna's *Physics* is a part of his philosophical encyclopaedia *Book of the Healing*, which was partially translated in Toledo in the twelfth century (Book I, II, III.1, and the beginning of III.2) and in Burgos between 1275 and 1280 (Book III.2–10).[4] Chapter 11 of Book III, dealing with the infinity of motion and time, the subsequent chapters of Book III, and Book IV were never translated in the Latin Middle Ages. Averroes's Long Commentary, in contrast, had been fully translated by Michael Scot, as Dag Nikolaus Hasse has shown,[5] and therefore Albert was able to read it.

One of Hossfeld's many articles on Albert the Great's natural science analyses the commentary on Book VIII of the *Physics* and records also other sources as well. Hossfeld observes that although Moses Maimonides is mentioned only twice, his influence on Albert is 'astonishingly large'.[6] However, even though Maimonides wrote many of his works in Arabic, his thought is without doubt different from that of the Islamic thinkers.

Since Albert was the first Scholastic thinker to write a commentary on Aristotle's *Physics*, even before it had become an official part of the university curriculum at Paris, I shall focus here on his interpretation of Aristotle's template in light of his relationship to one of his Arabic sources.

---

3 Albertus Magnus, *Physica, libri V–VIII*, ed. by Hossfeld (hereafter *Physica*); Albertus Magnus, *Physica: Translatio vetus*, ed. by Bossier and Brams.

4 Janssens, 'Reception of Avicenna's *Physics*', pp. 55–57.

5 Hasse, *Latin Averroes Translations.*

6 Hossfeld, 'Gott und die Welt', pp. 296–97.

## Perpetual Motion: Albert's Path into the Theme

In his commentary on the *Physics*, written around 1251, Albert does not start his discussion with Aristotle's template, but rather with one of his typical *digressiones*, a stylistic device by which Albert digressed from his text in order to explain philosophical difficulties in the source material at greater length:

> In this, our last book of the *Physics*, our intention is to discover whether there is any perpetual movement [*motus perpetuus*] that is like a cause of the perpetuity of movement in general [*in genere*]; for, at the end of Book VI of the *Physics* and above, we said in general that movement is generically perpetual and infinite on account of [its] nature, even though it is not necessary that rectilinear movement be perpetual on account of number.[7]

In his critical edition of Albert's paraphrase, Paul Hossfeld identifies this self-reference with two passages in Book VI in which we read arguments favouring the view that no change alone is by itself infinite.[8] Albert applies the phrase to changes in the world of coming-to-be and passing-away, and adds that circular movement cannot be eternal unless there is a source that produces continuous local movement and is itself eternal on account of its nature.[9]

Albert already diverges from Aristotle because, contrary to the Stagirite, he intends to enquire whether there is one specific and eternal movement that is the cause of all others. The reason for this enquiry, Albert notes, is to avoid falling into a vicious circle. If he could prove that movements in the sublunary world followed each other in a perpetual sequence, he would not have enough evidence to prove the existence of one eternal and first movement.[10] These remarks invoke the *De generatione et corruptione*; Albert is well aware that it follows the *Physics* chronologically, but this is not the issue because Albert's philosophical system is not conditioned by a strict chronology of works. As he notes, Aristotle ended the second book of his *De generatione et corruptione* with a discussion on relative and absolute necessity. For the Stagirite, coming-to-be must either go on forever or come to a halt. But, Albert continues, Aristotle also tells us that it has been demonstrated in his *Physics* that heavenly movement is eternal and that it is the cause for the eternal process of coming-to-be. On the basis of these premises, Albert proceeds to elucidate how the eternal character fits into the transient char-

---

7 *Physica*, VIII.1.1, p. 549, vv. 7–13: 'In ultimo autem hoc nostro libro physicorum nostra est intentio investigare, utrum sit aliquis motus perpetuus, qui est sicut causa perpetuitatis motus in genere; motum enim in genere diximus esse perpetuum et secundum naturam infinitum in FINE SEXTI LIBRI PHYSICORUM, SUPERIUS, licet secundum numerum non necessario sit aliquis rectus motus perpetuus'.

8 The passages indicated by Hossfeld are *Physica*, VI.3.3, p. 492, vv. 22 sq. and VI.3.5, p. 497, vv. 12 sq. See ibid., VI.3.5, p. 497, vv. 65 sq.: 'nulla mutatio [recta] est infinita'.

9 *Physica*, VIII.1.1, p. 549, vv. 15–16: 'si generans continuae sit allationis'.

10 Ibid., vv. 21–26.

acter of coming to be, finding the answer to this dilemma in cyclical succession. For Albert, absolute necessity and cyclical succession are bound together:

> For this reason, we wish to investigate here that from which we wish to proceed in the second book of *Peri geneseos*, namely: whether there is one perpetual first movement that is the cause of the continuous succession of movements in lower beings [immersed] in generable and corruptible matter.[11]

In this initial digression, Albert also guards against any suspicion of heterodoxy concerning the doctrine about God's creation of the universe, insisting he only intends to demonstrate that there has not been and will not be any time without movement and that he certainly does not maintain that God and motion are 'co-eternal'. In the same context, he invokes Boethius's argument that God precedes the world in eternity, not in time.[12] Albert suggests that Boethius intended to solve the conflict between God's omnipotence and man's free will, and to that end explained God's nature by defining eternity as 'the complete and perfect possession of an endless life'.[13] Eternity is also the key to understanding God's knowledge, which is all-embracing and ever-present, not subject to time.

In line with these ideas, Albert defines eternity as a kind of duration that does not have extension divisible into parts (*in partes simul stantes vel sibi succedentes*).[14] In Book IV of his *Physics*, Albert has already devoted one of the tractates to eternity,[15] which he defines in various ways. He considers the following definition to be the most appropriate:

> Eternity is space [*spatium*] without beginning and end, which does not have in itself any before and after, nor anything of succession, because it does not measure what becomes or what changes, but it rather is what always stays in the same way, never losing anything in the past and never acquiring anything in the future. And it is in this way that we shall talk about eternity.[16]

---

11 Ibid., vv. 33–37: 'Propter quod nos hic volumus investigare id, ex quo in SECUNDO PERI GENESEOS procedere volumus, hoc est, utrum sit aliquis unus motus primus perpetuus, qui est causa continuae successionis motuum in inferioribus in materia generabilium et corruptibilium'.

12 Ibid., vv. 42–45: 'secundum quem sensum etiam BOETHIUS in QUINTO CONSOLATIONIS PHILOSOPHIAE mundum dicit semper fuisse nec deum praecedere mundum tempore, sed aeternitate'.

13 Albert has already quoted the Boethian definition —'eternity is the complete and perfect possession of an endless life' — in *Physica*, IV.4.4.3, p. 296, vv. 13–14, where he writes: 'Propter quod aeternitatem in aeterno considerans BOETHIUS diffinit dicens, quod "est interminabilis vitae possessio tota simul et perfecta"'. This is a reference to Boethius's *Consolation of Philosophy*, V.6, trans. by Relihan, p. 145. In addition to Boethius, Isaac Israeli is another of Albert's sources on the issue (*Physics*, IV.4.2, p. 295, v. 29), although of less importance. See Levin, Walker, and Sadik, 'Isaac Israeli'.

14 *Physica*, VIII.1.1, p. 549, vv. 53–54.

15 Ibid., IV.4.1–5, p. 293, v. 23–p. 299, v. 74.

16 Ibid., IV.4.1, p. 294, vv. 21–26: '*aeternitas sit spatium principio et fine carens*, non habens in se prius et posterius nec aliquid successionis, eo quod non mensurat id quod fit vel mutatur, sed potius id quod stat uno modo semper, nihil amittens in praeterito et nihil acquirens in futuro, et hoc modo locutio nostra erit de aeternitate'.

A critical reader may see some inconsistency in applying this definition to God. Although *spatium* is used in a metaphorical sense, God's essence is not material at all, and eternity 'in this way' suggests materiality. Albert and his predecessors used the definition to overcome the difficulties arising from the Aristotelian doctrine of an eternal movement. Based on this definition, Albert responds to those who object by asking why God was idle or waited so long to create the world, but also to those who object that His infinite potency must cause an infinite effect in all aspects, above all in duration (*in duratione*).[17] The latter objectors do not realize, Albert explains, that God acts with all His potency, but that there is a ranking or order between cause and effect. Because of this order, the world has to start at a given time and its duration has a beginning.[18]

*Ordo* is another key concept introduced by Albert to explain the need for a beginning. As far as the first objection is concerned, Albert answers that since there is neither long nor short extension in eternity, no possible comparison to time is possible. With this argument, he seizes the opportunity to refute those 'who affirm God's will to delay in creating the world',[19] according to whom will is a cause which is such that it can delay its effect. Albert denies this claim on the grounds that eternity cannot precede time by a delay. We shall come back to his criticism of a delaying or postponing will later in the paper.

For now, let us turn to Albert's examination of the Aristotelian arguments in favour of the eternity of movement in Book VIII, Chapter 3, which is concerned with the 'Arguments of the Peripatetics concerning the Perpetuity of Movement' (*De probationibus Peripateticorum, quod motus sit perpetuus*).[20] This examination occurs after he has summarized the views of Plato, Anaxagoras, and Empedocles on the finite or perpetual nature of movement in Chapter 2, and it follows an initial definition of movement that distinguishes movement as perfection of the mobile from the perfection of the mover. Incidentally, this distinction was made, though not underlined, by Aristotle in his *Physics* (251a15–16); for Albert, in contrast, it takes centre stage. Both 'movements' are prior to movement as such, and priority is of two kinds in the subsequent changes:

> In all movements that are toward a form, such as growth, decrease, and alteration, the mobile must be prior to movement according to a twofold

17 See ibid., VIII.1.1, p. 550, vv. 66–72.

18 Ibid., vv. 74–76: 'Sed ordo, qui necessario ponitur inter causam et causatum, exigit, quod mundus duratione inceperit aliquando'. Let us note that on the latter issue, Albert diverges from Averroes, who distinguishes between an accidentally and an essentially related eternal sequence; see below, at note 60.

19 Ibid., vv. 14–15: '*Ex his etiam elucescit, quod falsum dicunt*, qui dicunt voluntatem dei esse dilatoriam in creando mundum'.

20 Ibid., VIII.1.3, p. 553, vv. 50–52.

> power [*potentia*], one of which is toward being [*esse*] and the other toward movement.[21]

In local movement, Albert suggests, priority toward being is redundant. In virtue of a thing's having its substantial form, its locomotion may ensue, if nothing hinders it. By contrast, what grows has the power of growth in its being (*potentia essendi*), and it begins to be itself when it begins to grow. That which has the potency of local movement need not acquire a substantial form. If nothing hinders it, it will move or cause to move. Albert wants to lead us to the following conclusion: If we were able to prove that local movement, which is the cause of all other motions is perpetual, then we could also prove that mover and mobile are eternal, because local movement does not require priority according to *esse*. Thus, the perpetuity of local movement has to be demonstrated, and Albert will do so by excluding its finiteness:

> If [mover and mobile] were not perpetual and therefore movement were not perpetual, the cause of this would have to be one of the two following: either they were said to be generated and made [*facta*] according to substance because they did not exist before, or they both existed perpetually, but the mover did not move the mobile because of some hindrance.[22]

Under the first hypothesis, the mobile comes to be from another substance, yet Albert reminds us that coming-to-be would be prior to the presupposed first movement. Moreover, coming-to-be is the end of a movement (*est autem generatio finis motus*), so that, if the first hypothesis is accepted, there would also be a movement prior to the coming-to-be, and so on — a conclusion that Albert considers 'impossible and unintelligible'.[23] Under the second hypothesis, mover and mobile existed perpetually, yet no movement took place because of some obstruction. For Albert, in the words of the Latin Aristotle, this situation 'seems irrational' (*videtur irrationale*) in the case of the first mover and mobile,[24] and he remarks that this view is shared by experts as well as the common people. Finally, he goes to great lengths to show the incoherence of the subtle arguments

---

21 Ibid., p. 554, vv. 23–26: 'in omnibus motibus, qui sunt ad formam, sicut est augmentum et diminutio et alteratio, oportet, quod ⌜mobile⌝ sit ante motum secundum duplicem ⌜potentiam⌝, quarum una est ad esse et altera ad motum'. Hossfeld refers the terms between small braces to Averroes's Long Commentary.

22 Ibid., vv. 43–49: 'Si enim non essent perpetua et ideo motus non esset perpetuus, huius causa esse non posset nisi altera duarum causarum, quarum una est, quod dicerentur ista generata et facta secundum substantiam, cum ante non essent, aut quod essent haec duo perpetua, sed movens non moveret ipsum mobile propter aliquod impedimentum'.

23 Ibid., p. 554, v. 50–p. 555, v. 6. In fact, adds Averroes, the alteration that precedes generation cannot be the first motion, since only circular motion can be eternal. Averroes, *Aristotelis De physico avditv libri octo cum Averrois Cordvbensis variis in eosdem commentariis*, ed. Venetiis (hereafter *CMPhys.*), VIII.7, fol. 343D; ed. by Schmieja, p. 20, vv. 13–14.

24 The Latin *videtur irrationale* (*Physica*, VIII.1.3, p. 555, v. 16) corresponds to the Greek *alogon*, 'unreasonable', which Aristotle often uses in this and similar contexts, e.g., at 251a 21.

(*subtiles rationes*) of those who sustain the view of Anaxagoras and Empedocles, and in refuting them, he often relies on Averroes's assistance.[25] In the following, therefore, I turn to a close comparison of Averroes's interpretation of Aristotle with Albert's borrowings from the Cordoban philosopher.

## Albert's Borrowing from Averroes: A Selective Approach

The famous Cordoban philosopher, jurist, and physician Averroes, whose long commentaries on Aristotle's *Physics*, *De caelo*, *Metaphysics*, and *De anima* were all translated into Latin in the early thirteenth century, explicitly identifies Aristotle's understanding of generic movement in *Physics* VIII, 251a20–28 with *motus caelestis primus*.[26] Averroes affirms that this movement is first not in time, but rather in nature, and that it is circular. He thus distinguishes sharply between movement of the heavenly bodies and progressive movement in the sublunary world. Moreover, Averroes distinguishes between creation *ex nihilo* and substantial coming-to-be in his approach to the issue in commentary 4, as we will see. Albert, in contrast, addresses local movement in general in line with Aristotle and mingles both views. But what does Albert have to say about creation *ex nihilo* and substantial coming-to-be? Will he follow Aristotle, or rather Averroes?

Averroes's commentary 8, which looks at Aristotle's 251a28–b5 (itself a parenthetical remark), begins with the assessment that Aristotle has established the cause of movement to be other than the cause of rest. On this basis, Averroes begins his argument where commentary 7 had left off, saying that if the mobile and the mover were eternally at rest, one or the other would need to undergo some alteration so that movement could start. He concludes from these premises that there would be some change prior to the supposed first change, but this would end in an infinite regress.[27]

When Averroes proceeds to comment on his template, he initially refers to Aristotle's observation that some powers can positively cause change in one direction but they can negatively cause change in the other direction. Fire heats, for instance, but if it is turned away, it cools. For this interpretation of Aristotle's text in Arabic, Averroes relies on the translation by Isḥāq ibn Ḥunayn, which opposes the two terms *mufradan* (Greek *monachōs*), and *jamīʿan* (all).[28] The Greek original and its Renaissance Latin translation, in contrast, do not contain

25 See *Physica*, VIII.1.3, p. 555, v. 18–p. 556, v. 41.

26 Averroes, *CMPhys.*, VIII.7, fol. 342L–M; ed. by Schmieja, p. 19, vv. 1–5.

27 Averroes's argument here is a summary of the argument of the text for comment 9, *Phys.* 251b5–10.

28 For *mufradan / jamīʿan*, see Aristotle, *Al-Ṭabīʿa*, ed. by Badawi, vol. 2, p. 806, vv. 10–12; *Aristotle's Physics VIII*, ed. by Arnzen (2021), p. 7, vv. 5–6: 'Some things are moving [*yataḥarrak*] only one by one, and some others produce two opposed motions one and all, for instance, fire heats and does not cool'. Besides this contradistinction, Ibn Ḥunayn's Arabic opposes moving itself and causing motion, but the Latin translation by Michael Scot, as well as Averroes's quotations, do not reflect that opposition.

the second adverb, and so invoke no opposition here. (Adding a note of caution, one should say that there is only one manuscript preserving the complete Arabic translation: MS Leiden, Warner 583). In his medieval Latin translation from the Arabic, Michael Scot translated these two terms with the Latin *singulariter* and *insimul*, the latter often meaning 'at once'. Thus, his Latin translation from the Arabic text of Aristotle reads: 'some powers cause only one movement, and some powers can cause two movements at once [*insimul*]',[29] though this meaning of *insimul* may not be the most common meaning of the Arabic *jamīʿan*.

In the Venice edition of the Long Commentary, Latin Averroes adds to the sentence 'et quaedam moveant duobus motibus insimul' the explanatory remark 'scilicet in horis diuersis'.[30] In a characteristic move, he introduces the distinction between essential and incidental, or accidental, and applies it to motion. Rational potencies can cause two contrary motions and they are essential actions, but natural powers cannot cause a duality except incidentally, as in the instance of fire. Fire heats and the action is essential, *per se*, whereas it cools incidentally, *per accidens*.[31] It is this interpretation by Averroes that Albert the Great follows in his own commentary.[32] He quotes 'quaedam autem movent et agunt secundum contrarios motus',[33] but adds a thorough explanation. Coldness accidentally heats if it is driven away and expelled, and heat binds and encompasses the core of the cold object.[34]

In the *Physics*, Aristotle sees rational powers as a cause of motion insofar as he considers them capable of acting in both the correct and the wrong way; he gives the instance of the expert using his knowledge to do mischief deliberately (*Phys.* 251a32–251b1). Describing rational powers is not Aristotle's purpose here, but Averroes, feeling the need to be more specific, says that 'two contrary actions exist essentially in the rational powers'.[35] Albert underwrites Averroes's interpretation and remarks, with Latin Aristotle, that the man who uses his knowledge to do evil 'sins wilfully' (*voluntarius peccat*).[36]

---

29 *CMPhys.*, VIII.8, fol. 343F; ed. by Schmieja, p. 20, vv. 17–18: 'Et quaedam movent singulariter, et quaedam movent duos motus insimul'.

30 Ibid., fol. 343I; ed. by Schmieja, p. 21, vv. 16–17 ('that is, at different times').

31 Ibid., fol. 343K; ed. by Schmieja, p. 22, vv. 5–7: 'altera actio est virtutum naturalium per se, et reliqua per accidens: in virtutibus autem rationabilibus duae actiones contrariae sunt essentialiter'.

32 See Albertus Magnus, *Physica*, VIII.1.3, p. 555, vv. 53–68.

33 The reading is very similar to the *translatio vetus*, quoted in ibid., p. 554, vv. 64–65.

34 Ibid., p. 555, vv. 62–68: 'Et aliquando quidem in naturalibus potentiis videtur aliquid esse simile potentiis sensibilibus et rationabilibus, quae dicuntur comprehensivae, quia frigidum aliquo modo calefacit per accidens, quando accipitur ut conversum et propellens et expellens ante se calidum et circumstans et ambiens ipsum et constringens in centro rei infrigidatae; sed hoc est per accidens'.

35 *CMPhys.*, VIII.8, fol. 343K; ed. by Schmieja, p. 22, vv. 6–7: 'In virtutibus autem rationalibus duae actiones contrariae sunt essentialiter'.

36 Albertus Magnus, *Physica*, VIII.1.3, p. 555, vv. 68–72: 'sed hoc est per accidens, sicut etiam sciens, licet ipse per se consideret recte, tamen voluntarius peccat aliquando, quando nec operatur secundum eam quae est in ipso, scientiam, sed potius revertitur ab ipsa et utitur ea ad contrarium'.

There is no doubt that the enquiry into the cause of movement and the cause of rest is the main issue, and Averroes is aware of this. Aristotle made the parenthetical remark, though immediately introducing an adversative clause:

> But at any rate, all things that are capable respectively of affecting and being affected, or of causing motion and being moved, are capable of it not under all conditions, but only when they are in a particular condition and approach one another.[37]

He concluded that movement only takes place under certain circumstances: if the mover and the mobile are near enough to each other and correctly predisposed, 'in a particular condition'.[38] Averroes notices this change, of course, moving away from the discussion of will and natural causes to the main discussion of the possibility of an infinite time of rest, and writes:

> Perhaps [Aristotle] brought in this [parenthetical passage] after that which he wanted to explain about the improbability which would occur if all things were supposed at rest for an infinite time, [and he did it] for one of these two reasons.[39]

The first reason Averroes suggests for this parenthesis in Aristotle is hinted by the adversative phrase *sed non omnia* (*all'oun hosa*, 'but at any rate', *Phys.* 251b1).[40] Averroes thinks that Aristotle wrote the clause to refute the possibility of an infinite rest preceding motion. If we were to assume that everything was at rest and then came into motion, we would have to posit the cause of motion to be different from that of rest. Admittedly, this only applies to principles that cause motion in just one direction. If some principles are capable of causing motion in one direction and also of reversing it, the cause must be identical.[41] The examples given by Aristotle for this objection — fire that accidentally causes coldness and a

---

37 *Phys.* 251b1–3. English translation in *The Works of Aristotle*, vol. 2., *Physica*, trans. by R. P. Hardie and R. K. Gaye, p. 800. Translation Bessarione, *CMPhys.*, VIII.8, fol. 343E: 'At vero quaecumque possunt facere, aut pati, aut movere, aut moveri, ea moueri non penitus possunt, sed si sic se habeant, et appropinquent sibi inuicem'; translation Michael Scot, *CMPhys.*, VIII.8, fol. 343G; ed. by Schmieja, p. 21, vv. 1–3: 'sed non omnia, quae possunt agere, aut pati, aut movere, aut moveri, necessario possunt hoc, sed quando fuerint talis dispositionis, aut quando fuerint vicinantia sibi ad invicem'. This follows Aristotle, *Al-Ṭabīʿa*, ed. by Badawi, vol. 2, p. 807, vv. 1–4; ed. by Arnzen, *Physics VIII*, p. 8, vv. 2–4: 'lākin laysa al-ashyā' kullu-hā', etc.

38 Michael Scot's translation reads *talis dispositionis*, faithfully rendering the Arabic *bi-ḥālin kadhā*. See *Al-Ṭabīʿa*, ed. by Badawi, vol. 2, p. 807, v. 3; *Physics VIII*, ed. by Arnzen, p. 8, v. 3. However, the Renaissance translation from Greek made by Cardinal Bessarione varies (*si sic se habeant*), which is closer to *all'hōdi echonta*. *CMPhys.*, VIII.8, fol. 343E. The basic meaning does not change.

39 *CMPhys.*, VIII.8, fol. 343M; ed. by Schmieja, p. 22, vv. 18–20.

40 Ibid., fol. 344B; ed. by Schmieja, p. 23, vv. 8–10: 'Sed non omnia, quae possunt, etc. et id est ponens hunc sermonem, et dans hanc rationem, non potest dicere hoc in omnibus rebus'.

41 Ibid., fol. 344A; ed. by Schmieja, p. 23, vv. 4–5: 'illud, quod agit quietem in istis virtutibus, est illud, quod agit motum'.

man using his science perversely[42] — lead Averroes to propose that some natural powers and most rational powers are of this kind. But most causes are not of this kind. Consequently, the possibility of coming in motion after an infinite rest cannot be admitted.[43]

Averroes favours a second explanation for Aristotle's parenthesis.[44] The second explanation he produces requires a third factor, namely the circumstances of spatial proximity and suitability that comprise all powers, both rational and natural.[45]

Since the Arabic version that Averroes read translated the Greek *hōdi echonta* by 'such a state' or 'such a condition' (*bi-ḥālim mā*),[46] Averroes may have been led to look for an internal cause in the soul of the mover. For instance, Averroes suggests by example (again anticipating Chapter 2) that when an animal moves after being at rest, some appetite (*appetitus*) must have come to pass in it. This appetite is caused by an alteration in the subject, and the alteration is produced by something external to the soul (*ab aliquo ente extra animam*). Humans seem to act similarly: they cannot act by their power of will unless they receive some external stimulus. This stimulus brings forth an image in the soul, which in turn generates knowledge and an appetite.

Through these examples and his argumentation in general, Averroes criticizes the Ashʿarite theologians (*loquentes nostrae legis*),[47] who, according to Averroes's view of their theology, affirmed that will 'is dependent on the action of an event [*ḥādith*] at a given time'.[48] Although Averroes tried to understand their discourse, he was seemingly unable to do so,[49] and thus concluded from his interpretation of Aristotle's *Physics* that even if there are powers producing rest and movement together (*insimul*), an infinite regress is certain from the assumption of a first

42 Ibid., fol. 343I–L; ed. by Schmieja, p. 22, vv. 14–17.

43 David Twetten rightly reminds me that Averroes is fully aware of the weakness of the first argument, recognizing that 'still the same question remains': *CMPhys.*, VIII.8, 344C–D; ed. by Schmieja, p. 23, v. 23–p. 24, v. 7: 'Sed adhuc remanet eadem quaestio. Potest enim aliquis dicere, quod primus motor totius est ex habentibus virtutes rationales, cum habentia virtutes rationales incipiunt movere post quietem, et quiescere post motum, absque eo, quod deficiat dispositio, per quam hoc movet, et hoc movetur, et absque eo, quod deficiat appropinquatio in tempore quietis'.

44 Ibid., fols 344E–K; ed. by Schmieja, p. 24, v. 10–p. 25, v. 12.

45 Ibid., fol. 344F; ed. by Schmieja, p. 24, vv. 16–17: 'Sermo eius erit vniuersalis in omnibus virtutibus motiuis, scilicet rationalibus, et irrationalibus, et hoc manifestum est'.

46 Aristotle, *Al-Ṭabīʿa*, ed. by Badawi, vol. 2, p. 807, v. 3: *bi-ḥālim mā*; *Aristotle's Physics VIII*, ed. by Arnzen, p. 8, v. 3, *bi-ḥālin kādhā*.

47 The term Ashʿarite derives from Abū al-Ḥasan al-Ashʿarī (d. 935 or 936), the founder of the theological school recognized as orthodox in Sunni Islam. See Watt, 'al-Ashʿarī, Abu l-Ḥasan'.

48 *CMPhys.*, VIII.8, fol. 344I; ed. by Schmieja, p. 25, vv. 13–14: 'voluntatem dependere de actione entis novi in hac hora'.

49 Averroes considers the action of an event to be unable to act on the will unless desire (*concupiscentia*) arises in the agent at that time. Desire is caused by 'presence of time', *praesentia temporis*. I must say that I cannot easily guess the Arabic terms behind the Latin words, perhaps *ḥuḍūr al-waqt*. It seems clear to me that Averroes points out an alteration in the agent as the condition for the arising of desire, which leads to an infinite regress (*CMPhys.*, VIII.8, fol. 344K; ed. by Schmieja, p. 25, vv. 19–24).

motion in time. Hence, Averroes saw no other way out than to affirm the perpetual character of movement.

Unlike Averroes, Albert had no reason to argue against the conclusion of the Ash'arite theologians, taken in itself and outside the domain of physics, but he nonetheless echoed Averroes's distinction between natural and rational powers.[50] Natural powers can only cause movement or rest when they suffer alteration by various dispositions.[51] Rational powers, in contrast, seem to cause movement or rest only by will — but Albert considers this position to be false, since 'various conceptions would have to be in them: when they would have to move and when they would have to be at rest'.[52] On this point, Albert clearly reads Aristotle with Averroes's guidance, considering conceptions and desires to be produced by temporal changes alone.

The Arabic concept of *bi-ḥālin kadhā* and its Latin equivalent *talis dispositionis esse* led both philosophers to the same interpretation as Aristotle's, that no motion can start after an infinite rest. In contrast to Averroes, however, Albert intended to defend two tenets: that there was no time without movement; and that the world is created with a first moment in time and not eternal. In order to do so, he uses, once again, the stylistic device of a *digressio* in Book VIII, tr. 1, c. 4 of his *Physics*, and introduces arguments for and against the eternity of movement and the universe.[53] He first objects to the absolute validity of the principle that 'nothing may come to be without qualification from not being' (*Phys.* I.9, 191b36–192a1) and restricts it to that which comes to be by means of generation, diverging from Averroes.[54]

At the end of this digression, Albert concludes that the universe could not have come to be by means of generation and that no time occurred in the past without movement. Albert does not see any contradiction between the two tenets. For him, creation brings the mobile and movement into existence simultaneously and out of nothingness, and the mobile, too, does not precede movement in time but only in nature.

Albert also considers himself to have sufficient evidence for the eternity of movement. Given the existence of mover and mobile, movement must necessarily take place as well. But the issue of how the mobile is created gives rise to some differences with Averroes. Although Albert agrees with Averroes in rejecting the possibility of a 'postponing will' (*voluntate dilatoria*), he diverges from him in the explanation of creation. Averroes admits that God does not precede the world in time; His priority is in causation alone. This class of priority is the priority

50 Albertus Magnus, *Physica*, VIII.1.3, p. 555, vv. 53–62.

51 See ibid., vv. 76–79. Albert limits his discussion, as he had set at the outset, to natural agents.

52 Ibid., vv. 83–85.

53 Ibid., VIII.1.4, p. 556, vv. 42–45: 'Et est digressio declarans, qualiter declarant praeinductae rationes motum et mundum esse perpetuum et qualiter non'.

54 Ibid., vv. 67–68: 'dico ego, quod ex nihilo nihil fit per generationem, sed ex nihilo aliquid fit per actum creantis'.

of the unchanging, timeless existent to an existent in time.[55] By contrast, Albert insists that the world had a beginning. Averroes says that to posit that mover and movement exist in an infinite time but without motion — i.e., actionless — and then to posit that they start moving at a given time (*in aliqua hora*) would be contrary to nature.[56] For Averroes, such a view would have been closely associated with that of the Muslim theologians who posited the mover moving by His will.

Albert identifies the voluntary mover with God when he reproduces Averroes's arguments, an argumentative move that seems to bring him even closer to the position of Averroes's opponents. Indeed, in the first discussion of the *Incoherence of the Incoherence*, which was not available to Albert but is reminiscent of his arguments, Averroes contests Algazel's words: 'The world was temporally created by an eternal will that decreed its existence at the time in which it came to be'.[57] In his commentary on the *Physics*, Averroes reiterates his point initially and remarks that those who posit the mover causing movement by His will are in a stronger position than those who posit the mover causing movement naturally because the order is not determined. Nonetheless, he still finds their position unsatisfactory. Averroes cautions his readers against understanding God's willpower in a human way, pointing out the following dilemma:

> 1) Should we assert a willpower in which it is possible to postpone the willed [*volitum*], this [the postponement] would happen because of the existence of something which was not yet existent, i.e., for want of some cause or of some disposition, and there would be neither time nor anything, because we have already posited that this will is first of [all] movements, so it is evident that the willed must be [simultaneous] with the will.
> 2) Should we assert that the will is eternal [*antiquam*], the willed[58] would also be eternal.[59]

Averroes thus accepts the second horn of the dilemma but introduces a distinction between essential and accidental dependence on a prior infinity.[60] Whereas for the Ashʿarite theologians, an eternal world would entail that one event depend essentially on a prior infinite series, for Averroes this dependence is *per accidens*, and therefore possible. In his eyes, this further evidences the incoherence of the Ashʿarite theologians, who mistake the accidental for the essential nature of

55 Averroes, *Tahafot at-Tahafot*, ed. by Bouyges, pp. 67–68; *The Incoherence of the Incoherence*, trans. by van den Bergh, vol. 1, p. 39.

56 *CMPhys.*, VIII.15, fol. 349I; ed. by Schmieja, p. 43, vv. 1–4 (paraphrasing 252a14–16): 'Aliud autem, quod est extra naturam, in hac opinione, est ponere motum, et motorem existentia tempore infinito, et motorem non movere, et motum non moveri, et post incipere in motu in aliqua hora absque eo, quod illic fuit causa, per quam moveret in illa hora, et non prius'.

57 Algazel, *Tahafot al-Falasifat*, ed. by Bouyges, p. 26, vv. 2–3; *The Incoherence of the Philosophers*, trans. Marmura, p. 15; cf. Averroes, *Tahafot at-Tahafot*, ed. by Bouyges, p. 7, vv. 6–7.

58 *Voluntatum*, instead of *volitum*, *CMPhys.*, VIII.15, fol. 349I; ed. by Schmieja, p. 44, vv. 5–6.

59 *CMPhys.*, VIII.15, fol. 349M; ed. by Schmieja, p. 43, v. 19–p. 44, v. 6.

60 *CMPhys.*, VIII.15, fols 350C–E; ed. by Schmieja, p. 43, vv. 2–21.

movement. The arguments here in his *Physics* strongly reflect the earlier polemic of the *Incoherence of the Incoherence*.

Albert's view differs from Averroes's in that he holds that there is no necessity for this kind of proportionality between first cause and effect, meaning that God's infinite potency can cause limited effects, and may infinitely exceed what originally exists outside Himself, namely, nothing.[61] In light of his opening digression and his definition of eternity,[62] Albert subsequently suggests that those who sustain that God's will can postpone its effect also assume that His eternity is divisible into successive parts and that He postponed the creation of the world to the future as if there were a future prior to the world; yet a future did not exist before the world.[63] Albert continues rebutting possible objections of dialectical nature;[64] then he proclaims that God's eternity is indivisible and that it 'precedes' creation in a specific sense: *ordo*, the ranking of eternity with regard to time, that of the cause with regard to the caused.

Albert ends his initial digression by admitting that from the viewpoint of natural science or physics, movement did not begin in the past.[65] This does not, however, entail the commonly expected consequence that the world did not have a beginning. Indeed, Albert argues against this consequence in the chapters to follow and, at their end, concludes that the doctrine of the creation of the world as having a beginning is neither a subject matter of physical science, nor can it be proved in physics.[66] By way of physical arguments (*per rationes physicas*), Albert suggests, we can only reach the conclusion that movement is perpetual — but there are other ways to understand creation, namely through metaphysical speculation.

In short, Albert agrees with Averroes in one respect, and disagrees in another. For Albert, Averroes is quite right to say that a) God, whose intellect created, or made, the world (*mundum operantem*), did not precede the latter in time, and b) God does not act with a 'postponing will'. However, he disagrees with Averroes, who considers it 'unintelligible' that God makes or creates the world in time according to His actual intellect from only one concept (*unius ideae*) and from eternity.[67]

It would be convenient to adduce a related statement in those of Averroes's works that were known to Albert, but I can find such a statement only in the

61 See Albertus Magnus, *Physica*, VIII.1.4, p. 557, v. 64–p. 558, v. 29; cf. VIII.1.1, p. 550, vv. 21–30, 66–72.
62 See ibid., VIII.1.1, p. 549, vv. 53–54.
63 Ibid., p. 550, vv. 45–46: 'ante mundum non fuit futurum quod continuaretur ad praesens instans temporis'.
64 Ibid., p. 550, vv. 48–57.
65 Ibid., p. 551, vv. 6–22.
66 Ibid., VIII.1.14, p. 579, vv. 43–46: 'Sed inceptio mundi per creationem nec physica est nec probari potest physice, et ideo hanc viam putatur Aristoteles tacuisse in physica'.
67 Ibid., VIII.1.4, p. 557, vv. 29–32: 'Quod deus ab aeterno unius ideae secundum intellectum existens operatus est mundum vel aliquam rem mundi in tempore, quod est post aeternitatem'.

*Incoherence of the Incoherence*, which he did not know (as already mentioned). The First Discussion in this work concerns the eternity of the world and is divided into several 'proofs'. The second of these is about causal, not temporal, priority of God to the world, and Averroes concludes

> that God never ceases to have power for action, and that it is impossible that anything should prevent His act from being eternally connected with His existence; and perhaps the opposite of this statement indicates the impossibility better still, namely, that He should have no power at one time but power at another.[68]

Albert would probably have feared the consequences of this bold and daring statement if he had been able to access it. Lacking that access, he understood Averroes to be opposed to temporal creation *ab aeterno* — a position that he wanted to defend. For Albert, God acted 'by His will and knowledge' (*per voluntatem et scientiam*), and neither His will nor His knowledge were subject to change. Albert thus amended Averroes's statement as follows:

> It is quite understandable that any intellect can abide by its thought and desire to produce a thing not immediately but later. And when the thing is made by it later, the change [*variatio*] that is the beginning of that thing is with regard to the produced thing and not with regard to the operating intellect. And we say that it is in this way with God, Who is the cause that prepossesses [*praehabens*] every created thing in an ideal way [*idealiter*], and He abides by His thought alone according to which His knowledge is the cause of things, as Aristotle teaches in Book XI of his *First Philosophy*.[69]

Albert's reference to the *Metaphysics* here is crucial. Apart from the fact that for him, and for Averroes as well, Book XI was our Book XII, we read a very similar statement in this parallel treatment of the first causes: 'A separated substance which is the cause of things, possesses beforehand the things as their principle and cause'.[70] Paul Hossfeld, the editor of *Physica* for the Editio Coloniensis, describes *Metaphysics* XII, 1072b26–30 as the Aristotelian passage of reference in the anonymous *translatio media*.[71] Albert was also acquainted with Averroes's

---

68 Averroes, *Tahafot al-Tahafot*, ed. by Bouyges, p. 95; trans. by van den Bergh, vol. 1, p. 56.

69 *Physica*, VIII.1.4, p. 557, vv. 32–40: 'Est enim satis intelligibile quemlibet intellectum posse stare in conceptu et appetitu uno faciendi rem non modo, sed postea, et quando postea ab eo fit res, variatio, quae est inceptio rei, est circa rem factam et non circa intellectum operantem; et ita dicimus esse in deo, qui est causa praehabens idealiter omnem rem creatam et uno modo stat in conceptu suo, secundum quod sua scientia est causa rerum, sicut docet ARISTOTELES in UNDECIMO PRIMAE PHILOSOPHIAE'.

70 Albertus Magnus, *Metaphysica*, XI.2.19, ed. by Geyer, p. 506, vv. 85–87: 'Diximus enim, quod substantia separata, quae est causa rerum, praehabet res sicut principium et causa'. See also the bilingual edition *Albert le Grand, Métaphysique*, ed. and trans. by Moulin, pp. 212–13.

71 In his comments on v. 40. *Physica*, VIII.1.4, p. 557, with further references to the *translatio media*: Aristotle, *Metaphysica*, ed. by Vuillemin-Diem, p. 214, vv. 19–23.

commentary on this passage in comment 39.[72] Aristotle's original, in the Latin translation available to Albert, begins by describing thought as one of God's activities,

> since its actuality is also pleasure. (And for this reason waking, sensation and thinking are most pleasant, and hopes and memories are pleasant because of them). Now thinking [*intelligentia*] in itself is concerned with that which is in itself best, and thinking in the highest sense with that which is in the highest sense best. And thought thinks itself through participation [*metalepsis, transumptio*] in the object of thought; for it becomes an object of thought by the act of apprehension and thinking, so that thought and the object of thought are the same.[73]

Averroes did not point to this passage of the *Metaphysics* in his *Long Commentary on the Physics*; in his *Long Commentary on the Metaphysics*, he ended his interpretation of the passage Text 39, on *Met.* XII, 1072b16–30, with just a short reference to God as the First Mover, 'the mover of the universe'.[74]

Incidentally, Averroes's *Long Commentary on the Metaphysics* is known for its criticism of the Christian Trinity, which Michael Scot softened in his translation,[75] but Averroes also criticized the Ashʿarite doctrine of the divine attributes as entities added to the divine essence. His criticism is related to Aristotle's words defining the divine essence: 'the actuality of thought is life, and [God] is that actuality' (*Met.* XII, 1072b26–27). Averroes interprets the Islamic doctrine of God's life and knowledge as complying with the Aristotelian one.

---

72 Averroes, *Tafsīr mā baʿd aṭ-Ṭabīʿat*, ed. by Bouyges, pp. 1616–24; Latin version by Michael Scot, *Aristotelis Metaphysicorum*, XII.3, fols 322A–323D; French translation: *Averroès: Grand Commentaire de la Métaphysique d'Aristote*, ed. by Martin, pp. 236–44; English translation: *Ibn Rushd's Metaphysics*, ed. by Genequand, pp. 157–61.

73 Aristotle, *Metaphysics*, 1072b16–21, trans. by Tredennick, p. 147. Albert was reading Aristotle, *Metaphysica* (*media*) XII, ed. by Vuillemin-Diem, p. 214, vv. 7–14: 'quoniam voluptas actus est huius (et propter hoc evigilatio et sensus, intelligentia delectabilissimum, spes vero et memoriae propter ea). Verum intelligentia secundum se eius est quod secundum se optimum, et maxime eius quod est maxime. Eum autem intelligit intellectus secundum transumptionem intelligibilis, intelligibile fit ordinans et intelligens; quare idem intellectus et intelligibile'.

74 Averroes, *Tafsīr mā baʿd aṭ-Ṭabīʿat*, ed. by Bouyges, p. 1624, v. 4.

75 Ibid., p. 1620, vv. 4–6. See also p. 1623, vv. 5–12. See *Ibn Rushd's Metaphysics*, trans. by Genequand, p. 159: 'It is in this respect that the Christians were mistaken when they adopted the doctrine of the Trinity in the substance; it does not save them from it to say that it (i.e. the substance) is three and God one because if the substance is multiple, the compound is one in the sense of unity superimposed on the compound'. I suggest translating 'in the sense of unity superimposed' (*bi-maʿná wāḥid zāʾid*) as 'because of one added meaning'. For Michael Scot's modified translation, see Averroes, *Aristotelis Metaphysicorum libri XIIII*, XII.3. 39, fol. 322I: 'putauerunt Antiqui trinitatem esse in Deo in substantia, et voluerunt evadere per hoc, et dicere quia fuit trinus, et unus Deus, et nescierunt evadere: quia, cum substantia fuerit numerata, congregatum erit vnum per vnam intentionem additam congregato'.

Albert holds that God's intellect is the cause of His knowledge, which is the cause of everything, as this intellect is above time.[76] Thus, time comes only with motion. However, a piece is still missing in the discussion. This concerns Averroes's comment on the eternal, which he considered also to act eternally. Albert, in contrast, wished to introduce a beginning of such action. We find his explanation on this point in the initial digression in his *Physics* paraphrase. Having rejected the view of the postponing will, he faced the issue that God's infinite power must also act in an infinite way. But if God's power acts infinitely, this implies that movement must be infinite too. Instead of the distinction between eternal and accidental eternity that we saw in Averroes, Albert draws a different distinction, one that focuses on the potencies of God and His creatures. God acts, of course, with all His infinite potency, but 'the necessary order existing between cause and effect necessitates the temporal beginning of the world'.[77] Albert here offers no further physical proofs for his explanation, referring us instead to Chapters 13–15 of Tract 1 and to his forthcoming *Metaphysica* (*c.* 1263).[78]

There is another doctrine on which Albert disagreed with Averroes. In an extended passage of his *Long Commentary on the Physics*,[79] Averroes explains Aristotle's words 'there must be something combustible before there can be combustion and something that can burn before there can be burning'; these were faithfully rendered by Michael Scot into Latin from the Arabic that he read.[80] By using this example of fire, Averroes once again opposed the Muslim theologians, who considered it possible for something to be generated out of nothing.[81] To uphold their claim, they argued that the subject of generation diminishes the nature of the agent (*diminutio agentis*), but God cannot be in a diminished state of functioning. In Averroes's eyes, these theologians were, in the first place, unable to understand that generation takes place only out of elements observable to our eyes, and in the second place, that there is no diminution at all when an action is impossible in itself.[82] This, I believe, is a tenet showing Averroes's deep rationalism.

---

76 *Physica*, VIII.1.4, p. 557, vv. 39–44: 'sua scientia est causa rerum, sicut docet ARISTOTELES in UNDECIMO PRIMAE PHILOSOPHIAE; et inceptiones rerum secundum ordinem sequuntur tempus, cum tamen intellectus, qui est causa scientiae, sit supra tempus, et tunc hoc modo dicimus illum intellectum esse causam motus in tempore'.

77 Ibid., VIII.1.1, p. 550, vv. 74–76, quoted in note 18 above.

78 See esp. Albertus Magnus, *Physica*, VIII.1.13, p. 576, vv. 44–50.

79 *CMPhys.*, VIII.4, fols 340I–341L; ed. by Schmieja, p. 10, v. 8–p. 14, v. 24. Averroes argues for the perfection existing potentially in the thing in which is accomplished and refers to Alexander of Aphrodisias's treatise *On the Principles of the Universe* to support his view.

80 *CMPhys.*, VIII.4, fol. 340I; ed. by Schmieja, p. 10, vv. 6–7: 'necesse est igitur ut res sit innata comburi prius antequam comburatur, et comburere, antequam comburat'. Arabic translation in *Al-Ṭabīʿa*, ed. by Badawi, vol. 2, p. 805, vv. 2–4; ed. by Arnzen, p. 5, vv. 10–11.

81 *CMPhys.*, VIII.4, fol. 341E; ed. by Schmieja, p. 12, v. 21: 'habent pro possibili aliquid generari ex nihilo'.

82 Ibid., fol. 341F, ed. by Schmieja, p. 13, vv. 7–8: 'sed dicere ipsum facere aliquid impossibile, et posse facere est deceptio'.

Albert, in contrast, held the opposite view, that God can and does act without underlying matter. Although he agrees with Averroes that there is no diminution when God does not enact what is impossible, he takes Averroes to be wrong (*decipitur*) in asserting that God causes generation only from some matter alone (*operatur per generationem ex aliqua materia tantum*). If this were the case, God's power (*virtus*) would be proportioned to the potency of His effect, which is nature. Albert quotes Aristotle to support his claim,[83] and reasons that God's power has no proportion to His effects. God can thus operate both with and without matter, but operating without is more appropriate to His excellence.[84] Moreover, matter is never deprived of form, but form cannot be produced out of matter: first, form must be created ex nihilo by the First Agent.[85]

My purpose here is not to discuss the validity of Albert's argument, but to point out that he once again tries to reconcile different views. He accepts Averroes's positions that God's power is not diminished when He acts with matter, that there is no past time without movement, and that the world did not come to be by means of generation.[86] But Albert rejects the position that the creation of the world is eternal.

In Chapters 13–15, Albert proclaims the doctrine of creation *ex nihilo* and the priority of the Creator to the created world in eternity, and defines the term *aeternitas* once again.[87] His arguments here intend to support the doctrine, but also to neutralize Aristotle's positive arguments contrary to it. And yet he is aware that none of his own arguments amount to scientific proofs, indeed that he has no means to obtain scientific proofs at all, as he acknowledges.[88]

Albert begins Chapter 13 with the words 'Now it is time to tell our view about the making of the world'.[89] His view can be summarized in a sentence, 'God is prior to the world in eternity', and he intends to confirm it not only on the basis of his faith, but also through rational arguments. Albert's reasoning in these chapters follows the *via perfectionis* and looks at material bodies, realizing that their perfecting cause, as a composite of matter and form, is found outside

---

83 In fact, he is quoting the pseudo-Aristotle of the *Liber de causis*. See *Liber de causis*, V.57, ed. by Pattin, p. 59 (as corrected by Taylor, 'The *Liber de Causis*', vol. 2, p. 451), edited for the Web by Zimmermann: 'Causa prima superior est narratione. Et non deficiunt linguae a narratione eius nisi propter narrationem esse ipsius'. Albert's quotation is quite exact. See *The Book of Causes*, trans. by Brand, p. 24: 'The First Cause is above every description. All tongues fail to describe it only because they are not able to describe its essence'. The *Liber de causis* was always essential to Albert's thought. See Krause and Anzulewicz, 'From Content to Method'.

84 *Physica*, VIII.1.4, p. 557, vv. 75–76: 'sed potius necessarium deum sine materia operari'.

85 Ibid., p. 558, vv. 22–23: 'ipsa per actum causae primae educitur de nihilo'.

86 Ibid., VIII.1.4, p. 558, vv. 34–39: 'Sed hoc quod est nobis hic sciendum, est, quod sciamus istas rationes, quae inductae sunt, non probare, nisi quod mundus non incepit per generationem et quod non fuit aliquod tempus in praeterito, in quo non fuerit motus. Et hoc quidem est verum et a nobis supra concessum'.

87 Ibid., VIII.1.13, p. 575, vv. 7–12.

88 Ibid., VIII.1.14, p. 579, vv. 44–45: 'nec probari potest physice'.

89 Ibid., VIII.1.13, p. 574, vv. 67–68.

of them. The process goes through higher beings up to the celestial bodies, where the cause of perfection is outside the universe. Once Albert has established the first agent of perfection, he enquires whether He acts by necessity or through choice and will.[90] Causation, Albert finds, can only be volitional because only a volitional agent can produce the great diversity found in the universe. Indeed, only a volitional agent can act freely, and only through freedom is variety possible: each cause is not determined to one and only one effect. Moreover, such an agent is not conditioned by a pre-existing matter, for God creates *ex nihilo*.

Here we see Albert defending a doctrine of faith by means of rational arguments that are not based on natural science (*scientia*). He cannot demonstrate *per rationes physicas*, and he is aware of that weakness. And yet he considers his arguments to be stronger than Aristotle's: *Videtur autem nobis ista ratio melior esse omnibus rationibus Aristotelis.*[91] To strengthen his arguments even further, Albert looks for another strategy, which aims to show that the positive Aristotelian arguments in favour of the eternity of the world are rather weak (Chapter 14). There is no need to posit either an eternal first matter or an eternal time. By virtue of creation, he thinks, he can give a more plausible explanation:

> The First Cause made [*fecit*] matter *ex nihilo* and also form *ex nihilo*, and He imprinted form in matter, and He acted [*faciendo*] upon matter, in the first movable, and in the whole universe, and He produced it *ex nihilo* with all the diversity existing in it. And we already proved this by probable reasoning [*probabiliter*].[92]

## Concluding Remarks

Albert is fully aware that his doctrine does not attain the status of demonstrative certitude and that his arguments are dialectical. However, he contends that they are more plausible than Aristotle's. After all, Aristotle was not God and could go wrong. Albert was considerably helped by Averroes in his argumentation, but he kept to his own path. Inheriting from Boethius a conception of eternity that was more useful to his purposes, he distanced himself from Averroes on this point. Yet he agreed with Averroes, and with Algazel,[93] on the point that God's priority to the created world is not temporal but causal, and he continually proclaimed that God preceded the universe in the way that eternity ranks above time and cause ranks above effect.[94]

Albert's doctrine of atemporal emanation echoes Avicenna's explanation of the creation of motion. Book III, Chapter 11 of Avicenna's *Physics* has the heading

90 Ibid., p. 576, v. 77: 'per electionem et voluntatem'.
91 Ibid., p. 577, vv. 44–45.
92 Ibid., VIII.1.14, p. 578, vv. 10–14.
93 Algazel, *The Incoherence of the Philosophers*, no. 81, trans. by Marmura, p. 31.
94 *Physica*, VIII.1.1, p. 550, vv. 3–4: 'ordinem aeternitatis ad tempus et causae ad causatum'.

'About the fact that nothing precedes motion and time save the essence of the Creator (may He be exalted) and that neither of the two has a first part in themselves'.[95] Avicenna argues from the standpoint of the possibility of existence that precedes any created thing, and affirms that time and motion are among the entities with an unstable existence (*ghayr qārrat al-wujūd*). This unstable existence is grounded upon 'exchange and transition from certain things to others', and therefore motion cannot have a first part. Avicenna establishes that only the divine essence is before it, but not temporally.[96] Avicenna's arguments were not known to Albert, since they are located in Chapter 11 of Book III and, as I have noted, the translation ended before Chapter 10, but there is a noteworthy agreement in a dependence on the divine at the level of motion.

In this paper, I have intentionally omitted a matter potentially related to Albert's thinking: Latin Averroism. Albert wrote his *Physics* around 1251–52,[97] and Latin Averroism was present at the University of Paris shortly afterwards, around 1255–60.[98] The Averroists insisted that rational arguments did not prove the creation of the world and that Aristotle's arguments based on the eternity of motion were stronger than those of his opponents. Albert, in contrast, intended to show that philosophy had strong arguments to prove the creation of the world, although they were found not in physics, but in metaphysics. Accordingly, Albert grounded his rational arguments on metaphysics, the speculative science *de substantia sensibili et immobili*, whereas the Latin Averroists accepted creation on grounds of the Christian faith alone.

In Albert's philosophical architecture, God is the First Cause but not the First Mover, as David Twetten has lucidly explained.[99] The distinction — and not identification — between God and the First Mover allows Albert to keep the eternity of motion apart from the eternity of the world, since the First Cause ontologically supersedes the First Mover, which remains constrained by the conditions of motion that cannot constrain God.

95 *Al-Shifā': al-ṭabī'īyāt*, ed. by Madkūr and Sa'īd, p. 232, vv. 12–13. For a detailed study of motion, see Hasnaoui, 'The Definition of Motion in Avicenna's *Physics*'.

96 Avicenna, *The Physics of 'The Healing'*, trans. by McGinnis, III.11, vol. 2, p. 364: 'Thus motion has no temporal beginning, but exists in the manner of an atemporal creation where nothing precedes it save the being of the Creator'.

97 According to Hossfeld in his editorial introduction, though hesitantly ('Non est vitiosum concludere'). *Physica*, Prolegomena, pp. v–vi. James A. Weisheipl gives a slightly earlier date, before 1250. Weisheipl, 'Albert's Works on Natural Science', p. 565.

98 Averroes and his reception by the Latin philosophers has been the subject of copious research. See Puig Montada, 'Averroïsme, une histoire chrétienne mais pas seulement'.

99 Twetten, 'Albert the Great on Whether Natural Philosophy'.

## Works Cited

### *Primary Sources*


Albertus Magnus, *Albert le Grand, Métaphysique. Livre XI, traités II et III*, ed. and trans. by Isabelle Moulin (Paris: Vrin, 2009)

———, *Metaphysica, libri VI–XIII*, ed. by Bernard Geyer, Editio Coloniensis, 16/2 (Münster: Aschendorff, 1964)

———, *Physica, libri V–VIII*, ed. by Paul Hossfeld, Editio Coloniensis, 4/2 (Münster: Aschendorff, 1993)

Algazel [al-Ghazālī], *Tahafot al-Falasifat*, ed. by Maurice Bouyges (Beirut: Imprimerie Catholique, 1927)

———, *The Incoherence of the Philosophers*, trans. with introduction and notes by Michael E. Marmura (Provo, UT: Brigham Young University Press, 1997)

Aristotle, *Al-Ṭabīʿa* [*Physics*], ed. by ʿAbd-ar-Raḥmān Badawi, 2 vols (Cairo: Al-Hay'a al-Miṣrīya al-ʿĀmma li-l-Kitāb, 1965)

———, *Aristotle: The Physics*, trans. by Philip H. Wicksteed and Francis M. Cornford, vol. 2 (London: William Heinemann, 1934)

———, *Aristotle's* Physics VIII, *Translated into Arabic by Isḥāq ibn Ḥunayn (9th c.)*, ed. with an introduction and glossaries by Rüdiger Arnzen, Scientia Graeco-arabica, 30 (Berlin: De Gruyter, 2021)

———, *Metaphysica lib. I–X, XII–XIV, translatio anonyma sive 'media'*, ed. by Gudrun Vuillemin-Diem, Aristoteles Latinus, 25/2 (Turnhout: Brepols, 1976)

———, *Metaphysics*, trans. by Hugh Tredennick, vols 17 and 18 of *Aristotle in 23 Volumes*, Loeb Classical Library, 271 and 287 (Cambridge, MA: Harvard University Press, 1989)

———, *Physica: Translatio vetus*, ed. by Fernand Bossier and Jozef Brams. Aristoteles Latinus, 7/1 (Leiden: Brill, 1990)

———, *The Works of Aristotle*, vol. 2, *Physica*, trans. by R. P. Hardie and R. K. Gaye (London: Oxford University Press, 1962)

Averroes, *Aristotelis De physico avditv libri octo cum Averrois Cordvbensis variis in eosdem commentariis* [Long Commentary on the *Physics*], trans. by Michael Scot, Aristotelis opera cum Averrois commentariis, 4 (Venice, 1562; repr. Frankfurt am Main: Minerva, 1962)

———, *Commentarium Magnum in Aristotelis Physicorum Librum Octavum.* In der lateinischen Übersetzung des Michael Scotus mit einer Einleitung herausgegeben von Horst Schmieja (Frankfurt am Main: Institute for the History of Arabic-Islamic Science, 2020)

———, *Aristotelis Metaphysicorum libri XIIII cum Averrois Cordubensis in eosdem commentariis* [Long Commentary on the *Metaphysics*], trans. by Michael Scot, Aristotelis opera cum Averrois commentariis, 8 (Venice, 1562; repr. Frankfurt am Main: Minerva, 1962)

———, *Averroès: Grand commentaire de la Métaphysique d'Aristote, Livre lam-lambda (= Tafsīr mā ba'd aṭ-Ṭabī'at)*, ed. by Aubert Martin (Paris: Les Belles Lettres, 1984)

———, *Averroes' Tahafut al-Tahafut (The Incoherence of the Incoherence)*, trans. with introduction and notes by Simon van den Bergh, 2 vols (London: Luzac, 1969)

———, *Ibn Rushd's Metaphysics: A Translation with Introduction of Ibn Rushd's Commentary on Aristotle's Metaphysics, Book Lâm*, trans. by Charles Genequand (Leiden: Brill, 1984)

———, *Tafsīr mā ba'd aṭ-Ṭabī'at* [Long Commentary on the *Metaphysics*] ed. by Maurice Bouyges, 2nd ed., 3 vols (Beirut: Dar El-Machreq, 1972)

———, *Tahafot at-Tahafot: Texte arabe*, ed. by Maurice Bouyges (Beirut: Imprimerie catholique, 1930)

Avicenna, *Al-Shifā': al-ṭabī'īyāt*, ed. by Ibrāhīm Madkūr and Zāyid Sa'īd (Cairo: al-Hay'ah al-Miṣrīyah al-'Āmmah li-l-Kitāb, 1983)

———, *The Physics of 'The Healing': A Parallel English-Arabic Text*, trans. with an introduction and notes by Jon McGinnis, 2 vols (Provo, UT: Brigham Young University Press, 2009)

Boethius, *Consolation of Philosophy*, trans. with an introduction and notes by Joel C. Relihan (Indianapolis, IN: Hackett, 2001)

*Liber de causis, The Book of Causes (Liber de causis)*, trans. with an introduction by Dennis J. Brand, Medieval Philosophical Texts in Translation, 25 (Milwaukee, WI: Marquette University Press, 1984)

———, in Adriaan Pattin, 'Le *Liber de causis*: Édition établie à l'aide de 90 manuscrits avec introduction et notes', *Tijdschrift voor filosofie*, 28 (1966), 90–203

## *Secondary Works*

Hasnaoui, Ahmad, 'The Definition of Motion in Avicenna's *Physics*', *Arabic Sciences and Philosophy*, 11 (2001), 219–55

Hasse, Dag Nikolaus, *Latin Averroes Translations of the First Half of the Thirteenth Century* (Hildesheim: Olms, 2010)

Hossfeld, Paul, 'Gott und die Welt: Zum achten Buch der Physik des Albertus Magnus (nach dem kritisch erstellten Text)', *Miscellanea Mediaevalia*, 21 (1991), 281–301

Janssens, Jules, 'The Reception of Avicenna's *Physics* in the Latin Middle Ages', in *O Ye Gentlemen: Arabic Studies on Science and Literary Culture*, ed. by Arnoud Vrolijk and Jan P. Hogendijk (Leiden: Brill, 2007), pp. 55–64

Krause, Katja, and Henryk Anzulewicz, 'From Content to Method: The *Liber de causis* in Albert the Great', in *Reading Proclus and the Book of Causes*, vol. 1: *Western Scholarly Networks and Debates*, ed. by Dragos Calma (Leiden: Brill, 2019), pp. 180–208

Levin, Leonard, R. David Walker, and Shalom Sadik, 'Isaac Israeli', in *The Stanford Encyclopedia of Philosophy* (Summer 2018 Edition), ed. by Edward N. Zalta, https://plato.stanford.edu/archives/sum2018/entries/israeli/

Puig Montada, Josep, 'Averroïsme, une histoire chrétienne mais pas seulement: L'influence d'Averroès sur les penseurs chrétiens et juifs', *Doctor Virtualis, rivista online di storia della filosofia medievale*, 13 (2015), 91–117

Taylor, Richard C., 'The *Liber de Causis* (*Kalām fī maḥḍ al-khair*): A Study of Medieval Neoplatonism', 2 vols, Ph.D. dissertation, University of Toronto, 1981

Twetten, David B., 'Albert the Great on Whether Natural Philosophy Proves God's Existence', *Archives d'histoire doctrinale et littéraire du Moyen Âge*, 64 (1997), 7–58

———, Steven Baldner, and Steven C. Snyder, 'Albert's Physics', in *A Companion to Albert the Great: Theology, Philosophy, and the Sciences*, ed. by Irven M. Resnick (Leiden: Brill, 2013), pp. 173–219

Watt, W. Montgomery, 'al-Ashʿarī, Abu 'l-Ḥasan', in *Encyclopaedia of Islam, Second Edition online*, ed. by P. Bearman, T. Bianquis, C. E. Bosworth, E. van Donzel, and W. P. Heinrichs (Leiden: Brill, 2012), http://dx.doi.org/10.1163/1573-3912_islam_SIM_0780

Weisheipl, James A., 'Albert's Works on Natural Science (*libri naturales*) in Probable Chronological Order', in *Albertus Magnus and the Sciences: Commemorative Essays 1980*, ed. by James A. Weisheipl (Toronto: Pontifical Institute of Mediaeval Studies, 1980), pp. 565–77

IRVEN M. RESNICK

# Chapter 6. Albert the Great's Treatment of Avicenna and Averroes on a Universal Flood and the Regeneration of Species

Albert the Great famously declared that 'You cannot be a complete philosopher without knowing both philosophies, Aristotle's and Plato's'.[1] Although little of Plato's work was available to him in Latin translation, beyond Calcidius's truncated version of the *Timaeus*,[2] Albert formed a picture of Plato's philosophy indirectly from Aristotle's remarks, from Galen, from the Neoplatonist Pseudo-Dionysius the Areopagite (on whose works Albert wrote extensive commentaries),[3] and from other late antique or medieval authors, including medieval Arab and Jewish writers whose works were then becoming available in Latin trans-

1 Albertus Magnus, *Metaphysica*, I.5.15, ed. by Geyer, p. 89, vv. 85–87. For the significance of this passage for Albert's views on Plato and Aristotle, see Franchi, 'Alberto Magno'.

2 Plato, *Timaeus*, ed. by Waszink. Just after the middle of the twelfth century, Plato's *Phaedo* and *Meno* had been translated from Greek by Henry Aristippus. I am not aware that Albert had access to these, but he does seem to be aware, indirectly, of some doctrines found in them. For Platonic influences upon Albert, see especially Anzulewicz, 'Die platonische Tradition bei Albertus Magnus'; Anzulewicz, 'Plato and Platonic/Neoplatonic Sources'; Anzulewicz, 'Albertus Magnus als Vermittler'. For Neoplatonic influence upon Albert, see also Pagnoni-Sturlese, 'A propos du néoplatonisme'.

3 Albert's commentaries on Pseudo-Dionysius's *De caelesti hierarchia* and *De ecclesiastica hierarchia* have appeared in Ed. Colon. 36/1 and in Ed. Colon. 36/2, and his commentaries on *De divinis nominibus* and the *Epistulas* have appeared in Ed. Colon. 37/1 and Ed. Colon. 37/2. Albert utilized the Latin translation produced by John the Saracen, who taught at Poitiers in the latter part of the twelfth century. For Pseudo-Dionysius's influence upon Albert, see Anzulewicz, 'Pseudo-Dionysius Areopagita und das Strukturprinzip'.

**Irven M. Resnick** (Irven-Resnick@utc.edu), Professor and Chair of Excellence in Judaic Studies at the University of Tennessee at Chattanooga, has edited or translated six volumes on Albert the Great.

*Albert the Great and his Arabic Sources*, ed. by Katja Krause and Richard C. Taylor, Philosophy in the Abrahamic Traditions of the Middle Ages, 5 (Turnhout: Brepols, 2024), pp. 167–194
BREPOLS PUBLISHERS 10.1484/M.PATMA-EB.5.136486

lation.[4] While there were no medieval Arabic translations of Plato's dialogues (but only epitomes of some of them), Albert received important works of Aristotle in Latin translations from Arabic — many completed by Gerard of Cremona in twelfth-century Spain[5] — and in four instances via the lemmata in Michael Scot's translations of Averroes's long commentaries. Ancient philosophical and scientific texts translated into Latin from Greek exemplars increased in number, moreover, during the twelfth and thirteenth centuries.[6] Albert the Great (who did not know Greek) received Aristotle's *Nicomachean Ethics,* whose ten books Bishop Robert Grosseteste of Lincoln had translated from the Greek in *c.* 1246–47, and he may also have had access to William Moerbeke's translation from the Greek of Aristotle's *De animalibus* (*c.* 1260). Because of persistent efforts from Late Antiquity to harmonize Plato and Aristotle, the differences that separated the two were initially muted, and this attempt at harmonization is reflected in the work of Albert the Great as well.

The philosopher writing in Arabic that Albert most admired and most often cited was Avicenna. Albert included Avicenna among 'the finest of philosophers',[7] and Albert's natural philosophy borrowed heavily from Avicenna's *Canon of Medicine,* his *Abbreviatio de animalibus,*[8] and his discussion of *De anima.* Although Averroes enjoyed a reputation among Scholastics as *the* Commentator upon Aristotle, Albert often prefers Avicenna's interpretation,[9] or seeks to reconcile the views of Avicenna and Averroes even though Averroes was perhaps Avicenna's

4 These would include not only Muslim, but also Jewish authors whose work appeared in Latin translation, e.g., Isaac ben Solomon Israeli (*c.* 855–*c.* 955), whose medical treatises were translated or adapted in Latin versions by Constantine the African in the eleventh century and whose *Book on the Elements* was also translated by Gerard of Cremona in the late twelfth century. Albert was also the first Parisian Scholastic author to cite a Latin translation of Maimonides' *Guide of the Perplexed* (*Dux neutrorum*), which possibly was translated in Paris at the priory of St James. See Hasselhoff, *Dicit Rabbi Moyses,* pp. 123–24. For discussion, see Maimonides, *Dux neutrorum vel dubiorum,* Part 1, ed. Di Segni, pp. 21–24.

5 As Charles Burnett noted, at least 116 works were translated from Arabic to Latin in Spain by known authors between 1116 and 1187. See Burnett, 'Some Comments on the Translating of Works'; Burnett, 'Arabic into Latin'; Burnett, 'Arabic Philosophical Works Translated into Latin'.

6 See esp. Hasse, 'Influence of Arabic and Islamic Philosophy'; D'Ancona, 'Greek Sources in Arabic and Islamic Philosophy'. For a list of medieval translations of Greek philosophical works, see also Dod, 'Greek Aristotelian Works Translated into Latin'; Trizio, 'Greek Philosophical Works Translated into Latin'; Gutas, 'Greek Philosophical Works Translated into Arabic'. Still useful are the lists in Grant, 'Translation of Greek and Arabic Science into Latin'. For the cultural significance of medieval translations from Greek exemplars, see Mavroudi, 'Translations from Greek into Latin and Arabic'.

7 Alongside Aristotle, Ptolemy, and Messellach (Māshā'allāh). Albert, *De generatione et corruptione,* ed. by Hossfeld, p. 206, vv. 32–33.

8 Avicenna's summary and commentary on Aristotle's zoological work, which became available in the 1230s in Michael Scot's translation. See d'Alverny, 'L'explicit du "De animalibus"'.

9 For example, Albert prefers Avicenna's definition of the science of First Philosophy or metaphysics over that of Averroes. See Albert, *Physica,* I.3.18, ed. by Hossfeld, p. 76, vv. 37–56. For discussion of Albert's debt to Avicenna, see esp. Bertolacci, 'Albert's Use of Avicenna'. But for discussion of Albert's evolving position in his later works, in which he begins to rely more clearly upon Averroes, see Bertolacci, '"Averroes ubique Avicennam persequitur"'.

fiercest philosophical opponent.[10] A good example of this reconciliation can be found in Albert's discussion of a universal flood and the potential for the regeneration of species thereafter by celestial power. Albert treated this topic in a lengthy digression in his commentary on the pseudo-Aristotelian *Book on the Causes of the Properties of the Elements*,[11] where he notes:

> There is, however, a great dispute between Avicenna and Averroes on these floods in their books, concerning what repairs the earth and replaces the animals on it after they have been extinguished and killed by a flood of water and of fire.[12]

The nature of their disagreement will be treated below, but Albert's attempt to harmonize the two authorities is clear from his remark that 'it seems to me, however, that one ought to agree with each of them in a certain respect'.[13]

## A Flood of Water and Fire

Before we can examine Albert's unique reconciliation, his reference to a flood not only of water but also of fire may require some explanation for the contemporary reader, for whom a flood is understood only in relation to water. His treatment of the floods is largely drawn from Avicenna's 'On Floods' (*De diluviis*),[14] which is an excerpt from a longer treatment of meteorology found in the second section treating natural philosophy in Avicenna's *Book on the Healing* (*Kitāb al-Shifāʾ*). Until Avicenna's larger discussion of meteorology was translated in its entirety by Johannes Gunsalvi, *c*. 1274–80, with the assistance of a Jew named Solomon, only two parts were available in Latin translation: *De diluviis* and *De mineralibus*.[15]

---

10 On Averroes's opposition to Avicenna, see Cerami, 'A Map of Averroes' Criticism'; Bertolacci, 'Averroes against Avicenna'. For development in Albert's attempts to harmonize or reconcile Avicenna and Averroes, see Bertolacci in this volume.

11 Albertus Magnus, *Liber de causis proprietatum elementorum*, ed. by Hossfeld (hereafter *De causis*). Hossfeld's edition contains both Albert's paraphrastic commentary and, at the bottom of each page, Gerard of Cremona's Latin translation of the text from the Arabic, after an edition established in Vodraska, 'Pseudo-Aristotle *De causis proprietatum et elementorum*'. For the translation, see *On the Causes of the Properties of the Elements*, trans. by Resnick (hereafter *On the Causes*). Albert likely completed this work between 1251 and 1254, while he was teaching in Cologne.

12 Albertus Magnus, *De causis*, I.2.13, p. 85, vv. 32–36; *On the Causes*, p. 90.

13 Albertus Magnus, *De causis*, I.2.13, p. 86, vv. 53–54; *On the Causes*, p. 93.

14 Found in chapter 2.6 of the meteorological part of Avicenna's *Book on the Healing*, which, once translated into Latin — probably by Michael Scot — circulated as an independent treatise with the title *De diluviis in Thimaeum Platonis*. On Michael Scot as the likely translator, see Hasse and Büttner, 'Notes on Anonymous Twelfth-Century Translations', esp. 344–49, 357. For Avicenna's treatment *de diluviis*, see Avicenna, *De diluviis*, ed. by Alonso Alonso (Latin text on pp. 306–08).

15 The section 'On Minerals', *De mineralibus*, had circulated in translation with glosses or commentary by Alfred of Sareshel (or Shareshill), but Alfred had inserted Avicenna's text at the end of Aristotle's *Meteorologica*, so that it appeared as an Aristotelian work. For Alfred's commentary, see *Alfred of Sareshel's Commentary on the* Metheora *of Aristotle*, ed. by Otte. For Alfred's role in the transmission

Avicenna's 'On Floods' provides a discussion prompted by certain passages in Plato's *Timaeus*. In the *Timaeus*, Critias recalls that Solon sought to introduce an Egyptian priest to figures from ancient Greek creation myths, for example to Phoroneus and Niobe, and to Pyrrha and Deucalion,[16] who lived 'after the deluge of the world [*inundatio mundi*]'[17] — a deluge, according to Albert the Great, that was actually the Biblical Flood described in Genesis (Gen. 6–8).[18] After reckoning the number of generations from the Flood, Solon attempted to calculate the age of the world. The priest rejected these calculations, however, based on his contention that the world has suffered many floods, both by fire and water.[19] An example of a flood by fire, he averred, is inscribed in the ancient myth of Phaeton, the child of the Sun, who caused his father's chariot to pass so close to earth that it burned its surface.[20] This, the priest alleged, conceals the scientific truth that a deviation in the world's orbit over a long period of time necessarily results in its destruction by fire, a destruction which Egypt alone has escaped due to its regular flooding by the waters of the Nile.[21]

Albert turns not only to Plato but also to Ovid for this doctrine of floods of water and fire.[22] Avicenna expands the notion and suggests that there may be a 'flood' (*diluvium*) stemming from each of the four elements. He defines a 'flood' generally as a catastrophic event that represents 'the victory of one of the [four] elements over the habitable quadrant [of the earth] or over one part [of it]'.[23] As a result, Avicenna allows for a flood not only of water or fire, but also of earth or air. In his commentary on *De caelo et mundo*, Albert may have had Avicenna in mind

---

of Avicenna's meteorology, see Mandosio, 'Follower or Opponent of Aristotle?' Albert the Great certainly knew Alfred's commentary, since he cited it in his discussion of the rainbow.

16 Deucalion is the Greek 'Noah': he and his wife Pyrrha, due to their piety, are chosen by the gods to survive the 'Universal Flood' of Greek mythology. The most common version of the tale is that of Ovid, *Metam.* I.262–415.

17 Plato, *Timaeus* 22A, ed. by Waszink, p. 13.

18 Albertus Magnus, *De causis*, I.2.9, p. 76, vv. 48–52; *On the Causes*, p. 71. Cf. Augustine, *De civ. Dei*, XVIII.8. Among Albert's contemporaries, Archbishop Rodrigo Jiménez de Rada (d. 1247) claims: 'At various times there have been other local floods, like the flood under Deucalion in Thessaly, which was about the time of Moses, and this flood was 1247 years after the flood of Noah'. *Breuiarium historiae catholicae*, I.24, ed. Fernández Valverde. Conrad de Mure (d. 1281) also identifies the flood of Deucalion as a local flood (*diluvium particulare*). See *Fabularius*, ed. by van de Loo, p. 408.

19 Plato, *Timaeus* 22D, ed. by Waszink, p. 14.

20 On Albert and the myth of Phaeton, see Albertus Magnus, *Politica*, II.6, ed. by Borgnet, p. 154b; *Summa theologiae*, pars 2, tr. 11, q. 52, m. 1, ed. by Borgnet, p. 552b; *De vegetabilibus*, VII.1.6, ed. by Meyer and Jessen, p. 607.

21 Plato, *Timaeus* 22D, ed. by Waszink, p. 14. For Albert's explanation of the regular flooding of the Nile, see *De homine*, I.1, ed. by Anzulewicz and Söder, p. 575, vv. 33–49.

22 See Albertus Magnus, *Summa theologiae*, pars 2, tr. 11, q. 52, m. 1, ed. by Borgnet, p. 552b; *Politica*, II.6, ed. by Borgnet, pp. 154b–155a; *De quindecim problematibus*, art. 6, ed. by Geyer, p. 38, vv. 54–65.

23 'Et est diluvium victoria unius elementorum super quartam habitabilem aut super unam partem'. Avicenna, *De diluviis*, ed. by Alonso Alonso, p. 306.

when he remarked that 'we found the chief men in philosophy speaking about a flood [*diluvium*] of any element upon another, as is clear in the enumeration of the floods which is found in the beginning of Plato's *Timaeus*'.[24]

## General and Local Floods and their Theological or Natural Causes

Albert also treats catastrophic events or 'floods' caused by elements other than fire and water in his commentary on the *Meteora*, in which he acknowledges that earthquakes and volcanoes may cast up soil or ash onto another place in a kind of flood. During an earthquake, moreover, the ground's movement and the power of winds (*venti*) beneath the earth may force water to the surface, producing a local flood (*diluvium particulare*).[25] In like manner, a marine earthquake may sunder the sea floor and result in a tidal wave that floods coastal areas; in just this fashion, Albert adds, several towns flooded in Normandy in his own day.[26] Similarly, changes in the revolutions or orbits of the planets may produce local flooding, as occurred in ancient Greece under King Dhuphilinus,[27] and they may bring about climate changes that dry out existing bodies of water or submerge other areas that once formed dry land. This, Albert remarks, is exactly what happened to the five cities of the Pentapolis,[28] which were placed at the bottom of the Dead Sea, and in Tungra Octavia,[29] whose fertile fields were once beneath the sea.[30]

Albert's treatment on the natural causes of floods of water and fire largely follows the noteworthy attempt of William of Conches (*c.* 1090–1154) to articulate them in his twelfth-century dialogue *Dragmaticon*. Unlike Albert's Dominican contemporary in Paris, Vincent of Beauvais, who frequently cites William of Conches by name in his massive *Speculum naturale*, Albert does not name William of Conches here, but he does seem to have known the *Dragmaticon*.[31] A watery

24 'nos invenimus praecipuos in philosophia viros loquentes de diluvio cuiuslibet elementi super aliud, sicut patet in enumeratione diluviorum, quae habetur in principio Timaei Platonis'. Albertus Magnus, *De caelo et mundo*, III.2.4, ed. by Hossfeld, p. 231, vv. 21–25.

25 Albertus Magnus, *Meteora*, III.2.17, ed. by Hossfeld, p. 146, vv. 22–24; cf. Avicenna, *De diluviis*, ed. by Alonso Alonso, p. 306.

26 Albertus Magnus, *Meteora*, III.2.17, ed. by Hossfeld, p. 146, vv. 39–41.

27 Ibid., II.2.15, p. 79, vv. 48–50.

28 That is, the five cities of Sodom, Gomorrah, Segor, Adama, and Seboim, mentioned at Wisd. 10.6 as having been destroyed by fire; cf. Gen. 19.24–25. It has often been thought that the Pentapolis were located around or even beneath the Dead Sea, as Albert himself bears witness. See also Andrew of St Victor, *Expositionem super Heptateuchum*, ed. by Lohr and Berndt, v. 1848; Petrus Cantor, *Summa quae dicitur Verbum adbreuiatum*, II.46, ed. by Boutry.

29 Perhaps the Belgian city of Tongres. See *De causis*, I.2.3, p. 66, vv. 50–56; *On the Causes*, p. 50 and n. 111.

30 Albertus Magnus, *Meteora*, II.2.15, ed. by Hossfeld, p. 79, vv. 30–35.

31 Silvia Donati, the editor of Albert's *De sensu et sensato*, identifies several instances in which Albert refers obliquely to the *Dragmaticon* and to William of Conches among 'some of the modern Latins'

flood, William remarks, may arise from increased moisture in the atmosphere, whereas a flood of fire may result when the earth absorbs so much moisture that the sun's heat then burns its surface. Or perhaps there are great rivers that flow beneath the earth; when these break through its surface, they cause a flood of water, but when they retreat further below the surface, the sun's heat causes a conflagration. Finally, William turns to an astronomical explanation:

> Some say that these things happen from the concurrent rising and setting of the planets. For if all the planets rise at the same time, being further removed than usual from the earth, they consume less moisture. This causes the moisture to increase, until it spreads over the earth, becoming a flood. But if only one planet or two or three rise without the others, the moisture does not increase to such an extent; for from what extent it increases from the rising of the one group, [to that same extent] it decreases from the setting of the other. If, however, all the planets set simultaneously, they burn the earth from their proximity to it and cause a conflagration. But if only two or three set, no conflagration occurs, because to the extent that the one group increases the probability of a conflagration from its proximity, [to the same extent] the other group decreases it from its remoteness. And notice that there is a general and a local flood. It is impossible to have two general floods but quite possible to have many local ones. And so Plato says that several floods have occurred before, which Augustine also endorses.[32]

Having distinguished a universal from a local flood, the *Dragmaticon* offered several natural explanations for both a flood of water and one of fire before the arrival of Avicenna's and Averroes's treatments in the Latin West. The possibility of explaining a theologically significant event described in the Bible by means of natural causality thus entered Albert's thought through a variety of sources. Yet William of Conches's *Dragmaticon* also had its argumentative limitations, particularly when compared to the Arabic material. While William rejected the possibility of more than one universal flood, he failed to make clear whether he considered that it is (logically) impossible to have two universal floods at the same time, or that it is impossible for one to follow after another in succession.

In his commentaries on Aristotle's *Meteora* and *De caelo et mundo*, Albert contributed to the expanding thirteenth-century discussion of universal and local floods and their causes, and at the same time paid testimony to a tension between theological and natural explanations.[33] In his *On the Causes of the Properties*

(*nonnuli Latinorum moderni*). See *De sensu et sensato*, I.1.5; I.3.2, ed. by Donati, p. 27, v. 26 and p. 98, v. 35 (*quidam dicunt*).

32 William of Conches, *A Dialogue on Natural Philosophy*, V.12.4, trans. by Ronca and Curr, p. 116. For the Latin text, see *Dragmaticon philosophiae*, V.12.4, ed. by Ronca, p. 176. At *Dragmaticon*, IV.1.3, William makes it clear that he neither confirms nor rejects this explanation that is attributed to unnamed others. For William's references to Plato and Augustine, see *Timaeus* 22C, ed. by Waszink, p. 14, and Augustine, *De civ. Dei*, XVIII.8.

33 See Schenk, 'Dis-Astri', esp. pp. 56–59.

*of the Elements*, Albert acknowledges that a proper investigation of a flood — whether universal or local in character, thus taking up William's distinction — will consider both theological and natural causes. Following William and using Avicenna and Averroes, Albert rejects entirely the criticism by theologians who eschew the search for natural causes:

> There are, however, some who attribute all these things [i.e., floods] to a divine disposition alone and who say that we should seek no cause for things of this sort other than the will of God. We agree with them in part, because we say that these things occur by the will of God, who governs the world, as a punishment for the evildoing of men. But we still say that God does these things on account of a natural cause, of which he who confers motion on all things is himself the first mover. However, we are not seeking causes of his will, but we are seeking the natural causes that are like certain instruments through which his will in such matters is brought into effect.[34]

For these theologians, then, the proximate cause of the universal flood is human sinfulness, while the ultimate cause is located in God's will. The universal flood is, moreover, not a natural but a supernatural event which God visited upon humanity only once in the past as a punishment.[35] For thirteenth-century theologians, the Noahide flood was often understood to be a divine punishment for sodomy or sexual crimes against nature.[36] In his commentary on Aristotle's *Meteora*, Albert also remarks that among the ancient theologians, the 'Hesiodists' as he calls them,[37] the rainbow became a sign that Providence will never again allow a universal flood of either water or fire to destroy the world.[38]

Albert's references to ancient 'Hesiodists' here likely point to an ongoing debate over theological or natural causes of a universal flood that can also be traced in his sources. Whereas Avicenna suggested the natural possibility of

---

34 Albertus Magnus, *De causis*, I.2.9, p. 76, v. 75–p. 77, v. 2; *On the Causes*, p. 71.

35 Vincent of Beauvais calculates that the Noahide flood occurred precisely in 1656 *Anno Mundi*. See his *Speculum naturale*, 32, cap. 27, in *Speculum quadruplex*, ed. Douai, vol. 1, pp. 2419–20.

36 See Peter of Poitiers, *Summa de confessione*, ed. by Longère, p. 65; Diekstra, 'Robert de Sorbon's *Cum repetes*', p. 139.

37 *At Liber de natura et origine animae*, II.7, ed. by Geyer, p. 30, v. 17, Albert incorrectly identifies Hesiod as an Epicurean philosopher (ibid., II.11, p. 35, vv. 25–26 and p. 36, vv. 36–43), and he identifies 'Hesiodists' as Epicureans at *Physica*, II.2.20, ed. by Hossfeld, p. 128, vv. 85–86. Hesiod was not properly speaking a philosopher at all, but rather related a comprehensive version of Greek creation myths in his *Theogony*. In *De praedicamentis*, VII.4, ed. by Borgnet, p. 279b, Albert identifies Hesiodists as those who pursue a 'theologizing philosophy', and at *Physica*, II.2.10 and V.3.3, ed. by Hossfeld, p. 114, v. 30 and p. 431, vv. 67–68, Albert refers to *Hesiodistae theologi*. Albert also mentions Hesiod and his followers at *De animalibus*, VI.3.1.99, ed. by Stadler, vol. 1, p. 483, v. 23; Albertus Magnus, *Albertus Magnus On Animals*, trans. by Kitchell and Resnick, vol. 1, p. 572 and n. 206; and *Ethica*, I.5.2, ed. by Borgnet, p. 58b. Pseudo-Albert identifies Seneca, among other ancients, as a Hesiodist. See Ps.-Albertus Magnus, *Philosophia pauperum*, IV.20, ed. Borgnet, p. 495a. The identity of these 'Hesiodists' remains uncertain.

38 Albert, *Meteora*, III.4.6, ed. by Hossfeld, p. 180, v. 69–p. 181, v. 1. Cf. Gen. 9.13.

a flood stemming from any of the four elements, Maimonides (d. 1204), for example, reproves those who 'deceive and delude others to believe that the Deluge in the time of Noah was merely due to a concentration of water, and was not a divine punishment for the immorality of the time'.[39] Contrary to Maimonides and Latin theologians, who seek all explanations for the flood in God and His will, the philosophers seek to explain floods by terrestrial and celestial causes.

Albert attempted a path of reconciliation between these two models, suggesting that God, as the first mover, utilizes natural causes to produce all effects in the world, effects which include local and universal floods of water and fire. Albert promised to examine the celestial causes in his proposed but unwritten *Astronomica disciplina*.[40] The Latin Scholastic debates over whether a universal flood can have a natural cause or only a supernatural explanation are also reflected in the 219 propositions of the Parisian Condemnation of 1277. Proposition 182 — *Quod possibile est quod fiat naturaliter universale diluvium ignis* — was condemned, and with it the claim that a universal flood of fire could arise from a natural cause.[41] For many Latin thinkers to follow, this condemnation implied, analogously, that a universal flood of water could not occur *secundam viam naturae*. An anonymous commentary on the Parisian Condemnation from the circle of the fifteenth-century Parisian Albertist John de Nova Domo clearly establishes this linkage:

> The reason for this article's error is that, although a universal flood of water would be produced, nonetheless such a flood of water was not produced naturally, since it was produced through a declaration, because God omnipotent had told Noah that he had to build an ark (Gen. 6.14), and because such a flood, moreover, was produced on account of man's sins committed in Sodom and Gomorrah, and so too in other cities. Therefore, certainly such [a flood] was not produced naturally, because nature does not punish sins, but only glorious God. Also, [it was not natural] because the water rose above the mountains,[42] which cannot occur naturally, since every light [element] is moved up and every heavy one is moved down toward the centre.[43] If, then, the flood of waters was not produced naturally, then therefore a flood of fire also will not occur naturally, because otherwise the

39 Maimonides, *Epistle to Yemen*, ed. by Twersky, p. 455.

40 Albert, *Meteora*, II.2.15, ed. by Hossfeld, p. 79, vv. 63–68.

41 This proposition receives number 182 in Piché and LaFleur, *La condemnation parisienne de 1277*, as well as in the *Chartularium Universitatis Parisiensis*, ed. by Denifle, vol. 1, p. 553. Mandonnet assigned the number 193 to this same proposition. Mandonnet, *Siger de Brabant et l'Averroïsme latin au XIIIe siècle*, part 2, p. 190.

42 See Gen. 7.20.

43 Aristotle, *Phys.* IV.4, 212a25–26; cf. Albertus Magnus, *Physica*, IV.1.12, ed. by Hossfeld, vol. 1, p. 224, vv. 34–42.

> principles pertaining to physics would be changed, namely that something light would ascend.[44]

Present-day scholarship has debated whether Albert was himself a target of the condemnation of proposition 182, a notion based in part on later remarks by Conrad of Megenberg and an anonymous fifteenth-century commentary known by its incipit as *Quod Deus*.[45] As Roland Hissette has shown, Conrad of Megenberg reproaches those who considered a universal flood to be a natural possibility.[46] Conrad remarks:

> Some of the masters in natural philosophy like Albert in *De proprietatibus elementorum* think that a universal catastrophe [*diluvium universale*] of water will occur and be caused by nature, if all the planets are together in diametrical opposition in the sign of Aquarius; and that a universal catastrophe of fire will occur if they are standing in diametrical opposition in the sign of the Lion.[47]

Albert the Great, however, influenced by Avicenna, had defended the position of natural causes in his *On the Causes of the Properties of the Elements*, and identified four celestial causes for a universal flood:

> I say, therefore, that the cause of the universal flood is comprised of four causes, of which one was a true seven-planet conjunction. The second is that all or many of them were in the lower part of their revolutions. The third is that the conjunction was such that it began in the sign of Aquarius near the four stars that are called the Water Pot of Aquarius [*Hydria Aquarii*] and which some call the Out-pourer of Water [*Effusor aquae*], because it was discovered that they have a special, wondrous effect in moving the waters [...].[48] The fourth and last cause is, however, that the moon was strengthened in its powers at the hour of conjunction, such that it was itself ascending from the circle of the hemisphere,[49] and that this conjunction was directly over the

---

44 'Ratio falsitatis illius articuli est, quia, licet factum sit universale diluvium aquarum, tamen tale diluvium aquarum non est factum naturaliter, cum est factum per intimationem, quia omnipotens deus dixerat Noe, quod deberet construere archam. Et etiam quia factum est tale diluvium propter enormia peccata hominum perpetrata in Sodoma et Gomorra, et sic de aliis civitatibus. Ergo utique tale non est factum naturaliter, quia natura non vindicat peccata, sed solus deus gloriosus. Etiam quia aqua ascendebat super montes, quod naturaliter non potuit fieri, cum omne leve sursum et omne grave deorsum ad centrum movetur. Si ergo diluvium aquarum non est factum naturaliter, ergo etiam diluvium ignis non erit naturaliter, quia alias principia physicalia mutarentur, scilicet quod leve ascenderet'. Wels, *Aristotelisches Wissen und Glauben im 15. Jahrhundert*, p. 101.

45 Lafleur, Piché, and Carrier, 'Le statut de la philosophie dans le décret parisien de 1277'.

46 See esp. Hissette, 'Albert le Grand et l'expression "Diluvium ignis"'; Hissette, 'Les recours et allusions à Albert le Grand'.

47 Gottschall, 'Conrad of Megenberg and the Causes of the Plague', p. 325. The 'sign of the Lion' is that of Leo in the zodiac. Aries, Leo, and Sagittarius form a fiery triplicity. See *Summa theologiae*, pars 2, tr. 11, q. 52, m. 1, ed. by Borgnet, p. 552b.

48 On the *Effusor aquae*, see *De causis*, I.2.6, p. 71, v. 93.

49 See Albertus Magnus, *De caelo et mundo*, II.2.5, ed. by Hossfeld, p. 135, vv. 12–39.

> water, and that it was at the hour and day of the moon. For then without doubt the moon had within itself whatever light was in all the planets.[50] And the moon moves with all that light according to the nature of the moon, and for this reason the water did not then advance gradually but leapt forth, as it were, toward the moon from the deepest bowels of the earth.[51]

In addition to these four celestial causes, Albert identified a fifth universal cause in the sublunary sphere:

> But another cause was in the lower bodies, it too universal, since it was necessary that on account of the motion of the water there be at that time many vapors in the air and that the power of the moon prevail in it, and for this reason the rains poured forth that converted both the vapors and much of the air into water. And this is one part of the lower cause. A second, however, was that there were many thick and strong vapors in the earth, which burst forth from the solid earth into the waters and cast out the waters from the depths of the abyss.[52]

When these causes are present with a diminished power or intensity, they produce only local or particular floods. In a significant endorsement of the science of astronomy, Albert adds that the 'discovery of the time of a flood, however, and its size and precise location, can only be known by the science of the movement of the stars'.[53] In his commentary on the *Sentences*, in contrast, he had limited astronomy's predictive power to a local flood, and did not extend it to a universal one — certainly not to the flood of fire expected at the End Time. For this reason,

> the end of the world cannot be known from the motion of heaven by the proper science of the astrologers [*ex propria scientia mathematicorum*]; but an end of the habitation of a kingdom's city and of our habitable earth, but not of the entire earth, can be known by the proper science of the astrologers; because when one quadrant of the habitable [earth] becomes uninhabitable now owing to the short length of the planet's diameters, and especially that of the sun, then for that very reason in another quadrant the length of another part of their same diameters will be longer; and the [earth] becomes habitable as an inhabitant crosses from one habitable quadrant into another. And nothing can be known more completely by the proper science of the astrologers, just as everyone knows who has come to know anything perfectly from art.[54]

---

50 See ibid., II.3.15, p. 178, vv. 7–14.

51 Albertus Magnus, *De causis*, I.2.9, p. 78, vv. 56–65 and 76–85; *On the Causes*, p. 76. For a useful discussion of Albert's understanding of the power of celestial conjunctions to cause floods, see Rutkin, 'The Natural Philosophical Foundations for Astrological Revolutions'.

52 Albertus Magnus, *De causis*, I.2.9, p. 78, v. 86–p. 79, v. 2; *On the Causes*, p. 76.

53 Albertus Magnus, *De causis*, I.2.9, p. 79, vv. 35–37; *On the Causes*, p. 77.

54 Albertus Magnus, *Super IV Sententiarum*, dist. 43, art. 7, ed. by Borgnet, p. 517b.

Perhaps based on this passage, Pseudo-Albert the Great's *De secretis mulierum* concludes mistakenly that Albert contradicts Avicenna concerning the natural possibility of a universal flood, and not concerning its astronomical predictability, since

> Aristotle says in the first book on *Meteorology* that a universal flood would be impossible in nature either by fire or by water. Albert gives the reason for Aristotle's view: that a flood is caused by a humidifying constellation, therefore if such a constellation were to control one part of the earth, then another constellation, that is, one that dries things out, would have power over another part of the earth.[55]

Based on this explanation, the anonymous late medieval Commentary A to the text of *De secretis mulierum* insists that the sort of planetary conjunction that could produce a universal flood is naturally impossible; natural causes can produce only a local flood, whereas a universal flood has occurred only miraculously as a punishment for human sin.[56] This text may reflect an attempt to shield Albert from criticism following the Parisian condemnation of 1277, just as Conrad of Megenberg suggested that Albert was not *really* discussing a universal flood in his *On the Causes of the Properties of the Elements*, perhaps because Albert also acknowledged there that 'Averroes says that there will never be a flood that is so universal that there is no evasion and no escape'.[57] Albert *does* dispute the claim of some philosophers and astrologers that it is possible to determine the end of the world in a flood of fire by predicting the appearance of the Great Year;[58] not even the saints have certain knowledge of when the End will occur. Although philosophers may be able to predict the appearance of local floods based on the movements of the celestial bodies, 'none determines the Day of Judgment'.[59]

In summary, then, Albert maintains that it is possible for natural philosophy to explain both universal and local floods according to natural causes. Although

---

55 Ps.-Albertus Magnus, *Women's Secrets*, cap. 4, ed. and trans. by Lemay, p. 97. Lemay prints two commentaries with her translation of the text of *De secretis mulierum*: Commentary A and Commentary B. These anonymous late medieval commentaries exist in many of the manuscripts and were often printed with, or even introduced into, the text itself. Lemay's transcription of Commentary A is based on the Lyons 1580 edition of *De secretis mulierum*. See *Women's Secrets*, p. 2. For a list of printed editions containing Commentary A, see ibid., pp. 181–82. Barragán Nieto has identified eighty-eight manuscripts containing *De secretis mulierum* and lists them in Ps.-Albertus Magnus, *El 'De secretis mulierum'*, ed. and trans. by Barragán Nieto, pp. 97–153; for the printed editions, see pp. 153–89.

56 Ps.-Albertus Magnus, *Women's Secrets*, ed. and trans. by Lemay, p. 97. For the judgment that a universal flood is impossible from natural causes, see also Raymund Llull, *Declaratio Raimundi*, ed. by Pereira and Pindl-Büchel, pp. 384–85.

57 Albertus Magnus, *De causis*, I.2.13, p. 86, vv. 78–80; *On the Causes*, p. 93. The reference is to Averroes, *Meteora*, II.1 (fols 29D–F), ed. Venetiis, p. 427A; *Metaph.*, XI(12), comm. 18 (fols 325Bff.).

58 The Stoics understood the Great Year to be the time when all the planets would return to their original places. It would signify the destruction and renewal of the universe.

59 Albertus Magnus, *Super IV Sententiarum*, dist. 43, art. 7, ed. by Borgnet, p. 518b.

the universal Noahide flood also delivered God's judgment and punishment upon human sinfulness, even that flood will be subject to a natural explanation. This apparently led some later readers of his *On the Causes of the Properties of the Elements* — for example, Johannes Stöffler and Jakob Pflaum — to predict according to celestial movements the next great biblical flood for the year 1524,[60] despite Albert's caveat in his *Commentary on the Sentences* (completed several years before *On the Causes of the Properties of the Elements*),[61] which limited astronomy's predictive capacity to local floods. Later defenders also sought to shield his reputation following the condemnation of 1277 by suggesting that Albert likewise did not believe that one can discover the natural causes for a universal flood, or by reaffirming that a universal flood, as Averroes said, is impossible according to natural causation and rather that every flood will be limited and localized. Such defences, however, seem to be the product of a historical revisionism and fail to recognize Albert's conviction that natural causation can be established for a universal flood.

## Christian Eschatology and the 'Flood of Fire'

One may easily understand allusions to a universal flood of water as a reference to the Biblical Flood, and appreciate the tension generated by competing theological and natural explanations. The flood of fire also gradually found a place in Christian theology, this time in eschatology. When commenting upon the actions of Lot's daughters at Gen. 19.30–36, for instance, Albert remarks that the Gloss (that is, the *Glossa ordinaria*) excuses them on the ground that they had learned that just as there was a judgment in the past by water (viz., the Biblical Flood), so too there will be a future judgment by fire.[62] Indeed, from the late thirteenth century, it was increasingly common to depict the Last Judgment as a flood of fire,[63] as Albert himself had done in his treatise *On the Resurrection*:

> Since, then, there are two judgments of the world, one in a qualified way [*secundum quid*] when some were punished, it was fitting for this to be accomplished by water, which is defined by the cold. The other [judgment], however, is absolute [*simpliciter*], and it is fitting that it be accomplished by fire, which is active in an absolute way. Besides, this occurs according to what is appropriate for sin. For at the time of the Flood the sin of carnal concupiscence, whose tinder [*fomes*] exists in the fire of lust, was at its highest.

---

60 See Rutkin, 'The Natural Philosophical Foundations for Astrological Revolutions', p. 100.

61 For the chronology of Albert's works, see especially Anzulewicz, 'Zeittafel'.

62 Albertus Magnus, *Summa theologiae*, pars II, tr. 18, q. 121, m. 1, art. 4, part. 2, sol., ed. by Borgnet, p. 392a.

63 Later, Martin Luther did so; see his *Enarrationes in Genesin*, cap. 1–4, 7, ed. by Schmid, pp. 64 and 115. Some early modern Christian theologians even sought additional support in rabbinic sources. See Losius and Salzmann, *Dissertationem philologicam*, p. 44.

> And for this reason it was fitting that judgment be rendered by means of a contrary, by means of a cold element. At the end, however, 'the charity of many will grow cold' [Matt. 24.12] as is said in the Gospel, and this from the icy frost of avarice. And for this reason that judgment will be rendered by means of a contrary: by means of the heat that exists in fire.[64]

Just as Albert had sought to identify the natural causes for a flood of water, so too he sought the causes for a flood of fire. In his *On the Causes of the Properties of the Elements*, he explained:

> There is a flood of fire, however, when the fire that has been called forth by the sun's light dries out and burns the hot climes and does not temper but inordinately warms the cold climes. And this occurs sometimes through a universal cause and sometimes through a cause that is partial. And sometimes a flood of fire is universal across the earth, and sometimes it is particular, as we said concerning a flood of water. However, the ancients did not know the true cause of a universal flood. For Plato said that an orbital deviation of the sun and the planets was the cause of a flood of fire, and he introduces the myth of Phaeton, which Ovid took from the Greeks and brought into Latin, and Plato says that although it seems to be a myth, it is nonetheless a true story.[65]

Albert dismisses Plato's explanation in the *Timaeus* that, as the Phaeton myth suggests, a deviation in the sun's orbit causes it to pass too close to the earth, burning the earth's surface.[66] The sun never leaves its appointed path or orbit, Albert insists, and therefore the ancients were in error. Instead, Albert remarks,

> I say [...] that the cause of the fire is gathered from five causes, that is, from the gathering [*congregatio*] of the sun and Mars and of Jupiter, and from the place of the gathering, that is, so that it occurs in Cancer, between Leo's heart and the Dog, and from the diameter of the sun and Mars, that is, so both of them and also Jupiter is in the lower part of their orbits; [...] the fourth [cause] is that the sun and the hot planets are not blocked by the cold planets, like Saturn and the moon and Venus, and especially by Saturn. And the fifth is that this conjunction occurs with the change of the triplicity of Saturn and Jupiter.[67] For at that time it will produce great events. But although all seven [planets] are perhaps conjoined, if the three that have been mentioned possess [their] powers, they still turn all the others to their properties, [...] But this happens very rarely, and for this reason a flood of fire occurs very rarely, and this is what the ancients called a long orbital deviation [...], and I think that

---

64 Albertus Magnus, *De resurrectione*, tr. 2, q. 10, art. 3, sol. 7, ed. by Kübel, p. 293, vv. 72–84. For the English translation, see *On Resurrection*, trans. by Resnick and Harkins, p. 170.

65 Albertus Magnus, *De causis*, I.2.12, p. 83, vv. 18–30; *On the Causes*, p. 85. See Ovid, *Metam.* II.179.

66 Plato, *Timaeus* 22C–D, ed. by Waszink, p. 14, vv. 6–12.

67 For the definition of a triplicity as three zodiacal signs in which successive conjunctions occur, see Albertus Magnus, *De causis*, I.2.2, p. 64, v. 82–p. 65, v. 17; *On the Causes*, pp. 46–47.

> this is the true cause of the flood of fire. For that flood has no cause among the lower [elements] as the flood of water did, because the lower elements cannot move fire nor is their vapor material for fire, but rather it is opposed to fire.[68]

In the fourteenth century, a debate once more ensued over whether this final conflagration will have natural causes or only a supernatural cause. Yet even though the assertion that a universal flood of fire could be subject to a natural explanation had been condemned in Paris in 1277 as proposition 182, some fourteenth-century thinkers interpreted the appearance of the plague as a universal flood and recapitulated the earlier debate over its causes.[69]

Nonetheless, new aspects entered the debates as well. Based on the assumption that the world will undergo a general conflagration in a flood of fire in a Final Judgment, and that this flood is subject to natural causation, numerous fourteenth-century thinkers sought to calculate precisely when the celestial phenomena already mentioned would produce that effect. As Laura Smoller has shown, in early fourteenth-century Paris the Alfonsine tables — prepared by Jewish astronomers in 1272 for King Alfonso X of Castile to provide the astronomical data necessary to predict eclipses and other heavenly phenomena — stimulated astrological calculations concerning the end of the world in a flood of fire.[70] In the second half of the fourteenth century, the Oxonian John of Eschenden (or Ashenden) criticized a certain master, perhaps John Aston, who lectured on the Bible at Merton College and who

> publicly asserted and determined in the Schools that there was a certain and determined number of years between the first Flood of water in the time of Noah and the second Flood of fire which is to come, namely 7900 years. And he said, as I gather, that he would show this from prophecy and Scripture by proofs of astronomy and philosophy.[71]

About the middle of the fourteenth century, also in Oxford, John of Eschenden composed his *Summa astrologiae iudicialis de accidentibus mundi*, in which he repeatedly cited Albert the Great's *On the Causes of the Properties of the Elements*.[72] John closely followed Albert's discussion concerning universal and local floods of water or fire and endorsed Albert's contention that it is appropriate to investigate their causes not only in the divine will, but also in nature.[73] John explicitly

---

68 Albertus Magnus, *De causis*, I.2.12, p. 84, vv. 14–19 and 38–60; *On the Causes*, pp. 87–88.

69 Carmichael, 'Universal and Particular'.

70 Smoller, 'The Alfonsine Tables'.

71 'publice asseruit et determinavit in scolis certum et determinatum numerum annorum [esse] a diluvio universali aque, qui erat tempore Noe, usque ad diluvium ignis futurum, viz. 7900tos annos; quem quidem numerum annorum dixit se ostensurum, pro ut mihi erat intimatum, ex prophecιis et scriptura sacra per astronomiam et philosophiam'. In Robson, *Wyclif and the Oxford Schools*, p. 103, n. 1. Robson takes the text from Digby MS 176, fol. 39v. The text further identifies this as a Joachimite error (pp. 102–03).

72 Thorndike, *History of Magic and Experimental Science*, pp. 325–42.

73 John of Eschenden, *Summa astrologiae iudicialis de accidentibus mundi*, tr. 2, dist. 8, cap. 2, fol. 316.

acknowledges, for instance, that a universal flood of fire will occur at the Last Judgment, when all four elements and all those that are composed of them will be purged by fire,[74] providing as support numerous proof texts from Scripture and from patristic and medieval authorities. Like Albert, John concedes that astronomy will be unable to predict this final, universal flood of fire,[75] and this seems to be the basis for his criticism of the Mertonian who calculated the date for the flood of fire. Other late medieval thinkers, such as John of Lubeck, confidently sought to predict from astronomical portents both the advent of Antichrist and the celestial conjunction of Saturn and Jupiter that would produce this final flood of fire.[76] These calculations, however, had gone far beyond the spirit of Albert's explanations of the universal flood, which took an outspoken stand against such predictions.

## The Regeneration of Species Following a Flood

As we saw in the previous section, Albert notes a disagreement between Averroes and Avicenna on whether there can be a universal flood, since 'Averroes says that there will never be a flood that is so universal that there is no evasion and no escape',[77] whereas Avicenna admits the possibility of a universal flood. In addition, Albert finds a disagreement between them 'concerning what repairs the earth and replaces the animals on it after they have been extinguished and killed by a flood of water and of fire'.[78] In the absence of any individuals that constitute a species, the debated question was whether celestial causes, in particular, can bring about the natural and spontaneous regeneration of the species from matter following a universal flood. Just like the question of a universal flood, this topic would, as we will see, become a source of heated debate for centuries to come.[79]

It was well understood among the commentators on Aristotle's *libri naturales* that generation typically requires the presence of male and female of a species. Yet for exceptional cases, such as a universal flood, Albert notes,

> Avicenna says that the powers of the stars mixed with the powers of the elements form and perfect all things, and only need a female on account of place. And for this reason he says that a 'womb' is required for generation

74 Ibid., fol. 317: 'Diluvium ignis universale erit in finali iudicio quando omnia quattuor elementa et omnia composita ex eis per ignem erunt purgata'.

75 Ibid., fol. 322.

76 Thorndike, 'Three Astrological Predictions'.

77 Albertus Magnus, *De causis*, I.2.13, p. 86, vv. 78–80; *On the Causes*, p. 93.

78 Albertus Magnus, *De causis*, I.2.13, p. 85, vv. 33–35; *On the Causes*, p. 90.

79 On the origins of debate on spontaneous generation, see McCartney, 'Spontaneous Generation'. On later treatment, see Bertolacci, 'Averroes against Avicenna'; Bertolacci, 'The Matter of Human Spontaneous Generation'; Kruk, 'A Frothy Bubble'; Hasse, 'Spontaneous Generation'; Gaziel, 'Spontaneous Generation'.

> only for the sake of well-being, namely, so that it may be better formed;[80] and, because a perfect place for something's generation among the elements is rare, owing to sudden changes of the elements, for this reason nature provides a fixed place, which is the womb of females. But it still often happens that the place for the generation of an animal is at times tempered [*contemperatus*] in the elements, and then he asserts that the stars produce the form for this animal, whose semen has been tempered in the elements.[81]

In support of this exception to the rule, Albert provides four arguments, which he also attributes to Avicenna, to show that generation can occur by the formative power of the stars without an antecedent individual of that species. As examples, he remarks that mice may be generated spontaneously from the earth as well as from coition;[82] serpents may be generated from human hair, or by coition,[83] and scorpions are said to have been produced from certain types of putrefaction and then subsequently by coition;[84] monstrous progeny may be formed that combine remote and unrelated species, such as a human and a cow or pig;[85] and, last, defective or monstrous births may be caused when, for example, 'some stars [are] in the sign of Aries'. In sum, 'from all these [indications] and others of this sort Avicenna proves that the first substances of any animal can be produced by the stars and that then they can be reproduced by coition'.[86] Moreover, 'Plato seems to have understood this when he said that the God of gods, the sower of the universe, created being [*esse*] and handed it to the stars to embody and complete it'.[87] As a result, Albert supports Avicenna's view, linking it to Plato's *Timaeus*,

80 See Avicenna, *De diluviis*, ed. by Alonso Alonso, p. 307: 'Sed matrix faciet ad meliorationem.'; Albertus Magnus, *De caelo et mundo*, II.1.2 and II.3.3, ed. by Hossfeld, p. 107, vv. 49–59, and p. 147, vv. 69–77.

81 Albertus Magnus, *De causis*, I.2.13, p. 85, vv. 37–49; *On the Causes*, p. 90.

82 Albertus Magnus, *De causis*, I.2.13, p. 85, vv. 50–52; *On the Causes*, p. 90. See Avicenna, *De diluviis*, ed. by Alonso Alonso, p. 307. Cf. Albertus Magnus, *De animalibus*, V.1.1.3 and XXII.2.1.123 (80), ed. by Stadler, vol. 1, p. 408 and vol. 2, p. 1415; Avicenna, *De animalibus*, XV.1, ed. Venetiis, fol. 59va. For ancient roots to the doctrine of spontaneous generation, see the still useful McCartney, 'Spontaneous Generation'. On Albert, also Balss, *Albertus Magnus als Zoologe*, pp. 59–62.

83 Albertus Magnus, *De causis*, I.2.13, p. 85, vv. 53–57; *On the Causes*, p. 90. See Avicenna, *Liber canonis*, IV, fen. 6, 4.51, ed. Venetiis, fol. 479va–b; *De diluviis*, ed. by Alonso Alonso, p. 307. Ps.-Albert reports in his *De secretis mulierum* that Avicenna demonstrated that, if one buries the hairs of a woman in fertile soil, by spring serpents will be generated from them. See Ps.-Albertus Magnus, *El 'De secretis mulierum'*, ed. by Barragán Nieto, p. 330. Other editions specify hair from a menstruating woman. See Ps.-Albertus Magnus, *Women's Secrets*, ed. and trans. by Lemay, p. 96.

84 Albertus Magnus, *De animalibus*, XV.1.8.45, ed. by Stadler, vol. 2, p. 1010, vv. 1–3, citing Avicenna.

85 See Albertus Magnus, *De animalibus*, XVIII.1.6.48, ed. by Stadler, vol. 2, pp. 1215–16; see *Quaestiones super De animalibus*, XVIII.7, ed. by Filthaut, p. 300, v. 76–p. 301, v. 38.

86 Albertus Magnus, *De causis*, I.2.13, p. 86, vv. 4–6; *On the Causes*, p. 91. Cf. Albertus Magnus, *De caelo et mundo*, II.1.2, ed. by Hossfeld, p. 107, vv. 49–59 with n. 59.

87 Albertus Magnus, *De causis*, I.2.13, p. 86, vv. 18–21; *On the Causes*, pp. 91–92. See Plato, *Timaeus* 41C–D, ed. by Waszink, p. 36, vv. 9–10. See also Albertus Magnus, *Metaphysica*, I.2.6, ed. by Geyer, p. 34, v. 70 with note, and p. 64, v. 7.

which maintains that our earth can be restored when the power of the stars informs a mixture of the elements.

In contradiction, writes Albert, 'Averroes opposed this, arguing with many arguments that animals that have a great diversity in their members and are called perfect, cannot be restored by the stars and the elements alone'.[88] Here, Albert epitomizes Averroes's four counterarguments just as he does those of Avicenna. In the first argument, Albert reads Averroes as holding that — unlike Avicenna, who remarked that a place or 'womb' is not necessary for generation but is required only for the sake of well-being — a 'womb' is absolutely necessary for the generation of the perfect animals. Second, if the stars could generate or regenerate such animals apart from the womb and sexual intercourse, then empirical evidence would certainly have presented itself to philosophers by now. Third, 'nature proceeds along the most direct path'; but generation *without* intercourse between male and female appears to be a more direct path.[89] Since we do not observe generation without intercourse in perfect animals, sexual intercourse must therefore be necessary. Finally, Averroes rejects the possibility that the influence of the stars can regenerate species because, in that case, 'equivocal generation would be prior to univocal generation [...]. And this is contrary to reason, because the equivocal cannot exist in any way unless the univocal, to which it is reduced, exists first'.[90]

For Albert, as for Averroes, the term univocal implies 'having one name' and some common nature. Thus, in the order of causes, univocal generation implies that something is generated from one having the same form as itself; equivocal generation, in contrast, implies that an organism is generated from something unlike itself, as, Albert believes, serpents might be generated from hair. Averroes's complaint against Avicenna's argument for equivocal generation thus seems to be that if equivocal generation precedes univocal generation, one could say that 'a' is generated from 'b' and 'b' is generated from 'c' and so on *ad infinitum*. Thus, univocal generation must precede equivocal generation, setting a standard against which equivocal generation becomes intelligible.[91]

Following this summary, Albert concedes that 'one ought to agree with each of them' — that is, with both Avicenna and Averroes — 'in a certain respect'.[92] Certainly, in Albert's eyes, after a universal flood, the power of the stars is sufficient to mix or arrange the elements in order to regenerate vegetation.[93] In the same way, the influence of the stars should be sufficient to (re)generate lower

88 Albertus Magnus, *De causis*, I.2.13, p. 86, vv. 26–29; *On the Causes*, p. 92. Cf. Averroes, *Metaphysica*, XI(12), comm. 18 (fols 326D–E). For some discussion of the debate, see Takahashi, 'Interpreting Aristotle's Cosmos', pp. 159–64.

89 Albertus Magnus, *De causis*, I.2.13, p. 86, vv. 39–40; *On the Causes*, p. 92.

90 Albertus Magnus, *De causis*, I.2.13, p. 86, vv. 45–50; *On the Causes*, p. 92.

91 See Albertus Magnus, *On the Causes*, p. 92 n. 266.

92 Albertus Magnus, *De causis*, I.2.13, p. 86, vv. 53–54; *On the Causes*, p. 93.

93 See Albertus Magnus, *De vegetabilibus*, I.1.7, IV.2.1, V.1.7, VI.1.21, VII.1.9, ed. by Meyer and Jessen, pp. 25, 233, 315, 394, 618–22; *De animalibus*, XVI.1.8.49, ed. by Stadler, vol. 2, p. 1085.

animals with similar body types, such as 'snakes and worms and fish';[94] these have appeared in newly formed lakes, Albert remarks in *De animalibus*, seemingly generated from mud or slime and not from antecedent members of the species.[95] He suggests (following Avicenna) that this has also been observed with respect to the beaver 'in a certain stream called the Iacton in Arabic', and concludes: 'it is most probable that the beaver was procreated in that same place by the power of the stars'.[96] Animals 'born from putrefaction, which have no univocal generation' may likewise be produced by celestial influence alone. Albert adds that 'the stars have the power to produce animals that are not too dissimilar, such as mice and bats', because their morphology is so similar; indeed, he identifies the bat 'as a sort of flying mouse' in his commentary on Aristotle's *On Animals*.[97] 'But for the reproduction of perfect animals, like the lion and the ox and the human, they [the stars] seem to suffice in no way.[98] And it is clear that they do not suffice in the human because a rational soul is not educed from matter, but is given by the first giver [of forms] according to the philosophers[99] [...] [this] is also the case for other perfect animals'.[100]

Thus, Albert follows Averroes's stricter criteria, according to which perfect animals have additional requirements for generation or reproduction that include coition, whereas 'animals [that] are imperfect in comparison to the human [...] do not require many things that are necessary to the perfect animals'.[101] Indeed, if Avicenna asserted that because coition is a voluntary movement among animals,

94 Albertus Magnus, *De causis*, I.2.13, p. 86, v. 59; *On the Causes*, p. 93.

95 Albertus Magnus, *De animalibus*, VI.2.2.81–82, ed. by Stadler, vol. 1, pp. 475–76. Albert identifies there certain fish and eels that appear in lakes that had dried out, but whose waters have been replenished. They seem to be generated not from the eggs of other members of the species, and not from copulation, but from slime or putrescence. He also includes earthworms in this category. See also Albertus Magnus, *De causis*, I.2.13, p. 86, vv. 59–61; *On the Causes*, p. 93.

96 Albertus Magnus, *De animalibus*, XV.1.8.46, ed. by Stadler, vol. 2, p. 1010, vv. 18–20; *Albertus Magnus On Animals*, trans. by Kitchell and Resnick, vol. 2, p. 1105.

97 See Albertus Magnus, *De animalibus*, XXIII.1.24.142 (109), ed. by Stadler, vol. 2, p. 1512, v. 5; *Albertus Magnus On Animals*, trans. by Kitchell and Resnick, vol. 2, p. 1651.

98 Albertus Magnus, *De causis*, I.2.13, p. 86, vv. 70–71; *On the Causes*, p. 93.

99 Albertus Magnus, *De causis*, I.2.13, p. 86, vv. 72–74; *On the Causes*, p. 93. Cf. Albertus Magnus, *De caelo et mundo*, I.3.4, ed. by Hossfeld, p. 63, v. 92–64, v. 8, and p. 63 n. 92; *De anima*, III.2.4, ed. by Stroick, p. 183, vv. 32–43. Albert refers to the 'giver of forms' (*dator formarum*) also at *Physica*, I.3.15, 2.2.11, ed. by Hossfeld, vol. 1, p. 69, v. 15 and p. 118, v. 10. According to Ps.-Albert's *Philosophia pauperum*, Averroes understood the giver of forms in relation to universals; thus, 'Some thought that forms exist externally [*ab extrinseco*], namely from the giver of forms. And before they are given, they exist independently outside matter, positioned in relation to the stars, as the Platonists [say]. Thus they say that when a man is born under a certain constellation, its form is impressed [upon him] with respect to its causality or its virtue'. Ps.-Albertus Magnus, *Philosophia pauperum*, I.4, ed. by Borgnet, p. 449b.

100 Albertus Magnus, *De causis*, I.2.13, p. 86, v. 76; *On the Causes*, p. 93.

101 Albertus Magnus, *Quaestiones super De animalibus*, I, q. 12.2, ed. by Filthaut, p. 89, vv. 17–20; Albertus Magnus, *Questions Concerning Aristotle's 'On Animals'*, trans. by Resnick and Kitchell, p. 36.

it is not necessary for generation or reproduction,[102] Albert insists (following Averroes) that

> coition is necessary for the generation of some animals because the more perfect something is, the more things are required for its generation. Thus, for the generation of some imperfect animals a universal agent with properly disposed matter is sufficient, as is evident in the generation of those from putrefaction. But for the generation of perfect animals there is required an agent of the same species as well as a universal agent.[103]

By contrast, for lower animals 'sperm is not necessary for generation from putrefaction. It is necessary, however, for the generation of perfect animals'.[104] The sperm in question here seems not to be female sperm (*sperma muliebre*), which Albert elsewhere treats as a material principle for reproduction that introduces only a confused and inchoate form to matter, but the male principle or *virtus*.[105] In sum,

> it is correct to argue in this way: Just as the power in the elements of the world relate[s] to imperfect animals, so does the power in the semen relate to perfect animals. But the power in the elements of the world produces a form of the imperfect animals. Therefore, the power which is in the semen produces a form of the perfect animals.[106]

Thus, Albert agrees with Averroes that the power of the stars alone is insufficient to regenerate perfect animals and the human, which is the most perfect animal of all;[107] these can be generated neither from putrefaction nor from the power of the stars.[108] For perfect animals, coition, the womb, and the formative power present in the semen are necessary. In addition, the human being's 'rational soul is not educed from matter, but is given by the first giver [of forms] according to the philosophers, and for this reason the first human hypostases were created

---

102 Avicenna, *De diluviis*, ed. by Alonso Alonso, p. 308.

103 Albertus Magnus, *Quaestiones super De animalibus*, V, q. 1, ed. by Filthaut, p. 153, vv. 43–50; Albertus Magnus, *Questions Concerning Aristotle's 'On Animals'*, trans. by Resnick and Kitchell, p. 185.

104 Albertus Magnus, *Quaestiones super De animalibus*, XV, q. 18, ed. by Filthaut, p. 270, vv. 78–81; Albertus Magnus, *Questions Concerning Aristotle's 'On Animals'*, trans. by Resnick and Kitchell, p. 466.

105 For Albert on female sperm as an *inchoatio formae*, see esp. *De animalibus*, III.2.8.155, ed. by Stadler, vol. 1, pp. 345–46. For discussion of his position on female sperm, see esp. Jacquart and Thomasset, 'Albert le Grand et les problèmes de la sexualité'; Jacquart and Thomasset, *Sexuality and Medicine in the Middle Ages*, chap. 2; Asúa, 'War and Peace', pp. 275, 285–86, 289. For Avicenna on female sperm and the Galenic-Aristotelian debate, see also Musallam, 'The Human Embryo'.

106 Albertus Magnus, *Quaestiones super De animalibus*, XVI, q. 11, ed. Filthaut, p. 281, vv. 71–76; Albertus Magnus, *Questions Concerning Aristotle's 'On Animals'*, trans. by Resnick and Kitchell, p. 493.

107 Albertus Magnus, *De animalibus*, XXI.1.1.3, ed. by Stadler, vol. 2, p. 1322, v. 28.

108 Albertus Magnus, *De quindecim problematibus*, art. 6, ed. by Geyer, p. 38, vv. 54–65. The Parisian condemnation of 219 propositions also condemned the view (under proposition 188) that 'quod homo posset sufficienter generari putrefactione'.

and formed by God'.[109] For the lower or more imperfect animals, however, Albert believes that he, Avicenna, and Averroes are in general agreement: 'Averroes says that what is said about the stars is true in similar animals and he says that it is not true in dissimilar animals'.[110] But Albert rejects completely the argument that univocal must precede equivocal generation:

> although one may speak in this way by agreeing with each of these philosophers in part, it is still not necessary [to agree] for the reason that univocal generation is said to have to precede equivocal generation; for we know that the stars are equivocal generators and they are nonetheless the first generators,[111] because they do not generate insofar as they are equivocal, but rather insofar as they agree.[112]

In his attempt to reconcile Averroes and Avicenna on this point, then, Albert proposed that although the more perfect animals cannot be regenerated from the stars alone, the imperfect animals can.

In the early sixteenth century, Pietro Pomponazzi would identify this harmonizing solution as the one favoured generally by the Latins, and cited Albert's *On the Causes of the Properties of the Elements* explicitly.[113] He nonetheless returned to Albert's *On the Causes of the Properties of the Elements*, to re-examine the debate between Avicenna and Averroes on celestial influence, spontaneous generation, and the regeneration of species, and affirmed at least the plausibility of Avicenna's position. Albert's attempted harmonization may have become the preferred position among the Latins, but the debate between Avicenna and Averroes that Albert had epitomized continued to draw the attention of early modern natural philosophers seeking justifications for spontaneous generation.[114]

## Conclusion

Disagreement between Avicenna and Averroes concerning a universal flood, and the potential for a regeneration of species thereafter, prompted Albert the Great to provide a detailed discussion on the natural causes of a flood, the differences between a universal and local flood, and the power or influence of celestial conjunctions to produce a flood of water or fire here below. His treatment in *On the Causes of the Properties of the Elements* would have a significant impact upon later medieval and early modern thinkers. Some, as I have shown, attempted

109 Albertus Magnus, *De causis*, I.2.13, p. 86, vv. 72–75; *On the Causes*, p. 93.

110 Albertus Magnus, *De causis*, I.2.13, p. 86, v. 85–p. 87, v. 1; *On the Causes*, p. 93.

111 See Albertus Magnus, *Metaphysica*, VIII.2.3 and VII.2.9, ed. by Geyer, p. 404, vv. 90–95 and p. 351, vv. 51–52 with nn. 53–54.

112 Albertus Magnus, *De causis*, I.2.13, p. 87, vv. 7–13; *On the Causes*, p. 94.

113 See Compagni, 'Métamorphoses animales et géneration spontanée'.

114 See, for example, Hirai, 'Daniel Sennert'; Hirai, 'Atomes vivants'.

to shield Albert from criticism that followed the Parisian condemnation of 219 propositions in 1277 by, it seems, 'revising' Albert's own view on natural causation and a universal flood. Others seem to have ignored Albert's caveat concerning the predictive capacity of astronomy with respect to a universal flood, leading them to calculate a fixed date in the future for a universal flood of water or of fire. Still others turned to his text for its discussion of the regeneration of species via spontaneous generation. All of these uses, however, stem from Albert's attempt to harmonize Plato and Aristotle, and Avicenna and Averroes.

## Works Cited

### Primary Sources


Albertus Magnus, *Albertus Magnus On Animals: A Medieval Summa Zoologica*, trans. by Kenneth F. Kitchell Jr. and Irven M. Resnick, revised ed., 2 vols (Columbus: Ohio State University Press, 2018)

———, *De anima*, ed. by Clemens Stroick, Editio Coloniensis, 7/1 (Münster: Aschendorff, 1968)

———, *De animalibus libri XXVI. Nach der Cölner Urschrift*, ed. by Hermann Stadler, 2 vols, Beiträge zur Geschichte der Philosophie des Mittelalters, 15–16 (Münster: Aschendorff, 1916–20)

———, *De caelo et mundo*, ed. by Paul Hossfeld, Editio Coloniensis, 5/1 (Münster: Aschendorff, 1971)

———, *De generatione et corruptione*, ed. by Paul Hossfeld, Editio Coloniensis, 5/2 (Münster: Aschendorff, 1980), pp. 109–213

———, *De homine*, ed. by Henryk Anzulewicz and Joachim R. Söder, Editio Coloniensis, 27/2 (Münster: Aschendorff, 2008)

———, *De praedicamentis*, ed. by Auguste Borgnet, Editio Parisiensis, 1 (Paris: Vivès, 1890), pp. 305–72

———, *De quindecim problematibus*, ed. by Bernhard Geyer, Editio Coloniensis, 17/1 (Münster: Aschendorff, 1975), pp. 31–44

———, *De resurrectione*, ed. by Wilhelm Kübel, Editio Coloniensis, 26 (Münster: Aschendorff, 1958), pp. 237–354

———, *De sensu et sensato*, ed. by Silvia Donati, Editio Coloniensis, 7/2a (Münster: Aschendorff, 2017), pp. 1–112

———, *De vegetabilibus, libri VII*, ed. by Ernst Meyer and Karl Jessen (Berlin: Reimer, 1867)

———, *Ethica*, ed. by Auguste Borgnet, Editio Parisiensis, 7 (Paris: Vivès, 1891)

———, *Liber de causis proprietatum elementorum*, ed. by Paul Hossfeld, Editio Coloniensis, 5/2 (Münster: Aschendorff, 1980), pp. 47–106

———, *Liber de natura et origine animae*, ed. by Bernhard Geyer, Editio Coloniensis, 12 (Münster: Aschendorff, 1955), pp. 1–44

———, *Metaphysica, libri I–V*, ed. by Bernhard Geyer, Editio Coloniensis, 16/1 (Münster: Aschendorff, 1960)

———, *Metaphysica, libri VI–XIII*, ed. by Bernhard Geyer, Editio Coloniensis, 16/2 (Münster: Aschendorff, 1964)

———, *Meteora*, ed. by Paul Hossfeld, Editio Coloniensis, 6/1 (Münster: Aschendorff, 2003)

———, *On the Causes of the Properties of the Elements (Liber de causis proprietatum elementorum)*, trans. and annotated by Irven M. Resnick, Mediaeval Philosophical Texts in Translation, 46 (Milwaukee, WI: Marquette University Press, 2010)

———, *On Resurrection*, trans. by Irven M. Resnick and Franklin T. Harkins, Fathers of the Church, Mediaeval Continuation, 20 (Washington, DC: The Catholic University of America Press, 2020)

———, *Physica, libri I–IV*, ed. by Paul Hossfeld, Editio Coloniensis, 4/1 (Münster: Aschendorff, 1987)

———, *Politica*, ed. by Auguste Borgnet, Editio Parisiensis, 8 (Paris: Vivès, 1891)

———, *Quaestiones super De animalibus*, ed. by Ephrem Filthaut, Editio Coloniensis, 12 (Münster: Aschendorff, 1955), pp. 77–321

———, *Questions Concerning Aristotle's 'On Animals'*, trans. by Irven M. Resnick and Kenneth F. Kitchell, Jr., Fathers of the Church, Medieval Continuation, 9 (Washington, DC: Catholic University of America Press, 2008)

———, *Super IV Sententiarum*, ed. by Auguste Borgnet, Editio Parisiensis, 29 (Paris: Vivès, 1894)

———, *Summa theologiae [= Summa de mirabili Scientia dei]*, ed. by Auguste Borgnet, Editio Parisiensis, 32 (Paris: Vivès, 1895)

Alfred of Sareshel, *Alfred of Sareshel's Commentary on the 'Metheora' of Aristotle*, ed. by James K. Otte, Studien und Texte zur Geistesgeschichte des Mittelalters, 19 (Leiden: Brill, 1988)

Andrew of St Victor, *Expositionem super Heptateuchum*, ed. Charles Lohr and Rainer Berndt, Corpus Christianorum Continuatio Mediaevalis, 53 (Turnhout: Brepols, 1986)

Averroes, *Aristotelis Metaphysicorum libri XIIII cum Averrois Cordubensis in eosdem commentariis et epitome*, Aristotelis opera cum Averrois commentariis, 8 (Venice, 1560; repr. Frankfurt am Main: Minerva, 1962)

———, *Meteora*, in *Aristotelis De coelo, De generatione et corruptione, Meteorologicorum, De plantis cum Averrois Cordubensis variis in eosdem Commentariis*, Aristotelis opera cum Averrois commentariis, 5 (Venice, 1562; repr. Frankfurt am Main: Minerva, 1962)

Avicenna, *De animalibus* (Venice: Johannes et Gregorius de Gregoriis, 1500)

———, *De diluviis*, in Manuel Alonso Alonso, 'Homenaje a Avicena en su milenario: Las traducciones de Juan González de Burgos y Salomón', *Al-Andalus*, 14 (1949), 291–319

———, *Liber canonis* (Venice, 1507; repr. Hildesheim: Olms, 1964)

*Chartularium Universitatis Parisiensis*, ed. by Heinrich Denifle and Émile Châtelain, 4 vols (Paris: Delalain, 1889–91)

Conrad de Mure, *Fabularius*, ed. by Tom van de Loo, Corpus Christianorum Continuatio Mediaevalis, 210 (Turnhout: Brepols, 2006)

John of Eschenden, *Summa astrologiae iudicialis de accidentibus mundi* (Venice: Johannes Lucilius Santritter, 1489)

Losius, Johann Justus, and Johann Gottfried Salzmann, *Dissertationem philologicam qua Hebraeos veteres Christum Scripturae scopum studiose olim quaesivisse, ex Genesi... ostenditur* (Magdeburg: Zeitleri, 1709)

Luther, Martin, *Enarrationes in Genesin*, ed. by Heinrich Schmid, vol. 1 of *Exegetica opera Latina*, ed. by Christophorus S. Th. Elsperger (Erlangen: Heyder, 1829)

Moses Maimonides, *Epistle to Yemen*, in *A Maimonides Reader*, ed. by Isidore Twersky (New York: Behrman House, 1972), pp. 437–62

———, *Moses Maimonides Dux neutrorum vel dubiorum*, Part 1, ed. by Diana Di Segni, Recherches de Théologie et Philosophie médiévales, Bibliotheca, 17/1 (Leuven: Peeters, 2019)

Peter of Poitiers, *Summa de confessione: compiliatio praesens*, ed. by Jean Longère, Corpus Christianorum Continuatio Mediaevalis, 51 (Turnhout: Brepols, 1980)

Petrus Cantor, *Summa quae dicitur Verbum adbreuiatum*, ed. by Monique Boutry, Corpus Christianorum Continuatio Mediaevalis, 196 (Turnhout: Brepols, 2004)

Plato, *Timaeus*, trans. with commentary by Calcidius, ed. by Jan H. Waszink (Leiden: Brill, 1962)

Pseudo-Albert the Great, *El 'De secretis mulierum' atribuido a Alberto Magno: Estudio, edición crítica y traducción*, ed. and trans. by José Pablo Barragán Nieto (Porto: Fédération Internationale des Instituts d'Études Médiévales, 2012)

———, *Philosophia pauperum sive Isagoge in libros Aristotelis physicorum, de coelo et mundo, de generatione et corruptione, meteorum et de anima*, ed. by Auguste Borgnet, Editio Parisiensis, 5 (Paris: Vivès, 1895)

———, *Women's Secrets: A Translation of Pseudo-Albertus Magnus's 'De Secretis Mulierum' with Commentaries*, ed. and trans. by Helen Rodnite Lemay (Albany: State University of New York Press, 1992)

Pseudo-Aristotle, 'Pseudo-Aristotle *De causis proprietatum et elementorum*: Critical Study and Edition', ed. by Stanley Luis Vodraska (unpublished doctoral dissertation, University of London, 1969)

Raymund Llull, *Declaratio Raimundi*, ed. by Michela Pereira and Theodor Pindl-Büchel, Corpus Christianorum Continuatio Mediaevalis, 79 (Turnhout: Brepols, 1989)

Rodrigo Jiménez de Rada, *Breuiarium historiae catholicae, libri I–V*, ed. by Juan Fernández Valverde, Corpus Christianorum Continuatio Mediaevalis, 72A (Turnhout: Brepols, 1992)

Vincent of Beauvais, *Speculum quadruplex: sive, Speculum majus: naturale, doctrinale, morale, historiale* (Douai, 1524; repr. Graz: Akademische Druck- u. Verlagsanstalt, 1964)

William of Conches, *A Dialogue on Natural Philosophy* (*Dragmaticon Philosophiae*), trans. by Italo Ronca and Matthew Curr (Notre Dame, IN: University of Notre Dame Press, 1997)

———, *Dragmaticon philosophiae*, ed. by Italo Ronca, Corpus Christianorum Continuatio Mediaevalis, 152 (Turnhout: Brepols, 1997)

### *Secondary Works*

Anzulewicz, Henryk, 'Albertus Magnus als Vermittler zwischen Aristoteles und Platon', *Acta Mediaevalia*, 18 (2005), 63–88

———, 'Plato and Platonic/Neoplatonic Sources in Albert', in *A Companion to Albert the Great: Theology, Philosophy, and the Sciences*, ed. by Irven M. Resnick (Leiden: Brill, 2013), pp. 595–601

———, 'Die platonische Tradition bei Albertus Magnus: Eine Hinführung', in *The Platonic Tradition in the Middle Ages: A Doxographic Approach*, ed. by Stephen Gersh and Maarten J. F. M. Hoenen (Berlin: De Gruyter, 2002)

———, 'Pseudo-Dionysius Areopagita und das Strukturprinzip des Denkens von Albert dem Grossen', in *Die Dionysius-Rezeption im Mittelalter*, ed. by Tzotcho Boiadjiev, Georgi Kapriev, and Andreas Speer (Turnhout: Brepols, 2000), pp. 251–96

———, 'Zeittafel', in *Albertus Magnus und sein System der Wissenschaften*, ed. by Albertus-Magnus-Institut (Münster: Aschendorff, 2011), pp. 28–31

Asúa, Miguel de, 'War and Peace: Medicine and Natural Philosophy in Albert the Great', in *A Companion to Albert the Great: Theology, Philosophy, and the Sciences*, ed. by Irven M. Resnick (Leiden: Brill, 2013), pp. 269–97

Balss, Heinrich, *Albertus Magnus als Zoologe* (Munich: Verlag der Münchner Drucke, 1928)

Bertolacci, Amos, 'Albert's Use of Avicenna and Islamic Philosophy', in *A Companion to Albert the Great: Theology, Philosophy, and the Sciences*, ed. by Irven M. Resnick (Leiden: Brill, 2013), pp. 601–10

———, 'Averroes against Avicenna on Human Spontaneous Generation: The Starting Point of a Lasting Debate', in *Renaissance Averroism and Its Aftermath: Arabic Philosophy in Early Modern Europe*, ed. by Anna Akasoy and Guido Giglioni (Dordrecht: Springer, 2013), pp. 37–54

———, '"Averroes ubique Avicennam persequitur": Albert the Great's Approach to the *Physics* of the *Šifā'* in Light of Averroes' Criticisms', in *The Arabic, Hebrew and Latin Reception of Avicenna's Physics and Cosmology*, ed. by Dag Nikolaus Hasse and Amos Bertolacci (Berlin: De Gruyter, 2018), pp. 397–431

———, 'The Matter of Human Spontaneous Generation According to Avicenna', in *Renaissance Averroism and Its Aftermath: Arabic Philosophy in Early Modern Europe*, ed. by Anna Akasoy and Guido Giglioni (Dordrecht: Springer, 2013), pp. 37–54

Burnett, Charles, 'Arabic into Latin: The Reception of Arabic Philosophy into Western Europe', in *The Cambridge Companion to Arabic Philosophy*, ed. by Peter Adamson and Richard C. Taylor (Cambridge: Cambridge University Press, 2005), pp. 370–404

———, 'Arabic Philosophical Works Translated into Latin', in *The Cambridge History of Medieval Philosophy*, ed. by Robert Pasnau and Christina Van Dyke (Cambridge: Cambridge University Press, 2010), vol. 2, Appendix B, pp. 814–22

———, 'Some Comments on the Translating of Works from Arabic into Latin in the mid-Twelfth Century', in *Orientalische Kultur und Europäisches Mittelalter*, ed. by Albert Zimmermann, Ingrid Craemer-Ruegenberg, and Gudrun Vuillemin-Diem (Berlin: De Gruyter, 1985), pp. 161–70

Carmichael, Ann G., 'Universal and Particular: the Language of Plague, 1348–1500', *Medical History*, 52, suppl. 27 (2008), 17–52

Cerami, Cristina, 'A Map of Averroes' Criticism against Avicenna: *Physics, De caelo, De generatione et corruptione*, and *Meteorology*', in *The Arabic, Hebrew and Latin Reception of Avicenna's Physics and Cosmology*, ed. by Dag Nikolaus Hasse and Amos Bertolacci (Berlin: De Gruyter, 2018), pp. 163–240

Compagni, Vittoria P., 'Métamorphoses animales et géneration spontanée: Développememts matérialistes après le *De immortalitate animae*', in *Pietro Pomponazzi entre traditions et innovations*, ed. by Joel Biard and Thierry Gontier (Amsterdam: Grüner, 2009), pp. 65–82

d'Alverny, Marie-Thérèse, 'L'explicit du "De animalibus" d'Avicenna traduit par Michel Scot', *Bibliothèque de l'école des Chartres*, 115 (1957), 32–42

D'Ancona, Cristina, 'Greek Sources in Arabic and Islamic Philosophy', in *The Stanford Encyclopedia of Philosophy* (Fall 2019 edition), ed. by Edward N. Zalta, https://plato.stanford.edu/archives/fall2019/entries/arabic-islamic-greek/

Diekstra, F. N. M., 'Robert de Sorbon's *Cum repetes (de modo audiendi confessions et interrogandi)*', *Recherches de théologie et philosophie médiévales*, 66 (1999), 79–153

[Dod, Bernard G.], 'Greek Aristotelian Works Translated into Latin', in *The Cambridge History of Medieval Philosophy*, ed. by Robert Pasnau and Christina Van Dyke (Cambridge: Cambridge University Press, 2010), vol. 2, Appendix B, pp. 793–97

Franchi, Alfredo, 'Alberto Magno e le origini della nozione di causalità efficiente: la teoria delle cinque cause nei "quidam" del V "Metaphysicorum"', *Sapienza*, 33 (1980), 178–85

Gaziel, Ahuva, 'Spontaneous Generation in Medieval Jewish Philosophy and Theology', *History and Philosophy of the Life Sciences*, 34 (2012), 461–79

Gottschall, Dagmar, 'Conrad of Megenberg and the Causes of the Plague: A Latin Treatise on the Black Death Composed ca. 1350 for the Papal Court in Avignon', in *La vie culturelle, intellectuelle et scientifique à la cour des Papes d'Avignon*, ed. by Jacqueline Hamesse (Turnhout: Brepols, 2006), pp. 319–32

Grant, Edward, 'The Translation of Greek and Arabic Science into Latin', in *A Source Book in Medieval Science*, ed. by Edward Grant (Cambridge, MA: Harvard University Press, 1974), pp. 35–41

Gutas, Dmitri, 'Greek Philosophical Works Translated into Arabic', in *The Cambridge History of Medieval Philosophy*, ed. by Robert Pasnau and Christina Van Dyke (Cambridge: Cambridge University Press, 2010), vol. 2, Appendix B, pp. 802–14

Hasse, Dag Nikolaus, 'Influence of Arabic and Islamic Philosophy on the Latin West', in *The Stanford Encyclopedia of Philosophy* (Spring 2020 edition), ed. by Edward N. Zalta, https://plato.stanford.edu/archives/spr2020/entries/arabic-islamic-influence/

———, 'Spontaneous Generation and the Ontology of Forms in Greek, Arabic, and Medieval Latin Sources', in *Classical Arabic Philosophy: Sources and Reception*, ed. by Peter Adamson (London: The Warburg Institute, 2007), pp. 150–75

———, and Andreas Büttner, 'Notes on Anonymous Twelfth-Century Translations of Philosophical Texts from Arabic into Latin on the Iberian Peninsula', in *The Arabic, Hebrew and Latin Reception of Avicenna's Physics and Cosmology*, ed. by Dag Nikolaus Hasse and Amos Bertolacci (Berlin: De Gruyter, 2018), pp. 313–69

Hasselhoff, Görge K., *Dicit Rabbi Moyses: Studien zum Bild von Moses Maimonides im lateinischen Westen vom 13. bis zum 15. Jahrhundert* (Würzburg: Königshausen & Neumann, 2004)

Hirai, Hiro, 'Atomes vivants, origine de l'âme et génération spontanée chez Daniel Sennert', in *Bruniana & Campanelliana*, 13 (2007), 477–96

———, 'Daniel Sennert on Living Atoms, Hylomorphism and Spontaneous Generation', in *Medical Humanism and Natural Philosophy: Renaissance Debates on Matter, Life and Soul*, ed. by Hiro Hirai (Leiden: Brill, 2011), pp. 151–72

Hissette, Roland, 'Albert le Grand et l'expression "Diluvium ignis"', *Bulletin de philosophie médiévale*, 22 (1980): 78–81

———, 'Les recours et allusions à Albert le Grand dans deux commentaires du "Symbolum Parisinum"', in *Nach der Verurteilung von 1277: Philosophie und Theologie an der Universität von Paris im letzen Viertel des 13. Jahrhunderts / After the Condemnation of 1277: Philosophy and Theology at the University of Paris in the Last Quarter of the Thirteenth Century*, ed. by Jan A. Aertsen, Kent Emery, and Andreas Speer (Berlin: De Gruyter, 2001), pp. 873–88

Jacquart, Danielle, and Claude Thomasset, 'Albert le Grand et les problèmes de la sexualité', *History and Philosophy of the Life Sciences*, 3 (1981), 73–93

———, *Sexuality and Medicine in the Middle Ages*, trans. by Matthew Adamson (Princeton, NJ: Princeton University Press, 1988)

Kruk, Remke, 'A Frothy Bubble: Spontaneous Generation in the Medieval Islamic Tradition', *Journal of Semitic Studies*, 35 (1990), 265–82

Lafleur, Claude, David Piché, and Joanne Carrier, 'Le statut de la philosophie dans le décret parisien de 1277 selon un commentateur anonyme du XV^e siècle: étude historico-doctrinale, édition sélective et synopsis générale des sources du Commentaire "Quod Deus"', in *Nach der Verurteilung von 1277: Philosophie und Theologie an der Universität von Paris im letzen Viertel des 13. Jahrhunderts / After the Condemnation of 1277: Philosophy and Theology at the University of Paris in the Last Quarter of the Thirteenth Century*, ed. by Jan A. Aertsen, Kent Emery, and Andreas Speer (Berlin: De Gruyter, 2001), pp. 931–1003

Mandonnet, Pierre, *Siger de Brabant et l'Averroïsme latin au XIII^me siècle*, 2nd part (Louvain: Institut Supérieur de Philosophie de l'Université, 1908)

Mandosio, Jean-Marc, 'Follower or Opponent of Aristotle? The Critical Reception of Avicenna's Meteorology in the Latin World and the Legacy of Alfred the Englishman', in *The Arabic, Hebrew and Latin Reception of Avicenna's Physics and Cosmology*, ed. by Dag Nikolaus Hasse and Amos Bertolacci (Berlin: De Gruyter, 2018), pp. 459–534

Mavroudi, Maria, 'Translations from Greek into Latin and Arabic during the Middle Ages: Searching for the Classical Tradition', *Speculum*, 90 (2015), 28–59

McCartney, Eugene S., 'Spontaneous Generation and Kindred Notions in Antiquity', *Transactions and Proceedings of the American Philological Association*, 51 (1920), 101–15

Musallam, Basim, 'The Human Embryo in Arabic Scientific and Religious Thought', in *The Human Embryo: Aristotle and the Arabic and European Traditions*, ed. by Gordon R. Dunstan (Exeter: University of Exeter Press, 1990), pp. 32–45

Pagnoni-Sturlese, Maria-Rita, 'A propos du néoplatonisme d'Albert le Grand', *Archives de Philosophie*, 43 (1980), 635–54

Piché, David, and Claude LaFleur, *La condemnation parisienne de 1277: Nouvelle edition du texte latin, traduction, introduction et commentaire* (Paris: Vrin, 1999)

Robson, John A., *Wyclif and the Oxford Schools: The Relation of the 'Summa de Ente' to Scholastic Debates at Oxford in the Later Fourteenth Century* (Cambridge: Cambridge University Press, 1961)

Rutkin, H. Darrel, 'The Natural Philosophical Foundations for Astrological Revolutions: Albertus Magnus's Commentary on the Pseudo-Aristotelian *De Causis Proprietatum Elementorum*', in *Sapientia Astrologica: Astrology, Magic and Natural Knowledge, ca. 1250–1800* (Cham: Springer International, 2019), pp. 93–115

Schenk, Gerrit J., 'Dis-Astri: Modelli interpretativi delle calamità naturali dal Medioevo al Rinascimento', in *Le calamità ambientali nel Tardo Medioevo Europeo: Realtà, percezioni, reazioni*, ed. by Michael Matheus, Gabriella Piccinni, Giuliano Pinto, and Gian Maria Varanini (Florence: Firenze University Press, 2010), pp. 23–76

Smoller, Laura, 'The Alfonsine Tables and the End of the World: Astrology and Apocalyptic Calculation in the Later Middle Ages', in *The Devil, Heresy and Witchcraft in the Middle Ages: Essays in Honor of Jeffrey B. Russell*, ed. by Alberto Ferreiro (Leiden: Brill, 1998), pp. 211–39

Takahashi, Adam, 'Interpreting Aristotle's Cosmos: Albert the Great as a Reader of Averroes' (unpublished doctoral thesis, Radboud University Nijmegen, 2017)

Thorndike, Lynn, *A History of Magic and Experimental Science: Fourteenth and Fifteenth Centuries* (New York: Columbia University Press, 1934)

———, 'Three Astrological Predictions', *Journal of the Warburg and Courtauld Institutes*, 26 (1963), 343–47

Trizio, Michele, 'Greek Philosophical Works Translated into Latin', in *The Cambridge History of Medieval Philosophy*, ed. by Robert Pasnau and Christina Van Dyke (Cambridge: Cambridge University Press, 2010), vol. 2, Appendix B, pp. 798–801

Wels, Henrik, *Aristotelisches Wissen und Glauben im 15. Jahrhundert: Ein anonymer Kommentar zum Pariser Verurteilungsdekret von 1277 aus dem Umfeld des Johannes de Nova Domo* (Amsterdam: B. R. Grüner, 2004)

ADAM TAKAHASHI

# Chapter 7. Against Averroes's Naturalism*

## *The Generation of Material Substances in Albert the Great's* De generatione et corruptione *and* Meteorologica *IV*

Historians of medieval philosophy have recently begun to address the Scholastic debate about the nature and composition of material substances.[1] While Scholastic authors directed much of their attention to logical, metaphysical, and theological themes, they also dealt with the nature of the material world.

Among the medieval thinkers who considered material substances and their generation, Albert the Great (*c.* 1200–80) can be singled out as the most important.[2] He commented on all of Aristotle's works that were available to him, thereby composing a complete scientific system that formed the basis for education in the Latin West. Albert discussed a broad range of natural phenomena in his explanations of the arguments in the Stagirite's natural books (*libri naturales*). Historians have examined Albert's theories about nature and the universe, but a number of his physical ideas still await more detailed analysis, not least that of the generation of material substances. How did Albert explain the nature and

* A large part of this article is based on the fourth and fifth chapters of my PhD thesis, 'Interpreting Aristotle's Cosmos'. I would like to express my gratitude to Hans Thijssen, Paul Bakker, Richard Taylor, and Katja Krause. Also many thanks to Jimmy Aames for making critical comments on an earlier draft and revising my English text, and to Kate Sturge for help with the editing. This study is also supported by JSPS KAKENHI Grant Number JP19K12934.

1 On the late medieval debate on the nature of material substances, see, among others, Maier, *An der Grenze von Scholastik und Naturwissenschaft*, pp. 3–140; Pasnau, *Metaphysical Themes*, pp. 17–175; Lagerlund, 'Material Substance'; Brower, *Aquinas's Ontology of the Material World*; Polloni, *Twelfth-Century Renewal of Latin Metaphysics*. See also Freudenthal, *Aristotle's Theory of Material Substance*.

2 On Albert's natural philosophy in general, see, among others, Weisheipl, *Albertus Magnus and the Sciences*; Hossfeld, *Albertus Magnus als Naturphilosoph*; Twetten, Baldner, and Snyder, 'Albert's Physics'.

**Adam Takahashi** (adam.takahashi@gmail.com) is an associate professor of philosophy at Kwansei Gakuin University (Nishinomiya, Japan).

*Albert the Great and his Arabic Sources*, ed. by Katja Krause and Richard C. Taylor, Philosophy in the Abrahamic Traditions of the Middle Ages, 5 (Turnhout: Brepols, 2024), pp. 195–224
BREPOLS PUBLISHERS 10.1484/M.PATMA-EB.5.136487

generation of material substances, including the four sublunary elements (fire, air, water, and earth)?[3]

An answer to this question can be found in his original theory on the topic, which tries to harmonize two different matter theories that he discerned in the works of Aristotle.[4] The first of these theories is hylomorphism, the notion that a substance is composed of form and matter. On this view, a particular thing, say a statue, is regarded as a compound of matter (for example, bronze) and form (or perceptible shape). The second is that of the four elements. According to this theory, a material substance is a thing composed of the elements fire, air, water, and earth. Each of the elements, which share an unchangeable substrate or prime matter, is in turn regarded as deriving from a combination of primary qualities (hot, cold, wet, and dry). Aristotle developed hylomorphism mainly in his *Physics*, while he advanced the theory of elements in *On the Heavens* III–IV, *On Generation and Corruption* (hereafter *GC*), and *Meteorology* IV. He did not, however, unify the two theories in his physical works. Since these two theories present divergent — if not contradictory — views on the nature and composition of material substances, the issue of how they could be reconciled was left as a pressing question for later commentators.

I will not discuss here whether medieval Aristotelians in general succeeded or failed in their attempts to reconcile the two matter theories. Instead, my aim is to show how Albert, in particular, tried to unify the hylomorphic analysis with the theory of the elements and primary qualities. To this end, I will examine Albert's paraphrases of *GC* and *Meteorology* IV, since these paraphrases are not only less studied than those of the *Physics* and *Metaphysics*, but also give us a picture of his general theory of the material world.[5] The theory presented in them offered a grand scheme under which, in his later works such as *On Minerals* (*De mineralibus*) and *On Animals* (*De animalibus*), Albert was able to give more particular explanations of natural substances.

What complicates the picture is that Albert developed his harmonization of the two theories through a critique of the Latin Averroes.[6] As I will show, Averroes (1126–98) had attempted to unify the theory of elements with hylomorphism in his Middle Commentaries on *GC* and *Meteorology* IV, both of which Albert was able to consult in translation. Although Averroes and Albert shared the same tendency of trying to unify Aristotle's two matter theories, there is also a clear difference in the ways that they approached the task. This difference revolves about the question: Are the substantial forms of the elements and more

3 To my knowledge, there is no comprehensive study of Albert's theory of material substances. See Weisheipl, *Albertus Magnus and the Sciences*.

4 See Maier, *An der Grenze von Scholastik und Naturwissenschaft*, pp. 3–140; also Lüthy, 'An Aristotelian Watchdog'; Petrescu, 'Hylomorphism'.

5 On Albert's paraphrases of Aristotle's *GC* and *Meteorology* IV, see notes 41 and 64.

6 Regarding the Latin translations of Averroes's works and their availability in the thirteenth century, see Hasse, *Latin Averroes Translations*.

complex material substances reducible to the operations of primary qualities? We will see that Averroes advanced a very naturalistic view, reducing the forms of the elements to primary qualities and giving a privileged status to the formative role of natural heat in sublunary generation.[7] Albert, in contrast, took a stand against Averroes by putting forward a non-naturalistic position in which the substantiality or essences of material substances, including the elements, are irreducible to the primary qualities.

In what follows, I first focus on Averroes's and Albert's arguments about the generation and corruption of elements in their commentaries on *GC*. I then turn to Albert's discussion of sublunary generation and his rejection of Averroes's view in his commentary on *Meteorology* IV.

## Albert's Theory of Elements and his Critique of Averroes in the Commentary on GC

### *Aristotle's Doctrine of Generation and Corruption in* GC *I.1–3 and II.1–4*

Before examining Averroes's and Albert's commentaries on *GC*, let me first summarize the arguments of Aristotle that formed the basis of Averroes's and Albert's views on the transmutation of elements.

Having explained the causes and principles of natural and celestial substances in *Physics* and *On the Heavens*, in *GC* Aristotle begins his discussion of the various kinds of change that sublunary entities can undergo.[8] He divides change into four kinds (*GC* I.4, 319b31–320a7): 'generation and corruption' (*genesis kai phthora*), 'alteration' (*alloiosis*), 'growth and diminution' (*auxesis kai phthisis*), and 'locomotion' (*phora*). This classification corresponds to four of the Aristotelian categories: generation and corruption are changes of substance, alteration is change of quality, growth and diminution are changes of quantity, and locomotion is change of place. Of these four kinds of change, I will examine Aristotle's theory

---

7 My reference to 'naturalistic' and 'non-naturalistic' or 'metaphysical' approaches found in the natural philosophy of the medieval and early modern Aristotelians relies on Kessler, 'Metaphysics or Empirical Science?'. According to Kessler, it was the emphasis on the role of primary qualities in the commentary tradition on *GC*, rather than the hylomorphic scheme of substantial bodies in the *Physics*, that gave a theoretical foundation to the naturalistic or materialistic approaches that we find in the works of Renaissance philosophers such as Pomponazzi, Cardano, and Telesio. Kessler argues: 'A new way of understanding Aristotle's natural philosophy was tried, which separated the "metaphysical" *Physics* from the "naturalistic" *De generatione et corruptione* and concentrated on the latter; and [...] this new way of reading Aristotle paved the way for a new reading of the book of nature itself' (p. 88).

8 All English translations of Aristotle's texts are taken from Aristotle, *Complete Works of Aristotle*, ed. by Barnes.

of generation and corruption, especially with regard to the four elements, by focusing on *GC* I.1–3 and II.1–4.[9]

In *GC* I.1–3, in order to define the notions of generation and corruption in their proper sense, Aristotle begins with a survey of the views of his predecessors. After criticizing major past theories including Democritus's atomism, he explains that generation and corruption take place when 'a thing changes from this to that as a whole' (*GC* I.2, 317a21–22). For him, generation and corruption have a special status, distinct from other kinds of change. Generation and corruption affect an entire substance, whereas the other changes only concern accidental properties such as quality, quantity, and place.[10] Aristotle also says that only generation and corruption of substances is entitled to be called 'unqualified' generation and corruption (*GC* I.3, 317b8–9). Therefore, we need to use other terms to signify changes of accidental properties. For instance, we might describe the process by which a hot thing changes into a non-hot or cold one as the generation of coldness and the corruption of hotness — but according to Aristotle, this is merely generation and corruption in one of the qualified senses; to be more precise, it should be called 'alteration', or change in quality.

In *GC* I.3, Aristotle further examines substantial change: how does a new substance come into being and how does an old one pass away? He answers this question by reflecting on the continuity of generation and corruption. There are two important points here: the reciprocal nature of generation and corruption, and the material cause of the continuity of substantial change. Regarding the first, Aristotle argues that generation and corruption are reciprocal processes that take place when one substance changes into another: 'in substances, the generation of one thing is always a corruption of another, and the corruption of one thing is always another's generation' (*GC* I.3, 319a20–22). If this claim applies to the four elements, as it appears to do in *GC* II.1–4, then it follows that the generation and corruption of the elements must be identical with their transformation into each other.[11]

Second, Aristotle addresses the material cause of the continuity of substantial change. Since generation and corruption is not the coming-to-be or passing-away of a substance from or into nothing, but rather the change of one substance into another, these two substances must share something that does not undergo any change. Otherwise, there would be no material continuity between what it generated and what is corrupted. It follows that the change of one substance into another requires an unchangeable material principle: the 'substratum'

9 For Aristotle's view of generation and corruption or substantial change in his *GC* I.1–3 and II.1–4, see Solmsen, *Aristotle's System of the Physical World*, pp. 321–52; Bostock, 'Aristotle on the Transmutation of the Elements'; de Haas and Mansfeld, *Aristotle: On Generation and Corruption*, pp. 25–121.

10 See Aristotle, *GC* I.2, 317a23–27: 'For in that which underlies the change there is a factor corresponding to the definition and there is a material factor. When, then, the change is in these factors, there will be coming-to-be or passing-away; but when it is in the thing's affections and accidental, there will be alteration'.

11 See Bostock, 'Aristotle on the Transmutation of the Elements'.

(*hypokeimenon*). Thus, Aristotle claims that 'the substratum is the material cause of the continuous occurrence' of generation (*GC* I.3, 319a19). It should be noted that what serves as the persisting subject of change is an important criterion distinguishing substantial change from alteration. In the case of substantial change, the persisting subject is not a substance, but a substratum that is neither perceptible nor exists actually, since substantial change only occurs when one substance changes as a whole into another. In the case of accidental change, substance serves as the persisting subject.

Aristotle's categorial classification of change is effective to the extent that it can sharply distinguish one kind of change from another kind, but his definition of the substantial change or a reciprocal change from one element to another raises a new problem with regard to the constituent principle of the substantiality or essence of material entities, especially in the case of the four elements. For, as we will see, one kind of element is distinguished from another only in terms of its primary qualities.

Aristotle's specific explanation of the generation and corruption of the elements can be found in *GC* II.1–4. There, he introduces two principles that are constitutive of the elements: a persisting substratum and the primary qualities. In *GC* II.1, Aristotle argues that there exists a material substratum underlying all the elements: 'a principle that is really first, the matter which underlies, though it is inseparable from, the contrary [qualities]' (*GC* II.1, 329a29–31). The substratum, then, serves as a material principle that, together with certain primary qualities, constitutes each element. Aristotle's idea of the substratum has traditionally been interpreted as 'prime matter', which in itself exists only potentially, and together with a combination of primary qualities constitutes an element.[12] It should be noted that Aristotle himself does not use this term in his works, whereas Averroes and Albert do, as we will see below.

In *GC* II.2–4, Aristotle examines the role that primary qualities play in substantial change. He views them as the essential principles that are at work in the transmutation of the elements. Primary qualities do not exist separately, but always come in certain combinations. Since a pair of contrary qualities (hot-cold or moist-dry) cannot be simultaneously present in the same subject, there can only be four possible combinations (hot-dry, hot-moist, cold-dry, and cold-moist). Aristotle argues that the four combinations of qualities correspond to the four elements: hot-dry corresponds to fire, hot-moist to air, cold-moist to water, and cold-dry to earth (*GC* II.3, 330a30–330b7). On the basis of this theory, he puts forward the idea that the substantial change of the elements (that is, the transmutation of one element into another) occurs as the result of the recombination of these qualities (see *GC* II.4, 331a26–b2). For instance, if one

---

12 It is still controversial whether Aristotle had in mind the idea of prime matter. The traditional positive view was formulated by Solmsen, 'Aristotle and Prime Matter'; Robinson, 'Prime Matter in Aristotle'; Aristotle, *Aristotle's 'De generatione et corruptione'*, trans. by Williams, pp. 211–19. Doubt has been cast on Aristotle's belief in prime matter by, among others, Gill, *Aristotle on Substance*, pp. 42–46.

combination of qualities (e.g., hot-dry) is changed into another (e.g., hot-moist), the corresponding element (fire) is changed into another element (air).

Aristotle's position can be summarized as follows:

1. Generation and corruption in the strict sense are a change of substance, not of accidental properties such as quality, quantity, and place.
2. When one of the four elements changes into another, there is a substratum that persists as an imperceptible underlying subject.
3. The substantial change of the elements takes place as a result of the recombination of the primary qualities inhering in the underlying substratum.

Aristotle's theory of the elements and of their change seems to blur the difference between substantial change and alteration, as well as that between substance and accidental qualities. Indeed, there has been much controversy as to whether the primary elements can properly be called substances in Aristotle's framework. Aristotle certainly regarded the four elements as substances in his *Metaphysics* (VII.2, 1028b81–83). Yet if these elements are distinguished from each other only in the way that the primary qualities are combined, then it would seem that the substantiality of the elements can be identified with a bundle of qualities.[13]

Later Aristotelians therefore faced the question of the extent to which the substantiality of the elements can be reduced to their accidental qualities. Once the theory of elements was combined with hylomorphism, this question was transformed into the following: Can the substantial or specific form of elements be regarded as an effect of the operation of primary qualities? Averroes and later Albert proposed different answers to this question in their attempts to harmonize the theory of elements with hylomorphism.

### *Averroes's Naturalistic Theory of Elements and its Doctrinal Background*

Moving on to Averroes's position, I will first discuss Averroes's commentary on *GC* I.1–3 in order to see how he responded to two of the three components of Aristotle's position: the categorial distinction between substantial change and other accidental changes, and the idea of a substratum as the material cause of the continuity of substantial change.[14] I then turn to Averroes's commentary on *GC* II.1–4 to examine his response to the third component of Aristotle's position,

13 On the unstable status of elements in Aristotle's matter theory, see Sokolowski, 'Matter, Elements and Substance in Aristotle'.

14 For Averroes's commentaries on *GC*, I use the following Latin texts: for the *Middle Commentary on GC*, I used *Averrois Cordvbensis Commentarivm Medivm in Aristotelis De Generatione et Corrvptione Libros*, ed. by Fobes (hereafter *Middle Commentary*); for the *Epitome of GC*, I used *Aristotelis opera cum Averrois commentariis*, ed. Venetiis, vol. 5 (hereafter *Epitome*). I also consulted *Averroes on Aristotle's 'De generatione et corruptione'*, trans. by Kurland. However, all the translations of Averroes's texts in this article are mine unless otherwise attributed. On these commentaries in general, see Puig Montada, 'Aristotle and Averroes on *Coming-To-Be and Passing-Away*'; Eichner, *Averroes (Abū l-Walīd Ibn Rušd) Mittlerer Kommentar*.

namely the idea that substantial change of the elements occurs as a result of a recombination of the primary qualities.[15]

Before going into any doctrinal details, let me make a preliminary remark on Averroes's commentaries on *GC*. Averroes wrote two commentaries on this treatise: an Epitome, or Short Commentary, and a Middle Commentary. The Middle Commentary, in particular, played a significant role in the Latin tradition. Translated probably by Michael Scot around 1230, it was regarded as the preeminent interpretation of Aristotle by medieval Latin commentators.[16] The Epitome, on the other hand, was not translated into Latin until the sixteenth century, and hence was not available to Albert. I will nevertheless consult this text as well, in cases where it helps to clarify Averroes's position.

In the *Middle Commentary on GC*, Averroes generally tends to simply reproduce Aristotle's teachings, and even in his interpretation of the substantial change of the elements, his comments are mostly repetitions of Aristotle's arguments. In fact, commenting on *GC* I.1–3, Averroes starts by paraphrasing Aristotle's survey of earlier views, then proceeds to discuss the nature of substantial change. He argues that unqualified generation takes place when one substance is totally changed into another, and claims that 'unqualified generation is a change of something as a whole from this into that, as from this water into that air'.[17] This statement clearly repeats Aristotle's claim that generation and corruption take place when 'a thing changes from this to that as a whole' (*GC* I.2, 317a21–22).

Up to this point, Averroes basically reproduces Aristotle's arguments when explaining the nature of substantial change or unqualified generation and corruption. But when he begins to analyse Aristotle's view on the continuity of substantial change, his account diverges from that of the Philosopher:

> The corruption of one and the same thing is [at the same time] the generation of another thing. For, given that the corruption of one thing is the generation of another thing, it is necessary that generation is never broken off, since through the succession of forms over a subject, which is matter, that from which generation comes in an unqualified sense, which is in potency, is not devoid of some being in act, which is form.[18]

Here, Averroes follows Aristotle's suggestion that the continuity of generation and corruption results from two principles: the reciprocal nature of substantial change

---

15 For Averroes's theory of the substantial change of elements and of the status of the form of elements, see Maier, *An der Grenze von Scholastik und Naturwissenschaft*, pp. 36–88; Eichner, *Averroes (Abū l-Walīd Ibn Rušd) Mittlerer Kommentar*, pp. 40–82 and 188–236.

16 See Caroti, 'Commentaries on Aristotle's *De generatione et corruptione*', p. 252.

17 Averroes, *Middle Commentary*, I.2, p. 21: 'Generatio simplex est transmutatio alicuius rei secundum totum ex hoc in hoc, ut hec aqua in hunc aerem'.

18 Ibid., I.3, pp. 26–27: 'Corruptio unius et eiusdem rei est generatio alterius; quoniam, cum corruptio alicuius fuerit generatio alterius, tunc necesse est ut generatio non abscindatur, quoniam per successionem formarum super subiectum, quod est materia, non denudatur illud ex quo generatio fit simpliciter, quod est in potentia, ab aliquo ente in actu, quod est forma'.

and the persistence of a material substratum. However, Averroes's account of substantial change diverges from Aristotle's in that he explains the transmutation of the elements by invoking hylomorphism, the view that bodies are composed of form and matter. Averroes's appeal to hylomorphism has its basis in *Physics* I.7–9, where Aristotle uses the hylomorphic scheme in order to explain substantial change in terms of form and matter.[19] Averroes regards substantial change as a change of form inhering in the same underlying matter. In his view, this change takes place when an underlying matter successively takes on new forms and loses old ones. Thus, he describes generation and corruption in terms of a 'succession of forms' (*successio formarum*).[20]

Furthermore, there is no doubt that Averroes has in mind the notion of 'prime matter' (*materia prima*) when he replaces Aristotle's term 'substratum' with 'matter'.[21] As mentioned above, Aristotle himself does not use the term in any of his works. Averroes, on the other hand, frequently uses 'prime matter' when referring to Aristotle's 'substratum' in his *Middle Commentary on GC*.[22] Rather than slavishly following Aristotle's distinction between substantial change and alteration, Averroes reformulates Aristotle's theory on the basis of hylomorphism, replacing his notion of 'substratum' with the term 'prime matter', which is more appropriate to a hylomorphic scheme.

Averroes's appeal to hylomorphism might be taken as a slight and insignificant modification of Aristotle's doctrine in *GC*. This hylomorphic scheme, however, leads him to develop a view different from the one that Aristotle presented in his treatise on the structure of each of the elements. Averroes's *Middle Commentary on GC* II.1–4 clarifies this difference. There, Averroes agrees with Aristotle's view on the transmutation of the elements:

> It is evident that every [element] can be generated from every other one, [...] for every one of these [elements] possesses some contrariety with every other one. [...] Since every one [of the elements] is contrary to any other one, it is necessary that every one is generated out of every other one.[23]

19 In his *Epitome*, Averroes says: 'We maintain that all generable and corruptible bodies are of two kinds, simple and compound; and each one of these kinds is composed of matter and form, as has been shown in an earlier work' (*Aristotelis opera cum Averrois commentariis*, ed. Venetiis, vol. 5, fol. 392I; trans. by Kurland, p. 124). In an editorial note to this passage, Kurland suggests that the 'earlier work' is Averroes's commentary on *Physics* I.7–9. For Aristotle's discussion in *Physics* I.7–9, see Bostock, *Space, Time, Matter, and Form*, pp. 1–18; also Cerami, 'Aristotelian Analysis of Generation'.

20 On the Averroist theory of the succession of forms, see Sylla, 'Medieval Concepts of the Latitude of Forms'.

21 On Averroes's idea of prime matter, see Hyman, 'Aristotle's "First Matter"'; Pasnau, *Metaphysical Themes*, pp. 60–66.

22 Averroes, *Middle Commentary*, pp. 33, 44, 70, 99.

23 Ibid., II.3, pp. 112–13: 'Et manifestum est quod omne ex omni potest generari, [...] quia unumquodque eorum habet aliquam contrarietatem cum unoquoque. [...] Et cum quodlibet est contrarium cuilibet, necesse est ut omne generetur ex omni'.

Averroes here merely paraphrases Aristotle's view that the transmutation of the elements takes place through a recombination of the primary qualities. He develops this doctrine using Aristotle's idea of the structure of the elements:

> He said: We have already declared that all these sensible bodies have matter that exists potentially and that is not devoid of one of the contrary [qualities], and that these four bodies [= four elements] are composed of this [matter] and of the contrariety that exists in it.[24]

Again, Averroes seems merely to follow Aristotle in regarding each element as consisting of one of the 'contrarieties' (*contrarietates*), or combinations of primary qualities, and the underlying substratum or prime matter. But in explaining the transmutation of the elements, he appeals not only to the combination of primary qualities, but also to the change of the form of the elements, adding:

> And since these four [qualities] are found only in combination, it is evident that the number of bodies, of which these [qualities] are forms, must be in accordance with the number of possible combinations of these primary qualities.[25]

Averroes regards the combination of the primary qualities of an element to be the form of that element.[26] He thus combines two views of material substances: hylomorphism, and the idea that the elements are composed of primary qualities and a substratum. Certainly, both views originate in Aristotle, but the Philosopher himself did not identify the form of an element with a combination of the primary qualities, whereas for Averroes, a recombination of the primary qualities in an element is identical with a change of its form.

Anneliese Maier has emphasized that Averroes's conception of the form of the elements as simply a complex of primary qualities is 'a genuinely Aristotelian concept of an element'.[27] One could argue that when Scholastic authors such as Albert later rejected Averroes's position regarding the forms of the elements, they diverged from Aristotle's doctrine and appealed to another metaphysical principle that gives substantiality or essence to each element, a principle not mentioned in GC.

---

24 Ibid., II.1, p. 98: 'Dixit nos autem declaramus quod omnia ista corpora sensibilia habent materiam existentem in potentia, non denudatam ab aliquo contrariorum, et quod hec corpora quatuor componuntur ex ea et ex contrarietate existente in ea'.

25 Ibid., II.3, p. 107: 'Et cum iste quatuor non inveniantur nisi coniuncte, manifestum est quod numerus corporum quorum iste sunt forme est secundum numerum compositionum possibilium istis qualitatibus primis'.

26 In his *Epitome*, Averroes says: 'The proximate matter of these [simple bodies or four elements] is prime matter, as has been shown, and their forms are the primary contraries, which are in them [...]' ('Materia propinqua eorum est materia prima, ut declaratum est. Forma autem eorum sunt prima opposita, quae sunt in eis'). *Aristotelis opera cum Averrois commentariis*, ed. Venetiis, vol. 5, fol. 392K; trans. by Kurland, p. 124.

27 Maier, *An der Grenze von Scholastik und Naturwissenschaft*, p. 43.

Before turning to Albert, I will sketch the theoretical background of Averroes's position, including the arguments of Alexander of Aphrodisias (fl. *c.* 200 CE) on the topic and Avicenna's critique of these.[28] Alexander's commentary on *GC* is important in understanding Averroes's position, because he is the most significant commentator named in Averroes's commentaries on *GC*.[29] Alexander recognizes the difficulty in distinguishing the substantial change of elements from alteration and the form of an element from a combination of the primary qualities. In a commentary on *GC* ascribed to Alexander, we find the following argument about the transmutation of the elements:

> That [i.e., the transmutation of the elements taking place as the generation of one element from another] is manifest owing to the fact that he [Aristotle] calls the reciprocal change of the primary bodies in respect of form, 'alteration'. This is evident from what he said: 'because alteration is only in respect of tangible affection'. [...] Not only do we observe that [bodies] change only in respect of tangible differences, but we say so because through these differences, [bodies] change in a manner that is known through sensation. These are their forms affected by a nature, and we observe that these forms belong to them for the sake of their change into one another, so that simple bodies may be generated into one another; otherwise there would be no alteration. Moreover, we observe that simple bodies are generated. For alteration is their change into one another.[30]

Here, as Emma Gannagé points out, Alexander suggests that according to Aristotle's definition, there is no clear distinction between substantial change and alteration in the transmutation of elements.[31] It is still a matter for further research whether the text just quoted is what Averroes was actually citing, but the view that the two changes of elements cannot be clearly distinguished can also be found in Alexander's other works, and in the work of other authors as well. For instance, in his treatise *De mixtione*, after declaring that 'earth and water, air and fire, have the same matter and are differentiated by their forms', Alexander explains what he means by 'form' in terms of contrary qualities.[32] In his commentary on Aristotle's *Meteorology* IV, too, he emphasizes that the primary qualities serve as the formal cause of the elements.[33] Furthermore, the sixth-century Aristotelian commentator Philoponus ascribes to Alexander the position that substantial change and alteration can be identified in the case of elements.[34]

---

28 Concerning Averroes's reliance on Alexander's commentary on *GC*, see Eichner, 'Ibn Rušd's Middle Commentary'.

29 Ibid., p. 287.

30 Alexander of Aphrodisias, *On Aristotle's On Coming-to-Be and Perishing*, trans. by Gammagé, pp. 60 and 104–05.

31 See ibid., pp. 59–63.

32 Alexander of Aphrodisias, *Alexander of Aphrodisias on Stoic Physics*, ed. by Todd, pp. 150–51.

33 Alexander of Aphrodisias, *On Aristotle's Meteorology 4*, trans. by Lewis, p. 65.

34 See Philoponus, *On Aristotle's On Coming-to-Be and Perishing*, trans. by Williams, pp. 142–43.

Alexander's position was influential in the Islamic world, as is attested in Avicenna's *De generatione et corruptione*, the third book of natural philosophy in his philosophical summa *The Book of the Healing* (*Kitāb al-Shifā'*).[35] There, Avicenna discusses the substantial change of elements by criticizing the opinion shared among Aristotle's commentators. In his view, the commentators failed to understand Aristotle's teaching:

> The commentators are confused about that, because there are those who make errors in distinguishing between [substantial] form and accidents by referring to the distinction between the natural form of bodies and their qualities, since they believe that all the qualities, or at least some of them, are [substantial] forms for these bodies.[36]

Avicenna criticizes the commentators for identifying the form of an element with its primary qualities. Given the background I have mentioned, Avicenna must have had in mind Alexander of Aphrodisias and his followers. Against Alexander's position, Avicenna argues that the substantial form of an element should be distinguished from its accidental properties. Shortly after the passage quoted above, he continues:

> Let us therefore say that each of the elements has a substantial form by which it is what it is, and consequent to the form are the perfections of quality, quantity, and place, and the warmness and coldness of each body belong to it due to the substantial form and the dryness and the moistness belong to it due to matter.[37]

Avicenna's claim here is that the substance or substantial form is ontologically prior to any accidental property, insofar as accidental properties can exist only in the substance.[38] In other words, accidental qualities are dependent upon the substantial form, and not vice versa. Therefore, if we were to suppose that the elemental forms arise from a combination of accidental qualities, we would be mistaking the ontologically prior for the posterior. In order to explain the origin

---

35 Avicenna, *Liber tertius naturalium De generatione et corruptione*, ed. by Van Riet (hereafter *De generatione et corruptione*). On Avicenna's theory of the substantial change of elements, see Stone, 'Avicenna's Theory of Primary Mixture'; McGinnis, *Avicenna*, pp. 84–88.

36 Avicenna, *De generatione et corruptione*, c. 6, p. 63: 'Commentatores autem vacillant in hoc, quia sunt titubantes in dividendo inter formam et accidentia quae ostendunt discretionem inter formam corporum naturalem et [iam] inter qualitates eorum, quia crediderunt quod omnes qualitates, vel saltem aliqua pars earum, erant formae istis corporibus [...]'. See Stone, 'Avicenna's Theory of Primary Mixture', p. 116; McGinnis, *Avicenna*, p. 87.

37 Avicenna, *De generatione et corruptione*, c. 6, p. 66: 'Dicamus ergo quod quodlibet elementorum habet formam substantialem cum qua est illud [et] quod est, et sequuntur formam istam complementa qualitatis et quantitatis et ubi, et caliditas et frigiditas cuiuslibet corporum approprianttur ex parte formae substantialis, et siccitas et humiditas ex parte materiae'. See Stone, 'Avicenna's Theory of Primary Mixture', p. 117; McGinnis, *Avicenna*, p. 87.

38 As for Avicenna's distinction between substance and accidental properties, see also Lammer, *Elements of Avicenna's Physics*, esp. pp. 114–21.

of the substantial form, Avicenna elsewhere invokes a metaphysical principle called the 'giver of forms' (*dator formarum*).[39] This notion occupies an important place in Avicenna's worldview. Supposing that the intelligences (along with the celestial orbs that are moved by them) emanate from the first cause, or God, Avicenna designates the lowest of these intelligences as the 'active intellect'. It is the active intellect that he also calls the 'giver of forms', since he supposes that the active intellect plays a formative role in the sublunary generation. Once sublunary matter is properly disposed to receive a certain form, the giver of forms bestows upon it a substantial form.[40]

### Albert's Critique of Averroes's Naturalism

Averroes's position raised serious questions about the identity and essence of the elements. Since he clearly identified the form of the elements with a combination of primary qualities, one might expect the substantial forms of the elements to be determined solely by material qualities. In what follows, I show that Albert was not satisfied with this, and put forward a non-naturalistic view of the substantial forms of the elements.[41]

Commenting on Aristotle's *GC* I.1–3, Albert tends to repeats Aristotle's theory of substantial change and his ideas of a substratum and prime matter.[42] According to Albert, some ancient philosophers identified 'unqualified generation' (*generatio simplex*) with 'alteration' (*alteratio*), while others distinguished between generation and alteration.[43] Following Aristotle, he supports the latter position and adds that unqualified generation in the present context is identical with 'change in(to) substance' (*mutatio in substantiam*).[44]

After reproducing Aristotle's survey of ancient opinions, Albert enters into a detailed discussion of substantial change. In his paraphrase of *GC* I.3, rather than simply reformulating Aristotle's arguments, he presents his own views by addressing the questions of whether generation and corruption really take place, what kind of entity serves as the subject of generation and corruption, and what

---

39 See, among others, Hasse, 'Avicenna's "Giver of Forms"'.

40 See Stone, 'Avicenna's Theory of Primary Mixture', esp. pp. 117–19.

41 On Albert's *De generatione et corruptione*, see Cadden, 'Medieval Philosophy and Biology of Growth'; Hossfeld, 'Grundgedanken in Alberts des Großen Schrift'; Caroti, 'Note sulla parafrasi del *De generatione et corruptione*'. See also Thijssen and Braakhuis, *Commentary Tradition*.

42 For Albert's view of the distinction between (substantial) change and (accidental) motion, see Baldner, 'Albertus Magnus and the Categorization of Motion'.

43 Albertus Magnus, *De generatione et corruptione*, ed. by Hossfeld (hereafter *De generatione*), I.1.2, p. 112, vv. 20–23: 'Quia ergo hi antiqui diversificati sunt circa genus generationis et alterationis, ponentes eas in genere uno vel diversis, ideo primo dicatur de opinione eorum'.

44 Ibid., vv. 15–20: 'Dico autem simplicem generationem, quae pure et vere est universalis generatio, mutatio scilicet in substantiam universaliter, de qua intendimus hic, sive etiam simplicem generationem, quae est generatio elementi, de qua in secundo huius voluminis loquemur'.

kind of generation is continuous and infinite.[45] All three of these questions clearly derive from Aristotle's work. Nonetheless, Albert's selection of the topics shows that he recognizes their centrality in the discussion of substantial change, and also the need to explicate Aristotle's ideas.

Albert does not tackle all three of these questions to an equal extent. In the case of the first issue, whether there is a substantial change distinct from other changes, he simply repeats Aristotle's argument.[46] Taking for granted the existence of both generation and corruption as distinct from alteration, Albert directs his attention to the second question: What serves as the 'subject' (*subiectum*) of generation and corruption?[47] He explains the idea of a subject or substratum by appealing to the notion of matter, also addressed in his paraphrase of the *Physics*.[48]

Albert's conception of matter is not the same as Aristotle's and Averroes's. The difference in their views on this issue becomes clear if we examine Albert's discussion on the nature of matter in his treatment of the third question, what kind of generation is continuous and infinite.[49] In his paraphrase of Aristotle's axiom that 'the generation of one thing is the corruption of another', Albert argues:

> Thus [one being] is under one form actually and under another potentially. And the cause of its restlessness is the infinite desire of matter, since [matter] does not desire only one form but every form successively, because it cannot have them at the same time. But this desire is the germ of form in matter, which is drawn out from [matter] itself.[50]

Albert explains the nature of matter by appealing to a particular notion that Averroes did not use: the 'germ of form' (*incohatio formae*).[51] Albert employs this notion to describe matter as a quasi-active principle that successively desires different forms. In attributing an active principle to matter, he clearly departs from previous Aristotelian commentators. But it should also be noted that, like

---

45 Ibid., I.1.19, p. 126, vv. 38–42: 'Determinatis autem his postea videndum, utrum generatur simpliciter aut corrumpitur aliquid, ut prima nostra quaestio sit, an sit generatio; et secunda, quid subicitur generationi; et tertia, quare continua et infinita sit generatio'.

46 Ibid., p. 126, v. 35–p. 127, v. 32.

47 Ibid., I.1.20, p. 127, v. 33–p. 128, v. 14.

48 See ibid., p. 127, vv. 36–41: 'Quod tamen etiam his determinatis mirabilem habet quaestionem [= utrum subiectum generationis sit aliquid ens actu vel nihil], quam ideo mirabilem voco, quia conducit ad altiorem considerationem, eo quod determinari vix potest, nisi cognoscatur materia, inquantum est principium substantiae compositae, quod utique pertinet ad primum physicum'. Cf. Albertus Magnus, *Physica (libri I–IV)*, I.3.3, p. 40, v. 70–p. 44, v. 86.

49 Albertus Magnus, *De generatione*, I.1.22, p. 129, v. 4–p. 130, v. 32.

50 Ibid., p. 130, vv. 23–29: 'Est igitur actu sub una forma et potentia sub altera, et causa inquietudinis eius est materiae desiderium in infinitum, quod non desiderat formam unam tantum, sed omnem formam successive, cum simul eas habere non possit. Hoc autem desiderium formae incohatio est in materia, quae educitur de ipsa'.

51 For Albert's idea of *incohatio formae*, see Nardi, *Studi di filosofia medievale*, pp. 69–101; Snyder, 'Albert the Great'.

Averroes, Albert explains the continuity of substantial change by means of the hylomorphic scheme.[52] He understands substantial change as a change of the form inhering in matter. Again following in Averroes's footsteps, Albert describes the continuity of substantial change in terms of a succession of forms. Although there is a divergence between Albert and Averroes concerning their conception of prime matter, we should not overlook the profound agreement between the two authors in their interpretation of the nature of the substantial change of the elements.

Now let us turn to Albert's paraphrase of *GC* II.1–4, where he deals with the generation and corruption of elements.[53] As we saw above, in *GC* II.1–4 Aristotle explains the substantial change of the elements through two principles: the substratum, or prime matter, and the combination of the primary qualities. Averroes adheres to hylomorphism, arguing that the substratum serves as the matter, the primary qualities as the form of the elements. Albert, in turn, seems to follow Averroes's hylomorphic account, for he also suggests that the substratum and primary qualities function as the 'material principle' (*principium materiale*) and the 'formal principle' (*principium formale*) of the elements respectively.[54] Thus, he explains the transmutation of the elements in terms of a change of the form inhering in matter:

> For these elements have a common matter and contrary [qualities], since they are transmuted into each other, and matter, which was under the form of one [element], is brought under the form of another, as fire becomes water or something else.[55]

Clearly, Albert regards this change as a change of form inhering in the same matter. This passage also indicates that he considers the transmutation of the elements to be the result of a recombination of the primary qualities. Indeed, in the subsequent paraphrase of *GC* II.4, he claims that 'intensified attributes [= qualities] change their [the elements'] substance into one other'.[56]

---

52 On Albert's hylomorphism, see Twetten, Baldner, and Snyder, 'Albert's Physics', esp. pp. 173–82.

53 Albert considered Aristotle's *GC* II to be foundational for his own analysis of the generation and corruption of more complex substances such as stones, metals, plants, and animals. See Albertus Magnus, *De generatione*, II.1.1, p. 177, vv. 13–18: 'Relinquitur autem in hoc secundo libro considerare de corporibus, quae vocantur elementa, antequam consideremus in particularibus libris de generatione et corruptione corporum specialium sicut lapidum et metallorum et plantarum et animalium'.

54 Albertus Magnus, *De generatione*, II.1.5, p. 179, vv. 37–42: 'Si autem habemus dicere principium materiale elementorum, antequam loquamur de eorum transmutatione ad invicem, non minus etiam habemus dicere principium formale ipsorum, hoc est qualitates distinguentes materiam elementi, secundum quas illa corpora qualia sunt'.

55 Ibid., II.1.4, p. 179, vv. 24–28: 'Haec enim elementa habent materiam communem et contrarietates, quia ad invicem transmutantur, et materia, quae fuit sub unius forma, efficitur sub forma alterius, sicut ignis fit aqua vel aliud'.

56 Ibid., II.2.1, p. 186, vv. 11–12: 'passiones intensae transmutant substantiam eorum ad invicem'.

However, Albert eventually draws a sharp distinction between the form of the elements and the qualities, and rejects the reduction of the formal cause of the elements to their primary qualities. His position is presented in a special digression entitled 'The primary qualities are not the substantial forms of the elements'.[57] Here, Albert speaks of the irreducibility of the elemental form to a combination of the primary qualities. He admits that the primary qualities differentiate the underlying common matter into the four elements,[58] but insists on the distinction between the form of the elements and the primary qualities, arguing that these qualities are 'not their [the elements'] substantial forms, because Aristotle says that substantial form is neither active nor passive'.[59] According to Albert, each of the primary qualities has either an active or a passive nature: hot and cold are active qualities, dry and wet are passive. And since the elemental form, unlike the primary qualities, is neither active nor passive, it cannot be identified with the primary qualities. Despite maintaining that the transmutation of the elements results from a recombination of primary qualities, Albert does not accept the identification of the form of the elements with their primary qualities.

Insofar as Albert insists on the distinction between the substantial form and accidental qualities, his view comes closer to that of Avicenna. Yet Albert was not able to read Avicenna's *De generatione et corruptione*, because the treatise was not available in the Latin West at the time Albert was writing his paraphrase of Aristotle's *GC*.[60] He may, though, have learned about Avicenna's view through another of his works: the *Metaphysics* or *Philosophia prima*.[61] There, Avicenna appeals to the ontological priority of substance to accidents. Substance, he argues, 'is constitutive of the existence of accidents and is not constituted by accidents; therefore substance is prior in existence'.[62] Indeed, as we will see below, Albert argues in his paraphrase of *Meteorology* IV that 'Avicenna and Algazel conceded that substantial form does not derive from active qualities'.[63] Thus, when Albert

57 Ibid., II.2.7, p. 190, vv. 15–16: 'primae qualitates non sunt formae substantiales elementorum'.

58 Ibid., vv. 17–21: 'licet primae qualitates distinguant elementorum materiam et dicantur primae differentiae elementorum, ita quod elementa differant et in numero ponantur penes coniunctiones ipsarum'.

59 Ibid., vv. 21–23: 'non tamen sunt substantiales formae eorum, quia dicit Aristoteles, quod forma substantialis nec activa est nec passiva'.

60 On the Latin reception of Avicenna's *De generatione et corruptione*, see Van Riet, 'Le *De generatione et corruptione* d'Avicenne'.

61 On Albert's reception of Avicenna's *Metaphysics* or *Philosophia prima*, see, among others, Bertolacci, 'Albert's Use of Avicenna'.

62 Avicenna, *Liber de Philosophia prima*, I.2.1, ed. by Van Riet, p. 66: 'Unde substantia est constituens esse accidentis, nec est constituta ab accidente; igitur substantia est praecedens in esse'.

63 Albertus Magnus, *Meteora*, IV.1.4, ed. by Hossfeld (hereafter *Meteora*), p. 215, vv. 2–4: 'Avicenna et Algazel concedebant formam substantialem non esse a qualitatibus activis'. This also suggests a possibility that Albert reached Avicenna's idea not through directly reading Avicenna's treatise, but through Algazel (al-Ghazālī, *c.* 1058–1111) and his *Intentions of the Philosophers* (*Maqāsid al-Falāsifa*), which is an exposition of Avicenna's work. The *Intentions* circulated in Latin under the title *Metaphysica* among Scholastic authors and was mistakenly considered to represent Algazel's own

adheres to the categorial distinction between substantial change and alteration, he was possibly inspired by Avicenna's view: a substance is ontologically prior to its accidental properties. Albert does not, however, adopt Avicenna's theory of the giver of forms, a point I will explore in the next section.

By focusing on their discussion about the substantial change of elements, I have shown how Averroes and Albert, in their commentaries on *GC*, tried to harmonize hylomorphism with the theory of elements. Averroes introduced hylomorphism into the discussion of the generation and corruption of elements, advancing a naturalistic position by identifying the form of the elements with a combination of primary qualities; Albert, while supporting the unification of hylomorphism with the theory of elements, put forward a non-naturalistic view in arguing for the irreducibility of the substantial forms of elements to their qualities. Albert's critique of Averroes's naturalistic position is found not only in his paraphrase of Aristotle's *GC*: it also appears in his paraphrase of *Meteorology* IV, where Albert reasoned similarly regarding spontaneous generation.

## The Debate on Spontaneous Generation in *Meteorology* IV

Having considered the opposition between Averroes and Albert on the substantial change of elements, this final section briefly examines Albert's paraphrase of *Meteorology* IV.[64] In *Meteorology* IV, Aristotle deals not with meteorological phenomena, as he does in the other three books of the treatise, but with the generation and formation of material substances in the sublunary region.[65] In the commentary tradition of the *Meteorology*, the fourth book was therefore regarded as following Aristotle's account of the generation and corruption of elements developed in *GC*.[66] Among the various issues addressed in the book, I will highlight Aristotle's discussion of spontaneous generation — the generation of living beings (such as worms and insects) without there being parents — and Albert's paraphrase of that discussion.[67] Although specialists in Albert's philosophy

philosophy. On Albert's and the Latin reception of Algazel's *Metaphysica*, see Salman, 'Algazel et les latins'; Janssens, 'Latin Translation of al-Ġazālī's *Maqāṣid al-Falāsifa*'; Minnema, 'Algazel Latinus'.

64 On Albert's *Meteorologica*, see Hossfeld, 'Der Gebrauch der aristotelischen Übersetzung in den *Meteora*'; Hossfeld, 'Das zweite Buch der *Meteora*'; Ducos, 'Théorie et pratique'. For the broader context of the reception of Aristotle's *Meteorology* IV in the Latin West, see Martin, 'Interpretation and Unity'.

65 On the special character of Aristotle's *Meteorology* IV, see Düring, *Aristotle's Chemical Treatise Meteorologica, Book IV*; Furley, 'The Mechanics of *Meteorologica* IV'; Martin, 'Interpretation and Unity', pp. 99–142.

66 See Alexander of Aphrodisias, *On Aristotle's Meteorology 4*, trans. by Lewis, p. 65: 'The book entitled the fourth book of the *Meteorologica* is by Aristotle but not part of the work of meteorology. For, those things said in it are not common to meteorology: rather what comes from these lectures ought to follow *On Generation and Corruption*'.

67 On Albert's discussion of spontaneous generation, see also Resnick in this volume.

have acknowledged the importance of the theory of spontaneous generation for understanding his theory of sublunary generation, the discussion of this issue in Albert's paraphrase of *Meteorology* IV has not received sufficient attention.[68]

## *An Overview of the Matter Theory in* Meteorology *IV*

The most important feature of Aristotle's matter theory in *Meteorology* IV is that it is very naturalistic or materialistic. Aristotle here completely dispenses with hylomorphism, and discusses natural phenomena solely in terms of the operation of primary qualities. He divides the four primary qualities into two classes, active (hot and cold) and passive (moist and dry),[69] and holds sublunary generation to be the result of the effects of the active qualities acting on passive ones. Such generation takes place, he writes, 'when the hot and the cold are masters of the matter [= passive qualities]' (*Meteor.* IV.1, 379a1). By this, Aristotle means that substantial bodies can come about only when active qualities maintain a proper ratio or proportion with passive ones. Otherwise — when the ratio between active and passive qualities does not remain stable — the bodies become putrefied. However, among the active qualities, he privileges the active role of hot or heat over that of cold, for the operation of cold can be witnessed only 'as far as heat is absent' (*Meteor.* IV.8, 384b27–28).

In his paraphrase of *Meteorology* IV, Albert basically reproduces Aristotle's view. Like Aristotle, he argues that natural phenomena should be understood in terms of the operations of active qualities on passive ones: 'We must consider all the operations of these qualities, which the active [principles] perform in a compound body and which happen in the specific forms of passive qualities'.[70] The active qualities transform and combine things, while the passive ones receive the operations of the active qualities.[71] He then adds a new explanation for the privileged status of hot over cold. On the premise that 'the formal [principle] is more active than the material one',[72] he argues that hot should be considered more active because it functions as the formal principle, whereas cold is less active because it serves as the material principle:

68 See Hasse, 'Spontaneous Generation'; Takahashi, 'Nature, Formative Power and Intellect'; Bertolacci, 'Averroes against Avicenna'.

69 Aristotle, *Meteor.* IV.1, 378b10–13: 'We have explained that the causes of the elements are four, and that their combinations determine the number of the elements to be four. Two of the causes, the hot and the cold, are active; two, the dry and the moist, passive'.

70 Albertus Magnus, *Meteora*, IV.1.1, p. 211, vv. 39–41: 'debemus assumere omnes operationes istarum qualitatum, quas efficiunt activa in corpore mixto et quae fiunt in speciebus qualitatum passivarum'.

71 See Aristotle, *Meteor.* IV.1, 378b21–25: 'Hot and cold we describe as active, for combining is a sort of activity; moist and dry are passive, for it is in virtue of its being acted upon in a certain way that a thing is said to be easy to determine or difficult to determine'.

72 Albertus Magnus, *Meteora*, IV.1.2, p. 212, vv. 24–25: 'Formale autem plus est activum quam materiale'.

> Even in minerals and metals, which seem to exist by [virtue of] the cold, the reception of form occurs by [virtue of] the hot, which mixes the moist [= water] with the dry [= earth] and brings it to form and species.[73]

Minerals seem to be shaped by the power of the cold. But for Albert as well as for Aristotle, they must be affected by the power of the hot before being subject to the operation of the cold. It is heat that leads minerals to attain their forms. According to Albert, heat works as the formative principle in producing any natural substance: 'the hot is the agent in generation in general and in every particular instance of generation'.[74]

This gives rise to the question of whether, for Albert, the forms of material substances are simply the effects of the operation of primary qualities. Albert goes beyond Aristotle by asking a question that the latter did not raise: 'What is it that produces that [specific] form in the nature of the mixture?' (*quid producat formam illam in naturam mixti?*).[75] Albert's intent in posing this question is to examine whether heat alone is sufficient to produce the substantial form of material substances.

In order to answer it, Albert first looks at the theories of Arabic philosophers such as Avicenna, Algazel, and the 'Peripatetic philosophers' (*peripatetici*), by whom he means especially Averroes, as I will show below. Albert claims that there is disagreement among the Arabic philosophers regarding the principle that gives the form to material substance. According to one theory he ascribes to Avicenna and Algazel, the substantial form of the mixture cannot be reduced to some bundle of the four primary qualities.[76] As we saw above, insofar as Albert adheres to the distinction between substance and accidental qualities, he is probably following Avicenna's non-naturalistic position. This does not, however, mean he accepts all of Avicenna's theories. According to Albert, Avicenna and Algazel argued that the primary qualities receive the substantial forms from a supralunar intelligence, the giver of forms.[77] Albert rejects Avicenna's theory of the giver of forms, because he heeds Aristotle's rejection of the Platonic idea of separate principles that can exist without matter.

Next, Albert turns to the theory that he ascribes to the 'Peripatetic philosophers'.[78] These Aristotelians, he writes, claimed that 'natural heat' (*calor naturalis*)

---

73 Ibid., vv. 34–37: 'etiam in mineralibus et metallicis, quae videntur constare per frigidum, acceptio formae est per calidum commiscens cum sicco humidum et deducens ipsum ad formam et speciem'.

74 Ibid., vv. 46–47: 'calidum est agens in universali generatione et particulari generatione'.

75 Ibid., IV.1.4, p. 214, vv. 76–77.

76 Ibid., p. 215, vv. 2–4: 'Avicenna et Algazel concedebant formam substantialem non esse a qualitatibus activis'.

77 Ibid., vv. 4–8: 'Sed potius dicebant qualitates primas disponere materiam alterando et commiscendo proportionaliter unicuique naturae et speciei et tunc ab intelligentiis sive primis substantiis, quae dant omnes formas, dari formas'.

78 Ibid, v. 12.

generates the forms of sublunary things.[79] Although Albert does not reveal his doctrinal source, he is probably relying on Averroes's *Middle Commentary* on Aristotle's *Meteorology* IV. Averroes had advanced a naturalistic position by identifying the formal cause of a material substance with heat itself, and claimed that 'generation takes place by natural heat' (*generatio fit per calorem naturalem*).[80]

Albert agrees with Averroes that natural heat serves as an indispensable principle in the formation of material substances, but he does not argue that natural heat by itself can produce substantial forms. Thus, he claims: 'The substantial form of the compound body, whether it is of a mineral, a plant, or an animal, is not introduced by means of an alteration [= a change in quality]'.[81] But how is the substantial form introduced, if not by heat? Regarding the formative principle, Albert says:

> For in this manner there is in that [natural heat] the power of the heaven and the celestial movers, and the power of the complexion or form of the nature [of the substance] that it forms, just as in the heat of the seed of plants or animals there is a power that is called formative of the animal or the plant. And it is the same way in minerals and their heat. And so the work of heat is to draw out form from matter, which exists potentially in it, and to fashion organs and arrange the location of the members and parts under the form that determines the matter.[82]

According to Albert, it is not natural heat itself, but rather the celestial powers involved in the heat that function as the formative principle of sublunary things. He compares the celestial power in natural heat with the 'formative' power in the seminal heat of animals.[83]

In the context of explaining Aristotle's theory of concretion or coagulation in *Meteorology* IV.5, Albert returns to the question of whether natural qualities suffice to produce the substantial forms of sublunary things. Concretion is a process

---

79 Ibid., vv. 30–35: 'Per hoc autem quod est calor naturalis simul, habet digerere humidum et commiscere cum sicco proportionaliter naturae, quae est forma et species mixti, ut humidum sit in quantitate et subtilitate et commixtione debita, quae exigitur ad speciem hanc vel illam, quam format calor, in quantum est naturalis'.

80 Averroes, *Middle Commentary on Meteorology* IV, *Aristotelis opera cum Averrois commentariis*, ed. Venetiis, vol. 5, fol. 461L: 'forma transmutans materiam ad recipiendum aliquam formam, et separare faciens primam, est de necessitate calor'; ibid., fol. 461M: 'Iam igitur declaratum est ex hoc quod calor est duplex, naturalis et extraneus: et quod generatio fit per calorem naturalem, corruptio autem per extraneum'.

81 Albertus Magnus, *Meteora*, IV.1.4, p. 214, vv. 54–55: 'per alterationem autem non inducatur forma substantialis mixti, sive sit mineralis sive plantae sive animalis'.

82 Ibid., p. 215, vv. 20–28: 'Sic enim est in eo vis caeli et motorum caelestium et vis complexionis sive formae illius naturae, quam format, sicut in calore seminis plantae vel animalis est vis, quae vocatur formativa animalis vel plantae. Et eodem modo est in mineralibus et calore eorum. Et sic opus caloris est educere formam de materia, quae potentia est in ipsa, et figurare organa et ordinare situm membrorum et partium sub forma terminante materiam'.

83 On Albert's notion of the formative power, see Takahashi, 'Nature, Formative Power and Intellect'.

in which a thing turns hard and becomes endowed with its own boundaries, distinguishing it from other things. Aristotle had argued that concretion takes place through the process of drying, which in turn occurs when natural heat operates on the passive qualities.[84] In his paraphrase of this section of Aristotle's text, Albert draws a sharp distinction between the form that heat brings about and the substantial form in the proper sense. In his view, there are two kinds of forms: 'specific form' (*forma specificata*) and 'material form' (*forma materialis*).[85] Specific form is the substantial form of a compound body; it gives a compound substance its proper species which distinguishes it from other substances. Material form, on the other hand, is just an effect, an 'impression' (*impressio*) received by passive qualities that are subject to the operation of natural heat. Natural heat, by itself, conveys only the material form, and not the substantial form, to material substances.

## Spontaneous Generation: Avicenna, Averroes, and Albert

Bearing in mind Aristotle's explanation of sublunary generation in terms of the operation of qualities and Albert's critical reception of that theory, let us turn to Albert's discussion of spontaneous generation. When Albert explains Aristotle's theory of spontaneous generation, he takes into account the views of Avicenna and Averroes, who had also presented their own opinions on the topic.

Avicenna discusses spontaneous generation in the meteorological part of *The Book of the Healing*.[86] This part was translated into Latin and circulated as a quasi-independent work among Scholastic authors under the title *On Floods* (*De diluviis*). In this short treatise, Avicenna presents a radical position: not only worms, but all animals can be generated without parents.[87] Albert refers to Avicenna's theory in his treatise *On the Causes of the Properties of the Elements* (*De causis proprietatum elementorum*). There, he deals with the effects of planets on the sublunary elements, and devotes one section to the question of whether, after floods, animals can be restored with the help of celestial power.[88] According to Albert, Avicenna claimed that 'the powers of the stars, which are commixed

84 Aristotle, *Meteor.* IV.5, 382b16–18: 'it is always a process of heating or cooling that dries things, but the agent in both cases is heat, either internal or external'.

85 Albertus Magnus, *Meteora*, IV.2.4, p. 250, vv. 11–21: 'Nos enim non habemus hic dicere de forma specificata corporum, quoniam tractatus de illa secundum se considerata pertinet ad philosophum primum; [...] Et ideo nos non habemus hic dicere nisi efficientem et passionem, quam imprimit efficiens qualitas in passivo, quae forma materialis est passivarum qualitatum in corpore mixto. Et vocatur haec passio sive materialis forma proprie impressio'.

86 On Avicenna's theory of spontaneous generation, see Kruk, 'A Frothy Bubble'; Hasse, 'Spontaneous Generation', esp. pp. 155–58; Bertolacci, 'Averroes against Avicenna'.

87 See Hasse, 'Spontaneous Generation', esp. pp. 155–56; Avicenna, *De diluviis*, ed. by Alonso Alonso.

88 Albertus Magnus, *De causis proprietatum elementorum*, I.2.13, ed. by Hossfeld, p. 85, v. 29–p. 87, v. 23. This chapter is devoted to the question 'from where the restoration of animals is made after some flood' ('unde fit restitutio animalium post aliquod diluviorum').

with the powers of the elements, form and bring about all things', and that every 'habitation' (*habitatio*) can be repaired by celestial power.[89] Albert agrees with Avicenna that celestial power is necessary for the existence of sublunary entities, but he ultimately rejects Avicenna's position, arguing that 'animals [...] cannot be restored only from stars and elements'.[90]

Averroes develops his own theory of spontaneous generation in his *Long Commentary on the Metaphysics* XII.[91] As Charles Genequand has rightly pointed out, Averroes deviates from the Philosopher by addressing 'the generation of some insects from putrescent matter or corpses'.[92] In his explanation of Aristotle's axiom that 'man is begotten by man, each individual by an individual' (*Met.* XII.3, 1070a28–29), Averroes seeks the principle of spontaneous generation in the operation of cosmic heat:

> This is why Aristotle said that man is begotten by man and by the sun. And this heat is made in earth and water by solar heat mixed with the heat of the other stars. For this reason, the sun and the other stars are the principle of life for each living being in nature. Therefore, the heat of the sun and the stars, diffused in water and earth, generates the animals born from putrefaction.[93]

Averroes claims that animal generation requires not only the sublunary elements or elemental qualities, but also stellar or cosmic heat. In his view, the sun and its heat function as a formative principle that determines the species of animals that are generated without parents.

Keeping in mind Avicenna and Averroes's positions, let us return to Albert's account of spontaneous generation in his paraphrase of *Meteorology* IV. A detailed analysis can be found in his paraphrase of the following statement by Aristotle:

> Animals too are generated in putrefying bodies, because the heat that has been expelled, being natural, organizes the particles thrown out with it.[94]

---

89 Ibid., p. 85, vv. 37–38: 'Dicit enim Avicenna virtutes stellarum commixtas viribus elementorum omnia formare et perficere'; ibid., p. 86, vv. 22–25: 'Istae igitur sunt rationes et similes his moventes Avicennam ad hoc quod dixit omnem habitationem ex stellarum viribus cum commixtione elementorum posse reparari'.

90 Ibid., p. 86, vv. 27–29: 'animalia [...] ex solis stellis et elementis reparari non posse'.

91 On Averroes's theory about spontaneous generation, see Averroes, *Ibn Rushd's Metaphysics*, trans. by Genequand, pp. 24–32; Davidson, *Alfarabi, Avicenna, and Averroes*, pp. 220–57; Hasse, 'Spontaneous Generation', esp. pp. 158–62; Bertolacci, 'Averroes against Avicenna'.

92 Averroes, *Ibn Rushd's Metaphysics*, trans. by Genequand, p. 24.

93 Averroes, *Long Commentary on Metaphysics*, *Aristotelis opera cum Averrois commentariis*, ed. Venetiis, vol. 8, fol. 143b: 'Unde Aristoteles dixit quod homo generatur ex homine et sole. Et factus est ille calor in terra et aqua ex calore solis admixto cum calore aliarum stellarum. Et ideo sol et alie stelle sunt principium vite cuiuslibet vivi in natura. Calor igitur solis et stellarum divisus in aqua et terra generat animalia ex putrefactione nata'. Cf. Averroes, *Ibn Rushd's Metaphysics*, trans. by Genequand, p. 111.

94 Aristotle, *Meteor.* IV.1, 379b6–8. In addition to this passage, Aristotle referred to the issue of spontaneous generation at two other places in *Meteorology* IV: 'It is not true that animals are generated in the concoction of food, as some say. Really they are generated in the excretion which

Albert first examines the position that he ascribes to Averroes, then presents his own reading.[95] Averroes, he writes, thought 'the power of the stars' (*virtus stellarum*) to be the cause of spontaneous generation. Albert seems to agree with the Commentator when he, too, states that all sublunary generation is caused by the stars.[96] He points out, however, that Averroes fails to present a satisfactory account of spontaneous generation, for if this type of generation took place only with the assistance of celestial power, it would occur in accordance with celestial motion. In reality, however, animals are not generated from putrefying bodies in accordance with the celestial cycles. Albert claims that the position he ascribes to Averroes cannot properly explain the cause of this kind of generation, and concludes that 'what Averroes said is imperfect' (*Dictum autem Averrois est imperfectum*).[97]

It should be noted that the position Albert ascribes to Averroes is not the one the Commentator himself advanced in his *Long Commentary on the Metaphysics*. Rather, the position sounds like Avicenna's, for it was Avicenna who argued that celestial power is able to restore all beings in the terrestrial region. Thus, although Albert refers to Averroes's position, he does not represent it accurately.

After criticizing the view ascribed to Averroes, Albert introduces another position, that spontaneous generation is caused by cosmic heat. As we saw above, it is this position that is close to Averroes's own view in his *Long Commentary on the Metaphysics*. According to Albert, this position, too, fails to offer a satisfactory account of the phenomenon. Since, Albert suggests, heat is of a single nature, it cannot by itself generate animals so diverse in size, shape, colour, and other properties.[98] Albert then clarifies his own view:

> For just as in animal semen there is the formative power of the animal according to the nature of that semen, which is a power in the foamy and natural spirit of the semen and works by means of natural heat, so the power in the genus of imperfect animals remains in the subtler moisture of the putrefied thing [...] Then the heat that is in it dissolves something from the subtler

---

putrefies in the lower belly, and they ascend afterwards' (IV.3, 381b9–12); and 'This explains the generation of animals in putrefying bodies: the putrefying body contains the heat which destroyed its proper heat' (IV.11, 389b5–6).

95 Albertus Magnus, *Meteora*, IV.1.11, p. 223, v. 26–p. 224, v. 65.

96 Ibid., p. 224, vv. 24–26: 'It is true that every generation is caused by stars, insofar as [the stars] are conveyed commonly in the oblique circle' ('Verum enim est quod omnis generatio causatur ex stellis, secundum quod feruntur in obliquo circulo communiter').

97 Ibid., p. 224, vv. 23–24.

98 Ibid., IV.1.11, p. 223, vv. 63–74: 'Quid constare facit humidum eductum in pellem et corpus animalis? Et cum ipsum humidum effluens ex partibus putrefacti corporis sit continuum: Quare inde non generatur unum animal, sed multa? Et cum ipsum sit unius naturae: Quare quaedam generata ex ipso sunt longa et stricta et quaedam lata et brevia et quaedam stricta, parva et brevia? Et unde venit diversitas colorum illorum animalium? Et quare ex humido illo non generatur planta, praecipue cum corpus putrefactum fuerit planta sicut arbor vel herba? Cum enim facilioris sit generationis planta quam animal, videtur forte alicui potius ex putrefactis debere generari plantas quam animalia'.

> moisture and makes spirit from it, which beats under the skin in a constant circuit and becomes the vehicle of the formative power that is in the moisture. And from it is drawn the form of the animal.[99]

Albert's view of spontaneous generation seems to be closer to that of Averroes, who regarded heat as the most important principle of spontaneous generation. But Albert supposes that spontaneous generation requires an additional principle, one that determines the particular features of animals. In his view, this special kind of generation takes place with the assistance of a formative power analogous to that which exists in human semen. This power operates on the putrefied matter and transforms it into a matrix from which diverse animals eventually emerge without parents.

## Conclusion

Aristotelians found in Aristotle's natural philosophy two different theories that explain the nature and change of material substances: hylomorphism and the theory of elements. Averroes and later Albert attempted to unify these two theories, but their strategies were quite different. In his commentary on Aristotle's *GC*, Averroes respected Alexander's interpretation and advanced the naturalistic position that the forms of the elements are identical with a combination of primary qualities. He did not identify substantial change with alteration even in the case of elements, but neither did he posit any immaterial principle distinct from the primary qualities that could distinguish one element from another. Albert rejected Averroes's naturalism and suggested that the substantial forms of the elements cannot be identified with the primary qualities. In this respect, his position comes closer to Avicenna's, even though Albert never adopted the latter's notion of the giver of forms.

The commonalities and differences between Averroes and Albert are even clearer in their treatment of Aristotle's theory of spontaneous generation. In his discussion of Aristotle's argument in the *Metaphysics* Book Lambda, Averroes claimed that cosmic heat plays a formative role in the spontaneous generation of living beings. Albert admitted that the activity of heat is crucial in sublunary generation, but he again was not satisfied with Averroes's naturalistic account. He added that spontaneous generation further requires the operation of a formative power analogous to that which exists in human semen.

---

99 Ibid., p. 224, vv. 11–21: 'Sicut enim in semine animalis est vis formativa animalis secundum naturam illius seminis, quae vis est in spiritu spumoso et naturali seminis et operatur per calidum naturale, ita vis in genere animalis imperfecti remanet in subtiliori humido putrefacti [...]. Deinde calidum, quod est in ipso, dissolvit aliquid de subtiliori humido et facit inde spiritum, qui tympanizat infra pellem constantem in circuitu et efficitur vehiculum virtutis formativae, quae est in humido; et ex illa educitur forma animalis'.

Although the main aim of this article has been to clarify Albert's theory of material substances, I have also highlighted Averroes's naturalistic position, which has not been as fully studied as his theory of the human soul and intellect. As I have argued, the Commentator did not hesitate to identify the forms of elements with the primary qualities, and he privileged the formative role of natural heat. My study has suggested that his natural philosophy may have offered later philosophers and theologians an alternative matter theory, different from the standard Scholastic theory, in which the hylomorphic scheme played a central role in the analysis of material substances.[100] For Albert and his programme of natural science, this alternative theory would have been too naturalistic (in Kessler's sense), in that it attempts to explain the nature and change of material substances too narrowly, on the basis of overly restricted materialist principles.

Moreover, Albert's matter theory, as developed in his paraphrases of *GC* and *Meteorology* IV and discussed in this paper, is crucial if we are to attain a proper understanding of his natural scientific programme as a whole. This is because his conception of the elements and sublunary generation, as presented in these works, served as a theoretical foundation upon which he developed more specific accounts of material substances and their change in his later works on mineralogy and zoology. Indeed, the phenomenon of animal generation was a critically important issue for Albert as well as for medieval and early modern Aristotelians in general. What is striking is that Albert presented hints of his own ideas on animal generation in his paraphrase of *Meteorology* IV before seriously tackling Aristotle's *Generation of Animals*, particularly with respect to his notion that a formative power is indispensable for the generation of animals to take place.

---

100 See Kessler, 'Metaphysics or Empirical Science?'; Sylla, 'Averroes and Fourteenth-Century Theories of Alteration'; Hirai, 'Telesio, Aristotle and Hippocrates on Cosmic Heat'.

## Works Cited

### *Primary Sources*

Albertus Magnus, *De causis proprietatum elementorum*, ed. by Paul Hossfeld, Editio Coloniensis, 5/2 (Münster: Aschendorff, 1980)

———, *De generatione et corruptione*, ed. by Paul Hossfeld, Editio Coloniensis, 5/2 (Münster: Aschendorff, 1980)

———, *Meteora*, ed. by Paul Hossfeld, Editio Coloniensis, 6/1 (Münster: Aschendorff, 2003)

———, *Physica, libri I–IV*, ed. by Paul Hossfeld, Editio Coloniensis, 4/1 (Münster: Aschendorff, 1987)

Alexander of Aphrodisias, *Alexander of Aphrodisias on Stoic Physics: A Study of the 'De mixtione' with Preliminary Essays, Text, Translation and Commentary*, ed. and trans. by Robert B. Todd (Leiden: Brill, 1976)

———, *On Aristotle On Coming-to-Be and Perishing* 2.2–5, trans. by Emma Gannagé (London: Bloomsbury, 2005)

———, *On Aristotle's Meteorology* 4, trans. by Eric Lewis (Ithaca, NY: Cornell University Press, 1996)

Aristotle, *Aristotle's De generatione et corruptione*, trans. with notes by C. J. F. Williams (Oxford: Oxford University Press, 1982)

———, *The Complete Works of Aristotle: Revised Oxford Translation*, ed. by Jonathan Barnes, 2 vols (Princeton, NJ: Princeton University Press, 1984)

Averroes, *Averrois Cordvbensis Commentarivm medivm in Aristotelis De generatione et corrvptione libros* [Middle Commentary on *GC*], ed. by F. H. Fobes (Cambridge, MA: The Mediaeval Academy of America, 1956)

———, *Averroes on Aristotle's 'De generatione et corruptione': Middle Commentary and Epitome*, trans. by Samuel Kurland (Cambridge, MA: The Mediaeval Academy of America, 1958)

———, *Aristotelis opera cum Averrois commentariis*, 14 vols (Venice, 1562–74; repr. Frankfurt: Minerva, 1962)

———, *Ibn Rushd's Metaphysics: A Translation with Introduction of Ibn Rushd's Commentary on Aristotle's Metaphysics, Book Lām*, trans. by Charles Genequand (Leiden: Brill, 1986)

Avicenna, *De diluviis*, in Manuel Alonso Alonso, 'Homenaje a Avicena en su milenario: Las traducciones de Juan González de Burgos y Salomon', *Al-Andalus*, 14 (1949), 291–319

———, *Liber de Philosophia prima sive Scientia divina, I–IV. Édition critique de la traduction latine médiévale*, ed. by Simone Van Riet, introduced by Gérard Verbeke (Leuven: Peeters, 1977)

———, *Liber tertius naturalium: De generatione et corruptione. Édition critique de la traduction latine médiévale et lexiques*, ed. by Simone Van Riet, introduced by Gérard Verbeke (Leuven: Peeters, 1987)

Philoponus, *On Aristotle's On Coming-to-Be and Perishing 1.6–2.4*, trans. by C. J. F. Williams (Ithaca, NY: Cornell University Press, 1999)

### Secondary Works

Baldner, Steven, 'Albertus Magnus and the Categorization of Motion', *The Thomist*, 70 (2006), 203–35

Bertolacci, Amos, 'Albert's Use of Avicenna and Islamic Philosophy', in *A Companion to Albert the Great: Theology, Philosophy and the Sciences*, ed. by Irven M. Resnick (Leiden: Brill, 2013), pp. 601–11

———, 'Averroes against Avicenna on Human Spontaneous Generation: The Starting-Point of a Lasting Debate', in *Renaissance Averroism and its Aftermath*, ed. by Anna Akasoy and Guido Giglioni (Dordrecht: Springer, 2013), pp. 37–54

Bostock, David, 'Aristotle on the Transmutation of the Elements in *De Generatione et Corruptione* I. 1–4', *Oxford Studies in Ancient Philosophy*, 13 (1995), 217–29

———, *Space, Time, Matter, and Form: Essays on Aristotle's 'Physics'* (Oxford: Clarendon Press, 2006)

Brower, Jeffrey E., *Aquinas's Ontology of the Material World: Change, Hylomorphism, and Material Objects* (Oxford: Oxford University Press, 2014)

Cadden, Joan, 'The Medieval Philosophy and Biology of Growth: Albertus Magnus, Thomas Aquinas, Albert of Saxony and Marsilius of Inghen on Book I, Chapter V of Aristotle's *De generatione et corruptione*' (unpublished doctoral dissertation, Indiana University, 1971)

Caroti, Stefano, 'Commentaries on Aristotle's *De generatione et corruptione*', in *Encyclopedia of Medieval Philosophy*, ed. by Henrik Lagerlund (Dordrecht: Springer, 2011), pp. 251–56

———, 'Note sulla parafrasi del *De generatione et corruptione* di Alberto Magno', in *Albert le Grand et sa réception au Moyen Âge: Hommage à Zénon Kaluza*, ed. by Francis Cheneval, Ruedi Imbach, and Thomas Ricklin, Separatum de *Freiburger Zeitschrift für Philosophie und Theologie*, 45 (1998), 6–30

Cerami, Cristina, 'The Aristotelian Analysis of Generation: *Physics* A and *Metaphysics* Z', *Documenti e studi sulla tradizione filosofica medievale*, 15 (2004), 1–38

———, *Géneration et substance: Aristote et Averroès entre physique et métaphysique* (Berlin: De Gruyter, 2015)

Davidson, Herbert A., *Alfarabi, Avicenna, and Averroes, on Intellect: Their Cosmologies, Theories of the Active Intellect, and Theories of Human Intellect* (Oxford: Oxford University Press, 1992)

de Haas, Frans, and Jaap Mansfeld, eds, *Aristotle: On Generation and Corruption, Book I* (Oxford: Oxford University Press, 2004)

Ducos, Joëlle, 'Théorie et pratique de la météorologie médiévale: Albert le Grand et Jean Buridan', in *Le temps qu'il fait au Moyen Âge*, ed. by Joëlle Ducos and Claude Thomasset (Paris: Presses de l'Université de Paris-Sorbonne, 1998), pp. 45–58

Düring, Ingemar, *Aristotle's Chemical Treatise Meteorologica, Book IV with Introduction and Commentary* (Göteborg: Elander, 1944)

Eichner, Heidrun, *Averroes (Abū l-Walīd Ibn Rušd) Mittlerer Kommentar zu Aristoteles' 'De Generatione et Corruptione': Mit einer einleitenden Studie versehen, herausgegeben und kommentiert* (Paderborn: Schöningh, 2005)

———, 'Ibn Rušd's Middle Commentary and Alexander's Commentary in Their Relationship to the Arab Commentary Tradition on the *De generatione et corruptione*', in *Aristotele e Alessandro di Afrodisia nella tradizione araba*, ed. by Cristina D'Ancona and Giuseppe Serra (Padua: Il Poligrafo, 2002), pp. 281–97

Freudenthal, Gad, *Aristotle's Theory of Material Substance: Heat and Pneuma, Form and Soul* (Oxford: Oxford University Press, 1995)

Furley, David, 'The Mechanics of *Meteorologica* IV: A Prolegomenon to Biology', in *Cosmic Problems: Essays on Greek and Roman Philosophy of Nature*, ed. by David Furley (Cambridge: Cambridge University Press, 1989), pp. 132–48

Gill, Mary Louise, *Aristotle on Substance: The Paradox of Unity* (Princeton, NJ: Princeton University Press, 1989)

Hasse, Dag Nikolaus, 'Avicenna's "Giver of Forms" in Latin Philosophy, Especially in the Works of Albertus Magnus', in *The Arabic, Hebrew and Latin Reception of Avicenna's Metaphysics*, ed. by Dag Nikolaus Hasse and Amos Bertolacci (Berlin: De Gruyter, 2012), pp. 225–49

———, *Latin Averroes Translations of the First Half of the Thirteenth Century* (Hildesheim: Olms, 2010)

———, 'Spontaneous Generation and the Ontology of Forms in Greek, Arabic, and Medieval Latin Sources', in *Classical Arabic Philosophy: Sources and Reception*, ed. by Peter Adamson (London: The Warburg Institute, 2007), pp. 150–75

Hirai, Hiro, 'Telesio, Aristotle and Hippocrates on Cosmic Heat', in *Bernardino Telesio and the Natural Sciences in the Renaissance*, ed. by Pietro Daniel Omodeo (Leiden: Brill, 2019), pp. 51–65

Hossfeld, Paul, *Albertus Magnus als Naturphilosoph und Naturwissenschaftler* (Bonn: Albertus-Magnus-Institut, 1983)

———, 'Der Gebrauch der aristotelischen Übersetzung in den *Meteora* des Albertus Magnus', *Mediaeval Studies*, 42 (1980), 395–406

———, 'Grundgedanken in Alberts des Großen Schrift "Über Entstehung und Vergehen"', *Philosophia Naturalis*, 16 (1976), pp. 191–204

———, 'Das zweite Buch der *Meteora* des Albertus Magnus', in *Albertus Magnus: Zum Gedenken nach 800 Jahren: Neue Zugänge, Aspekte und Perspektiven*, ed. by Walter Senner (Berlin: De Gruyter, 2001), pp. 416–26

Hyman, Arthur, 'Aristotle's "First Matter" and Avicenna's and Averroes' "Corporeal Form"', in *Harry Austryn Wolfson Jubilee Volume*, English Section, ed. by Saul Lieberman (Jerusalem: American Academy for Jewish Research, 1965), pp. 385–406

Janssen, Jules, 'The Latin Translation of al-Ġazālī's *Maqāṣid al-Falāsifa*', in *Encyclopedia of Medieval Philosophy*, ed. by Henrik Lagerlund (Dordrecht: Springer, 2011), pp. 387–90
Kessler, Eckhard, 'Metaphysics or Empirical Science? The Two Faces of Aristotelian Natural Philosophy in the Sixteenth Century', in *Renaissance Readings of the 'Corpus Aristotelicum': Proceedings of the Conference held in Copenhagen 23–25 April 1998*, ed. by Marianne Pade (Copenhagen: Museum Tusculanum Press, 2001), pp. 79–101
Kruk, Remke, 'A Frothy Bubble: Spontaneous Generation in the Medieval Islamic Tradition', *Journal of Semitic Studies*, 35 (1990), 265–310
Lagerlund, Henrik, 'Material Substance', in *The Oxford Handbook of Medieval Philosophy*, ed. by John Marenbon (Oxford: Oxford University Press, 2012), pp. 468–85
Lammer, Andreas, *The Elements of Avicenna's Physics: Greek Sources and Arabic Innovations* (Berlin: De Gruyter, 2018)
Lüthy, Christoph, 'An Aristotelian Watchdog as Avant-Garde Physicist: Julius Caesar Scaliger', *The Monist*, 84 (2001), 542–61
Maier, Anneliese, *An der Grenze von Scholastik und Naturwissenschaft* (Rome: Edizioni di Storia e Letteratura, 1952)
Martin, Craig, 'Interpretation and Unity: The Renaissance Commentary Tradition on Aristotle's *Meteorologica* IV' (unpublished doctoral dissertation, Harvard University, 2002)
McGinnis, Jon, *Avicenna* (Oxford: Oxford University Press, 2010)
Minnema, Anthony H., 'Algazel Latinus: The Audience of the *Summa theoricae philosophiae*, 1150–1600', *Traditio*, 69 (2014), 153–215
Nardi, Bruno, *Studi di filosofia medievale* (Rome: Edizioni di Storia e Letteratura, 1960)
Polloni, Nicola, *The Twelfth-Century Renewal of Latin Metaphysics: Gundissalinus's Ontology of Matter and Form* (Toronto: Pontifical Institute of Mediaeval Studies, 2020)
Puig Montada, Josep, 'Aristotle and Averroes on *Coming-To-Be and Passing-Away*', *Oriens*, 35 (1996), 1–34
Pasnau, Robert, *Metaphysical Themes, 1274–1671* (Oxford: Clarendon Press, 2011)
Petrescu, Lucian, 'Hylomorphism versus the Theory of Elements in Late Aristotelianism: Péter Pázmány and the Sixteenth-Century Exegesis of *Meteorologica* IV', *Vivarium*, 52 (2014), 1–26
Robinson, Howard M., 'Prime Matter in Aristotle', *Phronesis*, 19 (1974), 168–88
Salman, Dominique, 'Algazel et les latins', *Archives d'histoire doctrinale et littéraire du Moyen Âge*, 10–11 (1935–36), 103–27
Snyder, Steven C., 'Albert the Great, *Incohatio Formae*, and the Pure Potentiality of Matter', *American Catholic Philosophical Quarterly*, 70 (1996), 63–82
Sokolowski, Robert, 'Matter, Elements and Substance in Aristotle', *Journal of the History of Philosophy*, 8 (1970), 263–88
Solmsen, Friedrich, 'Aristotle and Prime Matter: A Reply to Hugh R. King', *Journal of the History of Ideas*, 19 (1958), 243–52
———, *Aristotle's System of the Physical World* (Ithaca, NY: Cornell University Press, 1960)
Stone, Abraham D., 'Avicenna's Theory of Primary Mixture', *Arabic Science and Philosophy*, 18 (2008), 99–119

Sylla, Edith D., 'Averroes and Fourteenth-Century Theories of Alteration', in *Averroes' Natural Philosophy and Its Reception in the Latin West*, ed. Paul J. J. M. Bakker (Leuven: Leuven University Press, 2015), pp. 141–92

———, 'Medieval Concepts of the Latitude of Forms: The Oxford Calculators', *Archives d'histoire doctrinale et littéraire du Moyen Âge*, 40 (1973), 223–83

Takahashi, Adam, 'Interpreting Aristotle's Cosmos: Albert the Great as a Reader of Averroes' (unpublished doctoral thesis, Radboud University Nijmegen, 2017)

———, 'Nature, Formative Power and Intellect in the Natural Philosophy of Albert the Great', *Early Science and Medicine*, 13 (2008), 451–81

Thijssen, Johannes M. M. H., and Henk A. G. Braakhuis, eds, *The Commentary Tradition on Aristotle's 'De generatione et corruptione': Ancient, Medieval and Early Modern* (Turnhout: Brepols, 1999)

Twetten, David, Steven Baldner, and Steven C. Snyder, 'Albert's Physics', in *A Companion to Albert the Great: Theology, Philosophy and the Sciences*, ed. by Irven M. Resnick (Leiden: Brill, 2013), pp. 173–219

Van Riet, Simone, 'Le *De generatione et corruptione* d'Avicenne dans la tradition latine', in *The Commentary Tradition on Aristotle's 'De generatione et corruptione': Ancient, Medieval and Early Modern*, ed. by Johannes M. M. H. Thijssen and Henk A. G. Braakhuis (Turnhout: Brepols, 1999), pp. 69–77

Weisheipl, James A., ed., *Albertus Magnus and the Sciences: Commemorative Essays 1980* (Toronto: Pontifical Institute of Mediaeval Studies, 1980)

LUIS XAVIER LÓPEZ-FARJEAT

# Chapter 8. Albert the Great's Use of Averroes in his Digressions on Human Intellectual Knowledge (*De anima* III.3.8–11)*

In *De anima* III.5, 430a15, Aristotle makes the famous distinction between an intellect that 'is what it is by virtue of becoming all things' and another 'which is what it is by virtue of making all things'. The first intellect is the passive or potential intellect, while the second is the active or agent intellect. The function of the agent intellect is analogous to light in the sense that light 'makes potential colours into actual colours', and it is described as separable (*chôristos*), separate (*chôristheis*), impassible (*apathês*), unmixed (*amigēs*), pure activity (*têi ousiai energeia*), immortal (*athanaton*), and eternal (*aidion*).

As is well known, Aristotle's *De anima* was one of the most commented-upon treatises among philosophers of Late Antiquity and, later, once it was translated from Greek into Arabic and Latin, among medieval philosophers. *De anima* III.5 led to intense interpretive debate, and Albert the Great was one of the many philosophers participating in that discussion. He dealt with human intellectual knowledge in several works, always engaging in an active and critical dialogue with Greek and Arabic Peripatetic sources in Latin translation — Alexander, Themistius, Theophrastus, Alfarabi, Avempace, Algazel, Avicenna, and Averroes.

---

* An earlier and shorter version of this paper, focusing on Avempace and Albert, was published in the *ACPA Proceedings 2012*. I would like to express my gratitude to the American Catholic Philosophical Association for letting me use a portion of that paper and to Prof. Steven Baldner for his comments on it. I have radically modified several parts of that version in order to enhance clarity and correct some inaccuracies. I am grateful to Richard C. Taylor, Katja Krause, Kate Sturge, José Molina, Jörg Tellkamp, Amos Bertolacci, Max Steinwandel, and Jules Janssens for their helpful comments and criticism.

**Luis Xavier López-Farjeat** (llopez@up.edu.mx), professor at the Universidad Panamericana, Mexico City, writes on classical Islamic philosophy. He is editor of *Tópicos, Journal of Philosophy*.

*Albert the Great and his Arabic Sources*, ed. by Katja Krause and Richard C. Taylor, Philosophy in the Abrahamic Traditions of the Middle Ages, 5 (Turnhout: Brepols, 2024), pp. 225–252
BREPOLS PUBLISHERS 10.1484/M.PATMA-EB.5.136488

In most of the works where Albert discusses human intellectual knowledge, he directly expresses his disagreement with Alfarabi, Avempace, Algazel, and Avicenna concerning their doctrines on the intellect. The main reason for this disagreement is that these philosophers conceived the agent intellect as something separated from the human soul, thus connecting natural human intellectual knowledge with a separate, immortal, and eternal substance that needs to be understood from an ontological perspective. By contrast, Albert's position throughout his career was that the agent intellect is a part of the individual human soul.

What has drawn the attention of some scholars is that Albert frequently claims the authority of Averroes to support his position. For instance, in a recent paper, Richard C. Taylor analysed Albert's use of Avicenna and Averroes in his *De homine*, a treatise completed around 1242. Taylor shows that in this early work, Albert misunderstood Averroes's teachings on the intellect and came to the false conclusion that, for Averroes, both the possible intellect (what Averroes calls the 'material intellect') and the agent intellect are powers of the individual human soul.[1] Certainly, in later works Albert corrected this interpretation, and clearly realized that Averroes also thought there was not only one separate intellect, but two separate intellects: the material and the agent intellects.

In his *Super Ethica* (*c.* 1250–52), Albert was already aware of Averroes's real doctrine of human intellectual knowledge.[2] The Cordoban thus joined the list of those Arabs with whom Albert disagreed. However, even after correcting his misunderstanding, Albert continued to invoke the authority of Averroes and refer to him as someone with whom he had philosophical affinities.

One brief line in Albert's *De anima* exemplifying this has been much interpreted, in different senses: *Nos autem in paucis dissentimus ab Averroe*.[3] In his classic work on Islam and the *Divine Comedy*, published in 1919, the Spanish scholar Miguel Asín Palacios adduces this line to show that Albert himself recognized the superiority of the Arabic philosophers over the Latins, even on thorny matters such as the nature of the agent intellect.[4] Later on, Gérard Verbeke was puzzled to find Albert writing that his doctrine of the intellect differs only in a few points from that of Averroes. Though Albert was certainly aware of the differences between his own teachings and those of Averroes, Verbeke notes, he does not insist on that point, but limits himself to defending the correct interpretation of Aristotle.[5]

---

1 See Taylor, 'Remarks on the Importance of Albert the Great's Analyses' and Taylor in this volume.

2 See Albertus Magnus, *Super Ethica*, I.15, ed. by Kübel, p. 79, vv. 76–80; p. 80, vv. 52–57; II.4, p. 106, vv. 71–82; VI.8, p. 451, vv. 30–59.

3 Albertus Magnus, *De anima*, III.3.11, ed. by Stroick, p. 221, v. 9.

4 Asín Palacios, *Islam and the Divine Comedy*, p. 505, n. 636.

5 Verbeke, 'L'unité de l'homme', pp. 222–23.

In a more recent publication, Pilar Herráiz Oliva returned to the discussion on the meaning of Latin Averroism in the thirteenth century.[6] She argues that it is impossible to explain 'Averroism' by associating it with a particular doctrine. Instead, different doctrinal positions on Aristotle and Averroes coexisted: the conservative, the moderate, the radical. Albert and Aquinas, in her view, are examples of moderate Averroism. Herráiz Oliva observes that having rejected Averroes's teachings on the intellect in *De unitate intellectus contra Averroem* (1256), in *De anima* (*c.* 1260–61) Albert writes that his own teachings differ to only a small extent from Averroes's views. She interprets the comment as suggesting a sort of affinity between Albert and Averroes and, of course, in favour of her interpretation of Albert as a moderate Averroist.[7]

Other scholars, such as Dominique Salman, Robert Miller, Alain de Libera, and Richard Taylor, have offered enlightening remarks to clarify the agreements and disagreements between Albert and Averroes.[8] Nevertheless, the ways in which Albert sometimes refers to Averroes have led, and continue to lead, to confusing and controversial conclusions. In the following, I look at Albert's *De anima* and revisit his use of Averroes's arguments against some of the other 'Arabs'. Albert's closeness to Averroes here, and the statement that his position differs only a little from that of Averroes, require careful revision. As I will show, in his *De anima*, Albert is not exactly an Averroist. My modest contribution is the reconstruction of Albert's arguments against the Arabs and the clarification of his peculiar use of Averroes to define his own position on the nature of human intellectual knowledge.

In *De anima*, Book III, treatise 3, chapters 6–11, Albert sets about a systematic debate with the Peripatetics and their view of human intellectual knowledge. He discusses their opinions largely chronologically, starting in chapters 6–7 with Greek Peripatetics such as Alexander, Themistius, and Theophrastus,[9] then engaging in chapters 8–9 with Alfarabi, Avempace, Avicenna, and Algazel.[10] Finally, in chapter 10, he deals with some Latin philosophers of the thirteenth century, and in chapter 11 provides what he takes to be the correct understanding of intellectual knowledge.[11] In this paper, I analyse chapter 8, where we find Albert's criticism of Alfarabi's and Avempace's doctrines of the intellect; chapter 9, where he discusses Algazel's and Avicenna's views on the same matter; and chapter 11,

---

6 Herráiz Oliva, 'Towards a New Methodology for Natural Philosophy', pp. 132–37.

7 Herráiz Oliva also remarks that Thomas Aquinas was the first to attack the Averroists directly, in his *De unitate intellectus contra Averroistas* (1270). What she does not mention is that in his *Commentary on the Sentences* (*c.* 1252–56), Aquinas already rejected Averroes's teaching on the intellect. See Aquinas, *In 2 Sent.*, d. 17, q. 2, a. 1.

8 See Salman, 'Albert le Grand et l'averroïsme latin'; Miller, 'An Aspect of Averroes' Influence on St Albert'; de Libera, *Métaphysique et noétique*, pp. 265–328; Taylor, 'Remarks on the Importance of Albert the Great's Analyses'.

9 Albertus Magnus, *De anima*, III.3.6, ed. by Stroick, pp. 214–16; III.3.7, p. 217.

10 Ibid., III.3.8, pp. 217–19; III.3.9, pp. 219–20.

11 Ibid., III.3.10, pp. 220–21; III.3.11, pp. 221–23.

where he offers a solution to the problem that seems at first sight to be very similar to that proposed by Averroes.[12]

My intention is, first, to reconstruct Albert's argumentation and highlight the philosophical reasoning on the basis of which he thinks that the doctrines of the intellect of the Arab Peripatetics are unable to provide proper solutions concerning human intellectual knowledge. Second, I show that Albert follows Averroes's criticism of the previous Arab Peripatetics quite closely, to the extent that in *De anima* III.3.11, he solves the problem of how we know intelligible forms by 'imitating' some aspects of Averroes's teachings on the intellect. Given that, as I will argue, Albert is aware how problematic Averroes's position is, *De anima* III.3.11 needs some clarification: To what extent is Albert following Averroes and in which aspects does he depart from him?

## Albert's Digression on Alfarabi and Avempace

In *De anima* III.3.8 Albert discusses mainly Avempace's conception of intellectual knowledge, which he regards as being closely connected to Alfarabi's. Whereas some of Alfarabi's philosophical works — among them *De intellectu* (*Fī al-ʿaql*) — were translated into Latin, none of Avempace's works were.[13] Like the rest of the Latin tradition, therefore, Albert knew Avempace's philosophical standpoint only through Averroes's *Long Commentary on the De anima*, which mentions three treatises authored by Avempace: the *Book on the Soul* (in the Latin Averroes, *Liber de anima*; in its original Arabic, *Kitāb al-nafs*), *On the Conjoining of the Intellect with Human Beings* (in Latin, *Continuationem Intellectus cum Homine*; in Arabic *Ittiṣāl al-ʿaql bi al-insān*), and *Letter of Farewell* (in Latin, *Expeditionis*; in Arabic, *Risālat al-wadāʿ*).[14]

In all these works, Avempace puts forth the view that ultimate happiness consists in the conjunction (*ittiṣāl*) of the material intellect with the agent intellect.[15] In general terms, Avempace builds his argument on the claim that intellection is most proper to human beings, but is imperfect because it is constrained by matter.

12 In fact, in these digressions Albert closely follows Averroes's analysis of the Greek Peripatetics and of Alfarabi and Avempace. The exceptions are Avicenna and Algazel, two thinkers he read in some Latin translations.

13 Though none of Avempace's works were translated in the Middle Ages (and hence were not accessible to Albert), the *Farewell* was translated in the Renaissance by Abraham de Balmes, as *Epistola expeditionis*. See Burnett, 'Arabic into Latin', p. 397.

14 In this paper, I use the following editions of Avempace's works. For the *Kitāb al-nafs* (*Book on the Soul*): *Ibn Bajjah's ʿIlm al-Nafs*, ed. by Maʿsumi; *Libro sobre el alma [Kitāb al-nafs]*, ed. by Lomba. For *Ittiṣāl al- ʿaql bi al-insān* (*On the Conjoining*): 'Tratado de Avempace', ed. by Asín Palacios; *Carta del adiós y otros tratados*, ed. by Lomba, pp. 83–104; 'Discours sur la conjonction de l'intellect', ed. by Genequand. For *Risālat al-wadāʿ* (*Of Farewell*): 'La *Carta del adiós* de Avempace', ed. by Asín Palacios; *Carta del adiós y otros tratados*, ed. by Lomba, pp. 19–72; 'Épître de l'adieu', ed. by Genequand.

15 On the complexity of the Latin translation of the term *ittiṣāl*, see Burnett, 'Coniunctio–Continuatio'.

Its perfectibility thus consists in the development of the aptitude to receive forms separated from matter, and its culmination is precisely the attainment of these forms through a kind of divine intervention.[16]

In his digression, Albert engages with Alfarabi and Avempace regarding this question, and presents the reasons why he considers both philosophers to be wrong. Initially, Albert depicts Avempace's view as assuming, first, that the material intellect is corruptible and generable, and is not part of the rational soul; and, second, that its action corresponds to imagination because this is a power, found in humans, which is united to the cogitative power. Because of his materialist view of the material intellect, in Albert's reading, Avempace answered this question in the same fashion as Alfarabi and 'Abubacher', assuming that the agent intellect is natural to humans but separated (*intellectus agens est natura hominis et est separatus*).[17] Hence, when the intellect is perfected by its operations, such as the creation (*creare*) and production (*facere*) of cognitive contents, it attains the intelligible forms, experiencing a sort of liberation (*quasi liberatus*).[18] This summary closely mirrors Avempace's position as found in his treatise *On the Conjoining of the Intellect with Human Beings*,[19] where he describes an ascending process that starts with the apprehension of the sensible forms and culminates in conjunction with the agent intellect and the attainment of perfection.

As Albert subsequently explains, Avempace added to the positions of Alfarabi and Abubacher that humans have two faculties related to their intellectual operations. The first is human and constituted by phantasms (*unam humanam, quam*

16 And in fact, this aptitude for receiving forms separated from matter is explained in *Kitāb al-nafs*, ch. 11, ed. by Lomba, pp. 122–35, ed. by Maʿsumi, pp. 117–21; *Risālat al-wadāʿ*, ch. 24, ed. by Asín Palacios, pp. 78–79 (Arabic p. 35); *Ittiṣāl al-ʿaql bi al-insān*, ch. 14, ed. by Asín Palacios, pp. 37–38 (Arabic pp. 17–18). Although Avempace was not translated into Latin until much later, I will provide some references to the treatises in Arabic where some of the doctrines to which both Averroes and Albert allude can be found.

17 Albertus Magnus, *De anima*, III.3.8, ed. by Stroick, p. 217, v. 85–p. 218, v. 5. Albert adds the name Abubacher to the discussion. Taylor believes that Albert wrongly thought this 'Abubacher' to be Abū Bakr Muḥammad Ibn Zakarīyah al-Rāzī. See Taylor in Averroes, *Long Commentary*, p. lxxxix n. 163. However, since, in his *Long Commentary*, Averroes often refers to Avempace as Abubacher, the possibility that Albert is referring to Avempace should not be easily discarded. The sentence 'intellectus agens est natura hominis et est separatus' is confusing here. How can the agent intellect be natural to humans and at the same time separated? In my view, Albert is assuming Alfarabi's emanative cosmology and metaphysics, according to which the separate agent intellect is involved in the human cognitive process. See Albertus Magnus, *Ethica* (*c.* 1262/1263), IX.3, ed. by Borgnet, p. 585a; *De natura et origine animae* (1262/1263), II.13, ed. by Geyer, p. 38, v. 40–p. 39, v. 8. This is also the teaching of Averroes. However, as will be shown in this article, Albert uses the term *separatus* in a different sense to that of the 'Arabs'. He regards the intellect and its powers as immaterial and separate from body. In this sense, inspired by Aristotle's *Nicomachean Ethics* IX.4, for Albert it is the nature of a human being to be intellect. See Albertus Magnus, *De anima*, I.1.1, ed. by Stroick, p. 2, v. 32–33.

18 Albertus Magnus, *De anima*, III.3.8, ed. by Stroick, p. 218, vv. 5–6.

19 Avempace, *Ittiṣāl al-ʿaql bi al-insān*, chs 13, 19, 20, ed. by Asín Palacios, pp. 37 and 43–45 (Arabic pp. 17 and 20–21).

*habet, inquantum colligatur phantasmatibus*); the second is a divine power and a vestige of the separate intelligence (*alteram autem divinam, quam habet, secundum quod est vestigium intelligentiae separatae*). Through the first, we understand forms abstracted from matter; through the second, we are able to understand forms separated from matter.[20] This position is presented in Avempace's *Letter of Farewell*.[21]

Regardless of Avempace's original positions, Albert unmistakably follows Averroes's explanation of Avempace in the *Long Commentary on the De anima*. There, Averroes initially refers to two treatises — *On the Conjoining of the Intellect with Human Beings* and *On the Soul* — in which Avempace proposes that the human intellect can attain intelligible forms by itself.[22] Averroes explains Avempace's main argument as follows: (1) the theoretical intelligibles have been produced; (2) everything that is produced has a quiddity; (3) for everything having a quiddity, the intellect is naturally able to extract that quiddity; (4) therefore, the intellect extracts the quiddities and intelligibles naturally.[23]

Averroes points out that Avempace drew this view from Alfarabi's *On Intellect and the Intelligible* (*De intellectu*) and later offered two similar presentations of it.[24] According to Averroes, Avempace argues in his treatise *On the Soul*

> that multiplicity does not accrue for the intelligibles of things except in virtue of the multiplication of spiritual forms with which they will be sustained in each individual. According to this, the intelligible of horse in me will be different from its intelligible in you. From this it follows by conversion of the opposite that for every intelligible not having a spiritual form by which it is sustained, that intelligible is one in me and in you.[25]

---

20 Albertus Magnus, *De anima*, III.3.8, ed. by Stroick, p. 218, vv. 9–13.

21 The distinction appears in *Risālat al-wadāʿ*, ch. 30, ed. by Asín Palacios, p. 84 (Arabic pp. 38–39).

22 'Dicamus igitur quod Avempeche multum perscrutabatur in hac questione et laboravit in declarando hanc continuationem esse possibilem, in sua epistola quam vocavit Continuationis Intellectus cum Homine; et in libro de Anima et in aliis multis libris videbitur quod ista questio non recessit ab eius cogitatione, neque per tempos nutus unius oculi'. Averroes, *Commentarium magnum*, Book III, comment 36, ed. by Crawford, p. 487 (hereafter *Commentarium magnum*); Averroes, *Long Commentary on the 'De Anima'*, ed. by Taylor (hereafter *Long Commentary*, pp. 388–89.

23 '[…] primo enim posuit quod intellecta speculativa sunt facta; deinde posuit quod omne factum habet quiditatem; deinde posuit quod omne habens quiditatem, intellectus innatus est extrahere illam quiditatem; ex quibus concluditur quod intellectus innatus est extrahere formas intellectorum et quiditates eorum'. Averroes, *Commentarium magnum*, Book III, comment 36, p. 490; *Long Commentary*, p. 391.

24 See Avempace, *Ittiṣāl al-ʿaql bi al-insān*, ch. 5, ed. by Asín Palacios, p. 31 (Arabic pp. 12–14); Alfarabi, *Risālah fī alʿaql*, ed. by Bouyges, pp. 3–7.

25 'In libro autem de Anima coniunxit huic quod intellectis rerum non contingit multitudo nisi per multiplicationem formarum spiritualium cum quibus sustinebuntur in unoquoque individuo, et per hoc fuerit intellectum equi apud me aliud quam intellectum eius apud te. Ex quo consequitur secundum conversionem oppositi quod omne intellectum non habens formam spiritualem a qua sustentatur, illud intellectum est unum apud me et apud te'. Averroes, *Commentarium magnum*, Book III, comment 36, p. 491; *Long Commentary*, p. 392. A similar argument appears in Avempace, *Kitāb al-nafs*, ch. 11, ed. by Lomba, pp. 128–30. Averroes himself followed this position in early works

Averroes then invokes an argument, supposedly put forth by Avempace, that since the quiddity of the intelligible and its form have no individual spiritual form upon which they can be sustained, and since the quiddity of the intelligible is not the quiddity of a singular individual, be it spiritual or bodily (it has indeed been explained that the intelligible is not an individual), it is therefore natural for the intellect to understand the quiddity of an intelligible belonging to an intellect that 'is one for all human beings, and what is such as this is a separate substance'.[26]

Similarly, Averroes presents a second argument, this time from Avempace's treatise *On the Soul*, that is built on the intellect's inability to extract quiddities to infinity.[27] This argument can be summarized as follows: (1) it is impossible to find a given quiddity which the intellect is not naturally constituted to extract from a lower quiddity (for that intellect would not then be called 'intellect' except equivocally, since it was asserted that the intellect is naturally constituted to separate the quiddity insofar as it is a quiddity); (2) it is impossible for the intellect to attain something which neither has a quiddity nor is a quiddity, for if something neither has a quiddity nor is a quiddity, it is absolute privation; (3) therefore, if the intellect attains a quiddity itself (that is, a quiddity not having a quiddity), this quiddity must be a separate form (*forma abstracta*).[28]

Averroes disagrees with the position he thus attributes to Avempace, questioning whether the term 'quiddity' can be said univocally of quiddities of material things and quiddities of separate intellects. Indeed, Averroes maintains against

---

such as his *Epitome of the De anima* and the *Epitome of the Parva naturalia*. However, as can be seen here, he rejected it later on in the *Long Commentary*.

26 'Deinde coniungit huic quod quiditas intellecti et forma eius non habet formam spiritualem individualem cui sustentatur, cum quiditas intellecti non est quiditas individui singularis, neque spiritualis neque corporalis; intellectum enim declaratum est quod non est individuum. Ex quo consequitur ut intellectus sit innatus intelligere quiditatem intellecti cuius intellectus est unus omnibus hominibus; et quod est tale est substantia abstracta'. Averroes, *Commentarium magnum*, Book III, comment 36, p. 491; *Long Commentary*, p. 392.

27 Actually, this argument appears in *On the Conjunction*; see Avempace, *Ittiṣāl al-ʿaql bi al-insān*, ch. 8, ed. by Asín Palacios, p. 33 (Arabic p. 15).

28 'Et si non fuerit concessum nobis quod ista quiditas est simplex et quod ens ex ea est idem cum intellecto, continget in ea quod contingit in prima, et est quod etiam habeat quiditatem factam. Et necesse est tunc aut ut hoc procedat in infinitum, aut ut intellectus secetur ibi. Sed quia impossibile est hoc procedere in infinitum (quia faceret quiditates et intellectus infinitos diversos in specie esse, scilicet secundum quod quidam eorum sunt magis liberati a materia quam quidam), necesse est ut intellectus secetur. Et cum secabitur, tunc aut perveniet ad quiditatem que non habet quiditatem, aut ad aliquid habens quiditatem sed intellectus non habet naturam extrahendi illam, aut ad aliquid non habens quiditatem neque est quiditas. Sed impossibile est invenire quiditatem quam intellectus non est innatus extrahere a quiditate, quoniam ille intellectus tunc non diceretur intellectus nisi equivoce (cum sit positum quod intellectus innatus est abstrahere quiditatem in eo quod est quiditas). Et impossibile est etiam ut intellectus perveniat ad aliquid non habens quiditatem neque est quiditas; hoc enim quod neque est quiditas neque habet quiditatem est privatio simpliciter. Remanet igitur tertia divisio, et est quod intellectus perveniat ad quiditatem non habentem quiditatem; et quod est tale est forma abstracta'. Averroes, *Commentarium magnum*, Book III, comment 36, p. 492; *Long Commentary*, pp. 392–93.

Avempace that if 'quiddity' were predicated equivocally, his arguments would be ineffective and knowledge of separate substances on the basis of quidditative abstractions would seem impossible. But the onus to prove univocity lies with Avempace: if 'quiddity' is said in a univocal way, he will need to explain how it is possible for the corruptible (he conceives the material intellect as something corruptible) to understand what is not corruptible.[29] As we will see, this is the core of the problem later remarked upon by Albert, following Averroes.

Averroes continues his objection by asking why, since it has been asserted that it is in the nature of the material intellect to understand separate things, one should not regard the intellection of separate things as 'analogous [*currit cursu*] to' the intellection of material forms, in such a way that to understand separate things would be part of the speculative sciences, and hence one of the problems dealt with by speculative science.[30] According to Averroes, Avempace does not take up a definitive position in this respect. On the one hand, in his *Letter of Farewell,* he mentions the possibility of two different sorts of intellection, the natural and the divine, concluding, in Averroes's account, that the knowledge of the separate intellect requires divine intervention. Albert, too, alludes to this point in his *De anima*. On the other hand, Avempace's *On the Conjoining of the Intellect with Human Beings* posits — as Averroes shows — an 'ascendant intellection', which starts from the common abstraction of material forms and culminates in the attainment of separate substance. The philosopher is the one able to attain the separate substance. From this, argues Averroes, it seems that in this treatise, Avempace considers the issue to be part of the speculative sciences. This solution does not convince Averroes either, because it does not show clearly enough how it is possible for human beings to acquire such ascendant intellection. Averroes asks whether we are naturally constituted to acquire this science or whether it is acquired through some kind of learning.

In Albert's presentation of Avempace's view, we find striking similarities with Averroes's own presentation — not surprisingly, since, as mentioned, Albert simply paraphrases the *Long Commentary on the De anima*. Unlike Averroes, however, Albert reads Avempace (and, for that matter, also Alfarabi and 'Abubacher') as putting forth a single argument that Albert considers to be true: (1) every intellectual content existing in the soul is in act after being in potency; (2) every

29 Averroes also refers to those who claim that the material intellect is not generable or corruptible. In this case, they would have to explain how what is naturally constituted to understand the quiddities in the future and the past can understand by virtue of a new intellection. Averroes, *Commentarium magnum*, Book III, comment 36, pp. 493–94; *Long Commentary*, p. 394.

30 'Et etiam, si posuerimus quod intelligere res abstractas est in substantia et in natura intellectus materialis, quare igitur ista intellectio non currit cursu intellectionum materialium nobis, ita quod hoc intelligere sit pars partium scientiarum speculativarum, et erit unum quesitorum in scientia speculativa?' Averroes, *Commentarium magnum*, Book III, comment 36, p. 494; *Long Commentary*, p. 394.

object known by the intellect has a quiddity; and (3) every quiddity in potency is actualized through the intellect.[31]

According to Albert, Avempace, Alfarabi, and 'Abubacher' also maintain that every quiddity of a composed and singular substance is potentially separable and that, when this quiddity is actualized, the agent intellect produces a pure quiddity. The problem that Albert points out in this regard is the same one we found in Averroes when he discussed Avempace's attempts to explain how human beings are able to understand a pure quiddity (that is, a separate substance): it remains unclear how the human intellect would be able to transform the quiddity it has attained from material objects into a separate and absolute quiddity.

Albert concludes, following Averroes, that Avempace (and thus Alfarabi and Abubacher) does not give a satisfactory answer to the question of how human beings are able to understand a pure quiddity. They build their argumentation on an ascendant abstraction that starts with the cognition of material forms, proceeds to phantasms, and ends in the highest level of abstraction, at which the light of the agent intellect makes it possible to attain separate quiddities — yet this position seems unable to account either for the difference between the abstraction of quiddities from matter, or for the quiddities of the separate intellects. Hence, Albert levels the same objection against Avempace as Averroes does: if the sense of 'quiddity' here is taken equivocally, the argumentation is erroneous; if it is taken univocally, the argumentation is also erroneous, and Avempace would need to explain how it is possible for a corruptible material intellect to conjoin with the incorruptible.[32]

In contrast to Averroes, Albert contemplates a third option for interpreting the term 'quiddity', one in which it is understood neither univocally nor equivocally, but somehow between the two. He finds, however, that this option of thought would imply the existence of several variations of quiddity in the intellectual faculty, yet according to Book VI of Aristotle's *Nicomachean Ethics*, congruence between the one who knows and that which is known (*congruentiam fit cognoscentis et cogniti*) is required.[33] Indeed, a faculty cannot apprehend an object that is not capable of being apprehended by its powers: the intellect is not able to apprehend sensible objects or the senses to apprehend intelligible objects.

To summarize, given that Avempace thought the material or possible intellect to be corruptible, both Averroes and Albert find that he fails to offer a fitting solution to the problem of how it can conjoin with something incorruptible. Paraphrasing Averroes, Albert points out how problematic it would be to conceive

31 Albertus Magnus, *De anima*, III.3.8, ed. by Stroick, p. 218, vv. 15–24.

32 'Si autem dicatur intellectus possibilis esse corruptibilis, tunc quaeretur secundum preadicta, qualiter corruptibile unitur incorruptibili'. Ibid., vv. 94–96. Aquinas's interpretation of Alfarabi and Avempace is very similar to that of Albert. This is not surprising since both are following Averroes's *Long Commentary*. See Krause, *Thomas Aquinas on Seeing God*, pp. 87–90; Wirmer, 'Avempace – "ratio de quidditate"'.

33 Albertus Magnus, *De anima*, III.3.8, ed. by Stroick, p. 218, vv. 82–87. See Aristotle, *Nicomachean Ethics*, VI.1, 1139a10–12.

of the possible intellect as incorruptible, since that would make it difficult to explain how it is possible for the intellect to have a new intellection that could entail the unity of form and matter and, in consequence, the understanding of a new quiddity.

## Albert's Digression on Avicenna and Algazel

In his interpretation of Avicenna and Algazel, Albert does not follow Averroes. Averroes made virtually no comment on Algazel in his *Long Commentary on the De anima* and paid scant attention to Avicenna. But as is well known, several of Avicenna's works were popular among Latin thinkers through the translations of Avendauth and Gundissalinus, especially his *De anima*, known in Latin as the *Liber sextus de naturalibus*.[34] Albert knew Avicenna very well and saw the Persian philosopher as one of the main philosophical authorities, which is why he is sympathetic to Avicenna's psychology in his earlier works, such as the *De homine*. In some instances, however, Albert also openly criticizes problematic theories entertained by Avicenna.[35]

One such criticism is found in Albert's *De anima*, III.3.9.[36] According to Albert's report, Avicenna and 'his follower' Algazel considered the nature of the possible intellect in us to be a sort of *tabula rasa*, that is, having the disposition to receive intelligible forms. Avicenna and Algazel (in Albert's reading) maintained that when we learn, we acquire intelligible forms because the agent intellect abstracts them and unites them to the possible intellect. All intellection consists in the conjunction of the possible intellect with the agent intellect and when this happens perfectly, knowledge of intelligible forms is attained. As a result, Albert took Avicenna and Algazel to assume that learning consists in acquiring a perfect disposition, by means of which the material intellect is directed to the agent intellect. Through this disposition, the possible intellect is actualized, that is, the forms become forms in act. Albert describes this process as consisting in a directedness of the possible intellect to the agent intellect and an emanation of the forms contained in the agent intellect into the soul.[37]

---

34 Avicenna, *Shifāʾ*: *al-Nafs*: Avicenna Latinus, *Liber de anima*, ed. by Van Riet; Avicenna, *Avicenna's De anima*, ed. by Rahman.

35 For the influence of Avicenna on Albert, see Hasse, *Avicenna's 'De anima'*, pp. 60–69.

36 Albert refers to Algazel as a follower of Avicenna. The only work by Algazel translated into Latin by Albert's times was the *Doctrines of the Philosophers* (*Maqāsid al-falāsifa*), known as *Summa theoricae philosophiae* or as *Logica et philosophia Algazelisi*. The Latin translation omitted the introduction where Algazel explains that he is reporting the teachings of the philosophers. Thus, several Latin philosophers, among them Albert and even Aquinas, thought Algazel was a faithful follower of Avicenna's philosophy. See Minnema, 'Algazel Latinus'; Minnema, '*Cave hic*: Marginal Warnings'; Minnema, 'A Hadith Condemned'.

37 Albertus Magnus, *De anima*, III.3.9, ed. by Stroick, p. 219, vv. 39–56: 'Huic etiam quaestioni satisfacere intendit Avicenna et insecutor eius Algazel dicentes, quod in veritate intellectus possibilis

Albert offers a synthetic account of the route by which Avicenna and Algazel conclude that the forms are contained in the agent intellect and not in the possible intellect. Before reaching this conclusion, he explains, both thinkers discussed two other alternatives. According to the first of these, a human being can possess the forms whenever he wants because those forms are in the rational soul (the possible intellect) as in a treasure trove; according to the second alternative, those forms are in the rational soul in such a way that a human being is able to possess them whenever he wants, even if they are not in the soul as in a subject. The first alternative is false because it would entail a kind of memory in which the soul would be able to remember everything. Furthermore, 'what is in the soul is abstracted from time, but all that is remembered is remembered at a given time' (*quae sunt in anima, sunt abstracta a tempore et ominis memorans memoratur sub differentia temporis determinati*).[38] The second alternative is also false, because if the forms were in the soul, it would be unnecessary to perceive in act or to learn something through science; put simply, all sensible forms would already be in the soul, that is, in the possible intellect.

Since these two options are false, Albert continues, Avicenna and Algazel conclude that a third alternative is the correct one: the forms are not in the rational soul (in the possible intellect) but in the agent intellect. Therefore, a disposition through which the soul learns to direct itself to the agent intellect is necessary in order to explain how the possible intellect attains these forms. Albert presents four arguments against this position:

(1) It is not clear how it is possible to acquire this disposition through the understanding of objects that are not permanent, that is, material objects.
(2) According to Avicenna and Algazel, the agent intellect conjoins with the possible intellect immediately after the acquisition of any item of scientific knowledge. Yet given that the agent intellect is incorporeal and separate, it would seem that this conjunction happens before the acquisition of full scientific knowledge, and if this is so, it would seem that the two intellects are always united.
(3) Claiming that the agent intellect is the giver of forms is not an explanation of why the forms differ in genus and species.
(4) According to Avicenna and Algazel, forms emanate from the agent intellect into the possible intellect, but if the agent intellect manifests itself

---

est in nobis primo sicut tabula rasa et planata, quemadmodum supra expositum est, quia cum addiscit homo, acquirit formas intellectas per hoc quod agens denudat eas et coniungit eas intellectui possibili; et cum quaelibet illarum formam intellectualitatis accipiat ab agente, oportet, quod possibilis in qualibet illarum convertatur ad agentem; et cum perfecta fuerit conversio eius ad agentem, tunc coniungitur intellectui agenti ut formae, et tunc per ipsum intelligit separata. Et ideo addiscere nihil aliud est nisi acquirere perfectam habitudinem, qua materialis ad agentem convertatur; quam cum acquirit, tunc omnia scit actu ex hoc quod convertitur ad agentem largientem ei formas intellectas, quia ex ipso emanant in animam omnes formae intellectae'.

38 Ibid., vv. 65–67.

in a specific way, then the forms that emanate from it into matter manifest themselves in a specific manner. This argument does not consider that we can have new intellections, or, in other words, the forms emanating from the agent intellect at this time are not necessarily the same forms that will emanate in the future.[39]

Albert thus examines Avicenna's doctrine of the intellect with regard to both its coherence and its difficulties. He first refers to Avicenna's conception of the material intellect (*intellectus possibilis / ʿaql hayūlānī*), suggesting that this intellect is in us as a *tabula rasa* and is a receptacle of forms. Avicenna therefore rejects — and in this respect Albert agrees with him — the possibility of the forms being in the individual material intellect innately. Then Albert refers to the acquisition of the intelligible forms through a disposition. This seems to be what Avicenna calls intellect *in habitu* (*ʿaql bi al-malaka*). But in contrast to Avicenna, Albert does not clearly distinguish between the intellect *in habitu* and the actual intellect or intellect *in effectu* (*ʿaql bi al-fiʿl*), that is, the intellect with the presence of primary intelligibles (*maʿqūlāt ūlā*) and the intellect with the acquisition of secondary intelligibles.[40] Indeed, he does not mention Avicenna's assertion that the primary intelligibles (or first principles, such as 'the whole is greater than the part') are acquired by the intellect *in habitu*, a rather puzzling element of Avicenna's doctrine.

Modern scholars have discussed this matter, since it is not clear whether these primary intelligibles emerge spontaneously through the material intellect or are some sort of *a priori* principles provided by the agent intellect.[41] Probably Albert assumed that the primary intelligibles arise from the agent intellect, because he directly raises the problem of the acquisition of intelligible forms, which is actually what Avicenna tries to resolve when he explains the intellect *in effectu* (*ʿaql bi al-fiʿl*) and the acquired intellect. Avicenna's explanation of the intellect *in effectu* is scanty, at least in his *De anima*,[42] where he mentions that this intellect

---

39 Ibid., v. 5. See ibid., p. 220, vv. 5–24: 'Contra autem istud dictum sunt praecipue quattuor rationes. Quarum prima est, quia adhuc quaestio est ut prius, qualiter possibilis per intellecta non remanentia in ipso acquirat aptitudinem coniungendi se agenti; de hoc autem nullam assignat causam. Secunda autem est, quia secundum hoc immediate agens continuatur possibili post scientiam, et cum sit incorporeus et separatus, videtur, quod eadem ratione continuetur ei etiam ante scientiam, et sic semper continuatur ei, ut videtur. Tertia autem est: si solum agens largitur formas intellectas, cum ipsae differant genere et specie, deberet assignare Avicenna, quae esset causa differentiae. Quarta autem est, quia cum convertitur possibilis ad agentem, tunc fluunt formae intellectae in ipsum possibilem; cum ergo agens se habet uno modo, formae fluentes ab ipso receptae in materiali erunt uno modo. Et hoc est falsum, quia modo fluunt formae quaedam et cras aliae, et sic de aliis'.

40 Avicenna, *Shifāʾ: al-Nafs*, I.5: *Liber de anima*, vol. 1, ed. by Van Riet, pp. 96–100; *Avicenna's De anima*, ed. by Rahman, pp. 48–50.

41 See Gutas, *Avicenna and the Aristotelian Tradition*, pp. 170–77; Davidson, *Alfarabi, Avicenna, and Averroes*, pp. 85–87; Hasse, *Avicenna's 'De anima'*, pp. 179–80.

42 For a wider explanation of these secondary intelligibles in other works by Avicenna, see Hasse, *Avicenna's 'De anima'*, pp. 180–83.

acquires secondary intelligibles (definitions, genera, species, differences, etc.) but without yet using them.[43] Primary and secondary intelligibles are necessary for the actualization of the intelligibles — a process that takes place in what Avicenna calls the acquired intellect (*ʿaql al-mustafād*), that is, the stage that is reached when we are able to actively think the intelligible forms. This intellect is what Albert calls the 'speculative intellect'. The main difficulty arises when Avicenna holds that the actualization of intelligible forms somehow needs help from a separate intellect, namely the agent intellect (*ʿaql faʿʿāl*).

To put it more simply, the actualization of the possible intellect — what has been explained here as the receptive material intellect in Albert's interpretation — needs assistance from the agent intellect in order to become speculative intellect (acquired intellect, in Avicenna's words: the intellect which is able to understand intelligible forms). Practically all of Albert's Peripatetic sources, from Alexander to the Arabs, held the assistance of the agent intellect to be necessary for understanding intelligible forms. The problem is how this assistance takes place. As we have seen, Albert disagrees with the explanations of Avempace, Alfarabi, Abubacher, and Avicenna.

In the four arguments listed above, Albert also criticizes Avicenna's conception of the agent intellect as providing the intelligible forms. However, it should be noted that Avicenna's doctrine of the intellect is more deeply problematic than the portrait painted by Albert in this digression. On the one hand, Avicenna describes a process of abstraction in which the role of the internal faculties is essential in order to explain how it is possible to abstract intelligible forms. In this process, the agent intellect helps to transform particular images stored in the imagination into intelligible forms. On the other, Avicenna seems — at least in Albert's view — to suggest that the agent intellect provides the intelligible forms that emanate (*fāḍa*) from it.[44]

---

43 See Avicenna, *Shifāʾ: al-Nafs*, I.5: *Liber de anima*, vol. 1, ed. by Van Riet, pp. 96–97; *Avicenna's De anima*, ed. by Rahman, p. 49.

44 Avicenna can be regarded as a philosopher who conceives of knowledge as a naturalistic process in which the human intellect is able to abstract intelligible forms by itself; on the other hand, there are strong reasons to hold that his conception of human cognition requires an external agent — the agent intellect — in order to complete the cognitive process, so that his approach depends on a cosmological-metaphysical model and does not make sense without it. These two approaches have led to divergent interpretations of Avicenna's view of human cognition. Some have emphasized the importance of abstraction (Hasse, 'Avicenna on Abstraction'; Gutas, 'Intuition and Thinking'; Gutas, 'Empiricism of Avicenna'; Gutas, 'Avicenna's Philosophical Project'), others the role of the agent intellect (Davidson, *Alfarabi, Avicenna, and Averroes*; Black, 'Psychology: Soul and Intellect'; Black, 'How Do We Acquire Concepts?'; Taylor, 'Al-Fārābī and Avicenna' — though more recently he has reconsidered his position; Lizzini, *Fluxus (fayḍ)*; Hasse, 'Avicenna's Epistemological Optimism'). Other recent interpretations have argued that the two perspectives should complement each other (Alpina, 'Intellectual Knowledge'; McGinnis, 'Making Abstraction Less Abstract'; Taylor, 'Avicenna'; Ogden, 'Avicenna's Emanated Abstraction'). More recently, Gutas's interpretation of Avicenna as an empiricist has been discussed in Kaukua, 'Avicenna's Outsourced Rationalism'.

This, indeed, is the point where Albert appears most dissatisfied with Avicenna's account of the way the agent intellect acts upon the possible intellect. The core of arguments (1) and (2) is that Avicenna does not explain how the material intellect is united to the immaterial and separate agent intellect; the core of arguments (3) and (4) is that the agent intellect cannot be the 'giver of forms'. For Albert, both sets of arguments thus entail reasons to reject the theory of emanation.

It may be worth pointing out here that Albert does not allude to Avicenna's analogy in which the agent intellect is to knowledge as the sun is to vision. This analogy is key to understanding what Avicenna intended to express when he argued that the agent intellect acts upon the human intellect. In his *De anima*, V.5, Avicenna explains that the action of the agent intellect occurs through the mediation of illumination: *mediante luce intelligentiae agente / bi-tawassuṭi ishrāq ʿaql faʿʿāl*.[45] The role of Avicenna's agent intellect is thus not exactly what Albert understood it to be: the agent intellect is not the provider of the intelligible forms (*largitur formas intellectas*), as Albert holds in argument (3) as such, but rather the mediator that enables the process of abstraction by the human intellect through the illumination of the objects of abstraction. This analogy might have enabled Albert to give a different account of Avicenna's explanation on how the agent intellect acts in human knowledge. For some reason, however, Albert does not mention it.[46]

45 'Cuius comparatio ad nostras animas est sicut comparatio solis ad visus nostros, quia sicut sol videtur per se in effectu, et videtur luce ipsius in effectu quod non videbatur in effectu, sic est dispositio huius intelligentiae quantum ad nostras animas. Virtus enim rationalis cum considerat singula quae sunt in imaginatione et illuminatur luce intelligentiae agentis in nos quam praediximus, fiunt nuda a materia et ab eius appendiciis et imprimuntur in anima rationali, non quasi ipsa mutentur de imaginatione ad intellectum nostrum, nec quia intentio pendens ex mulits (cum ipsa in se sit nuda considerata per se), faciat similem sibi, sed quia ex consideratione eorum aptatur anima ut emanet in eam ab intelligentia agente abstractio. Cogitationes enim et considerationes motus sunt aptantes animam ad recipiendum emanationem, sicut termini medii praeparant ad recipiendum conclusionem necessario, quamvis illud fiat uno modo et hoc alio, sicut postea scies. Cum autem accidit animae rationali comparari ad hanc formam nudam mediante luce intelligentiae agentis, contingit in anima ex forma quiddam quod secundum aliquid est sui generis, et secundum aliud non est sui generis, sicut cum lux cadit super colorata, et fit in visu ex illa operatio quae non est similis ei ex omni parte'. Avicenna, *Shifāʾ: al-Nafs*, V.5: *Liber de anima*, vol. 2, ed. by Van Riet, pp. 127–28; *Avicenna's De anima*, ed. by Rahman, p. 235.

46 Although Albert does not use this particular analogy of Avicenna's, in his solution to the problem in chapter 11 he applies a similar analogy drawn from Averroes. Now, if this analogy is essential in order to understand that Avicenna did not intend to conceive the agent intellect as the provider of forms, then why did Albert not pay attention to the sense given to it by Avicenna, instead choosing Averroes's sense? McGinnis notes that Avicenna never explicitly identifies the agent intellect with the provider of forms (McGinnis, *Avicenna*, p. 120). Hasse points out that the term *wāhib aṣ-ṣuwar* is used by Avicenna in his doctrine of creation but not in his epistemology, and suggests that Latin thinkers attributed this expression to Avicenna because Algazel, in his *Maqāṣid* (*Metaphysica* in Latin), used it four times where Avicenna did not use it (Hasse, *Avicenna's 'De anima'*, p. 188). I draw attention to Albert's treatment of Avicenna and Algazel in this digression as if they held the same doctrine. It should be said, however, that Hasse has more recently argued that the agent intellect is an epistemological principle (Hasse, 'Avicenna's Epistemological Optimism').

## Albert's Use of Averroes in *De anima* III.3.11

After rejecting a reading of Avicenna's doctrine as he saw it, Albert turns to the last of his Arabic sources, Averroes. Having criticized Alfarabi, Avempace, Avicenna, and other philosophers, Albert explains the true cause and the manner in which the conjunction with the agent intellect takes place in us. In *De anima*, III.3.11, he begins his solution by saying, as mentioned at the beginning of this article, that he dissents in only a few respects from Averroes's *Long Commentary on the De anima*.[47] This is quite surprising given that just a little earlier, in III.2.3 and III.2.7, he has pointed out Averroes's mistakes. Now, however, in III.3.11, he summarizes Averroes's standpoint in three premises: (1) Averroes conceives of the agent intellect as separated from the human soul, as do most of the other Peripatetic philosophers (Avempace, Alfarabi, Avicenna, and Algazel); (2) he claims that the agent intellect is united in some way to the possible intellect and that it is necessary to explain the cause of this union; and (3) he thinks that the union between the possible and the agent intellect is formal, and he explains why.

In his *Long Commentary on the De anima*, Averroes certainly subscribed to the notion that the agent intellect is an eternal, separately existing substance. The relationship between the agent intellect and the material intellect ('possible intellect', in Albert's terminology) is raised at several points in Book III.[48] Averroes clearly states that 'a human being is intelligent in act only owing to the conjoining of the intelligible with him in act'.[49] In a long and detailed explanation of how this conjunction takes place, Averroes sets out six different intellects: agent, material, acquired, dispositional or *in habitu*, theoretical, and passible. The acquired intellect is the intellect at the precise moment when it receives the intelligibles in act; the intellect *in habitu* is the intellect already disposed with knowledge which we can access at will; the theoretical intellect is the intellect that already contains the intelligibles. These three intellects are part of the soul and they are different stages of the intellective act, whereas the 'passible intellect' comprises the internal powers. Therefore, the passible intellect is called 'intellect' only equivocally, since these powers provide images and intentions, but not intelligibles.

Averroes devotes long paragraphs to the 'material intellect', which he conceives of as a unique receptive intellect, separate from matter and shared by all human beings.[50] According to Averroes, Aristotle himself refers to it when he distinguishes between two parts of the intellect: a receptive intellect and the agent

47 'Nos autem in paucis dissentimus ab Averroe, qui inducit istam quaestionem in Commento Super Librum De Anima'. Albertus Magnus, *De anima*, III.3.11, ed. by Stroick, p. 221, vv. 9–11.

48 Averroes, *Commentarium magnum*, Book III, comments 18–20, pp. 437–54; *Long Commentary*, pp. 349–63.

49 'Dicamus igitur quod manifestum est quod homo non est intelligens in actu nisi propter continuationem intellecti cum eo in actu'. Averroes, *Commentarium magnum*, Book III, comment 5, p. 404; *Long Commentary*, pp. 319–20.

50 Averroes, *Commentarium magnum*, Book III, comments 4–5, pp. 383–413; *Long Commentary*, pp. 300–29.

intellect.[51] The receptive or material intellect is 'that which is in potency all the intentions of universal material forms and is not any of the beings in act before it understands any of them';[52] the agent intellect 'makes the intentions which are in the imaginative power to be movers of the material intellect in act after they were movers in potency [...]. [Also from Aristotle it is apparent, says Averroes,] that these two parts are neither generable nor corruptible and that the agent is related to the recipient as form to matter'.[53] Averroes thus conceives of the agent and material intellects as two ontologically distinct substances.

When commenting on *De anima*, III.5, 430a14–17, Averroes once again distinguishes between the intellect that becomes everything, the intellect that makes the human intellect understand everything, and the intellect that understands everything as a positive disposition and is described as a 'light'.[54] These are the material intellect, the agent intellect, and the intellect *in habitu*. Whereas the agent intellect has no receptivity in itself and actually does not understand anything from the material world, the material intellect is characterized by its receptivity.[55]

As Richard Taylor has pointed out in several publications, Averroes rejects the conception of the material intellect as an individual entity in his *Long Commentary on the De anima*, since this would imply that the received intelligibles are restricted to the individual nature of the material intellect.[56] For Averroes, the material intellect thus has to be a separate and unique entity, capable of receiving in potency the intentions of universal material forms without contracting them into individuals. The nature of the material intellect is so unusual that Averroes says it is not matter, nor form, nor a composite of these, but a *quartum genus esse* (a fourth kind of

---

51 'Et ideo opinandum est, quod iam apparuit nobis ex sermone Aristotelis, quod in anima sunt due partes intellectus, quarum una est recipiens, cuius esse declaratum est hic, alia autem agens'. Averroes, *Commentarium magnum*, Book III, comment 5, p. 406; *Long Commentary*, p. 321.

52 'Idest, diffinitio igitur intellectus materialis est illud quod est in potentia omnes intentiones formarum materialium universalium, et non est in actu aliquod entium antequam intelligat ipsum'. Averroes, *Commentarium magnum*, Book III, comment 36, p. 387; *Long Commentary*, p. 304.

53 'Et ideo opinandum est, quod iam apparuit nobis ex sermone Aristotelis, quod in anima sunt due partes intellectus, quarum una est recipiens, cuius esse declaratum est hic, alia autem agens, et est illud quod facit intentiones que sunt in virtute ymaginativa esse moventes intellectum materialem in actu postquam erant moventes in potentia, ut post apparebit ex sermone Aristotelis; et quod hee due partes sunt non generabiles neque corruptibiles; et quod agens est de recipienti quasi forma de materia, ut post declarabitur'. Averroes, *Commentarium magnum*, Book III, comment 5, p. 406; *Long Commentary*, p. 321.

54 Averroes, *Commentarium magnum*, Book III, comment 18, pp. 437–54; *Long Commentary*, pp. 349.

55 Averroes's conception of the material intellect differs in his three commentaries on the *De anima*: in the Epitome, the material intellect is an individual disposition of the forms of the imagination; in the Middle Commentary, it also exists in each individual, but as a disposition provided by the agent intellect at birth for the reception of the intelligibles when mature human beings are able to abstract through the actualization of the agent intellect; in the Long Commentary, there is a new and quite puzzling conception, namely the material intellect as a separate and unique receptive intellect shared by all human beings.

56 See Taylor, 'Cogitatio, Cogitativus and Cogitare'; Taylor, 'Separate Material Intellect'; Taylor, 'Agent Intellect'.

being).[57] The material intellect is separate from the human soul's internal powers, but requires from the soul the attainment of the intentions of material forms; at the same time, it is a substance distinct from the agent intellect, but requires the agent intellect's intervention for the actualization of the intelligibles. Given its privileged relationship with both the human and agent intellects, the material intellect is able to understand both material forms of the world and separate forms that come to exist in itself.

In Book III, comment 36, Averroes treats of the same question that Albert poses in his *De anima*, III.3.11, concerning human intellectual knowledge and the conjunction with the agent intellect.[58] Comment 36 is crucial (1) for understanding Averroes's position on intellectual human knowledge (a highly relevant opinion for Albert), (2) for understanding the agent intellect as both the final and formal cause of our knowledge, and (3) for understanding how Averroes conceives of the conjunction between the material and the agent intellects.

Concerning (1), Averroes writes:

> For us who have asserted that the material intellect is eternal and the theoretical intelligibles[59] are generable and corruptible [...] and that the material intellect understands both, namely, the material forms and the separate forms, it is evident that the subject of the theoretical intelligibles and of the agent intellect in this way is one and the same, namely, the material [intellect]. Similar to this is the transparent medium which receives color and light at one and the same time; and light is what brings color about.[60]

Later on, I show how Albert takes up this view. With regard to (2), Averroes claims that

> it will necessarily happen that the intellect which is in us in act be composed of theoretical intelligibles and the agent intellect in such a way that the agent intellect is as it were the form of the theoretical intelligibles and the theoretical intelligibles are as it were matter. In this way we will be able to generate intelligibles when we wish. Because that in virtue of which something carries out its proper activity is the form, and we carry out our proper activity in virtue of the agent intellect, it is [therefore] necessary that the agent intellect be form for us.[61]

---

57 Averroes, *Commentarium magnum*, Book III, comment 5, p. 409; *Long Commentary*, p. 326.

58 See Taylor, 'Abstraction and Intellection in Averroes'.

59 This is the phrase that Taylor uses to translate 'intelligibilia speculativa'.

60 'Nos autem cum posuerimus intellectum materialem esse eternum et intellecta speculativa esse generabilia et corruptibilia eo modo quo diximus, et quod intellectus materialis intelligit utrunque, scilicet formas materiales et formas abstractas, manifestum est quod subiectum intellectorum speculativorum et intellectus agentis secundum hunc modum est idem et unum, scilicet materialis. Et simile huic est diaffonum, quod recipit colorem et lucem insimul; et lux est efficiens colorem'. Averroes, *Commentarium magnum*, Book III, comment 36, p. 499; *Long Commentary*, p. 398.

61 'Quoniam hoc posito, continget necessario ut intellectus qui est in nobis in actu sit compositus ex intellectis speculativis et intellectu agenti ita quod intellectus agens sit quasi forma intellectorum

Finally, regarding (3) Averroes explains that

> when the theoretical intelligibles are united with us through forms of the imagination and the agent intellect is united with the theoretical intelligibles (for that which apprehends [theoretical intelligibles] is the same, namely the material intellect), it is necessary that the agent intellect be united with us through the conjoining of the theoretical intelligibles.[62]

With this in mind, I turn back to Albert's *De anima*, where he affirms that he will 'imitate' Averroes concerning aspects of agreement between Averroes and Aristotle. Albert accepts something of Averroes's conception of the agent intellect as the cause of the actualization of the material intellect (possible intellect, in Albert's terminology). He also adopts something of Alfarabi's position on the agent intellect as efficient and formal cause of our knowledge and on the fact that our knowledge comes through the agent intellect's union with us.[63] And concurring with Averroes, Avempace, and partially with Alfarabi, Albert considers it necessary to clarify how that union occurs. He proceeds to explain how the 'Peripatetic philosophers' understood the role of the agent intellect in human knowledge acquisition as follows:

> (1) We acquire some cognitive content naturally. This means that we do not apprehend this content through sensation or through someone's teachings, nor do we attain it through inquiry. This content amounts to the first principles of demonstration, and the agent intellect infuses intelligibility into them.
> (2) There is other cognitive content that we apprehend voluntarily, as when we learn something with the assistance of a teacher or through our inquiry. When we attain this content in act, the agent intellect is joined to us as efficient cause.
> (3) The possible intellect continuously receives the light of the agent intellect, and when it attains all cognitive content, this light is possessed as an adherent form. Given that this light is the essence of the agent intellect, it adheres to the possible intellect as form to matter. The result of this union is what the Peripatetic philosophers call 'acquired and divine intellect'.[64]

---

speculativorum et intellecta speculativa sint quasi materia. Et per hunc modum poterimus generare intellecta cum voluerimus. Quoniam, quia illud per quod agit aliquid suam propriam actionem est forma, nos autem agimus per intellectum agentem nostram actionem propriam, necesse est ut intellectus agens sit forma in nobis'. Averroes, *Commentarium magnum*, Book III, comment 36, pp. 499–500; *Long Commentary*, pp. 398–99.

62 'Quoniam, cum intellecta speculativa copulantur nobiscum per formas ymaginabiles, et intellectus agens copulatur cum intellectis speculativis (illud enim quod comprehendit ea est idem, scilicet intellectus materialis), necesse est ut intellectus agens copuletur nobiscum per continuationem intellectorum speculativorum'. Averroes, *Commentarium magnum*, Book III, comment 36, p. 500; *Long Commentary*, p. 399.

63 Albertus Magnus, *De anima*, III.3.11, ed. by Stroick, p. 221, vv. 34–35.

64 Ibid., p. 221, vv. 69–91 and p. 222, vv. 4–14.

In sum, according to Albert the Peripatetic philosophers hold that there are three ways in which the agent intellect, essentially separated in its nature and substance, conjoins to us: (1) as some potency or faculty of the soul, (2) as efficient cause when it actualizes the cognitive content, and (3) as a form. Of these three, the third is the one that, according to Albert, really describes how the most perfect union with the agent intellect takes place. The cause of this union is the speculative intellect ('intellect *in habitu*', in Averroes's terminology). This, Albert notes, is why the speculative intellect needs to be prior to the acquired intellect — in other words, why the existence of intentions of universal material forms is prior to the actualizing reception of them by the acquired intellect.[65] In the same fashion, Albert argues that the possible intellect ('material intellect', in Averroes's terminology though, as we shall see, not understood in the same way as Averroes does) is prior to the intellective act. This consists in the actualization of the intelligibles through the intervention of the agent intellect, which Albert says 'comes from outside', here using an explicit reference to Aristotle's *De generatione animalium*.[66]

After these remarks, Albert proceeds to resolve the difficulties that have arisen from the Peripatetic views he discussed in the previous chapters. He starts with two: first, how it is possible to unite the corruptible and generable with the incorruptible, that is, the possible intellect with the agent intellect; and second, how it can be that some philosophers (Averroes in particular) understood the union between the agent intellect, the speculative intellect (or intellect *in habitu*), and the possible intellect (or material intellect) to take place as a union between form and matter.

Regarding the first question, Albert responds that, unlike the Peripatetics, '[we do not say] that the possible intellect (or material intellect) is corruptible, but that it is separated and incorruptible, just like the agent intellect'.[67] This statement may seem to be close to Averroes's view. However, it needs clarification. Earlier, in *De anima* III.2.3, Albert has explained that according to Aristotle, the possible intellect is without mixture (*immixtus*), separated (*separatus*), imperishable (*impassibilis*), receptive of all things (*receptibilis omnium*), and not individual (*non hoc aliquid*).[68] Albert says that these characteristics must be accepted as true. At first sight, then, Albert, like Averroes, seems to accept that the possible intellect is separated and not a *hoc aliquid*. He explains that it is separated in the sense that it has no specified and individualizing characteristics, as is also the case for universals.[69] Now, if the possible intellect is not individual, as Averroes thought,

65 Ibid., p. 222, vv. 15–28.

66 Aristotle, *De generatione animalium* XVI.3, 736b27–28.

67 Albertus Magnus, *De anima*, III.3.11, ed. by Stroick, p. 222, v. 44–47.

68 Ibid., III.2.3, p. 179, v. 94–p. 180, v. 3.

69 Ibid., p. 180, vv. 45–73: 'Separatus autem ponitur ab omnibus Peripateticis intellectus possibilis, secundum quod separatum duo importat. Quorum unum est, sicut separatum dicimus denudatum a specificantibus et individuantibus, sicut prima universalia dicimus esse separata. Secundum autem est, sicut dicimus separatum id quod est potentia aliquid, antequam habeat illud. [...]. Ratio autem

then it is one shared by all mankind.[70] However, a few lines later, Albert considers that assertion to be madness and nonsense: if it were the same intellect for all mankind, then when someone acquires scientific knowledge, all other human beings would also acquire it, and this is absurd. Later, in *De anima* III.2.7, Albert again strongly rejects Averroes's conception of the possible intellect as one for all mankind, concluding that the possible intellect is an individual intellect in each human being.[71] So in what sense could it be said that the possible intellect is *separatus* and a *hoc aliquid*? And if the possible intellect is separated just as the agent intellect is, in what sense is the agent intellect separate?

Before turning to these two queries, let us recapitulate the second question posed above: how it can be that some philosophers, such as Averroes, understood the union between the agent, speculative, and possible intellects to take place as a union between form and matter. Albert responds that although the speculative intellect (*in habitu*) is generated from a potency or disposition, the possible intellect does not exert any transformation upon it. Therefore, it is not the case that the three intellects become only one intellect, as happens when matter and form are united.[72] In other words, the three intellects (possible, agent, and speculative) are functionally different.

Once again, Albert *seems* to follow Averroes here. He promptly explains the conjunction of the three intellects using a well-known analogy found in Averroes: the agent intellect unites with the possible intellect as light with the transparent medium, and the speculative intellect unites with the possible intellect as colours with the illuminated transparent medium.[73] In this analogy, both the agent intellect and the possible intellect are separated (they are the light and the transparent medium), while the speculative intellect is the product of the union between the light and the medium. Everything seems to indicate that Albert somehow agrees with Averroes's conception of the possible and the agent intellects as separate and incorruptible. Yet more careful attention to Albert's position is required. Although his position in *De anima* III.3.11 might appear to be close to Averroes, if we keep

demonstrativa, quae induxit Aristotelem, quod intellectum possibilem posuit separatum primo modo separationis, illa fundatur super duo necessaria. Quorum unum est, quod omnis cognitio animae secundum congruentiam aliquam existit ei; et ideo congruentiam harmonicam oportuit esse sensus ad sensibile et imaginationis ad imaginabile. Sic igitur etiam cognitio universalis erit secundum congruentiam aliquam animae. Specificatum autem et individuatum non habet aliquam congruentiam ad universale, sed potius oppositionem in modo; cum ergo cognitio universalis sit secundum possibilem intellectum, oportet ipsum esse separatum hoc modo quo universale est separatum'.

70 Ibid., p. 181, vv. 55–57: 'Et videtur sequi ex isto, quod intellectus possibilis unus est numero in omnibus hominibus, qui sunt et fuerunt et erunt'.

71 Ibid., III.2.7, p. 186, v. 59–p. 188, v. 6.

72 Ibid., III.3.11, p. 222, vv. 44–59.

73 Ibid., v. 55. See Black, 'Conjunction and the Identity'. This is the well-known light analogy that Averroes sets out in comment 36 of the *Long Commentary*: just as light is constitutive of colours, the agent intellect is constitutive of the intelligibles. I will not discuss in depth here how problematic this analogy is within Averroes's rational psychology.

in mind what he has argued previously — in *De anima* III.2.3 and III.2.7 — it becomes obvious that this is not exactly the case.

I have said that in *De anima* III.2.3 and III.2.7, Albert rejects Averroes's conception of the possible intellect as being ontologically one for all mankind. From these two passages, we can conclude that for Albert the possible intellect is *hoc aliquid* such that it is an individual intellect belonging to each individual human being. In *De anima* III.3.11, however, he holds that the possible intellect is separated and incorruptible. If it is separated, it is distinct from an individual material subject; if it is incorruptible, it is not mixed with matter and hence it is immaterial. If the possible intellect is not an individual subject insofar as it is immaterial, then it seems to be, as Averroes thought, *non hoc aliquid*. So how should we interpret *De anima* III.3.11?

As mentioned, in *De anima* III.2.7, Albert argues that if the possible intellect were one for all mankind, then the scientific knowledge acquired would be the same for all of us. Experience shows that this is false, since then we would all have the same knowledge. Thus, Albert concludes that the possible intellect is individual (*hoc aliquid*) and part of the soul (*pars animae*). Earlier in *De anima*, at III.2.3, he has explained that for Aristotle (and also for Averroes) the possible intellect is without mixture, separated, imperishable, purely receptive, and not something individual (*non hoc aliquid*).[74] Quickly detecting how troublesome it would be to adopt the Averroist position, Albert argues that the possible intellect is a *hoc aliquid*. The key to understanding how the possible intellect can be a *hoc aliquid* and at the same time separated and incorruptible is to keep in mind that in *De anima* III.2.7, Albert has argued that the possible intellect is *pars animae*. This means it is part of the individual soul, but is separated from the body and, as such, incorruptible for each individual human being.

At the beginning of *De anima* III.3.11, Albert argues that according to Aristotle, the soul — like any other nature — has a passive principle and an active principle.[75] There is thus no doubt, Albert says, that the agent intellect is a part and a potency of the soul. In other words, the agent intellect is an active part of the soul whose function is to actualize the possible intellect, that is itself the passive part of the intellectual soul. Both in III.3.11 and previously in III.2.18–19, Albert explains the nature of the agent intellect. In III.3.11 he agrees with the Peripatetics in general that the agent intellect is more separated than the possible intellect. Both intellects are, as mentioned, separated and incorruptible. This means, for Albert, that these intellects are not part of the body but immaterial parts of the soul.

When Albert says that he disagrees 'only a little' with Averroes, and that he will imitate Averroes concerning those aspects where the Cordoban agrees with

74 Albertus Magnus, *De anima*, III.2.7, ed. by Stroick, pp. 186–88. For this ambiguous characterization of the possible intellect, see Tellkamp, 'Why Does Albert the Great Criticize Averroes?' See also Tracey, 'Albert the Great'.

75 Aristotle, *De anima* III.5, 430a10–13.

Aristotle, we therefore need to specify the aspects in which he does follow Averroes. Certainly, Albert's criticism of the Peripatetics is built upon Averroes's views. We have seen that Albert disagrees with the positions of Alfarabi, Avempace, Avicenna, and Algazel concerning human intellectual knowledge: the possible intellect is not corruptible, as Avempace thought, and it is not a mere receptacle of forms provided by the separate agent intellect, as Avicenna thought in Albert's interpretation. Albert, like Averroes, argues that the possible intellect is separated and incorruptible. On the other hand, whereas Averroes regards both intellects as separated from the individual soul, Albert conceives of them both as parts of the soul. Hence, he agrees with Averroes regarding the operations of the possible and agent intellects, but disagrees on how we should understand their ontological separability and incorruptibility.

## Conclusion

At the beginning of *De anima* III.3.11, Albert follows Averroes (and Aristotle) exclusively regarding the operation of the two intellects: the agent intellect actualizes the potential cognitive content of the possible intellect, but both — against Averroes — are complementary parts of the soul, distinguished from each other by their operations. Consequently, after following Averroes very closely in his explanations and criticism of some Greek and Arab Peripatetics, Albert departs from Averroes in a crucial matter: the nature of the possible and agent intellects.

Albert's digressions on the teachings of philosophers of the Arabic tradition illustrate the philosophical complexities that arise when he discusses, corrects, and sometimes partially follows their views. In the three digressions I have explored, Albert takes on board the arguments of Alfarabi, Avempace, Avicenna, Algazel, and Averroes on the operation of the possible and the agent intellects. However, he rejects Avempace's conception of the possible intellect as corruptible, Avicenna's conception of the separate agent intellect as 'giver of forms', and Averroes's conception of both the material and the agent intellects as separated from the human soul. At first sight, it would seem that Albert 'differs only a little' from Averroes, and this could suggest that he is a sort of moderate Averroist. However, as I have shown, he distances himself from Averroes on several key points.

Despite his departure from 'the Arabs', in short, it is clear the Albert builds his own doctrine of the intellect in intense dialogue with and careful interpretation of thinkers of the Arabic Peripatetic tradition. Albert's digressions show that he constantly engages with 'the Arabs', finding their positions both very attractive and at the same time quite problematic. In the complex context of the discussions of Averroes and the other Arab Peripatetics in his *De anima,* Albert's position amounts to a reaffirmation of his own understanding of the soul and its powers as found in his early *De homine,*[76] with, however, a profound revision and reassessment of his own interpretation of Averroes's teachings.

76 See Taylor, 'Remarks on the Importance of Albert the Great's Analyses', pp. 139–47.

## Works Cited

### *Primary Sources*

Albertus Magnus, *De anima*, ed. by Clemens Stroick, Editio Coloniensis, 7/1 (Münster: Aschendorff, 1968)

———, *De natura et origine animae*, ed. by Bernhard Geyer, Editio Coloniensis, 12 (Münster: Aschendorff, 1955)

———, *Ethica*, ed. by Auguste Borgnet, Editio Parisienis, 7 (Paris: Vivès, 1891)

———, *Super Ethica*, ed. by Wilhelm Kübel, Editio Coloniensis, 14/1 and 14/2 (Münster: Aschendorff, 1968–72)

Alfarabi, *Risālah fī alʿaql*, ed. by Maurice Bouyges, Bibliotheca Arabica Scholasticorum, 8/1 (Beirut: Dār el-Mashriq, 1938)

Aquinas, Thomas, *Scriptum super libros Sententiarum*, ed. by Pierre Mandonnet (Paris: P. Lethielleux, 1929)

Averroes, *Averrois Cordubensis Commentarium magnum in Aristotelis De anima libros*, ed. by F. Stuart Crawford (Cambridge, MA: The Mediaeval Academy of America, 1953)

———, *Long Commentary on the 'De Anima' of Aristotle*, trans. with an introduction and notes by Richard C. Taylor, with Thérèse-Anne Druart (New Haven, CT: Yale University Press, 2009)

Avempace, 'La *Carta del adiós* de Avempace' [*Risālat al-wadā*], trans. by Miguel Asín Palacios, *Al-Andalus*, 8 (1943), 1–87

———, *Carta del adiós [Risālat al-wadā] y otros tratados filosóficos*, ed. and trans. by Joaquín Lomba (Madrid: Trotta, 2006)

———, 'Discours sur la conjonction de l'intellect avec l'homme adressé à l'un de ses fréres' [*ʿIttiṣāl al-ʿaql bi-l-insān*], in *La conduite de l'isolé et deux autres épîtres*, ed. and trans. by Charles Genequand, Textes et Traditions, 19 (Paris: Vrin, 2010), pp. 183–203

———, 'Épître de l'adieu', in *La conduite de l'isolé et deux autres épîtres*, ed. and trans. by Charles Genequand, Textes et Traditions, 19 (Paris: Vrin, 2010), pp. 88–121

———, *Ibn Bajjah's 'Ilm al-nafs'*, trans. with notes by M. S. Hasan Maʿsumi (Karachi: Pakistan Historical Society, 1961)

———, *Libro sobre el alma [Kitāb al-nafs]*, ed. and trans. by Joaquín Lomba (Madrid: Trotta, 2007)

———, 'Tratado de Avempace sobre la unión del intelecto con el hombre' [*ʿIttiṣāl al-ʿaql bi-l-insān*], trans. by Miguel Asín Palacios, *Al-Andalus*, 7 (1942), 1–47

Avicenna, *Avicenna's De anima: Being the Psychological Part of Kitāb al-Shifā'*, ed. by Fazlur Rahman (London: Oxford University Press, 1959)

Avicenna Latinus, *Liber de anima seu Sextus de naturalibus: Édition critique de la traduction latine médiévale*, ed. by Simone Van Riet, introduced by Gérard Verbeke, vol. 1 (books IV–V), vol. 2 (books I–III) (Leiden: Brill 1968–72)

### Secondary Works

Alpina, Tommaso, 'Intellectual Knowledge, Active Intellect and Intellectual Memory in Avicenna's *Kitāb al-Nafs* and Its Aristotelian Background', *Documenti e studi sulla tradizione filosofica medievale*, 25 (2014), 131–83

Asín Palacios, Miguel, *Islam and the Divine Comedy*, trans. by Harold Sutherland (London: Routledge, 2008)

Black, Deborah, 'Conjunction and the Identity of Knower and Known in Averroes', *American Catholic Philosophical Quarterly*, 73 (1999), 159–84

———, 'How Do We Acquire Concepts? Avicenna on Abstraction and Emanation', in *Debates in Medieval Philosophy: Essential Readings and Contemporary Responses*, ed. by Jeffrey Hause (New York: Routledge, 2014), pp. 126–44

———, 'Psychology: Soul and Intellect', in *The Cambridge Companion to Arabic Philosophy*, ed. by Peter Adamson and Richard C. Taylor (Cambridge: Cambridge University Press, 2005), pp. 308–26

Burnett, Charles, 'Arabic into Latin: The Reception of Arabic Philosophy into Western Europe', in *The Cambridge Companion to Arabic Philosophy*, ed. by Peter Adamson and Richard C. Taylor (Cambridge: Cambridge University Press, 2005), pp. 370–404

———, 'Coniunctio–Continuatio', in *Mots médiévaux offerts à Ruedi Imbach*, ed. by Iñigo Atucha, Dragos Calma, Catherine König-Pralong, and I. Zavattero (Turnhout: Brepols, 2011), pp. 185–98

Davidson, Herbert A., *Alfarabi, Avicenna, and Averroes, on Intellect: Their Cosmologies, Theories of the Active Intellect, and Theories of Human Intellect* (New York: Oxford University Press, 1992)

de Libera, Alain, *Métaphysique et noétique: Albert le Grand* (Paris: Vrin, 2005)

Gutas, Dimitri, *Avicenna and the Aristotelian Tradition: Introduction to Reading Avicenna's Philosophical Works* (Leiden: Brill, 1988)

———, 'Avicenna's Philosophical Project', in *Interpreting Avicenna: Critical Essays*, ed. by Peter Adamson (Cambridge: Cambridge University Press, 2013), pp. 28–47

———, 'The Empiricism of Avicenna', *Oriens*, 40 (2012), 391–436

———, 'Intuition and Thinking: The Evolving Structure of Avicenna's Epistemology', in *Aspects of Avicenna*, ed. by Robert Wisnovsky (Princeton, NJ: Markus Wiener, 2001), pp. 1–38

Hasse, Dag Nikolaus, 'Avicenna on Abstraction', in *Aspects of Avicenna*, ed. by Robert Wisnovsky (Princeton, NJ: Markus Wiener, 2001), pp. 39–72

———, *Avicenna's 'De anima' in the Latin West: The Formation of a Peripatetic Philosophy of the Soul, 1160–1300* (London: The Warburg Institute, 2000)

———, 'Avicenna's Epistemological Optimism', in *Interpreting Avicenna: Critical Essays*, ed. by Peter Adamson (Cambridge: Cambridge University Press, 2013), pp. 109–19

Herráiz Oliva, Pilar, 'Towards a New Methodology for Natural Philosophy: Latin Averroism Revisited', *Mediterranea*, 6 (2021), 131–55

Kaukua, Jari, 'Avicenna's Outsourced Rationalism', *Journal of the History of Philosophy*, 58 (2020), 215–40

Krause, Katja, *Thomas Aquinas on Seeing God: The Beatific Vision in his Commentary on Peter Lombard's 'Sentences' IV.49.2* (Milwaukee, WI: Marquette University Press, 2020)

Lizzini, Olga, *Fluxus (fayḍ): Indagine sui fondamenti della metafisica e della fisica di Avicenna* (Bari: Edizioni di Pagina, 2011)

McGinnis, Jon, *Avicenna* (New York: Oxford University Press, 2010)

———, 'Making Abstraction Less Abstract: The Logical, Psychological, and Metaphysical Dimensions of Avicenna's Theory of Abstraction', *Proceedings of the American Catholic Philosophical Association*, 80 (2006), 169–83

Miller, Robert, 'An Aspect of Averroes' Influence on St Albert', *Mediaeval Studies*, 16 (1954), 57–71

Minnema, Anthony H., 'Algazel Latinus: The Audience of the *Summa theoricae philosophiae*, 1150–1600', *Traditio*, 69 (2014), 153–215

———, '*Cave hic*: Marginal Warnings in Latin Translations of Arabic Philosophy: A Case Study of Algazel', *Manuscripta*, 61 (2017), 72–140

———, 'A Hadith Condemned at Paris: Reactions to the Power of Impression in the Latin Translation of al-Ghazālī's *Maqāṣid al-falāsifa*', *Mediterranea*, 2 (2017), 145–62

Ogden, Stephen, 'Avicenna's Emanated Abstraction', *Philosophers' Imprint*, 20 (2020), 1–26

Salman, Dominique, 'Albert le Grand et l'averroïsme latin', *Revue des sciences philosophiques et théologiques*, 24 (1935), 38–65

Taylor, Richard C., 'Abstraction and Intellection in Averroes and the Arabic Tradition: Remarks on Averroes, *Long Commentary on the De Anima* Book 3, Comment 36', in *Sujet libre: Pour Alain de Libera*, ed. by Jean-Baptiste Brenet and Laurent Cesalli (Paris: Vrin, 2018), pp. 321–25

———, 'The Agent Intellect as "Form for Us" and Averroes Critique of al-Fārābī', *Tópicos*, 29 (2005), 29–51

———, 'al-Fārābī and Avicenna: Two Recent Contributions', *Middle Eastern Studies Association Bulletin*, 39 (2005), 180–82

———, 'Avicenna and the Issue of the Intellectual Abstraction of Intelligibles', in *Philosophy of Mind in the Early and High Middle Ages*, ed. by Margaret Cameron (London: Routledge, 2019), pp. 56–82

———, 'Cogitatio, Cogitativus and Cogitare: Remarks on the Cogitative Power in Averroes', in *L'élaboration du vocabulaire philosophique au Moyen Âge: Actes du Colloque international de Louvain-la-Neuve et Leuven, 12–14 septembre 1998*, ed. by Jacqueline Hamesse and Carlos Steel (Leuven: Peeters, 2000), pp. 111–46

———, 'Remarks on the Importance of Albert the Great's Analyses and Use of the Thought of Avicenna and Averroes in the *De homine* for the Development of the Early Natural Epistemology of Thomas Aquinas', in *Die Seele im Mittelalter: Von der Substanz zum funktionalen System*, ed. by Günther Mensching and Alia Mensching-Estakhr (Würzburg: Königshausen & Neumann, 2018), pp. 131–48

———, 'Separate Material Intellect in Averroes' Mature Philosophy', in *Words, Texts and Concepts Cruising the Mediterranean Sea: Studies on the Sources, Contents and Influences of Islamic Civilization and Arabic Philosophy and Science*, ed. by Rüdiger Arnzen and Jörn Thielmann (Leuven: Peeters, 2004), pp. 289–309

Tellkamp, Jörg A., 'Why Does Albert the Great Criticize Averroes' Theory of the Possible Intellect?', in *Via Alberti: Texte – Quellen – Interpretationen*, ed. by Ludger Honnefelder, Hannes Möhle, and Susana Bullido del Barrio (Münster: Aschendorff, 2009), pp. 61–78

Tracey, Martin J., 'Albert the Great on Possible Intellect as *locus intelligibilium*', in *Raum und Raumvorstellung im Mittelalter*, ed. by Jan A. Aertsen and Andreas Speer (Berlin: De Gruyter, 1998), pp. 267–303

Verbeke, Gérard, 'L'unité de l'homme: saint Thomas contre Averroès', *Revue Philosophique de Louvain*, 58 (1960), 220–49

Wirmer, David, 'Avempace – "ratio de quidditate": Thomas Aquinas's Critique of an Argument for the Natural Knowability of Separate Substances', in *Wissen über Grenzen: Arabisches Wissen und lateinisches Mittelalter*, ed. by Andreas Speer and Lydia Wegener (Berlin: De Gruyter, 2006), pp. 569–90

JÖRN MÜLLER

# Chapter 9. Is There an Intellectual Memory in the Individual Human Soul?*

## *Albert the Great between Avicenna and Aquinas*

In her seminal study on philosophical conceptions of memory in ancient and medieval times, Janet Coleman drew attention to the close link that exists between memory and knowledge: 'Memory cannot be treated separately from a more inclusive theory of knowing. [...] [A]ncient and medieval theories of memory are intricately linked to an epistemology'.[1] Consequently, any theory of memory which disregards the epistemological background is bound to be incomplete. It seems reasonable to suggest that this equation also holds the other way round: any convincing theory of knowledge has to deal with memory and its contribution to cognition and thought. But the extent to which this second statement is really justified needs clarification, especially in the light of the thirteenth century's debates on intellectual knowledge. There, we encounter a very lively debate about the connection of memory and intellectual knowledge, in which Albert the Great's position seems to be unique, in that it is somehow situated between Avicenna and Thomas Aquinas. This controversy is highly illuminating, also on a larger scale, because it involves some fundamental tenets of epistemology and anthropology that shaped the intellectual physiognomy of the thirteenth century. Furthermore, it testifies to the profound influence of the legacy of Arabic philosophy on the Scholastic era in the Latin West, especially in the case of Albert the Great.

---

* I would like to thank Dag Nikolaus Hasse, Richard Taylor, and Katja Krause for their valuable comments on this paper (although they might still disagree with my reading of Avicenna and Aquinas).

1 Coleman, *Ancient and Medieval Memories*, p. 231.

**Jörn Müller** (joern.mueller@uni-wuerzburg.de) is professor of ancient and medieval philosophy at the Julius-Maximilians-Universität Würzburg. His research focuses on ethics, anthropology, and philosophical psychology.

*Albert the Great and his Arabic Sources*, ed. by Katja Krause and Richard C. Taylor, Philosophy in the Abrahamic Traditions of the Middle Ages, 5 (Turnhout: Brepols, 2024), pp. 253–282
BREPOLS PUBLISHERS 10.1484/M.PATMA-EB.5.136489

The kernel of this debate may be summarized as follows. Whereas Avicenna denies the existence in the human soul of an intellectual memory in which intelligible species are stored, Aquinas vigorously defends the idea of such an intellectual memory. Albert, meanwhile, seems to sit happily on the fence between them, neither subscribing to one of the alternatives wholeheartedly nor rejecting them out of hand.

This question, and especially its development in the thirteenth century, has been somewhat neglected in the scholarly literature on the subject.[2] Avicenna's denial of intellectual memory has certainly been noted, since it figures in the highly contested issue of 'abstractionism' versus 'emanationism' in the interpretation of his epistemology. But the debates in Avicennian scholarship mainly focus on the controversial role and contribution of the transcendent agent intellect as regards the acquisition and subsequent use of intellectual knowledge.[3] Rather than reviewing this thorny issue once again, my purpose here is to ask how Avicenna's outright denial of intellectual memory was received and discussed in thirteenth-century Latin philosophy. I will not go back to the original Arabic sources, but only to their Latin translations and especially their interpretation and critical assessment by Albert the Great and Thomas Aquinas. Thus, Avicenna will mainly be seen through the lens of his later Latin readers. My main concern is the question of how the different reactions of Albert and Aquinas to Avicenna's stance may ultimately be explained. In particular, I am interested in offering a possible explanation for Albert's seemingly wavering attitude vis-à-vis Avicenna's denial of an intellectual memory in the individual soul.

In the following, I will retrace the most important arguments at the heart of this debate concerning intellectual memory and its wider philosophical background, with special emphasis on the epistemological issues involved. I will do so

---

2 Hasse, *Avicenna's 'De anima'*, pp. 186–90, who offers the most comprehensive account of Avicenna's psychology and its influence on the Latin West, outlines the basic positions involved in this debate and notes that it deserves closer study (p. 190). Coleman, *Ancient and Medieval Memories*, pp. 432–35, offers some observations on Aquinas's stance vis-à-vis Avicenna, but does not take into account Albert's 'intermediate' position, which is also crucial for understanding Aquinas's criticism. For an analysis of Aquinas's criticism of Avicenna's denial of intellectual memory, see Müller, 'Memory as an Internal Sense', on which I draw in the second section of this paper.

3 The 'traditional' interpretation (which is present in the medieval Latin tradition and still maintained in recent scholarship, e.g., by Black, 'How Do We Acquire Concepts') sees the direct emanation of forms from the transcendent agent intellect, or the conjoining of the individual human soul with it, as the cornerstone of Avicennian epistemology. The causal role of the agent intellect in effecting intellectual knowledge in the human soul is rather strong in this reading. Criticizing this model, Dag Nikolaus Hasse and Dimitri Gutas have advanced a more naturalistic account, in which the abstraction of intelligible content from sense-data is mainly based on discursive reasoning. In this rather empiricist understanding of Avicennian epistemology, the agent intellect plays a smaller role with regard to the content of intellectual understanding and is reduced to being the ontological locus of intelligibles. For critical surveys of recent developments in this discussion, see Alpina, 'Intellectual Knowledge', pp. 135–42; Taylor, 'Avicenna', pp. 59–72. Despite its promising title, Alpina, 'Intellectual Knowledge', pp. 171–73, has nothing really illuminating to say on the problem of intellectual memory.

in three steps. First of all, I describe the precise core of the debate as delineated in Albert's double-edged treatment of Avicenna's position in his early treatise *De homine*. I then turn to Aquinas's harsh criticism of Avicenna's denial of intellectual memory. Finally, I will boldly enter into more speculative territory and hint at a possible explanation of the different attitudes shown by Albert and Aquinas to Avicenna and the existence of an intellectual memory in the human soul. As becomes clear in the concluding section, these differences may be explained by the underlying epistemological models with which Albert and Aquinas operate.

## The Core of the Debate: Albert on Avicenna's Account of Memory

In his early work *De homine*, completed around 1242, Albert devotes a long section to memory in its two main functional aspects:

(i) as *memoria*, that is, roughly speaking, as the continuous storing of past sensory experiences by images which can be accessed by the mind at will in order to have them present again;
(ii) as *reminiscentia*, that is, as the conscious retrieval or recollection of some memory content which has been 'forgotten' and cannot be accessed directly.[4]

This division roughly corresponds to the ancient Greek notions of *mnēmē* and *anamnēsis*, discussed by Plato and Aristotle in various places.[5] For each concept, Albert wants to clarify as neatly as possible what its definition is, what its objects are, and how it operates. As usual in his oeuvre, he draws on a wide variety of sources from the Graeco-Latin tradition as well as from Arabic writings.

The first point to note here is that, in *De homine*, Albert judges Avicenna to be the best authority in this area, since he has given the most adequate definition of memory overall.[6]

Nonetheless, in the early stages of his project of establishing an accurate understanding of memory and its functions, Albert already hits a stumbling block. The authors whom he terms 'Peripatetic' follow Aristotle's treatise *De memoria et reminiscentia* closely and see memory as a power of the sensitive

4 See Albertus Magnus, *De homine*, ed. by Anzulewicz and Söder (hereafter *De homine*), p. 297, v. 1–p. 312, v. 25. For the main differences between *memoria* and *reminiscentia*, esp. p. 309, vv. 5–74.

5 For Plato, see mainly *Platonis Opera*, ed. by Burnet: *Philebus* 33c–34c; *Meno* 80d–86c; *Phaedo* 72e–77a. Aristotle's account is to be found in one of his *Parva naturalia*, *De memoria et reminiscentia*. For useful overviews of the medieval reception of Aristotle's *Parva naturalia*, see Brumberg-Chaumont, 'Le première reception'; De Leemans, 'Medieval Commentaries'. See also Müller, 'Memory in Medieval Philosophy' on the importance of the division between the Aristotelian and the (neo-)Platonic understanding of memory in the whole development of philosophical theories of memory in the Middle Ages.

6 See Albertus Magnus, *De homine*, p. 301, vv. 40–41: 'Magis propriam [sc. diffinitionem memoriae] autem omnibus his dat Avicenna in VI de naturalibus'.

soul.[7] Avicenna, for example, thinks that memory is the storing-place for the intentions of the estimative power. Both *memoria* and *reminiscentia* are among the five so-called 'internal senses', together with common sense, imagination, and the cogitative/imaginative faculty (in Avicenna's account).[8] These internal senses are post-sensory faculties of the soul, which work on the impressions received by the bodily organs and the five external senses. They are located in the three chambers of the brain (which is why they are sometimes also called 'cerebral senses'). Thus, their operation is dependent on the body; in the case of memory, this means that the representational images with which memory works are basically physical imprints with a cognitive content. That content is based on sense-perception and imagination, explaining why it is not exclusively found in human beings: some higher animals also share in memory.

This 'Peripatetic' picture of memory as an internal sense was very much shaped by the Arabic thinkers and their reading and further development of Aristotle's philosophical psychology. It discernibly clashes with the Augustinian tradition of memory as one aspect of the human mind (*mens*), alongside intellect and will.[9] According to Augustine, the human mind is an image of the Trinity in its three forms of *intelligentia, memoria,* and *voluntas*. This means that memory is a power of the mind, belonging to the highest part of the human being. If one takes Augustine seriously and transfers his conception to faculty psychology, as endorsed by many in the thirteenth century, memory must be a power of the rational or intellectual soul — and not of the sensitive soul, as 'Peripatetic' authors such as Avicenna suggest.

Indeed, the legacy confronting Latin authors at the dawn of Scholasticism was twofold: on the one hand, the Augustinian idea of memory as an integral aspect of a theory of mind; on the other, the Peripatetic (or Arabic) view of memory as a subject of natural philosophy, which included the Galenic physiological anatomy of memory. Memory thus seemed to be a kind of *duplex memoria*: sensitive and intellectual memory.[10] As can easily be imagined, the tendency to harmonize

---

7 In his *De memoria et reminiscentia*, II.1, ed. by Donati, p. 124, vv. 9–10, Albert names Alexander of Aphrodisias, Themistius, Alfarabi, Avicenna, and Averroes. On the programmatic character of this appeal in Albert's project of constructing a 'Peripatetic' system of natural sciences, see Müller, *Albertus Magnus über Gedächtnis*, pp. 13–18.

8 On the different internal senses and the complex history of their conceptualization, see the seminal paper by Wolfson, 'The Internal Senses', and, more recently, Di Martino, *'Ratio particularis'*, who focuses on the connection between Arabic and later Latin writers. Avicenna's basic account of memory as an internal sense is given in his *Liber de anima*, I.5, IV.1, and IV.3; see also *Avicenna's Psychology* [*Kitāb al-Najāt*], ch. 3, trans. by Rahman, pp. 30–31. For closer analyses of this account, see Gätje, 'Gedächtnis und Erinnerung'; Coleman, *Ancient and Medieval Memories*, pp. 341–62; Bloch, *Aristotle on Memory*, pp. 145–53.

9 See especially books VIII–XV of Augustine's *De Trinitate*. The most comprehensive account of memory is found in his *Confessiones*, XX.11–38. For useful overviews of Augustine's theory of memory, see Mourant, *Saint Augustine on Memory*; O'Donnell, 'Memoria'.

10 On the manifold notions of memory in medieval philosophy, see Di Martino, '*Memoria dicitur multipliciter*'; Müller, 'Memory in Medieval Philosophy'.

these conflicting strands of thought by talking indiscriminately of memory as a storing-place of both sensitive and intelligible forms was strong, and dominated approaches to memory in the first half of the thirteenth century.[11]

I cannot go into detail here on how this hidden clash of traditions is handled by Albert, who explicitly recognizes these two different conceptions of memory in his *De homine*.[12] The crux is that Albert does not harmonize them at all costs, as many of his predecessors did, but instead tries to separate the philosophical (Aristotelian) from the theological (Augustinian) discourse on memory as tidily as possible, because he thinks that the use of the notion 'memory' with regard to sensitive and intelligible forms is ultimately 'equivocal'.[13] In his view, the true meaning of memory (*memoria proprie dicta*) is the Aristotelian one, which is most accurately interpreted by the Arabs.[14] Consequently, Albert treats the Augustinian notion of memory, as an image of the Trinity, in a separate section of *De homine*.[15]

This split between Aristotle's and Augustine's notions of memory provides the general background to the discussion about the possibility of an intellectual memory in the Scholastic era. From an Augustinian perspective, such a storing-place for intelligibles is not only possible but indispensable. From a Peripatetic one, however, it is far from clear how a power of the sensitive soul can be the subject of intelligibles at all. Albert puts the point succinctly in a neat syllogism:

> *Major premise*: No power of the sensitive soul stores intelligibles (because these powers take up their species with material attachments, and thus do not deal with a purely intelligible content).
> *Minor premise*: Memory is a power of the sensitive soul (since some higher animals share it).
> *Conclusion*: Memory cannot store intelligibles.[16]

---

11 See Coleman, *Ancient and Medieval Memories*, chs. 16–17, and Müller, *Albertus Magnus über Gedächtnis*, pp. 48–50. Compare also Albert's highly critical judgement on his predecessors, which points in the same direction; see below, especially the citations in notes 71 and 73.

12 See Albertus Magnus, *De homine*, p. 302. The most comprehensive treatment of Albert's dealings with memory throughout his work is provided by Anzulewicz, '*Memoria* und *Reminiscentia*'; see also Coleman, *Ancient and Medieval Memories*, ch. 19. On Albert's innovative conception of artificial memory, see Yates, *Art of Memory*, pp. 62–69; Carruthers, *Book of Memory*, pp. 172–79 and 345–60.

13 Albertus Magnus, *Summa theologiae*, I.3.15, ed. by Siedler, p. 69, vv. 25–28: 'dicendum, quod memoria aequivoce dicitur [...] ad thesaurum formarum intelligibilium et thesaurum formarum sensibilium'.

14 See Albertus Magnus, *De ecclesiastica hierarchia*, III, ed. by Burger, p. 78, v. 40–p. 79, v. 9. Albert also praises the superiority of the Arabic approach to memory as an internal sense in the prefaces to the two treatises of his commentary on Aristotle's *De memoria et reminiscentia*, I.1, ed. by Donati, p. 113, v. 7–p. 115, v. 16; II.1, p. 124, v. 4–p. 125, v. 35.

15 See Albertus Magnus, *De homine*, p. 547, v. 37–p. 550, v. 47.

16 See Albertus Magnus, *De homine*, p. 298, vv. 27–33: 'Nulla virtus animae sensibilis est conservans intelligibile; memoria est virtus animae sensibilis; ergo non est conservans intelligibile. Prima patet per hoc quod omnis virtus animae sensibilis, ut dicit Avicenna, accipit speciem cum appendiciis materiae. Secunda vero probatur per hoc quod memoria invenitur in brutis, in quibus non est nisi anima sensibilis'.

Albert names Avicenna as the one who pinpointed this problem, but defers an explicit discussion of it to a later treatment in *De homine*.[17] The debate is then resumed in his discussion of the speculative intellect, when he asks how the intelligible species are present in the human soul. The background to this question is the famous claim by several Arabic interpreters of Aristotle's *De anima* that the intellect is one and the same for all human beings. This thesis of the 'unity of the intellect' (often somewhat misleadingly called 'monopsychism') is firmly rejected by Albert.[18] But although he is a staunch defender of the intrinsic nature of the intellect, he recognizes that there are some serious epistemological problems that have to be solved in this area. One of them directly touches upon the question of whether there is a memory of intelligibles. Its exact phrasing is: 'Does the habit of the speculative intellect [1] stay in it [*sc.* in the speculative intellect] after its consideration, or [2] does it stay in some form of memory which is a part of the rational soul, or [3] does it altogether not stay in the rational soul?'[19] The tripartite form of his question already presents the three alternatives to be discussed. Again, Albert references Avicenna as his basic source text, specifically the *Liber de anima*,[20] the arguments of which he reproduces faithfully. Let us take a closer look at them.

One major argument in Avicenna's position against the idea of memory as a depository for intelligibles is based on a feature of Aristotle's conception of the phenomenon. Aristotle believes that remembering always involves images of past experiences, which implies that every item present in memory necessarily has a more or less precise time index. Yet nothing that is universal is perceived as being at a determinate time. Since particular memories lack the universality that is characteristic of intelligibles, they cannot be intelligibles.[21] Hence, it comes as no surprise to find Avicenna rejecting the idea that sensitive memory might be a suitable location of intelligibles. This line of reasoning is also strengthened on

17 See Albertus Magnus, *De homine*, p. 299, vv. 28–30: 'Item, Avicenna in VI de naturalibus probat in rationali anima non esse memoriam. Et quia de hoc infra erunt quaestiones, rationes eius infra ponentur'.

18 See Albertus Magnus, *De homine*, p. 436, v. 46–p. 438, v.13 ('Utrum unus et idem numero intellectus speculativus sit in omnibus animabus rationalibus'). It should be noted that in *De homine*, Albert misunderstands Averroes and mainly criticizes Avicenna for positing a separate intellect outside the human soul; see Taylor, 'Remarks on the Importance', pp. 139–47.

19 Albertus Magnus, *De homine*, p. 439, vv. 45–49: 'Quinto et ultimo quaeritur, utrum habitus intellectus speculativi post considerationem manet in ipso, vel in memoria aliqua quae sit pars animae rationalis, vel omnino non manet in anima rationali. Haec est quaestio Avicennae in VI de naturalibus'.

20 See Avicenna, *Liber de anima*, V.6, ed. by Van Riet, pp. 146–50. Since the Latin translation is the source text used by Albert (and Thomas Aquinas), I cite from this edition and not from the Arabic text edited by Rahman.

21 See Albertus Magnus, *De homine*, p. 440, vv. 34–46: 'Item, tempus est una condicionum particularis; universale autem remotum est ab omnibus condicionibus particularis; ergo universale remotum est a differentia temporis. [...] Inde sic: Omnis memoria est cum differentia temporis determinata; nullum universale est cum differentia temporis; ergo nullum universale est in memoria. Intellectus autem tantum est universalis; ergo nulla memoria est in intellectu'.

another level. Nothing which is or involves a body is a candidate for the deposition of intelligibles. Since memory is a psychic function working in and through the posterior ventricle of the brain, it can be disqualified as a storing-place for intelligibles.

Avicenna also considers another candidate. The intelligibles might be stored in the essence of the soul, that is, the intellect, which does not operate through any bodily organ. Aristotle's famous dictum that the intellective soul is 'the location of forms' (*topos eidōn*) seems to be a clear hint to be taken up here.[22] But Avicenna is not satisfied with this solution either, pointing out that for an intelligible form, 'to be in the soul' means nothing other than to be known by the soul in actuality.[23] Consequently, any form that is not actually contemplated by the possible intellect is not truly present there. Therefore, at the very moment when the intellect turns away from the intellection of one thing towards another object, the intelligible form of the thing ceases to be actually in the soul. There is no depository for it in the sense of an intellectual memory to which it might be transferred after its actual cognition or when contemplation has ended. Intelligibles exist in the soul only when the soul thinks them in complete actuality.[24]

Nonetheless, Avicenna realizes, as does Albert in his reconstruction, that this solution prompts a counterquestion: If there is no depository for intelligibles in our intellect which is comparable to the memory of the sensitive soul, does this not mean that we have to learn everything afresh all the time? At one moment, I grasp the truth of a mathematical theorem by actually contemplating it, but the moment I turn my mind away from it, nothing seems to be left of it. Consequently, the next time I encounter this problem, I will have to go through the same stages of cognition as before. If learning means something like 'acquiring knowledge to have it readily at hand for future use', Avicenna's denial of intellectual memory seems to strand us in a situation not unlike the one suffered by Sisyphus: every time we have rolled the stone up the hill, it inevitably rolls down again the moment we switch our mind to something else — and we have to fetch it once more in the same manner as before. How is any intellectual progress by learning possible in this model? As Richard Taylor puts it, one of the major challenges to Avicenna's epistemology seems to be

> how, given Avicenna's denial of intellectual memory to the human soul, there can be a connection of the actualized human intellect with the separate Agent

---

22 See Aristotle, *De anima*, III.4, 429a27–29.

23 See Avicenna, *Liber de anima*, V.6, ed. by Van Riet, p. 148: 'formam enim intellectam esse in anima hoc est quod apprehendi eam'.

24 See ibid., p. 147: 'Impossibile est enim dici hanc formam esse in anima in effectu perfecte, et non intelligi ab ea in effectu perfecte; sensus enim de hoc quod eam intelligit non est nisi quia forma existit in ea; unde impossibile est corpus esse thesaurum eius. Et impossibile est etiam animae esse thesaurum eius: hoc enim quod est thesaurus eius nihil aliud est nisi quia forma intellecta existit in ea et nos sic intelligimus. Non est autem sic apprehensio formae; apprehendere etenim formam non est thesauri sed tantum retinere'.

> Intellect to allow human access to the ontologically distinct intelligibles in the Agent Intellect to permit the individual human soul the ability to recall abstracted intelligibles at will.[25]

Avicenna's answer to this challenge involves his basic model of different modes or stages of the intellect and their functioning in intellectual cognition. The possible intellect acquires its intelligible species by turning to the agent intellect, from which these forms emanate into it. This is not to be construed, as it has often been in the past, as a simple form of illumination, because it involves the abstraction of the intelligible content from its bodily attachments still present in the phantasms. Rather, looking at these particulars in the light of the agent intellect 'disposes the soul for something abstracted to flow upon it'.[26]

This intertwining of abstraction and emanation is too complex for me to do it full justice here,[27] but the gist of Avicenna's analysis seems to be that after the possible intellect has cognized an intelligible thing for the first time, it is more apt or suited towards possible iterations of that cognition. To learn something for the first time is like having one's eye cured from an illness so that it can afterwards look at the object without any effort whenever one wants.[28] Applied to the realm of intellectual learning, this means that when one knows the intelligibles, one's soul is in such a state that it can make their form present whenever one wishes. The human soul has acquired the perfect disposition for being united with the external agent intellect, from which the abstracted form easily flows into the possible intellect. Avicenna calls this stage 'the acquired intellect' (*intellectus*

---

25 Taylor, 'Avicenna', p. 74.

26 Avicenna, *Liber de anima*, V.5, ed. by Van Riet, p. 127.

27 For the divide in contemporary scholarship between the emanationist and abstractionist interpretation of Avicenna, see note 3 above. For a short review of the recent debate and an alternative view trying to avoid the potential pitfalls of this antagonism, see also Hasse, 'Avicenna's Epistemological Optimism'. Hasse's main idea, that emanationism is the response to the problem of intellectual memory in Avicenna, seems to me basically correct, but his overall account still faces some problems, as is rightly pointed out by Alpina, 'Intellectual Knowledge', pp. 141–42, and Taylor, 'Avicenna', pp. 68–70. Generally, I agree with Taylor that the idea of making the separate agent intellect into a kind of collector or storehouse of intelligible forms abstracted by individual human souls clashes with its description as a pure act without potency. Taylor, 'Avicenna', pp. 72–76, tries instead to resolve this issue by turning to Themistius, of whose interpretation of Aristotle's *De anima* Avicenna was probably aware.

28 See Avicenna, *Liber de anima*, V.6, ed. by Van Riet, p. 149: 'Aptitudo autem quae praecedit discere est imperfecta; postquam vero discitur est integra. Cum enim transit in mentem eius qui discit id quod cohaeret cum intellecto inquisito et convertit se anima ad inspiciendum [...], solet anima coniungi intelligentiae et emanat ab ea virtus intellectus simplicis, quem sequitur emanatio ordinandi. Si vero avertitur a primo, fiunt ipsae formae in potentia, sed in potentia proxima. Ergo primum discere est sicut curatio oculi, qui, factus sanus, cum vult, aspicit aliquid unde sumat aliquam formam; cum vero avertitur ab illo, fit illud sibi in potentia proxima effectui'.

*adeptus*),[29] an important concept for both Avicenna and Albert, and one to which I will return in some detail later in this paper.

This Avicennian model of learning and knowing differs considerably from our customary understanding of the process. We commonly think of intellectual learning as the acquisition of a habitual 'knowing that', a kind of propositional knowledge, which latently stays in us until we activate it from within. The focus here is on the actual or potential mode of being of the intelligible content. Avicenna's model, in contrast, concentrates on the power of the intellect. The intellect does not acquire a long-standing 'knowing that' but rather a kind of 'knowing how', namely the ability to connect easily with the agent intellect in order to get the intelligibles from there whenever the agent wishes.[30] Indeed, Avicenna distinguishes different stages of potentiality in the intellect itself, ranging from pure possibility to a power for actualization of intelligibles that is close to actual cognition of the intelligible content.[31] Through repeated processes of learning and knowing, one climbs up this ladder of different levels of the intellect, and this, in Avicenna's view, is sufficient to guarantee the fundamental difference between learning or acquiring knowledge for the first time and using that knowledge afterwards. If we conceive of learning as acquiring a dispositional 'knowing how' rather than as a habitual 'knowing that', there is no need for a depository in the human soul which contains or stores former cognitions in their propositional content. The human intellect just uses its newly acquired ability to turn to the transcendent and universal agent intellect, which functions here as a kind of external 'hard disk'

---

29 See ibid., pp. 148–50: 'Restat ergo ut ultima pars sit vera, et ut discere non sit nisi inquirere perfectam aptitudinem coniungendi se intelligentiae agenti, quousque fiat ex ea intellectus qui est simplex, a quo eminent formae ordinatae in anima mediante cogitatione. [...] Cum enim dicitur Plato esse sciens intelligibilia, hic sensus est ut, cum voluerit, revocet formas ad mentem suam; cuius etiam sensus est ut, cum voluerit, possit coniungi intelligentiae agenti ita ut ab ea in ipsum formetur ipsum intellectum, non quod intellectum sit praesens suae menti et formatum in suo intellectu in effectu semper, nec sicut erat priusquam disceret. Hic enim modus intelligendi in potentia est virtus quae acquirit anima intelligere cum voluerit; quia, cum voluerit, coniungetur intelligentiae a qua emanat in eam forma intellecta. Quae forma est intellectus adeptus verissime et haec virtus est intellectus in effectu'.

30 The explicit distinction between 'knowledge-how' and 'knowledge-that', which I use in a rather broad sense here, goes back to Gilbert Ryle and is one of the most interesting battlegrounds of contemporary analytic philosophy. See Fantl, 'Knowledge How', for further details on the current state of the debate. I use the expression 'knowing how' in the sense of the 'ability account' (one of the contenders in this area): to know how to do something — e.g., to ride a bicycle — does not denote a propositional knowledge about it, but simply the possession of the dispositional ability to do it. It is an open question in contemporary philosophy whether this kind of 'knowledge-how' is somehow dependent on 'knowledge-that' after all. In Avicenna's case, the acquisition and use of the 'acquired intellect' certainly involves intellectual cognition and perhaps even quasi-propositional content in the sensitive part of the soul (as Katja Krause has suggested to me in a personal communication). But this does not alter the fact that the acquired intellect itself does not consist in having propositional 'knowledge-that' but in the power to tap in easily to the agent intellect and get the intelligibles directly from it.

31 For Avicenna's discussion of potentiality in general, see *Liber de anima*, I.5, ed. by Van Riet, pp. 95–96.

already containing all intelligible forms; after the first successful act of abstraction, it can be accessed at will by the individual human soul and subsequently delivers the corresponding intelligible form 'on demand'.

It is quite telling that Averroes, although differing from Avicenna's theory of intellectual knowledge in several important respects, ultimately subscribes to this denial of intellectual memory in any individual human being.[32] If one assumes an external 'depository' of intelligibles — like the transcendent agent intellect in the Arabic interpretation of Aristotle's *nous* — there is, strictly speaking, no need for a storing-place for them in the human mind. The denial of intellectual memory in man is thus closely tied to the controversial thesis of the unity of the intellect that was so hotly debated by Latin thinkers in the thirteenth century.[33]

How, then, does Albert judge this Avicennian attempt to dispose of the habitual knowledge of acquired intelligibles? As emerges from his *De homine*, he is in complete agreement with Avicenna with regard to the rejection of intellectual memory in the literal meaning of the expression.[34] The rational soul does not possess a memory in the same sense that the sensitive soul does, because of the epistemological difficulties discussed above; one should therefore keep the Augustinian notion of memory as a part of our mind tidily apart from the Aristotelian concept under discussion here.[35] Albert agrees with Avicenna that memory in its basic Aristotelian understanding is not a suitable place for storing

32 See Averroes, *Commentarium magnum*, III.20, ed. by Crawford, pp. 449–51. For Averroes's account of memory and its epistemological background, see Coleman, *Ancient and Medieval Memories*, ch. 18.

33 This connection is already visible on the surface of some of the texts. In Albert's *De homine*, the discussion of Avicenna's position (p. 439, v. 40–p. 442, v. 30) follows closely upon his rejection of the unity thesis (p. 436, v. 46–p. 438, v. 13). The same pattern can be observed in Aquinas: *Summa contra gentiles*, II.73, ed. Leonina, pp. 459–63 ('Quod intellectus possibilis non est unus in omnibus hominibus'), is followed by his rejection of Avicenna's denial of intellectual memory in II.74 (ibid., pp. 469–70). See also Aquinas, *Summa theologiae* (hereafter *STh*) I.79, a. 5, ed. Leonina, p. 269 (unity of the intellect), and a. 6, pp. 270–01 (intellectual memory). On the general influence of Albert's *De homine* on the development of Aquinas's natural epistemology concerning the understanding of the intellect and the intelligibles in act, see Taylor, 'Remarks on the Importance'.

34 See Albertus Magnus, *De homine*, p. 441, vv. 65–72: 'Solutio: Sine praeiudicio aliorum dicimus quod anima rationalis proprie loquendo non habet memoriam. Et si Augustinus dicat quod memoria est pars imaginis, ipse accipit memoriam prout est praeteritorum, praesentium et futurorum, ut supra diximus [...]. Sententia autem Avicennae in hoc quod non est habere memoriam animam rationalem, est eadem nobiscum'. This is already anticipated in Albert's earlier discussion of whether memory belongs to the rational or the sensitive soul, where he states (ibid., p. 303, vv. 16–20): 'Dicatur ergo ad primum quod virtus animae sensibilis non est conservans intelligibile per se, sed per accidens. Ea vero quae sequuntur et probant memoriam esse partem sensibilis animae, concedimus'. This concession also explicitly pertains to Avicenna's position that there is no memory in the rational soul (see ibid., p. 299, vv. 28–30). In *De homine*, Albert even sees recollection as a power of the sensitive soul, differing from memory only in its use of reason, not in its psychological seat (*in subiecto*; ibid., p. 310, vv. 54–56). He also maintains this idea in his later commentary on Aristotle's *De memoria et reminiscentia*; see Müller, *Albertus Magnus über Gedächtnis*, pp. 38–42.

35 See Albertus Magnus, *De homine*, p. 300.

intelligible contents, but takes issue with his outright denial of the possibility that the intelligibles might be conserved in the determinate particular intellect itself.

This conclusion rests on a premise from Avicenna's *Liber de anima* not accepted by Albert, namely, that there is a fundamental difference between powers that are apprehensive and powers that are retentive. This is a pattern already visible in Avicenna's description of the five internal senses. There is always a correlation between a power that is apprehensive, that is, cognizing, and another that is retentive, that is, storing the cognitional content as a depository (*thesaurus*). Analogously, imagination is the depository of common sense and memory is the depository of estimation.[36] These corresponding powers display a kind of division of labour which defines their activities of apprehending and storing the forms. That division, though, also limits their functional capabilities. According to Avicenna, a retentive power is not able to apprehend its actual content on its own, while an apprehensive power is not able to store or conserve it in itself.

Albert, in contrast, thinks that this strict functional separation applies only to bodily powers or to psychic faculties which rely on the body, and does not apply to the human mind insofar as it is separated from the body. For Albert, the possible intellect is capable of receiving intelligible species (that is, of apprehending or knowing them) and, at the same time, of retaining or storing them as available knowledge. Contrary to in the physical realms, there is no contradiction here, since the intellect is not the location of the intelligible species in a literal — spatial — sense, but rather their subject in the sense of an underlying matter informed by the cognition of the intelligible thing.[37]

Albert's solution to the problem is not an argument designed to strike down Avicenna at all costs. Rather, it betrays a certain uneasiness with the Avicennian epistemology he has presented. Albert takes issue mainly with the status of the intelligible and the idea that it flows from the agent into the possible intellect, which means that the intelligible must somehow already exist in the agent intellect. Albert's initial question is whether the agent intellect brings this intelligible into existence by performing an act of abstraction. In that case, the agent intellect itself would have to receive something new; it would acquire this form afresh.

36 See Avicenna, *Liber de anima*, IV.1, ed. by Van Riet, pp. 8–10.

37 See Albertus Magnus, *De homine*, p. 441, v. 72–p. 442, v. 17: 'Et ipse distinguit inter thesaurum formarum et virtutem apprehensivam, dicens quod thesauri non est nisi retinere formam et non apprehendere, et propter hoc forma non est in thesauro sicut in virtute apprehensiva; sed virtutis apprehensivae est apprehendere formam et non retinere. Et propter hoc dicit quod species intelligibilis non retinetur in intellectu possibili, quia ipse est virtus apprehensiva, sed ex conversione sui ad intellectum agentem generatur in ipso, cum actualiter considerat. Nos autem dicimus quod manet in intellectu possibili, eo quod Aristoteles expresse dicat quod memoria et reminiscentia habent suos actus apprehensionis. Unde falsum est quod thesauri non sit apprehendere. In virtutibus enim corporalibus alterius quidem virtutis est recipere et alterius retinere; [...] Sed in intellectuali virtute eiusdem virtutis est recipere et retinere, eo quod oppositorum actus ibi non sunt oppositi, cum sint separata opposita a materia et potentia agendi et patiendi. Unde intellectus possibilis recipit formas intelligibilium et retinet eas'. On the significance of this passage, see also Taylor, 'Remarks on the Importance', pp. 146–47.

Yet this would mean that it was in potency towards this form beforehand, which clashes with the idea that the agent intellect is in no way passive or potential, but is universally active. So might the forms always exist actually in the agent intellect? This would amount to a model that comes close to a Platonic notion of preexistent knowledge only actualized by acts of sense cognition; learning is nothing other than remembering what one already knows — the crucial idea of Platonic *anamnēsis*. But in that case, there would be absolutely no need for any kind of abstraction of the intelligible species from the *phantasmata*, which seems to be a regular element of Avicennian epistemology. Thus, there seems to be an impasse here, since the ontological status of the intelligible species cannot be explained convincingly (at least according to Albert's view).[38]

To be sure, some of these problems arise from the fact that Albert regards the agent intellect as an internal power of the soul and thus fundamentally disagrees with Avicenna's model, which externalizes it as a source of knowledge accessible to all human beings. However, his reservations regarding Avicenna's outright denial of intellectual memory show that Albert spots some general problems in Avicenna's intermingling of an abstractionist and an emanationist view of the intelligibles. We will return to this issue in more detail below, in the context of Albert's later commentary on Aristotle's *De anima*.

To sum up the discussion thus far: In his early *De homine*, Albert shares the concerns voiced by Avicenna against making sensitive memory a storehouse for intelligibles, and generally argues against the conflation of Aristotelian and Augustinian notions of memory. He does not think that the 'knowing how' of Avicenna's possible intellect is a completely adequate substitute for the stored 'knowing that' of acquired intelligibles. But he also does not intend to completely reject Avicenna's denial of intellectual memory, and rather betrays a general dissatisfaction with some features of Avicennian epistemology on which this position rests. Thomas Aquinas is much more radical here. He takes, so to speak, the bull by the horns.

38 Albertus Magnus, *De homine*, p. 441, vv. 25–36: 'Sed contra [sc. Avicennam]: Illa intelligibilia quae emanant ab intelligentia agente, aut sunt abstracta a phantasmatibus, aut sunt semper in intellectu agente. Si primo modo, tunc oportet quod in aliquo recipiantur sicut in subiecto, cum abstracta sunt; hoc autem non est intellectus agens, eo quod ipse non sit in potentia; sed est agens universaliter; igitur addiscere non erit tantum acquirere aptitudinem convertendi se ad intellectum agentem, sed etiam acquirere formam intelligibilem. Si secundo modo, tunc nulla formarum intelligibilium acquiritur per intellectum, et sic nihil videntur prodesse phantasmata quae accipiuntur per sensus'.

## Aquinas's Criticism of Avicenna's Denial of Intellectual Memory

Aquinas acknowledges several times that Avicenna's denial of intellectual memory was taken up approvingly by some earlier authors,[39] but he criticizes this idea throughout his whole oeuvre, in his philosophical commentaries as well as his theological works. Aquinas's preoccupation with this matter is well reflected by the sheer number of treatments to be found in his writings. In at least ten different passages throughout his work, Aquinas carefully reconstructs and rigorously rebuts Avicenna's denial of intellectual memory, in some cases discernibly inspired by Albert's doubts concerning its consistency.[40] Aquinas's arguments can be summarized into three categories: Avicenna's position defies (1) reason, (2) Aristotle, and (3) the Catholic faith.

(1) First of all, Aquinas sees a kind of irrationality (or rather incoherence) in Avicenna's denial of an intellectual memory. The intellect is a higher potency than the faculties of the sensitive soul; above all, it is infinitely more stable and fixed than they are. This is aptly expressed in the fact that the intellect remains intact after our physical death whereas the sensitive powers perish. But, as we are constantly reminded by Aquinas, whatever is received, is received in the manner of the receiving subject. Therefore, the intelligibles are received in the intellect according to its nature, and this means in the case of the intellect: in a very stable and fixed way.[41] Is it not strange to credit some lower psychic powers such as imagination and memory with a strong continuity in the storing of their objects while denying it to the intellect, the highest faculty of the soul? In this area, Aquinas also takes up Albert's criticism of Avicenna that it is not necessary to split the apprehensive and the retentive powers in the case of the possible intellect. Rather, the possible intellect can receive and know the intelligible species as well as store them.[42] Aquinas points out that Avicenna's distinction between 'apprehending' and 'storing' powers makes sense as long as it is applied only to the

39 See especially Dominicus Gundissalinus, *De anima*, X, ed. by Muckle, pp. 93–98, who quotes at length Avicenna's *De anima*, V.6.

40 These are (1) Aquinas, *In IV Sent.*, L.1.2; (2) *Summa contra gentiles*, II.74; (3) *Super I ad Corinthos*, XIII.3; (4) *Sentencia libri de anima*, III.2; (5) *Sentencia libri de sensu et sensato*, II.2; (6) *De veritate*, X.2; (7) *De veritate*, XIX.1; (8) *Quodlibet* III.9.1; (9) *STh*, I.79.6; (10) *STh*, I–II.67.2. These are also the basic texts for the following reconstruction of his arguments, an abridged version of which is to be found in Müller, 'Memory as an Internal Sense'. For Aquinas's account of memory in general, see Coleman, *Ancient and Medieval Memories*, ch. 20.

41 See Aquinas, *Quodlibet*, III.9.1, ed. Leonina, p. 279: 'Quod enim species intelligibiles in intellectu possibili non conseruentur, est contra rationem; quod enim in aliquo recipitur, est in eo per modum recipientis; unde, cum intellectus possibilis habeat esse stabile et inmobile, species intelligibiles oportet quod in eo recipiantur stabiliter et inmobiliter'.

42 See Aquinas, *Summa contra gentiles*, II.74, ed. Leonina, p. 470: 'Apprehendere igitur et conservare, quae in parte animae sensitivae pertinent ad diversas potentias, oportet quod in suprema potentia, scilicet in intellectu, uniantur'.

powers of the sensitive soul which operate on a bodily basis,[43] though it fails to grasp the peculiarities of the intellect in its immaterial character.[44] But how does the possible intellect store the intelligibles?

(2) In order to answer that question, we have to turn to the second group of arguments that Aquinas hurls at Avicenna. They boil down to the accusation that Avicenna has not completely understood Aristotle with regard the ontological status of the intelligible species. Aquinas stresses the frequently repeated Aristotelian formula that the intellectual soul is the 'place of forms' and adds that this is to be understood in terms of potentiality and actuality.[45] When something is presently known, the form actualizes the possible intellect by being itself in an actual state; the moment the mind turns away from it, the intelligible species does not vanish completely — as Avicenna believed — but stays in the mind in a potential manner. In this context, Aquinas frequently emphasizes a distinction made by Aristotle in *De anima* (III.4, 429b5–9), according to which someone who has actually learnt something is in a different state of potentiality than she was before.[46] One has acquired a habit of knowledge (*habitus scientiae*) which is halfway between pure possibility and full actualization.[47] This stored knowledge is already a first actuality which is close to being realized fully, and the second realization is not dependent on something from outside the soul but can be accomplished within. The mind is now ready and able to operate through itself by means of the stored intelligibles. It is a pervasive line of thought in Aquinas that memory in all its forms (sensitive or intellectual) is to be understood not as a psychic power in its own right, but as a habit or disposition that is acquired

43 See Aquinas, *De memoria et reminiscentia*, II, ed. Leonina, pp. 109–10.

44 See Aquinas, *Summa contra gentiles*, II.74, ed. Leonina, pp. 469–70.

45 See Aquinas, *Super I ad Corinthos*, XIII.3, Index Thomisticus, p. 386: 'Secundo est contra auctoritatem Aristotelis in III de anima, qui dicit quod "cum intellectus possibilis est sciens unumquodque, tunc etiam est intelligens in potentia". Et sic patet quod habet species intelligibiles per quas dicitur sciens, et tamen adhuc est in potentia ad intelligendum in actu, et ita species intelligibiles sunt in intellectu possibili, etiam quando non intelligit actu. Unde etiam Philosophus dicit quod anima intellectiva est locus specierum, quia sc. in ea conservantur species intelligibiles'.

46 Aquinas combines this idea with a doctrine of abstraction of intelligibles from sensory perception, which is not explicit in Aristotle but rather is one interpretation of his text, first proposed by Alexander of Aphrodisias. This 'abstractionism' was conveyed through Alexander to the Arabic tradition and the Latin West; Aquinas just takes it for granted as an established tenet of Aristotle's psychology. For further background on abstraction, see Taylor, 'Epistemology of Abstraction'.

47 See Aquinas, *Sentencia libri De anima*, III.2, ed. Leonina, p. 209: 'Contra quod [sc. positionem Avicennae] hic manifeste Philosophus dicit quod intellectus reductus in actum specierum per modum quo sciencia actus est, adhuc est potencia intellectus; cum enim intellectus actu intelligit, species intelligibiles sunt in eo secundum actum perfectum, cum autem habet habitum scienciae, sunt species in intellectu medio modo inter potenciam puram et actum purum'. See also Aquinas, *In IV Sent.*, L.1.2 ad 5, Index Thomisticus, p. 704: 'Ad quintum dicendum, quod in intellectu nostro remanent formae intelligibiles, etiam postquam actu intelligere desinit; nec in hoc opinionem Avicennae sequimur. Remanent autem hujusmodi formae intelligibiles in intellectu possibili, cum actu non intelligit, non sicut in actu completo, sed in actu medio inter potentiam puram et actum perfectum, sicut etiam forma quae est in fieri, hoc modo se habet'.

by repetition and enables the same operations afterwards. Intellectual memory is therefore not a different power from the possible intellect, but nothing other than the intellect disposed in a certain way by the presence of potentially known intelligibles which may be activated again at will.[48] Aquinas calls this the 'habitual intellect' (*intellectus in habitu*).[49]

Aquinas's solution comes down to the idea that the possible intellect retains the intelligible species as second potentialities — that is, first actualities. Aquinas seems to wonder why Avicenna did not simply apply his own theory of different degrees of potentiality to the question at hand, since this would have led him in the right direction. Actually, Avicenna does so in a way, but he restricts his application to the knowing subject, that is, the possible intellect and its different forms of relation towards the knowable object; he does not use it to qualify the ontological status of the known intelligible in the human soul, as Aquinas suggests.

This limitation provokes another criticism by Aquinas, one that seems somewhat strange at first glance: he accuses Avicenna of sharing the Platonic view on the matter.[50] This comes as a surprise, because Avicenna himself discusses the Platonic position in his denial of intellectual memory in his *Liber de anima*,[51] but explicitly rejects it in favour of his own rival model. Aquinas recognizes this to a certain extent, but he sees a substantial epistemological agreement between Plato and Avicenna. In both cases, in every act of cognition by the possible intellect, the intelligible content is supplied from outside the human soul. In Plato's case, the outside source are several transcendent ideas, while in Avicenna's epistemology, these ideas are replaced by one agent intellect.[52] The immediate drawback of Avicenna's conception of intellectual memory as a realization of an acquired 'knowing how' instead of recalling a previous 'knowing that' also becomes apparent here. According to Aquinas, this conception means that by coming to know one science, we will have learned all of them, due to the unity of

48 See Aquinas, *STh*, I.79.7.

49 See Aquinas, *STh*, I.79.6, ad 3. For his reading of the Arabic theory of the four intellects, see *STh*, I.79.10.

50 See Aquinas, *Summa contra gentiles*, II.74, ed. Leonina, p. 469: 'Sed si diligenter consideretur, haec positio [scil. Avicennae], quantum ad originem, parum aut nihil differt a positione Platonis. Posuit enim Plato formas intelligibiles esse quasdam substantias separatas, a quibus scientia fluebat in animas nostras. Hic [scil. Avicenna] autem ponit ab una substantia separata, quae est intellectus agens secundum ipsum, scientiam in animas nostras fluere. Non autem differt, quantum ad modum acquirendi scientiam, utrum ab una vel pluribus substantiis separatis scientia nostra causetur'.

51 See Avicenna, *Liber de anima*, V.6, ed. by Van Riet, pp. 146–47. For a criticism of Plato's idea that learning is nothing else than remembering, see ibid., IV.3, p. 41.

52 See especially Aquinas, *STh*, I.84.4, ed. Leonina, pp. 319–21, where Aquinas acknowledges that Avicenna offers a rival account to Plato because he does not take the intelligibles to be subsistent entities, and also notes some further significant differences. Nevertheless: 'Et sic in hoc Avicenna cum Platone concordat quod species intelligibiles nostri intellectus effluunt a quibusdam formis separatis: quas tamen Plato dicit per se subsistere, Avicenna vero ponit eas in intelligentia agente' (ibid., p. 320).

the agent intellect as a common source of knowledge of intelligible forms for the human soul.[53]

It might certainly be asked whether Aquinas misunderstood one of them, Plato or Avicenna — or perhaps even both (or if he purposefully bent them to say what was convenient for him in order to have opponents). But his structural comparison between Avicenna's and Plato's handling of the intelligible and its ontological status clearly indicates what he himself is after when he asserts the existence of intellectual memory. In Aquinas's eyes, both Plato and Avicenna neglect or else cannot convincingly explain why the human intellect has to turn to the products of imagination (*phantasmata*) in every one of its operations. This *conversio ad phantasmata* is a basic tenet of Aristotelian natural epistemology as Aquinas understands it, and it is also present in Aristotle's *De memoria et reminiscentia*.[54] Aquinas discusses it elaborately in his own commentary on *De memoria et reminiscentia*, explicitly referring to Avicenna's denial of intellectual memory.[55]

This helps to explain why Aquinas thinks that Avicenna ultimately falls short of Aristotle's theory of the intellect. If the possible intellect overcomes the state of pure potentiality by receiving the intelligible form abstracted from the phantasm, in Avicenna's model it also gains the disposition to turn directly to the agent intellect at a later time in order to receive the intelligible species from there via emanation. There is no longer any need for them to be processed by abstraction, at least with regard to all the intelligibles which have already been known before at least once. But according to Aquinas, this raises the question (both for Avicenna and Plato) of why the human soul has been united to the body at all if it is by nature ultimately fitted to receive the intelligibles without bodily assistance.[56] His own anthropology points the other way: the union between the human soul and body is a natural one; thus the epistemology of natural cognition cannot simply dispense with the body or with sense experiences and the phantasms resulting from them.

---

53 As Richard Taylor pointed out to me, Aquinas's criticism of Avicenna as a kind of Platonist may be derived from his reading of Averroes's *Long Commentary on De anima*, where Themistius is accused of a similar 'Platonism'. In general, the question of whether and to what extent Aquinas draws on Averroes in order to criticize Avicenna's epistemology might prove interesting for the debate about the possibility of an intellectual memory, but this line cannot be followed here.

54 Aristotle, *De memoria et reminiscentia*, I.450a1, trans. by Sorabji, p. 49: 'Memory, even the memory of objects of thought, is not without an image'. See also Aristotle, *De anima*, III.7, 431a16–17; III.8, 432a8–9. See Cory, 'What Is an Intellectual "Turn"', for the Arabic influences on Aquinas's reading of this formula.

55 See Aquinas, *De memoria et reminiscentia*, II, ed. Leonina, pp. 107–09.

56 See Aquinas, *STh*, I.84.4, ed. Leonina: 'Si autem anima species intelligibiles secundum suam naturam apta nata esset recipere per influentiam aliquorum separatorum principiorum tantum, et non acciperet eas ex sensibus, non indigeret corpore ad intelligendum: unde frustra corpori uniatur'.

Now, Aquinas sticks to the *conversio ad phantasmata* even after the possible intellect has already received the intelligible forms.[57] But this idea might prove awkward for his own defence of an intellectual memory. Does it not simply render the concept of stored intelligibles superfluous after all, because it favours a model of intellectual cognition based on the memory-images of the sensitive soul? Is it not sufficient that we are able to produce all intelligibles at will by turning to our sense memory in order to recall the images from which we can then abstract the intelligible content? Aquinas does not think so. If one wants to remember some intelligible already acquired, the habitual knowledge of it already present helps the intellect to effectively sort out the phantasms through which the full actualization of the intellect can easily be reached at will. Whereas in the acquisition of knowledge, habitual knowledge is produced via abstraction from the phantasms, later remembering goes the other way around: from the stored intelligible species to the process of thinking by means of images. The existing intelligible species are 'applied to the phantasms'.[58] But this means that the intellect not only has the capability of retaining the intelligibles in a potential state after their first cognition; it must also be able to access them at will in order to facilitate actual thinking of everything that one already knows. Recollection (*reminiscentia*), which Aquinas likens to syllogistic reasoning, is still an operation of the sensitive soul, but the sensitive soul is 'nobler and stronger in humans than in other animals because of its connection with the intellect'.[59] Consequently, Aquinas stresses that recollection has a certain 'affinity and propinquity' to universal reason due to a 'flow-back' (*refluentia*) from it.[60] Thus, the idea of an intellectual memory has a fundamental role in the overall architecture of Aquinas's natural epistemology.

(3) Intellectual memory is also indispensable for Aquinas when it comes to his conception of how the *anima separata* — the soul after the death of the

57 See Aquinas, *De veritate*, X.2, ad 7, ed. Leonina, p. 302: 'Unde, cum phantasma hoc modo se habeat ad intellectum possibilem sicut sensibilia ad sensum, ut patet per Philosophum in III de anima, quantumcumque aliquam speciem intelligibilem apud se intellectus habeat, numquam tamen actu aliquid considerat secundum illam speciem, nisi convertendo se ad phantasma. Et ideo, sicut intellectus noster secundum statum viae indiget phantasmatibus ad actu considerandum antequam accipiat habitum, ita et postquam acceperit'. See also *Summa contra gentiles*, II.73, ed. Leonina, p. 462, where Aquinas stresses the reversal of causality involved: first the phantasms move the possible intellect, then the possible intellect uses them as a kind of tool (*quasi instrumento*) to produce the images needed for the abstraction of intelligible species.

58 Aquinas, *Super I ad Corinthios*, XIII.3, Index Thomisticus, p. 386: 'Indiget tamen in hac vita convertere se ad phantasmata, ad hoc quod actu intelligat, non solum ut abstrahat species a phantasmatibus, sed etiam ut species habitas phantasmatibus applicet'. See also Aquinas, *De memoria et reminiscentia*, 2, ed. Leonina, pp. 108–09: 'Non ergo propter hoc solum indiget intellectus possibilis humanus fantasmate ut acquirat intelligibiles species, set etiam eas quodam modo in fantasmatibus inspicit'.

59 Aquinas, *De memoria et reminiscentia*, VIII, ed. Leonina, p. 133 (translation from Carruthers and Ziolkowski, *Medieval Craft of Memory*, p. 188).

60 Aquinas, *STh*, I.78.4, ad 5, ed. Leonina, p. 257: 'Ad quintum dicendum quod illam eminentiam habet cogitativa et memorativa in homine, non per id quod est proprium sensitivae partis; sed per aliquam affinitatem et propinquitatem ad rationem universalem, secundum quandam refluentiam'.

natural body — can have intellectual knowledge. We now enter the third area in which he criticizes Avicenna, namely the theological consequences of the denial of intellectual memory, which concern Aquinas's supernatural epistemology. Interestingly enough, this seems to have been the original background of Aquinas's preoccupation with the subject. In the fourth book of his *Sentences* commentary and in his early questions *On Truth*, Aquinas discusses Avicenna's view under the heading of whether the separated soul still knows the intelligible species after death.[61]

The problem is quite straightforward. The sensitive faculties of the soul are disabled after our physical death, and acquisition of new knowledge via abstraction from phantasms is no longer possible in the absence of impressions from the imagination and from sensitive memory. Furthermore, the separate soul cannot make use of any sense-memories to think the intelligible forms that have already been acquired during our lifetime. The faculties of the sensitive soul, including imagination and memory, are lost with bodily death, as Aquinas stresses in accordance with Aristotle.[62] Thus, the natural mode of intellectual cognition via *conversio ad phantasmata* inevitably fails.

In order to counteract these postmortem pitfalls, Aquinas keeps to the idea that the intellect of the separated soul preserves direct access to the intelligibles acquired during our lifetime because they are still habitually stored in the intellectual memory. This turning back to intellectual memory cannot be the only mode of intellectual cognition accessible to the separated soul; otherwise stillborn babies would have no knowledge after death at all, as Aquinas remarks, because they did not have the opportunity of producing any intelligible species. Apart from this knowledge, therefore, there must be other and higher forms of knowledge. Aquinas sets out three different modes of cognition of the *anima separata*, two of which do not rely on stored intelligibles, whereas the third consists in the activation of intelligibles already acquired during our lifetime.[63] Thus, intellectual memory is still one valuable source of knowledge in the afterlife, which has to be defended against Avicenna's devastating claim that the intelligibles are only in the soul when they are perceived actually.

At least at some points, the target here is not so much Avicenna's theory itself, but rather the consequences that some later Latin authors drew from it. Using Avicenna's denial of an intellectual memory created during the earthly life, it is

---

61 See Aquinas, *In IV Sent.*, L.1.2, Index Thomisticus, p. 703 (Quaestio: 'Utrum anima per species quas nunc a corpore abstrahit, separata postmodum per eas aliquid intelligat'): 'Respondeo dicendum, quod circa hoc duplex est opinio. Una occasionatur ex opinione Avicennae, qui dicit, quod intellectus noster possibilis non servat aliquas formas postquam actu intelligere desinit; quia formae intelligibiles non possunt esse in intellectu possibili nisi ut in vi apprehendente'. See also Aquinas, *De veritate*, XIX.1 ('Utrum anima post mortem possit intelligere') and *Quodlibet*, III.9.1 and XII.8.1.

62 See Aristotle, *De anima*, I.4, 408b27–29.

63 For these three modes, see Aquinas, *De veritate*, XIX.1, ed. Leonina, p. 566. For the general differences between natural and postmortal cognition of the human soul, see Aquinas, *STh*, I.89.1, ed. Leonina, pp. 370–71.

possible to 'resurrect' the Platonic idea of recollection: The human intellect is endowed from its birth with connatural intelligible forms, the actualization of which is rather hindered by the union of the soul with the body. After being freed from this nuisance, the intellect in its after-worldly state can access all these forms at will.[64] Once more, Avicenna's argument seems to pave the way for a Platonic epistemology that Aquinas rejects without hesitation. The background is, again, his fundamental anthropological conviction that the union between the soul and the body is a natural and essential one and not an accidental one, as Platonizing philosophers and theologians would have it. According to Aquinas, the body cannot be any kind of obstacle to the achievement of intellectual cognition (or merely a functional instrument to be dispensed with later). Furthermore, Avicenna's argument that no intelligibles are left behind in the intellectual soul by the cognitions during our lifetime ultimately defeats the idea that man's natural striving for knowledge is realized by a gradual advance towards the contemplation of God, in which grace does not destroy nature but perfects it. In consequence, in Aquinas's model, our acquired intellectual memory of the natural world stays intact on a formal level after our physical death even though its material basis (the body and the external and internal senses) has been destroyed or disabled.[65]

As one can gather from all these different arguments, Aquinas is keen to refute Avicenna's argument and to install in its place a full-blown conception of intellectual memory that serves his philosophical and theological purposes both in disputes about natural and supernatural epistemology and in anthropological controversies. He is aware that he thereby gradually moves away from the original Aristotelian notion of memory as a power of the sensitive soul, but he still believes himself to be in basic agreement with Aristotle's theory of the intellect.[66] Whereas Albert tries to separate the philosophical (Aristotelian) notion of memory from the theological (Augustinian) one as clearly as possible, Aquinas tries to bring them as closely together as possible. This is not the only difference between teacher and pupil, as we will see in the concluding section.

---

64 See Aquinas, *In IV Sent.*, L.1.2.

65 See Aquinas, *STh*, I–II.67.2, ed. Leonina: 'Unde quantum ad ipsa phantasmata, quae sunt materialia in virtutibus intellectualibus, virtutes intellectuales destruuntur destructo corpore: sed quantum ad species intelligibiles, quae sunt in intellectu possibili, virtutes intellectuales manent. Species autem se habent in virtutibus intellectualibus sicut formales'.

66 There is a telling passage in his early *Sentences* commentary (Aquinas, *In IV Sent.*, XLIV.3.3b, ad 4, Index Thomisticus, pp. 649–50), where Aquinas first talks of sensitive memory in the Aristotelian sense and afterwards distinguishes it from an Augustinian understanding pertaining to the intellective part of the soul. He is quite clear that only this second form of memory is capable of remembering after death, when the first one is lost. His later position tends to blur this distinction and to ascribe to Aristotle the permanent storing of intelligibles in the possible intellect. Whereas Albert thinks that *memoria proprie dicta* is sensitive memory, Aquinas sees intellectual memory as the most appropriate understanding: 'vis qua mens nostra retinere potest huiusmodi intelligibiles species post actualem considerationem, memoria dicetur. Et hoc magis accedit ad propriam significationem memoriae' (Aquinas, *De veritate*, X.2, ed. Leonina, pp. 301–02).

## Why Does Albert Sit on the Fence?

Aquinas's continued battle against the denial of intellectual memory betrays a clear dissatisfaction with the Avicennian epistemology underlying that claim. As we saw above, he identifies in Avicenna's position some indications of a Platonizing position that he deems altogether unsuitable, especially with regard to its anthropological consequences.[67] Since the conception of an intellectual memory serves many philosophical and theological purposes in Aquinas, it is hardly surprising that he tries to remove a major epistemological obstacle to it by refuting Avicenna's denial every time the hydra raises its head.

It seems to me quite telling that Albert — at least to my knowledge of the texts — fell relatively silent on the issue of intellectual memory after his balanced judgement on Avicenna's position in the early work *De homine*. There was certainly no lack of opportunity for him to come back to the matter, for example in his commentary on Aristotle's *De memoria et reminiscentia* or in the treatise *De intellectu et intelligibili*, which concludes Albert's project of building a complete Peripatetic psychology based on a theory of the intellect. Yet Albert shows little inclination to return to the subject, apart from an interesting short passage in his commentary on *De anima*, to which I will turn below. The puzzling question here, put simply, is: Is there a reason why Albert refrains from further criticisms of Avicenna in the style of his pupil Aquinas?

One might suspect that Albert generally criticizes Avicenna with more restraint than Aquinas, at least in matters of psychology, despite his constant criticism of the 'externalization' or separation of the agent intellect beginning in his early *De homine*. This general tendency was already visible in his treatment of Avicenna's argument concerning the conservation of intelligibles, addressed above. Instead of making a blunt criticism, Albert carefully points out that Avicenna is basically right to deny the idea of an intellectual memory in the literal sense of the phrase. Despite Augustine's testimonies on behalf of such a notion, memory as a subject of psychology or natural philosophy (and not of theology) belongs to the sensitive powers.[68] This corroborates the great influence of Avicenna on

67 Aquinas spots the same 'lapse' into Platonism in Averroes's doctrine of the intellect as well; see Taylor, 'Remarks on the Importance', pp. 136–37.

68 This does not mean that Albert denies outright Augustine's notion of a kind of *a priori* intellectual memory that can be activated from within by a form of recollection only triggered by sense-perception and phantasms. But this is not to be confused with the discussion at hand, which pertains to intelligibles acquired from without (i.e., *a posteriori*). Albert distinguishes two kinds of intelligible species relevant to memory: *species mentis* (corresponding to Augustine's *a priori* inward memory) and *species rei* (abstracted *a posteriori* from external things). He specifies the contents of this *a priori* intellectual memory especially with a view to moral principles: it does not contain only general theoretical axioms (e.g., that the whole is bigger than its parts), which are intuitively evident once one has learned the meaning of the concepts involved, but also the general rules of natural law that guide our conscience in its applications to particular cases. These principles are truly inborn in our intellectual memory and not caused by any former experiences in the way that sensitive memories or intellectual cognitions are produced (see Albertus Magnus, *De homine*,

the theory of the soul in Albert's early work, clearly established by Dag Hasse.[69] In his later works as well, however, Albert continues to regard Avicenna as one of the leading authorities in psychology, especially concerning the investigation of memory in the Aristotelian tradition. This comes to the fore most evidently at the beginning of his commentary on *De memoria et reminiscentia*, when Albert states that he will begin his commentary not with the Aristotelian text, but with a digression on Avicenna's and Averroes's opinions on the matter.[70] The reason Albert offers for this rather unusual procedure is telling: he thinks that nearly all the Latin authors have simply misunderstood the faculties of memory and recollection,[71] so that he follows the Arabic understanding of these phenomena in many places. In particular, Albert stresses the Aristotelian formula that one can speak of 'remembering the intelligibles' only in an accidental and not in an essential way.[72] He even praises Avicenna and Averroes for their correct understanding of this basic insight, and states that one should not give credit to other, Latin, interpreters who want to turn Aristotelian recollection into a faculty of the intellectual soul.[73]

Apart from this general approval of the Arabic interpretations of Aristotelian memory — in marked contrast to the misunderstandings by the contemporary Latin doctors — there is, I think, also another reason why Albert does not dwell too much on his earlier criticism of Avicenna's position in *De homine*. In the absence of a smoking gun, we are about to enter some speculative territory. In brief, in the most refined version of his own theory of intellectual knowledge,

---

pp. 549–50; Anzulewicz, *Memoria* und *reminiscentia*, p. 173). However, Albert explicitly relegates this understanding of an *a priori* intellectual memory to theology because it is based on the idea of the human soul as a likeness of God. Albert certainly does not see any insuperable contradiction between the philosophical and theological understanding of memory, as Anzulewicz rightly stresses, but in fact he does not seriously try to place them in one coherent epistemological frame.

69 See Hasse, *Avicenna's 'De anima'*, pp. 60–69; Hasse, 'The Early Albertus Magnus'.

70 An English translation of this commentary is available in Carruthers and Ziolkowski, *Medieval Craft of Memory*, pp. 122–52. A comprehensive philosophical analysis of the work is offered in Müller, *Albertus Magnus über Gedächtnis*.

71 Albertus Magnus, *De memoria et reminiscentia*, I.1, ed. by Donati, p. 113, vv. 7–22: 'Et est digressio declarans sententiam Avicennae et Averrois de memoria. [...] Quia autem, ut mihi videtur, omnes fere aberraverunt Latini in cognitione harum virtutum quas memoriam et reminiscentiam appellamus, ut aestimo propter verborum Aristotelis obscuritatem, ideo primo volumus ponere planam de memoria sententiam Peripateticorum, antequam Aristotelis sententiam prosequamur'.

72 Compare Aristotle, *De memoria*, I, 449a13–14. Also ibid., 450a22–25, trans. by Sorabji, p. 49: 'It is apparent, then, to which part of the soul memory belongs, namely the same part as that to which imagination belongs. [...] things that are not grasped without imagination [i.e., objects of thought; J.M.] are remembered in virtue of an incidental association'. Albert offers subtle interpretations of this idea in his *De memoria et reminiscentia*, I.3, ed. by Donati, pp. 118–19, and II.2, p. 126.

73 See Albertus Magnus, *De memoria et reminiscentia*, II.1, ed. by Donati, p. 124, vv. 8–13: 'Ponemus igitur primo sententias Averrois et Avicennae et Alexandri et Themistii et Alfarabii, qui omnes concorditer dicunt quod reminiscentia nihil aliud est nisi investigatio obliti per memoriam. Non igitur credendum est eis qui dicunt reminiscentiam esse partis intellectualis animae secundum se'.

Albert has — contrary to Aquinas — only quite a restricted use for a habitual memory of acquired intelligibles.

This comes to light in his numerous digressions in Book III of his commentary on *De anima*, written between 1254 and 1257, in which he develops his own theory of the intellect in critical response to different 'Peripatetic' commentators on Aristotle's theory of the intellect.[74] There, he conspicuously turns against a modern 'Latin' reading of Aristotle which holds that after we have acquired knowledge, the corresponding intelligibles are in us in a habitual manner.[75] The reference is rather vague,[76] but one cannot help noting that Albert's criticism is also directed at the position taken by Aquinas in his construction of intellectual memory in terms of a habitual intellect, as discussed above. Indeed, Albert criticizes this idea of a habitual memory of the intelligibles with direct (and affirmative) reference to Avicenna. There are several points of criticism of the 'Latin' position, first and foremost that it cannot give an adequate explanation of whether (and if so, how) the human soul is ultimately capable of knowing the separate, immaterial, substances. Ultimately, Albert advises his contemporaries to look for a more convincing solution to the problem in the writings of the Peripatetic authors instead of inventing false principles of their own to support their alternative (and false) readings of Aristotle.[77]

In the digression immediately following this attack on contemporary Latin authors, Albert reveals his real objective, namely the Arabic theory of the four intellects with the pinnacle of the 'acquired intellect' (*intellectus adeptus*).[78] He

---

74 See Albertus Magnus, *De anima*, III.3, chs. 6–11, ed. by Stroick, p. 214, v. 83–p. 223, v. 38. For a comprehensive and convincing analysis of Albert's theory of knowledge as it is developed in his commentary on *De anima*, see Winkler, 'Zur Erkenntnislehre'.

75 See Albertus Magnus, *De anima*, III.3.10, p. 220, vv. 30–57: 'Forsitan autem dicit aliquis sequens dicta quorundam doctorum, qui Peripateticos non imitantur, quod intelligibilia sunt in nobis in habitu, cum acquisita est scientia. [...] Et haec est via, quam fere sequuntur omnes moderni Latinorum, sed isti in principiis non conveniunt cum Peripateticis. Si enim scientia sit qualitas in anima, tunc scientia non est universale, quod est in intellectu, quod est ubique et semper. Peripatetici autem concorditer dixerunt, quod universale secundum actum non est nisi in anima, et quod per universale continuatur cum intellectu separato; et haec est obiectio Avicennae, in quam omnes concordaverunt'.

76 One possible target is William of Auvergne, who in his *De anima* (written around 1240) closely links intellectual knowledge with the notion of *habitus* and also refers to *memoria*. See William of Auvergne, *De anima*, VII.8–9, *Opera omnia* II, pp. 214–15. On William's relationship with Avicenna in this respect, see Baumgartner, *Die Erkenntnislehre*, pp. 78–84, who erroneously argued that William basically followed the Arabic path. For a comprehensive account of William's noetics, see William of Auvergne, *De l'âme*, ed. by Brenet, pp. 13–71, with a French translation of his *De anima*, VII.1–9.

77 See Albertus Magnus, *De anima*, III.3.10, ed. by Stroick, p. 220, v. 94–p. 221, v. 5: 'Et ex omnibus his patet, quod difficillima quaestio est, quae supra est inducta, et Latini quidem huc usque neglexerunt illam quaestionem; et huius causa est, quia non convenerunt in positionibus suis cum dictis Peripateticorum, sed diverterunt in quandam alteram viam et secundum illam finxerunt alia principia et alias positiones'. The open question to which Albert alludes concerns the possibility of cognizing immaterial substances; see Aristotle, *De anima*, III.7, 431b17–19.

78 See Albertus Magnus, *De anima*, III.3.11, ed. by Stroick, p. 221, v. 6–p. 223, v. 38 ('Et est digressio declarans veram causam et modum coniunctionis intellectus agentis nobiscum').

sees the essence of this model as follows.[79] As a result of continuous study of the sciences, the human intellect can reach a new level of intellection at which it is finally capable of knowing the separate substances. In this state, the agent intellect no longer effects single cognitions of material things in the possible intellect via abstraction of the intelligible forms from phantasms; instead, the agent intellect becomes the form of the possible intellect, and in this state of 'conjunction' the intelligibles are transferred directly from the agent intellect to the possible intellect without the need for abstraction. This is important because the cognition of separate — that is, immaterial — substances can certainly not be achieved via any sense-images or phantasms. Albert calls the formal conjunction of the agent and the possible intellect the 'acquired intellect', following the Arabic tradition,[80] and he wholeheartedly embraces this view as the basis for his own conception of worldly happiness based on doing philosophy.

Although Albert refers mainly to Averroes and Alfarabi as his sources in this digression, his model is also indebted to Avicenna. The 'acquired intellect' in this sense plays a pivotal role in Albert's theory of the intellect, and supplies, in my opinion, the missing background to Albert's forgiving stance on Avicenna's denial of intellectual memory. In the present epistemological context, it is significant that this model of the acquired intellect as the formal conjunction of the possible and the agent intellect severely restricts the function of a habitual memory of intelligibles, at least once the intellect has reached its highest stage as *intellectus adeptus*. On the level of the acquired intellect, there is no more need for an epistemology in the Thomistic style, in which the stored intelligibles are necessary to sort out the appropriate phantasms and elicit actual thinking about what one already knows. Albert emphasizes that the acquired intellect has no need at all for any *conversio ad phantasmata* because the possible intellect is now 'plugged' directly into the agent intellect and acquires the intelligibles from there; he even quotes Avicenna to the effect that all sensitive powers, including imagination, are thus completely left behind, like a vehicle that is no longer of any use at the end of a journey.[81]

---

79 On this topic and the subsequent description of the acquired intellect in Albert, see Müller, 'Der Einfluss der arabischen Intellektspekulation'; also Sturlese, '"Intellectus adeptus"'.

80 Albertus Magnus, *De anima*, III.3.11, ed. by Stroick, p. 221, v. 81–p. 222, v. 9: 'Quaedam autem speculata fiunt in nobis per voluntatem, quia scilicet studemus inveniendo et audiendo a doctore, et haec omnia fiunt intellectu agente influente eis intelligibilitatem [...]; et ideo in omnibus his accipit continue intellectus possibilis lumen agentis et efficitur sibi similior et similior de die in diem. Et hoc vocatur a philosophis moveri ad continuitatem et coniunctionem cum agente intellectu; et cum sic acceperit omnia intelligibilia, habet lumen agentis ut formam sibi adhaerentem [...]. Et hoc sic compositum vocatur a Peripateticis intellectus adeptus et divinus; et tunc homo perfectus est ad operandum opus illud quod est opus suum, inquantum est homo, et hoc est opus, quod operatur deus, et hoc est perfecte per seipsum contemplari et intelligere separata'.

81 See Albertus Magnus, *De anima*, III.2.19, ed. by Stroick, p. 206, vv. 43–54: 'Amplius autem, adhuc aliud est consideratione dignum de istis intellectibus, quoniam in veritate, quando intellectus possibilis procedit de potentia ad actum, tunc utitur reminiscentia et sensu et imaginatione et phantasia, quoniam ex sensu accipit experientiam et ex experientiis memoriam et ex memoriis

So is Avicenna right after all to completely discard the idea of an intellectual memory? At long last, Albert returns to this topic. He agrees with Avicenna that the *intellectus adeptus* designates the possible intellect's acquired disposition to turn to the agent intellect in order to receive the intelligible forms directly (which is basically Albert's understanding of Avicennian emanation). He asks, however, how Avicenna can explain this disposition without assuming that the already acquired intelligibles somehow remain in the soul. One basic feature of Albert's *intellectus adeptus* is that it is acquired on the basis of comprehensive scientific studies; once we have come to know everything that can be gathered by abstraction from the sensible world, we are intellectually mature to the degree that we may turn directly to the cognition of separate substances. This is only possible by the formal conjunction of the possible and the agent intellect, both of which are conceived as powers of the individual human soul in Albert's epistemology.

Albert's criticism of Avicenna is based on this model. The intelligibles which have been acquired previously have to stay somewhere in the soul as a kind of 'medium' between the two intellects in order to enable their formal union.[82] This does not mean that they are actually used any longer in specific instances of intellectual cognition, as we saw above: abstraction paves the way for a direct intuition of intelligible forms (based on emanation, that is, direct transfer, from the agent intellect), but it no longer constitutes the epistemic mode once the intellect is fully 'acquired'. However, Albert still has a functional role in store for the intelligibles acquired before by means of abstraction — because if they simply vanished after their first cognition, there would be no permanent basis for the acquired intellect (since the permanent connection between the possible and the agent intellect in the *intellectus adeptus* is 'mediated' by or grounded in the collection of acquired intelligibles). They are needed in order to maintain the connection between the two intellects, so that the possible intellect is always in 'standby mode', ready to be tuned to the agent intellect immediately and at will.

Therefore, Avicenna cannot be right in his complete denial of any intellectual memory in the individual soul. Albert does, though, concede that the acquired intellect itself is defined neither by receiving nor by retaining these intelligibles.

---

universale. Cum autem iam habeat scientiam, vocatur intellectus adeptus, et tunc non indiget amplius virtutibus sensibilis animae, sicut qui quaerit vehiculum, ut dicit Avicenna, ad vehendum se ad patriam, cum pervenerit ad patriam, non indiget amplius vehiculo'. Cf. Avicenna, *Liber de anima*, V.3, ed. by Van Riet, p. 105, and Avicenna, *Avicenna's Psychology* [*Kitāb al-Najāt*], trans. by Rahman, p. 56, where the lower faculties of the soul are compared to an animal that becomes a hindrance for the rider after the destination has been reached.

82 See Albertus Magnus, *De anima*, III.3.11, ed. by Stroick, p. 222, v. 95–p. 223, v. 10: 'Quod autem quaeritur, utrum intellecta maneant apud intellectum, quando actu illa non considerat, sicut dicit Avicenna, per rationes superius inductas nos non iudicamus esse verum, quia cum medium coniunctionis possibilis ad agentem sint speculata, oportet ipsa manere, aut extrema dividerentur. Bene tamen concedimus, quod ipsa non est diffinita per tenere neque etiam per recipere sicut virtutes corporales, sed potius, cum anima intellectualis sit ut "locus specierum" universalium, remanent universalia speculata apud intellectum sicut in loco suae generationis'.

Their permanence is a kind of *conditio sine qua non* for the higher form of cognition; it is not essential to or constitutive of it. Again, Albert sits on the fence between criticism and praise of Avicenna in this area. Looking at the way Albert treats Avicenna's theory of the intellect in the digressions on *De anima* III in general, he does not really condemn him. He refers approvingly to Avicenna's characterization of the acquired intellect and only criticizes him for not having given the cause of the formal conjunction between possible and agent intellect.[83]

## Concluding Remarks

According to Albert, Avicenna was basically correct with regard to the *intellectus adeptus* and its mode of operation (which discards once and for all the previously necessary *conversio ad phantasmata* so cherished by Aquinas), although he went wrong when it came to the formal conjunction. On this, Albert carefully corrects him: intellectual memory is somehow needed for establishing and keeping up the acquired intellect even if it is not specifically used in the acquired intellect's single operations. In Albert's view, this lapse by Avicenna does not indicate a fundamental flaw in his epistemology that needs to be hammered home at every opportunity.

Albert's rather forgiving attitude towards Avicenna concerning the complete denial of intellectual memory contrasts sharply with Aquinas's repeated and severe criticism of the doctrine. This difference between the two scholars may be explained at least partially by the differing importance for their own projects of the idea of a habitual storing of intelligibles. For Albert, that idea does not seem as far-reaching as for Aquinas, who ultimately aims for a reconciliation of the Aristotelian and Augustinian traditions of memory in a theological perspective (whereas Albert tries to keep the two more separate). Thus, Albert's restatement of his position concerning intellectual memory in his commentary on *De anima* (written around 1254–57) may already be a reaction to his pupil's sharp criticism of Avicenna, which starts with Aquinas's early commentary (written around 1252–54) on Peter Lombard's *Sentences*.[84]

Their different attitudes to intellectual memory also indicates a divergence between teacher and pupil at a deeper level. Aquinas remains rather critical of the Arab theory of the four intellects, and especially of the acquired intellect, because he keeps closely to Aristotelian *conversio ad phantasmata* as a cornerstone of his natural epistemology; this is one of the main reasons why he denies the

83 See Albertus Magnus, *De anima*, III.3.9, p. 219, v. 35–p. 220, v. 27, esp. p. 220, vv. 5–9: 'Contra autem istud dictum [sc. Avicennae] sunt praecipue quattuor rationes. Quarum prima est, quia adhuc quaestio est ut prius, qualiter possibilis per intellecta non remanentia in ipso acquirat aptitudinem coniungendi se agenti; de hoc autem nullam assignant causam'.

84 See Aquinas, *In II Sent.*, d. 17, q. 2, a. 1. On this passage, see Taylor, 'Remarks on the Importance', pp. 132–38.

idea of a perfect happiness (*beatitudo perfecta*) during our lifetime. In contrast, Albert ultimately subscribes to the Arabic idea of the 'acquired intellect' as the crowning achievement of philosophical knowledge, which finally transcends the normal mode of human cognition and makes us truly happy, as humans rather than images of God, already in this life through the contemplation of the separate substances.[85] With regard to his mature theory of the intellect, which ultimately underlies this whole debate about intellectual memory, Albert is in many ways closer to Avicenna than to Aquinas. In this way, my analysis once more bears testimony to the tenet of Janet Coleman (quoted at the beginning of this chapter) that the topic of memory is always embedded into a more inclusive theory of knowledge, which has to be taken into account in some detail in order to solve intriguing problems in this area.

85 On this vital difference between Albert and Aquinas, see Müller, '*Duplex beatitudo*'. For a comprehensive take on Albert's theory of cognition, see Anzulewicz and Krause, 'Albert der Große'.

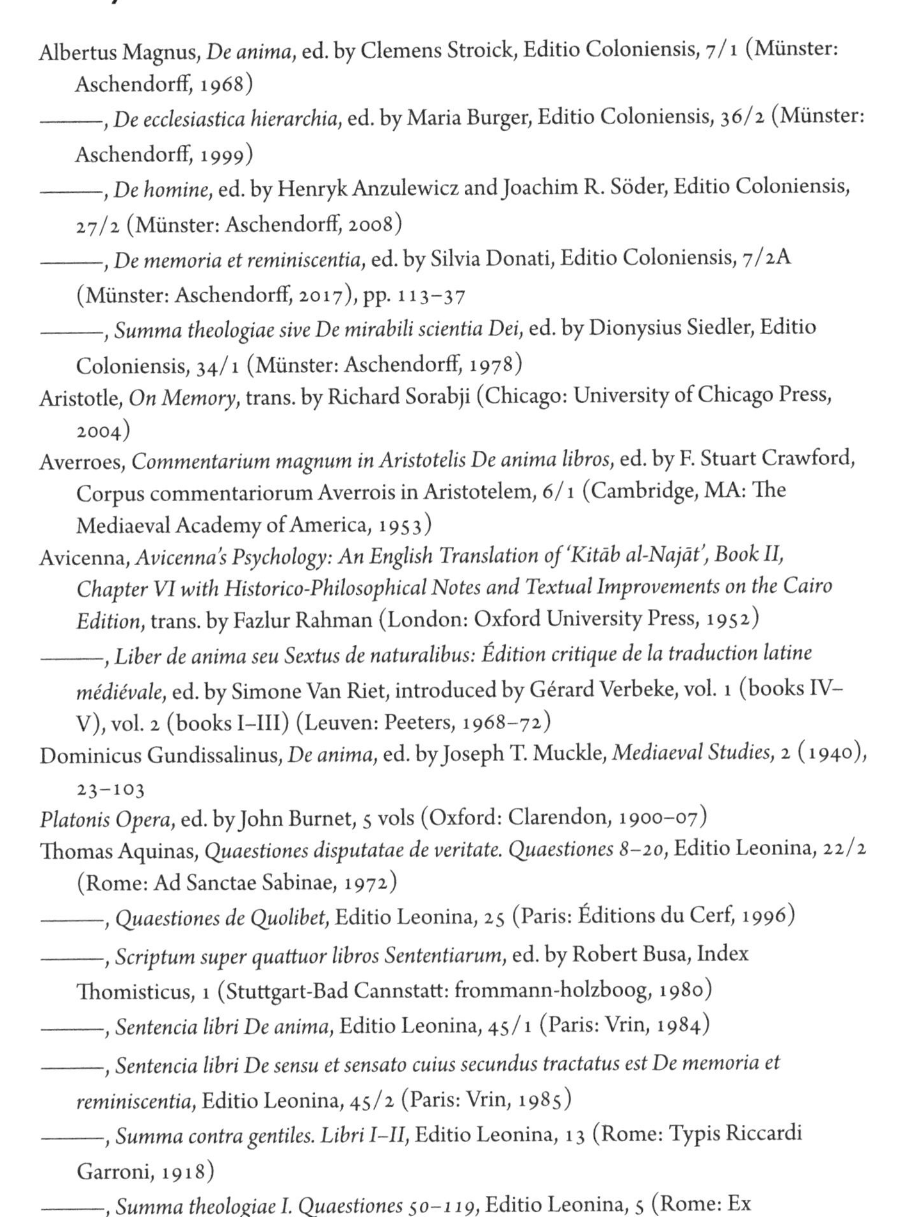

## Works Cited

### *Primary Sources*


Albertus Magnus, *De anima*, ed. by Clemens Stroick, Editio Coloniensis, 7/1 (Münster: Aschendorff, 1968)

———, *De ecclesiastica hierarchia*, ed. by Maria Burger, Editio Coloniensis, 36/2 (Münster: Aschendorff, 1999)

———, *De homine*, ed. by Henryk Anzulewicz and Joachim R. Söder, Editio Coloniensis, 27/2 (Münster: Aschendorff, 2008)

———, *De memoria et reminiscentia*, ed. by Silvia Donati, Editio Coloniensis, 7/2A (Münster: Aschendorff, 2017), pp. 113–37

———, *Summa theologiae sive De mirabili scientia Dei*, ed. by Dionysius Siedler, Editio Coloniensis, 34/1 (Münster: Aschendorff, 1978)

Aristotle, *On Memory*, trans. by Richard Sorabji (Chicago: University of Chicago Press, 2004)

Averroes, *Commentarium magnum in Aristotelis De anima libros*, ed. by F. Stuart Crawford, Corpus commentariorum Averrois in Aristotelem, 6/1 (Cambridge, MA: The Mediaeval Academy of America, 1953)

Avicenna, *Avicenna's Psychology: An English Translation of 'Kitāb al-Najāt', Book II, Chapter VI with Historico-Philosophical Notes and Textual Improvements on the Cairo Edition*, trans. by Fazlur Rahman (London: Oxford University Press, 1952)

———, *Liber de anima seu Sextus de naturalibus: Édition critique de la traduction latine médiévale*, ed. by Simone Van Riet, introduced by Gérard Verbeke, vol. 1 (books IV–V), vol. 2 (books I–III) (Leuven: Peeters, 1968–72)

Dominicus Gundissalinus, *De anima*, ed. by Joseph T. Muckle, *Mediaeval Studies*, 2 (1940), 23–103

*Platonis Opera*, ed. by John Burnet, 5 vols (Oxford: Clarendon, 1900–07)

Thomas Aquinas, *Quaestiones disputatae de veritate. Quaestiones 8–20*, Editio Leonina, 22/2 (Rome: Ad Sanctae Sabinae, 1972)

———, *Quaestiones de Quolibet*, Editio Leonina, 25 (Paris: Éditions du Cerf, 1996)

———, *Scriptum super quattuor libros Sententiarum*, ed. by Robert Busa, Index Thomisticus, 1 (Stuttgart-Bad Cannstatt: frommann-holzboog, 1980)

———, *Sentencia libri De anima*, Editio Leonina, 45/1 (Paris: Vrin, 1984)

———, *Sentencia libri De sensu et sensato cuius secundus tractatus est De memoria et reminiscentia*, Editio Leonina, 45/2 (Paris: Vrin, 1985)

———, *Summa contra gentiles. Libri I–II*, Editio Leonina, 13 (Rome: Typis Riccardi Garroni, 1918)

———, *Summa theologiae I. Quaestiones 50–119*, Editio Leonina, 5 (Rome: Ex Typographia Polyglotta, 1889)

———, *Summa theologiae I–II. Quaestiones 1–70*, Editio Leonina, 6 (Rome: Ex Typographia Polyglotta, 1891)

———, *Super I ad Corinthos XI–XVI (reportatio)*, ed. by Roberto Busa, Index Thomisticus, 6 (Stuttgart-Bad Cannstatt: frommann-holzboog, 1980)

William of Auvergne, *De l'âme (VII, 1–9)*, ed. by Jean-Baptiste Brenet, Sic et Non (Paris: Vrin, 1998)

———, *Opera omnia*, 2 vols (Paris: F. Hotot, 1674; repr. Frankfurt am Main: Minerva, 1963)

## *Secondary Works*

Alpina, Tommaso, 'Intellectual Knowledge, Active Intellect and Intellectual Memory in Avicenna's *Kitāb al-Nafs* and Its Aristotelian Background', *Documenti e studi sulla tradizione filosofica medievale*, 25 (2014), 131–83

Anzulewicz, Henryk, '*Memoria* und *Reminiscentia* bei Albertus Magnus', in *La mémoire du temps au Moyen Âge*, ed. by Agostino Paravacini Bagliani (Florence: Edizioni del Galluzzo, 2005), pp. 163–200

———, and Katja Krause, 'Albert der Große und sein holistisches Konzept menschlicher Erkenntnis', in *Veritas et subtilitas: Truth and Subtlety in the History of Philosophy*, ed. by Tengiz Iremadze and Udo Jeck (Amsterdam: John Benjamins, 2018), pp. 157–94

Baumgartner, Matthias, *Die Erkenntnislehre von Wilhelm von Auvergne* (Münster: Aschendorff, 1893)

Black, Deborah, 'How Do We Acquire Concepts? Avicenna on Abstraction and Emanation', in *Debates in Medieval Philosophy: Essential Readings and Contemporary Responses*, ed. by Jeffrey Hause (New York: Routledge, 2014), pp. 126–44

Bloch, David, *Aristotle on Memory and Recollection: Text, Translation, Interpretation, and Reception in Western Scholasticism* (Leiden: Brill, 2007)

Brumberg-Chaumont, Julie, 'La première reception du *De memoria et reminiscentia* au Moyen Âge Latin: Le commentaire d'Adam de Buckfield', in *Les 'Parva naturalia' d'Aristote: Fortune antique et médiévale*, ed. by Christophe Grellard and Pierre-Marie Morel (Paris: Presses universitaires de Paris-Sorbonne, 2010), pp. 121–42

Carruthers, Mary, *The Book of Memory: A Study of Memory in Medieval Culture*, 2nd, rev. ed. (Cambridge: Cambridge University Press, 2008)

———, and Jan M. Ziolkowski, eds, *The Medieval Craft of Memory: An Anthology of Texts and Pictures* (Philadelphia: University of Pennsylvania Press, 2002)

Coleman, Janet, *Ancient and Medieval Memories* (Cambridge: Cambridge University Press, 1992)

Cory, Therese Scarpelli, 'What Is an Intellectual "Turn"? The *Liber de Causis*, Avicenna, and Aquinas's Turn to Phantasms', *Topicós*, 45 (2013), 129–62

De Leemans, Pieter, 'Medieval Commentaries on Aristotle's *Parva Naturalia*', in *Encyclopedia of Medieval Philosophy: Philosophy between 500 and 1500*, ed. by Henrik Lagerlund, vol. 2 (Vienna: Springer, 2011), pp. 917–23

Di Martino, Carla, '*Memoria dicitur multipliciter*: L'apporto della scienza psicologica araba al medievo latino', in *Tracce nella mente: Teorie della memoria da Platone ai moderni*, ed. by Maria Michela Sassi (Pisa: Edizioni della Scuola normale superiore, 2007), pp. 119–38

———, *'Ratio particularis': Doctrines des sens internes d'Avicenne à Thomas d'Aquin* (Paris: Vrin, 2008)

Fantl, Jeremy, 'Knowledge How', *The Stanford Encyclopedia of Philosophy* (Fall 2017 Edition), ed. by Edward N. Zalta, https://plato.stanford.edu/archives/fall2017/entries/knowledge-how/

Gätje, Helmut, 'Gedächtnis und Erinnerung bei Avicenna und Averroes', *Acta Orientalia*, 49 (1988), 7–36

Hasse, Dag Nikolaus, *Avicenna's 'De anima' in the Latin West: The Formation of a Peripatetic Philosophy of the Soul, 1160–1300* (London: The Warburg Institute, 2000)

———, 'Avicenna's Epistemological Optimism', in *Interpreting Avicenna: Critical Essays*, ed. by Peter Adamson (Cambridge: Cambridge University Press, 2013), pp. 109–19

———, 'The Early Albertus Magnus and His Arabic Sources on the Theory of the Soul', *Vivarium*, 46 (2008), 232–52

Mourant, John A., *Saint Augustine on Memory* (Villanova, PA: Villanova University Press, 1980)

Müller, Jörn, *Albertus Magnus über Gedächtnis, Erinnern und Wiedererinnerung: Eine philosophische Lektüre von 'De memoria et reminiscentia' mit Übersetzung* (Münster: Aschendorff, 2017)

———, '*Duplex beatitudo*: Aristotle's Legacy and Aquinas's Conception of Happiness', in *Thomas Aquinas and the Nicomachean Ethics*, ed. by Tobias Hoffmann, Jörn Müller, and Matthias Perkams (Cambridge: Cambridge University Press, 2013), pp. 52–71

———, 'Der Einfluss der arabischen Intellektspekulation auf die Ethik des Albertus Magnus', in *Wissen über Grenzen*, ed. by Andreas Speer and Lydia Wegener (Berlin: De Gruyter, 2006), pp. 545–68

———, 'Memory as an Internal Sense: Avicenna and the Reception of His Psychology by Thomas Aquinas', *Quaestio*, 15 (2015), 497–506

———, 'Memory in Medieval Philosophy', in *Memory: A History*, ed. by Dmitri Nikulin (Oxford: Oxford University Press, 2015), pp. 92–124

O'Donnell, James J., 'Memoria', in *Augustinus-Lexikon*, ed. by Cornelius Mayer, vol. 3 (Basel: Schwabe, 2010), pp. 1249–57

Sturlese, Loris, '"Intellectus adeptus": L'intelletto e i suoi limiti secondo Alberto il Grande e la sua scuola', in *Intellect and Imagination in Medieval Philosophy*, ed. by Maria Candida Pacheco and José F. Meirinhos (Turnhout: Brepols, 2006), vol. 1, pp. 303–21

Taylor, Richard C., 'Avicenna and the Issue of the Intellectual Abstraction of the Intelligibles', in *Philosophy of Mind in the Early and High Middle Ages*, ed. by Margaret Cameron (London: Routledge, 2019), pp. 56–82

———, 'The Epistemology of Abstraction', in *The Routledge Companion to Islamic Philosophy*, ed. by Richard Taylor and Luis X. López-Farjeat (London: Routledge, 2016), pp. 273–84

———, 'Remarks on the Importance of Albert the Great's Analyses and Use of the Thought of Avicenna and Averroes in the *De homine* for the Development of the Early Natural Epistemology of Thomas Aquinas', in *Die Seele im Mittelalter: Von der Substanz zum funktionalen System*, ed. by Günther Mensching and Alia Mensching-Estakhr (Würzburg: Königshausen & Neumann, 2018), pp. 131–48

Winkler, Norbert, 'Zur Erkenntnislehre Alberts des Großen in seinem *De anima*-Kommentar als systematische Einheit von *sensus, abstractio, phantasma, intentiones, species, universalia* und *intellectus*', *Bochumer Philosophisches Jahrbuch für Antike und Mittelalter* 19 (2016), 70–173

Wolfson, Harry, 'The Internal Senses in Latin, Arabic and Hebrew Philosophical Texts', in *Harvard Theological Review*, 28 (1935), 69–133

Yates, Frances A., *The Art of Memory* (London: Routledge and Kegan Paul, 1966)

HENRYK ANZULEWICZ

# Chapter 10. What Makes a Genius?*

## *Albert the Great on the Roots of Scientific Aptitude*

In an anonymous question, *An anima racionalis sit mortalis*, which is preserved in a recently discovered thirteenth-century Albertus Magnus manuscript,[1] the author claims that the rational soul performs its essential activity, knowing, in accordance with the body's dispositions (*disposiciones corporis*). In support of this claim, the author invokes the natural-philosophical and medical authority of Aristotle and Galen. According to Aristotle, the functional efficiency of basic principles of life gives evidence of their dependence upon the body. Walking is Aristotle's paradigmatic example of such functional efficiency, where the feet serve as bodily instruments of the soul's disposition to move.[2] The anonymous author corroborates the interdependence of body and mind through, on the one hand, an appeal to the Aristotelian doctrine that human beings with delicate bodies possess especially gifted minds,[3] and, on the other, an appeal to Galen's affirmation of a causal connection between the quality of a person's brain matter and his cognitive

---

* Translated from the German by Martin J. Tracey, Benedictine University, Lisle, IL. My thanks go to Katja Krause, Richard C. Taylor, Steven Harvey, and Jules Janssens for their insightful comments on this paper, and to Kate Sturge for her help in the editing process.

1 The manuscript contains two of Albert's writings, *De fato* and *Super Ethica*, and belongs to a private collection. I wish to express my profound gratitude to Dr Francesco Siri (IHRT, Paris) and the codex's owner for having made it possible to inspect the codex and for making a reproduction of it available to the Albertus-Magnus-Institut.

2 Aristotle, *Gen. An.* II.3, 736b22–24; trans. by Michael Scot: *De animalibus* XVI.3, ed. by van Oppenraaij, p. 74. Cf. Albertus Magnus, *De homine*, ed. by Anzulewicz and Söder, p. 35, vv. 60–64; Albertus Magnus, *De animalibus*, XVI.1.12.67, ed. by Stadler, vol. 2, p. 1096, vv. 18–24.

3 Anon., *An anima racionalis sit mortalis*, MS lat. CP 439 fol. 14vb: 'probatur ex secundo *De anima* quod molles carne bene aptos dicimus mente'. See Aristotle, *DA* II.9, 421a25–26; trans. vetus: Albertus Magnus, *De anima*, II.3.23, ed. by Stroick, p. 132, v. 82 and p. 133, vv. 24–38; Albertus Magnus,

**Henryk Anzulewicz** (anzulewicz@albertus-magnus-institut.de), of the Albertus Magnus Institute, Bonn, has published critical editions and translations of Albert's works as well as books and essays on Albert the Great.

*Albert the Great and his Arabic Sources*, ed. by Katja Krause and Richard C. Taylor, Philosophy in the Abrahamic Traditions of the Middle Ages, 5 (Turnhout: Brepols, 2024), pp. 283–310
BREPOLS PUBLISHERS 10.1484/M.PATMA-EB.5.136490

performance.[4] These and similar statements by the anonymous author accurately reflect doctrinal views held by Albert the Great, and give the impression that Albert himself might be the author of the anonymous *Quaestio*.

In what follows, I present certain aspects of Albert the Great's views regarding the natural prerequisites of human scientific aptitude. My focus is on endogenous and exogenous factors influencing both the physical constitutions of human beings (specific and individual) and their cognitive dispositions. I examine Albert's views regarding what, in his opinion, are the most important internal or psychophysiological and environmental or ecological determinants of those constitutions and dispositions. Albert's writings on anthropology, natural philosophy, ethics, and metaphysics offer an enormous amount of material pertinent to this inquiry, and my essay can only address a limited number of representative texts and themes.

My argument proceeds in three steps. First, I show how Albert grounds scientific aptitude, what epigenetic state of the human being he presents as its prerequisite, and how he determines the epistemic subject matter of any such aptitude.[5] Next, I present Albert's views on the endogenous factors that influence and differentiate the individual's aptitude for science. The third part of my contribution attempts to reconstruct, on the basis of Albert's natural-philosophical writings, the external conditions, such as geographical regions and climates, that Albert regards as the natural prerequisites of an individual's cognitive capacities. My concluding remarks foreground and appraise Albert's systematically innovative attempt to develop a psychophysiological account — including both endogenous and exogenous factors — of the scientific aptitude of human beings at the level of the species and the individual.

## Albert's View of Human Beings' Scientific *aptitudo* from their Epigenetic State

In his early work *Summa de creaturis*, a philosophical-theological synthesis concerning the principles of creaturely actuality,[6] Albert investigates the human aptitude for science in a discussion of the theme of separate intelligences.[7] There, he identifies such aptitude as a human trait, one that both distinguishes human

---

*Quaestiones super De animalibus*, VIII.5–9, ed. by Filthaut, p. 190, vv. 14–16: 'Mores sequuntur complexiones. Probatio per Philosophum II *De anima*: "Molles carne sunt apti mente"'.

4 On Galen's approach, see Marechal, 'Galen's Constitutive Materialism'.

5 I use the term *epigenesis* in this context essentially in the Aristotelian sense and as a *via media* between a manifold determinism and the creative will of an individual. See Maienschein, 'Epigenesis and Preformationism'.

6 For Albert, those principles are formless matter (*materia prima*), time (*tempus*), heaven (*caelum*), and separated intelligence (*angelica intelligentia*).

7 Albertus Magnus, *De quattuor coaequaevis* (*Summa de creaturis* I), IV.61.4, ed. by Borgnet, pp. 655b–656a.

beings and is constitutive of them. More precisely, he considers the bodily and mental constitution to be constitutive of human beings as humans and of their essence, as well as of their rational — or rather, intellectual — endowment. Albert regards the intellectuality, rationality, and freedom of human beings as the unique characteristic that distinguishes them from all other animals. He connects intellectuality, understood in the sense of intuitive insight (*intellectus simplex*), to a human being's epistemic aptitude for any kind of science. At the same time, he distinguishes three kinds of science: general methodological sciences (*rationalia*), the practical sciences (*ethica*), and the theoretical sciences (*naturalia*, *mathematicalia*, and *divina*).[8] It is worth noting that this taxonomy represents Albert's earliest attempt to classify the sciences.

Closer examination of Albert's early statements about human nature shows that he grounds the human aptitude for science neither in discursive reason (*intellectus compositivus* or *ratio*) nor in free will (*liberum arbitrium*), but rather in intuitive insight (*intellectus simplex*). Nevertheless, Albert assigns the leading role in the cognitive process to reason and will, with their natural, specific, and individual properties. Reason and will, together with the human natural desire for knowledge, constitute an epigenetic predisposition that makes human beings capable of scientific knowing. With regard to the intellect, Albert agrees with Aristotle and the Greek and Arabic Peripatetics that intellect, as the principle of cognition, is entirely separate from the body.[9] He also agrees with Aristotle and his followers that every cognitive process originates in sense perception. The organic body, with its powers of sensation, is accordingly understood as an instrument of an individual human's intellect, precisely analogous to matter, since matter, on Albert's view, is the instrument of the separate intelligence that pervades all of nature's operations.[10]

The function of external and internal sensory perception, which collects individual sensory impressions and prepares them for higher cognitive operations — be it directly or in the form of an experience stored in the sensory *memoria* — is the basis for the generation of knowledge and thus for the human aptitude for science. Albert first points this out in *De homine* and later, among other places,

8 Ibid., p. 656a: 'Id autem quod praestat irrationabilibus animalibus, est triplex. Quorum primum inest secundum intellectum simplicem: quia licet minimum sit quoad sensibiles virtutes, in quibus convenit cum brutis: tamen omni virtute est potens secundum magnitudinem intellectus. Et dicitur omnis virtus intellectus virtus ipsius ad rationalia et ethica et naturalia et mathematicalia et divina. Secundum inest per intellectum compositivum, qui ratio est, secundum quod omni continuitate habet scientiam rationalem: continuat enim terminum termino in habitudine subjecti ad praedicatum, et habitudine propositionis ad propositionem secundum quamlibet speciem ratiocinationis. Tertium est in ratione liberi arbitrii, scilicet secundum naturam animae liberum et potentissimum'.

9 Actually, when Albert uses the term *Peripatetici*, he refers almost exclusively — apart from Aristotle, Porphyry, Theophrastus, Themistius, and Alexander of Aphrodisias — to the Arabic (including Jewish) tradition. See de Libera, 'Épicurisme, stoïcisme, péripatétisme', pp. 360–62.

10 Albert expresses this view with the axiom 'Opus naturae est opus intelligentiae'. See Weisheipl, 'The Axiom "Opus naturae est opus intelligentiae"'; Hödl, '*Opus naturae est opus intelligentiae*'.

in his commentary on the *Metaphysics*, in each instance invoking Aristotle's *Posterior Analytics*.[11] Clearly, it is this organically grounded framework for receptive-abstractive perception, expressing general principles anchored in human nature, that makes human beings able to conduct scientific inquiry of any kind. In his *De homine* and his commentaries on Aristotle's *De anima* and *De sensu et sensato*, as well as in his original work *De intellectu et intelligibili*, Albert explains more and more precisely different aspects of the question of how the senses, in their psychophysiological constitution and function, influence the human aptitude for science, from the reception of sensory data to the generation of concepts.

According to Albert, the will, by setting in motion the cognitive process, contributes decisively to the human capacity for scientific knowledge and to its realization, and in equal measure to the general psychological make-up of intellection and to the apperception of the sensory data. Together with reason, the will serves as the mover of the human being's other powers, owing to the nature of the will. This view accords with the account of the will's function developed by Anselm of Canterbury and is in harmony with some earlier authors including Aristotle, Augustine, and John of Damascus.[12] Indeed, Albert picks up Anselm's account and joins it to a strikingly voluntarist view of knowledge taken from Augustine. This underlies Albert's account, in his early works, of the cause of the connection between the agent intellect and the passive intellect — a connection that generates the act of cognition and thereby becomes a precondition for the cognitive process.[13] In this way, Albert makes it clear that the will, to the extent that it is a spontaneous and autonomous agency, is a power of motion and desire that is active alongside or within reason. It determines the activity of each psychological faculty — including the cognitive acts underlying scientific aptitude, such as the transformation of the cognitive power from potentiality to actuality and of the potential intellect from potency to act.[14]

---

11 Albertus Magnus, *De homine*, ed. by Anzulewicz and Söder, p. 1, v. 21–p. 2, v. 20; Albertus Magnus, *Metaphysica*, I.1.7, ed. by Geyer, p. 10, v. 35–p. 11, v. 48. Cf. Aristotle, *APo* II.19, 99b34–100a9; Albertus Magnus, *Analytica posteriora*, II.5.1, ed. by Borgnet, pp. 229b–230b.

12 Anselmus Cantuariensis, *Cur Deus homo*, II.10, ed. by Schmitt, p. 107, vv. 1–2: 'Omnis potestas sequitur voluntatem. Cum enim dico quia possum loqui vel ambulare, subauditur: si volo'. For the related contributions of Aristotle, Augustine, and John of Damascus, see Albertus Magnus, *De homine*, ed. by Anzulewicz and Söder, p. 489, v. 46–p. 492, v. 59.

13 Albertus Magnus, *De homine*, ed. by Anzulewicz and Söder, p. 491, vv. 51–54: 'voluntas prout est ad intrinseca, universalis motor est virium ad actum, non ita quod ipsa sit immixta eis, sed quia ceterae virtutes cum actibus suis sunt in ratione voliti et per hoc efficiuntur obiecta voluntatis'; ibid., p. 492, vv. 51–59: 'voluntas uno modo dicta est motor generalis omnium potentiarum ad actum; et propter hoc dicit Anselmus quod cum dico "possum loqui", subintelligitur "si volo", et similiter cum dicitur "possum ambulare" vel "possum intelligere", et sic de aliis. Et hoc est etiam quod dicit Augustinus quod nulla causa est, cum intelligibile est in anima, quare quandoque actu intelligit et quandoque non, nisi voluntas'.

14 Ibid., p. 442, vv. 23–29: 'Si autem quaeritur, quid sit reducens possibilem de habitu in actum, dicendum quod nihil nisi voluntas. Et hoc patet per diffinitionem habitus datam a Philosopho, scilicet quod habitus est, quo quis aliquid agit, quando voluerit. Voluntas enim tripliciter est in anima,

Unlike in his early works, in his *De anima* commentary Albert subordinates the emphasis he often, invoking Augustine, places on the role of the will in cognitive processes to an emphasis on the natural, subjective predispositions of the knower and to the connatural spontaneity of intellection. In his account of the cognitive capacity of individual human beings, he accords a greater role than Aristotle does to the will — chiefly inspired by Averroes, but also making free use of Avicenna and Algazel, appropriating all their theories to his own argumentative needs.[15] Aristotle makes no mention of the will in the key passage where he sets out his theory of intellect (*DA* III.5). In the Philosopher's view, the will seems to have no direct share in the cognitive process. By contrast, when explaining this passage of Aristotle, Albert introduces the will as the principle of desire and free choice, thus merging a Stoic conception with his Aristotelian sources.[16]

---

scilicet motor aliarum potentiarum ad actum et pars imaginis et pars liberi arbitrii'. See Anzulewicz, 'Vermögenspsychologische Grundlagen', pp. 114–15; Schönberger, 'Rationale Spontaneität', esp. pp. 228 sqq.

15 Albertus Magnus, *De anima*, III.3.11, ed. by Stroick, p. 221, vv. 71–84: 'convenimus cum Averroe in toto et cum Avempeche et in parte cum Alfarabio, dicentes, quod intellecta speculata dupliciter fiunt in nobis; quaedam enim fiunt in nobis per naturam, ita quod non accipimus ea per aliquid vel ab aliquo doctore nec per inquisitionem invenimus ea, sicut sunt "dignitates demonstrationum primae", quae sunt prima et vera, ante quae omnino nulla sunt, quae non scimus ex sensu, nisi inquantum terminos cognoscimus, notitia autem terminorum non facit notitiam principiorum nisi per accidens. Quaedam autem speculata fiunt in nobis per voluntatem, quia scilicet studemus inveniendo et audiendo a doctore, et haec omnia fiunt intellectu agente influente eis intelligibilitatem'; ibid., III.3.9, p. 219, vv. 57–61: 'Dicunt enim isti [sc. Avicenna et Algazel], quod cum formae intellectus speculativi sint in anima, quas homo intelligit, quando vult, quod aut sunt in anima sicut in thesauro aut sunt ita in anima, quod licet non sint in ipsa sicut in subiecto, habet tamen eas, quando vult'. Cf. Averroes, *Commentarium magnum in De anima*, III.36, ed. by Crawford, p. 496, vv. 488–93; Avicenna Latinus, *Liber de anima seu Sextus de Naturalibus*, V.6, ed. by Van Riet, p. 147, v. 16–p. 150, v. 67; Algazel, *Metaphysics*, II.4.5, ed. by Muckle, p. 175, vv. 10–18.

16 See Albertus Magnus, *De anima*, III.3.11, ed. by Stroick, p. 221, vv. 81–86: 'Quaedam autem speculata fiunt in nobis per voluntatem, quia scilicet studemus inveniendo et audiendo a doctore, et haec omnia fiunt intellectu agente influente eis intelligibilitatem, et faciendo haec intellecta secundum actum esse intellecta intellectus agens coniungitur nobis ut efficiens'. See Anzulewicz, 'Vermögenspsychologische Grundlagen', pp. 111–15. Albert does not here consider the theological account of will as a component of the divine likeness within the soul endowed with reason. That account is also missing from his typology of the human being's internal powers of motion, in which he supplements the Aristotelian conception (*DA* III.10) from the point of view of the Platonists and the theologians. See Albertus Magnus, *De anima*, III.4.10, ed. by Stroick, pp. 240–42; Anzulewicz and Anzulewicz, 'Einleitung', pp. 101–06. However, as Richard Taylor wrote me in an email on 20 February 2020, 'although there is disagreement, generally scholars of ancient philosophy hold that Aristotle did not have a developed concept of free will. Rather, this notion was developed in Stoicism. Both the Latin and Arabic traditions took over this notion and read it back into Aristotle. In the Arabic tradition al-Fārābī took free will and choice to be an essential presumption for human nature'. See al-*Farābī, Commentary and Short Treatise on Aristotle's 'De Interpretatione'*, trans. by Zimmermann, p. 77. These issues are discussed in Phillipson, *Aquinas, Averroes, and the Human Will*. Averroes mentions will (*iradah*, translated as *voluntas*) many times in his paraphrasing *Middle Commentary on the Nicomachean Ethics*. But for him, will is part of the bodily cogitative power, whereas for Aquinas, will is part of the immaterial power of intellect. See Taylor, '*Cogitatio*, *Cogitativus* and

Looking more closely at the text of *De anima* underlying Albert's commentary, it is striking that Aristotle ascribes a role to the will that is epistemically (but not voluntarily) similar to the one ascribed to it in *De homine* by Albert, following Anselm and Augustine. Aristotle, however, restricts the efficacy of the will within intellection, taking it explicitly into account only in connection with sense perception and rational desire. On the one hand, Aristotle suggests casually and indirectly that the will plays a certain vague role within the cognitive process; on the other, he claims that it is the leading principle in our striving for a way of life ruled by reason. Aristotle notes in *De anima* II.5 that the elements of cognition lie to a certain extent within the soul, which is why cognition underlies the volitional decision.[17] In *De anima* III.9, he underscores that practical intellect manifestly does not lead to action without the desire whose principle is the will.[18] When one examines the noetic significance of the two functions of the will, as laid down by Aristotle and interpreted by Albert, it becomes clear that, with respect to the soul's powers, a will that is active within the cognitive processes is one of the conditions underlying the human scientific aptitude and capacity for action. For according to Albert, cognitive activity and practical action not only presuppose the epigenetic predisposition of scientific aptitude and the naturally implanted desire for knowledge, but also underlie the voluntarist law 'si volo'.[19]

Within his general consideration of the ontological foundations and epigenetic presuppositions of scientific aptitude, Albert conceives of that aptitude's subject in an inclusive way. He expands the epistemic realm that the human being is able to access so as to include all sciences without distinction. I pointed out these features of Albert's thought and his attendant taxonomy of sciences at the start of this section. By contrast, the differentiation between epistemic subjects on the basis of individual predisposition depends, in Albert's view, on both endogenous and exogenous factors. Albert discusses those factors extensively in his

*Cogitare'*, pp. 138–42. I would like to thank Prof. Taylor for this important and interesting additional information.

17 Aristotle, *DA* II.5, 417b22–24; trans. vetus: Albertus Magnus, *De anima*, II.3.3, ed. by Stroick, p. 100, vv. 80–81: 'Causa autem est, quoniam singularium secundum actum sensus est, scientia autem universalium; haec in ipsa quodammodo sunt anima. Unde intelligere quidem in ipsa, cum velit'. Cf. ibid., vv. 41–60.

18 Aristotle, *DA* III.9, 433a15–25; trans. vetus: Albertus Magnus, *De anima*, III.4.5, ed. by Stroick, p. 233, vv. 86–87 and p. 234, vv. 89–92: 'Et appetitus propter aliquid omnis est. Non enim appetitus hic practici intellectus est, ultimum autem principium actionis est. Quare rationabiliter haec duo videntur moventia, appetitus et intelligentia practica. Appetitivum enim movet et propter hoc intelligentia movet, quia principium ipsius appetitivum est. Et phantasia autem cum moveat, non movet sine appetitu. Unum igitur quiddam est movens, quod est appetitivum. Si enim duo sunt, appetitus et intellectus, et secundum communem aliquam speciem motus movent. Nunc autem intellectus quidem non videtur movens sine appetitu. Voluntas enim appetitus est. Cum autem secundum appetitum movetur, et secundum voluntatem movetur. Appetitus autem movetur extra'. Cf. ibid., p. 233, v. 80–p. 235, v. 22 (comm. Alberti).

19 See notes 12 and 13 above.

natural-philosophical writings, particularly in *De natura loci* and *De animalibus*.[20] He also summarizes them in his commentary on the *Metaphysics*, where he succinctly discusses the phenomenon of a natural desire for knowledge and the object of that desire, as well as the foundation for the multiplicity of scientific disciplines.[21]

Before Albert examines these aspects of scientific aptitude in his commentary on the *Metaphysics* and gives psychophysiological reasons for the differences in their specific expressions, he asserts emphatically, within a digression in his *De anima* commentary echoing the famous passage opening Aristotle's *Metaphysics*, that a natural desire for knowledge is proper to all human beings and that the only life that is truly human — the only life it is worthy for human beings to lead — is a life of continuous intellectual perfection. In his early works, Albert already regards this capacity for knowledge and science, which conforms with the natural desire for knowledge, as being anchored in human nature and constitutive of it. (Albert usually distinguishes the capacity for knowledge from the aptitude for science, but sometimes conflates them when speaking of capacities as found within subjects.[22]) At the same time, he does not overlook enormous deficits on the part of some human beings with respect to the natural desire for knowledge. As Albert notes in his *De anima* commentary, he has observed that some people do not make good use of their natural predisposition or do right in respect of the *telos* of human life that arises from it. He likens a life that follows only the law of the senses to the life of wild animals (*bestiae*).[23] On Albert's view, the true, natural, and essential fulfilment of human life has the good of the intellect alone as its goal; human fulfilment lies in contemplative happiness (*felicitas contemplativa*), which consists in intellectual activity undertaken for its own sake.

---

20 Cf. Albertus Magnus, *Metaphysica*, I.1.5, ed. by Geyer, p. 8, vv. 28–31: 'De his autem et huiusmodi in *Physicis* determinatum est a nobis, ubi de organorum complexionibus et de natura locorum habitabilium locuti sumus'; Mayer, 'Die Personallehre in der Naturphilosophie von Albertus Magnus'; Anzulewicz, 'Zwischen Spekulation und Erfahrung'.

21 Albertus Magnus, *Metaphysica*, I.1.5, ed. by Geyer, p. 7, v. 41–p. 8, v. 33.

22 See note 24 below.

23 Albertus Magnus, *De anima*, III.3.11, ed. by Stroick, p. 222, vv. 88–94: 'iste solus in veritate modus est, quo "omnes homines natura scire desiderant", quia isto solo modo homo est homo et operatur, quae sunt hominis. Vide enim eos qui sic intellectum non sunt adepti, si dicas aliquid de contemplabilibus, non intelligunt plus quam bestiae, quae in singularium semper remanent cognitione'. Cf. Albertus Magnus, *De intellectu et intelligibili*, I.1.7, ed. by Borgnet, pp. 488b–489a: 'tripliciter homini unitur intellectus. Uno scilicet modo ut natura dans esse: et sic est individuus. Alio modo ut potentia per quam est operatio intelligendi: et sic est virtus universalis. Tertio modo ut forma acquisita ex multis intelligibilibus, sicut planius tractatum est de intellectu agente qui non unitur contemplativis ut agens tantum, sed ut beatitudo eorum est, quando perveniunt ad hoc quod in eis est ut forma: et secundo et tertio modis secundum prudentiam et sapientiam dictus intellectus non inest aequaliter omnibus hominibus, sed aliis plus, et aliis minus, et aliis fortassis nihil inest de intellectu'.

This view of Albert's is grounded in his radically intellectualist understanding of the human being as 'intellect alone' (*solus intellectus*).[24] In the passages of his *De anima* commentary and *De intellectu et intelligibili* that are my focus here, Albert does not answer the question of why some people do not follow this natural intrinsic *telos*, or why people follow it in varying ways. His answer must therefore be sought elsewhere in his work.

## The Influence of Endogenous Factors on Individual Aptitude for Science

For Albert, as for Aristotle, it is clear that, as a rule, all human beings naturally desire to know. Albert locates the cause and goal of this natural desire in the highest good, which belongs to the perfection of nature and is implanted within it according to the desire for it in an inchoate and imperfect way.[25] Since, on Albert's understanding, the human being as such is intellect alone, the natural desire for perfection as the highest good is a desire for intellectual perfection as the good of the intellect — more precisely, of the perfectible potential intellect (*intellectus possibilis*), which, like a blank slate (*tabula rasa*), is naturally ready for the reception of intelligibles. It is nevertheless observed, Albert notes, that the desire for knowledge comes to expression differently in many people and is completely absent in some. He explains this phenomenon in two ways. The first is that a moral and affective deterioration paralyses nature in its pursuit of the good, so that it is unable to realize its potentiality and in fact does not reach its goal.

24 See Albertus Magnus, *Liber de natura et origine animae*, II.6, ed. by Geyer, p. 29, vv. 23–38: 'Sed delectari in propria et connaturali operatione non impedita non convenit nisi secundum naturam. Si enim est propria, est essentialiter conveniens secundum formam, quae dat esse et rationem, et si est connaturalis, non est per accidens, et si est non impedita, non habet contrarium. Taliter autem felicitas contemplativa se habet ad intellectum. Est autem felicitas illa contemplatio divinorum, non cum continuo et tempore, sicut in scientia *De anima* probatum est et in scientia *De intellectu et intelligibili*, quae deo convenit et intellectibus divinis omnibus caelestibus et homini. Propter quod Aristoteles dicit, quod homo non est nisi intellectus, eo quod bonum intellectus solum sibi proprium est et connaturale'. Cf. ibid., II.13, p. 39, vv. 24–46.

25 On this point and the following, see Albertus Magnus, *Metaphysica*, I.1.5, ed. by Geyer, p. 7, vv. 44–64: 'Quaeret autem fortasse aliquis, quae illa natura in homine sit, qua omnes homines natura scire desiderant. Homo enim studet adipisci, quod desiderat; et si desiderat illud natura, vehementius erit in ipsum desiderium, et hoc vehementius accendet ad studium; et hoc in multis hominibus non videmus. Amplius, naturaliter non desideratur, nisi quod est de naturae perfectione, cum sit non habitum. Sicut enim in fine primi *Physicorum* traditum est a nobis, desiderium est imperfecti ad bonum et optimum et divinum, cuius incohatio est in ipso, sed a perfecto deficit. Non enim desideraret nisi per aliquod simile, quod est in eo, ad bonum et optimum, sicut turpe desiderat bonum et sicut femina masculum. Non enim desiderat inquantum turpe vel femina, quia sic sunt contraria masculo et bono et contrarium non desiderat contrarium; sed potius desiderant inquantum imperfecta; sic turpe aliquid habet boni et femina masculi. Et sic oportet, quod sit hoc, propter quod intellectus possibilis est, sicut tabula rasa ad scibilia praeparatus natura, qua omnes homines scire desiderant'.

This deterioration, as Albert assumes following Aristotle, may have physiological causes, insofar as a human being's moral habits result from his complexional constitution, that is, from a particular proportional mixture of the bodily humours blood, phlegm, yellow bile, and black bile (*sanguis, fleuma, colera rubea*, and *colera nigra* or *melancolia*).[26] His second explanation is that the desire of those who seek to attain the natural good comes to expression very differently due to certain psychophysiological characteristics.[27] What exactly does Albert have in mind in this second explanation?

According to Albert, two kinds of conditions determine the character and object of the desire for cognition and knowledge, corresponding to an individual's suitability to operate in this or that field of science. One kind is mental competence, which he calls 'intelligence' or 'mental power'; its capability depends on the physiological constitution of the brain and sensory organs and on their interaction. The other kind is the complexional constitution of the knower and his individual capacities, as well as the hierarchically differentiated disciplines, from metaphysics to rhetoric, to which those capacities correspond. Invoking the theory of complexion enables Albert to offer a natural-philosophical explanation of the endogenous factors underlying the capacity for cognition, which include primary qualities, metabolism and bodily fluids, heredity, fixed and mutable patterns of behaviour (*habitus*), and bodily dispositions (*dispositiones*) with regard to certain objects and natural stimuli.[28]

The theory of complexions is based on the theory of humours, which is preserved in rudimentary form in the *corpus Hippocraticum* and was developed by Galen, among others, with recourse to the Aristotelian conception of primary qualities and the theory of temperaments (that is, to the four-fluid theory mentioned above).[29] Avicenna, who ordered, corrected, and further developed the medical teachings of Galen, incorporated the theory of complexions into his *Liber canonis* as well as *De animalibus*, and became the theory's most important trans-

---

26 Albertus Magnus, *Quaestiones super De animalibus*, VIII.5–9, ed. by Filthaut, p. 190, vv. 14–16: 'Mores sequuntur complexiones. Probatio per Philosophum II *De anima*: "Molles carne sunt apti mente"'. See Lennox, 'Aristotle on the Biological Roots of Virtue'. Albert takes up the humoral theory deriving from Galen in many places in *De animalibus*; he devotes a digression chapter, with explicit reference to Galen, to the theory's causes, formation, and properties. Albertus Magnus, *De animalibus*, III.2.4, ed. by Stadler, vol. 1, pp. 330–34. The editor of Albert's *De animalibus*, Hermann Stadler, suggests that Albert's immediate source in this case is Avicenna, *Liber Canonis*, I.1.4.1–2, ed. Venetiis, fol. 4v–6v.

27 Albertus Magnus, *Metaphysica*, I.1. 5, ed. by Geyer, p. 7, vv. 65–70: 'Et quia corruptelae consuetudinum et affectionum a natura deducunt hominem, ut in X *Ethicorum* probatum est, ideo natura haec non semper potest movere ad studium, sed manet in potentia et virtute, contrariis habitibus naturale bonum corrumpentibus impedientibus, ne in actum procedat'.

28 Ibid., p. 8, vv. 9–28. See König-Pralong, '*Omnes homines natura scire desiderant*', p. 132; Schipperges, 'Das medizinische Denken bei Albertus Magnus', p. 285; Mayer, 'Die Personallehre in der Naturphilosophie von Albertus Magnus', pp. 209, 223, 235, 254.

29 On Galen's dependence upon Aristotle's natural-philosophical perspectives, see Lennox, *Aristotle's Philosophy of Biology*, pp. 119–23.

mitter for Albert.[30] On the criterion of complexion, which determines the organic basis of the perceptual and cognitive faculties with respect to their physiological constitution and functionality, Albert distinguishes four types of natural aptitude for different disciplines among four corresponding groups of people.

The highest rung on the hierarchical ladder of aptitude for science is occupied by human beings who possess a balanced bodily complexion.[31] This guarantees the purest form of cognition insofar as the cognitive faculty, the intellect, which is fully separated from internal sense perception, has as its object completely abstract entities — entities that Albert characterizes as divine, sublime, and subtle. According to Albert, this essentially complexional form of cognition enables human beings to engage in metaphysics. The prerequisite for such engagement, the 'unmixedness and purity' of the intellect, has a physiological basis. It stems from the complexion of the pneuma — that is, of the life-spirit understood medically as a sublime bodily substance — and of the primary sensory qualities of heat and luminous moisture, insofar as the complexion is not stiffened by freezing cold or disturbed by heat, which by nature leads to the mixing of primary qualities.[32]

On the second rung of the hierarchical ladder representing the natural aptitude for the sciences and its connection to particular scientific disciplines, Albert places humans with specially well-trained powers of imagination, whose complexional constitution makes them, above all, able to grasp geometrical figures. This characteristic, he explains, is generated on the one hand by the tempered dry and complexional, non-freezing cold of the imaginative organ, and on the other by the intellect, which relies on the organ of imagination. This kind of organic-complexional and mental constitution makes a human being especially capable of engagement with mathematical sciences.[33]

---

30 See Jacquart, '§ 52. Die Medizin als Wissenschaftsdisziplin', p. 1599. Albertus Magnus, *De animalibus*, XII.1.2.18–24, ed. by Stadler, vol. 1, p. 804, v. 31–p. 807, v. 15; Avicenna, *De animalibus*, XII.1, ed. Venetiis, fol. 44v–45r.

31 See Albertus Magnus, *De animalibus*, VIII.6.1.234, ed. by Stadler, vol. 1, p. 670, vv. 12–22: 'Adhuc autem est unum quod omnibus praedictis plus est attendendum, et hoc est aequalitas complexionis, et recessus ab excellentia contrariorum, quae est in complexionantibus, et accessus ad caeli aequalitatem, quoniam licet anima neque sit armonia complexionantium neque aliqua forma corporalis armoniam consequens: tamen quia perfectio est corporis et habet potentiam vitae organicam, multum cooperatur nobilitas complexionis nobilibus operationibus ipsius. Est autem quaedam nobilissima operatio ipsius quae cognoscere et iudicare de sensibilibus acceptis, et illa simplicior et formalior est in nobiliori complexione, et ignobilior et particularior et materialior est in ea quae est minus nobilis'.

32 Albertus Magnus, *Metaphysica*, I.1.5, ed. by Geyer, p. 8, vv. 9–14: 'Et horum quorumcumque intellectus quidem immixtus et purus est et complexio subtilis spiritus et caloris et humoris luminosi non constantis per frigidum congelans nec turbata per calidum commiscens, illi student bene et libenter divinis et magnis et subtilibus rebus'. Cf. Albertus Magnus, *De intellectu et intelligibili*, I.3.3, ed. by Borgnet, p. 501b. On the concept of pneuma (*spiritus*), see Bono, 'Medical Spirits and the Medieval Language of Life'; Meroni, 'The Doctrine of Spirit'.

33 Albertus Magnus, *Metaphysica*, I.1.5, ed. by Geyer, p. 8, vv. 14–19: 'Quorum autem organum imaginationis optime ad tenendum figuras per temperate siccum et complexionale non congelans

Other natural aptitudes arise, Albert supposes, when the organ of sense perception is very marrowy and pure and the pneuma is bright, unmixed, and not slowed down by freezing cold in its function of transmitting sensory data. If all these qualities come together in the imaginative organ, the imagination's ability to act will be small: its physiological permeability reduces its ability to securely hold its images. When, under such qualitative conditions, a human being's intellect is oriented towards sense perception, his theoretical interest tends to be directed towards the natural world. Thus, in human beings with an imaginative organ of this constitution, where there is cooperation of the cognitive power and sense perception, one finds a predisposition towards natural science.[34]

Just as for the preceding three levels, Albert's account of the ground and character of the natural aptitude for a form of *bios theoreticos* on the lowest level rests on the particular complexional constitution of the organs of inner perception and the role of the senses in the cognitive process. As is to be expected, Albert states that on this fourth level, the initial conditions for both sense perception and intellect have further deteriorated in quality. This applies to the pneuma, in that it freezes and darkens under the influence of cold. A complexional constitution of this kind, Albert thinks, binds perception and knowledge to external appearances and not to truth content, which is why it stands in the way of conducting more profound theoretical work and is sufficient only for a concern with rhetoric.[35]

The differentiation and brief explanation in Albert's commentary on the *Metaphysics* of the individual capacity for scientific knowledge based upon endogenous factors, and the corresponding classification of scientific disciplines at each level of the hierarchy, is neither Albert's sole nor his most important presentation of this complex problematic. The passage, alongside a parallel passage in *De intellectu et intelligibili*,[36] instead affords a concise overview of what Albert presents in detail in various contexts in his natural-philosophical works, especially in the two commentaries on *De animalibus* and in *De natura loci*.[37] His detailed explanations of the endogenous causes of the constitutional characteristics and dispositions of human beings are fruitful adaptations and elaborations of psychophysiological and biological-medical concepts and theories drawn from Aristotle, Galen, and Avicenna. In these reworkings, the Persian physician and philosopher plays the

frigidum praeparatum est et intellectus reflexus ad ipsum, hi doctrinalibus mathematicis gaudent studiis et huiusmodi'.

34 Ibid., vv. 19–24: 'Quorum autem medullosum et bene purum est organum sensus cum spiritu lucido non commixto et non pigro per frigidum congelans et organum imaginationis consequenter erit non bene tenens figuras et intellectus erit reflexus ad sensum, hi gratas habent speculationes naturales'.

35 Ibid., vv. 24–28: 'Quorum autem congelati sunt spiritus et non bene clari propter frigus inspissans, occupantur circa signa rethorica et detinentur in his nec profundantur in aliqua veri speculatione'.

36 Albertus Magnus, *De intellectu et intelligibili*, I.3.3, ed. by Borgnet, pp. 501a–502b.

37 See also Albertus Magnus, *Metaphysica*, I.1.5, ed. by Geyer, p. 8, vv. 28–31: 'De his autem et huiusmodi in *Physicis* determinatum est a nobis, ubi de organorum complexionibus et de natura locorum habitabilium locuti sumus'.

role of both an independent medical authority and an important transmitter of Galenic medicine.[38]

The same element of adaptation applies to the exogenous conditions of the human constitution, which Albert considers an essential determinant of each individual's aptitude for science. His foundational, detailed, and most extensive explanations, however, concern endogenous aspects of individual bodily constitution. These aspects include the theory of primary and and the theory of secondary qualities; the former is interpreted in older medicine as a thermodynamic theory of aggregate states.[39] Again, Albert attributes fundamental importance to the medical-physiological theory of complexion, which concerns an accidental quality, also called a complexion, that emerges from the mixing and interaction of the fluids and the secondary qualities.[40] Equally relevant for him is humoural theory (or humoural pathology, insofar as the theory concerns the aspect of disease), which accounts for the four digestions (that is to say, the four phases of digestion: *pepansis*, *omotes*, *epsesis*, and *molysis*); the fluids; and the formation of temperaments from the mixture of fluids and of homogeneous members from the connection between the fluids (complexion).[41]

In addition, Albert treats in *De animalibus* the physiological and psychosomatic aspects of the individual's constitution, such as the shape and interrelationship of their limbs (*membrorum figura* and *habitudo*). These in turn explain the natural tendencies of the individual's emotions and physiological 'rules of life'

38 On this dual role, see Jacquart, '§ 52. Die Medizin als Wissenschaftsdisziplin', p. 1599.

39 In the classic doctrine of elements, the primary qualities are warm (*calidum*), cold (*frigidum*), dry (*siccum*), and moist (*humidum*). As secondary qualities, which arise from the combination of primary qualities, Albert like Aristotle recognizes heavy-light (*grave-leve*), slippery-dry (*lubricum-aridum*), fine-gross (*subtile-grossum*), and hard-soft (*durum-molle*). Albertus Magnus, *De generatione et corruptione*, II.1.7, ed. by Hossfeld, p. 181, vv. 51–55, v. 57 (Arist.); Aristotle, *GC* II.2, 329b32–34; trans. vetus: Aristoteles Latinus 9/1, p. 55, vv. 9–11.

40 In his work on natural-philosophical foundations, the *Physics* commentary, Albert defines complexion as follows (*Physica*, II.2.1, ed. by Hossfeld, p. 98, vv. 24–26): 'est enim complexio qualitas una proveniens ex reciproca actione et passione qualitatum contrariarum in corporibus commixtis'. That definition is silently adapted from Avicenna's *Liber primus naturalium. De causis et principiis naturalium*, I.6, ed. by Van Riet, p. 62, v. 78–p. 63, v. 80, where it has the following nearly identical wording: 'complexio est qualitas veniens ex reciproca passione qualitatum contrariarum in corporibus sibi permixtis'. Cf. Albertus Magnus, *Physica*, VII.1.7, ed. by Hossfeld, p. 531, vv. 36–39: 'quae [sc. complexio] est qualitas resultans ex permixtione humorum et elementorum, ex eo quod plurimum cuiuslibet est cum plurimo alterius cuiuslibet, ubique alterans ipsum et alteratum ab ipso'.

41 Albertus Magnus, *De animalibus*, III.2.3–4, ed. by Stadler, vol. 1, pp. 320–34; Albertus Magnus, *Meteora* 4.1.12–26, ed. by Hossfeld, p. 224, v. 66–p. 242, v. 43. See Mayer, 'Die Personallehre in der Naturphilosophie von Albertus Magnus', pp. 203–06, 210–13; Barbado, 'La physionomie, le tempérament et le caractère', pp. 336–42; Balss, *Albertus Magnus als Zoologe*, pp. 52–53 [44–45]; Balss, *Albertus Magnus als Biologe*, pp. 203–04; Stubbe, *Albertus Magnus*, pp. 87–88. The humoral theory as well as the doctrines on complexion and temperament derive ultimately from Galen. See Marechal, 'Galen's Constitutive Materialism', p. 191 n. 2.

(*regimina vitae*).[42] In this context, Albert devotes much attention to questions regarding biological heredity, which he associates with the critique of Galen's view of semen. He locates the actual source of inheritance in a twofold effective power of the semen: one of these powers determines the embryo's membership in the human species (*virtus ad speciem inducendam*), while the other, which derives its efficacy from the properties of the seminal matter, generates the embryo's complexion, shape, and colour, and transmits where applicable the hereditary property, in modern terms the genetic material. Semen provides the proper material foundation for heredity, whereas menstrual blood plays no active role in procreation and heredity.

Noteworthy here is Albert's causal explanation of the possibility of physiologically different trajectories in heredity. On his view, hereditary traits either derive from the father, when the effective power of the semen is able to shape entirely the procreative matter (menstrual blood), or they derive from both parents, when the procreative material resists to a greater or lesser extent the shaping power that arises from the semen's properties and partially retains its original characteristics, stemming from the mother. If the procreative matter resists formation through the effective power of the seminal matter entirely, then there is inheritance of the mother's attributes alone. It is also possible for the effective power to be shaped by an attribute that derives neither from the father nor from the mother's procreative material. In such a case, the characteristics inherited are not from the parents but perhaps from the grandparents or more distant ancestors. Finally, Albert does not rule out the appearance of 'hereditary traits' that, on account of a multitude of opposed properties contained within the semen and procreative material, have nothing in common with the 'genetic' make-up of any ancestor.[43]

---

42 Albertus Magnus, *De animalibus*, I.2.2.126–30, ed. by Stadler, vol. 1, p. 46, v. 4–p. 47, v. 29. See Mayer, 'Die Personallehre in der Naturphilosophie von Albertus Magnus', p. 209.

43 Albertus Magnus, *De animalibus*, IX.2.2.98, ed. by Stadler, vol. 1, p. 713, v. 34–p. 714, v. 11: 'Sane autem intelligendum est quod dicimus de sanguinis menstrui virtute. Non enim damus ei secundum scientiam Perypatheticam aliquam virtutem operativam. Sed cum in spermate sit tota virtus operativa, sperma duplicem habet virtutem, unam quidem ad speciem inducendam, et quoad hanc non quaerit nisi materiam humanae speciei convenientem, et non attendit similitudinem aliquam, et hanc semper consequitur in propria materia sibi subiecta. Aliam autem habet virtutem ex qualitatibus materiae: et ex illa operatur complexionem et figuram et colores: et secundum illam aliquando oboedit [aliquando] materia in toto, et inducitur similitudo patris. Aliquando autem resistit ei in toto qualitas materiae, et tunc inducitur similitudo matris. Aliquando autem in parte vincit et in parte succumbit, et tunc inducitur similitudo utriusque in diversis membris conceptus. Et aliquando operatur secundum aliam qualitatem quae est in ipso: et tunc nullius parentum, sed forte avi aut proavi inducitur similitudo, et forte inducitur similitudo quae ad nullum est de tota progenie, quoniam multa sunt superiora et inferiora, quorum qualitates sunt in spermate et in sanguine menstruo'. Cf. ibid., XVIII.1.4–5 ('De causa similitudinis generati ad parentes vel avos praecedentes, et de causa dissimilitudinis eiusdem ad progenitores; De improbatione erroris eorum qui alias causas similitudinis et dissimilitudinis eorum quae generantur, assignaverunt'), ed. by Stadler, vol. 2, pp. 1205–14; Balss, *Albertus Magnus als Zoologe*, pp. 61–62 [53–54] and 66 [58]; Mayer, 'Die Personallehre in der Naturphilosophie von Albertus Magnus', p. 223.

Also noteworthy is Albert's view that physiognomy, physique, and age display certain psychical and mental characteristics. Invoking the authority of a physiognomist named Loxus,[44] he offers examples, drawn from many cases, of certain inferences concerning scientific aptitude that can be drawn from a person's body.[45] For example, collarbones far from the armpits, arranged on a wide and broad curve, indicate mental weakness; collarbones that are closed and pressed to the armpits and chest indicate a retardation of perception and a clumsiness of the mind; a more even arrangement of collarbones indicates optimal strength and intelligence. Similarly, an anatomically well-arranged chest, the individual parts of which are broad, is natural for human beings and therefore best. A thin chest reveals mental deficiency, while a fleshy one is typical of ineducable and ignorant human beings.[46]

Concluding this section, it should be noted that Albert repeatedly affirms the view that aging involves a clarification of the 'spirit'. The same claim is made by the anonymous author of *An anima racionalis sit mortalis*, whose positions mirror Albert's. This view of aging rests on the assumption that the gases produced in the digestion of food no longer arise, and so cause no disturbance of the animal pneuma. Lucidity of mind remains undimmed, supporting the sensory judgment and thus the cognitive ability.

## External Conditions as the Natural Prerequisites of an Individual's Cognitive Capacities

Exogenous factors form another aspect of the psychophysiological constitution and dispositions of human beings that Albert studies in detail, if less fully than the endogenous factors. In Albert's eyes, place and climate are the main such factors.

---

44 'Loxus' is an unidentified physician, likely of the third century BCE, whose physiognomic work is not preserved; Valentin Rose identifies him with Eudoxus of Cnidus: Rose, 'Die Physiognomia des Apuleius', pp. 82–83. See Thomann, 'Studien zum *Speculum physionomie*', pp. 5 and 10; Popović, *Reading the Human Body*, p. 88; Misener, 'Loxus, Physician and Physiognomist'; Evans, *Physiognomics in the Ancient World*, pp. 10–11.

45 Albertus Magnus, *De animalibus*, I.2.23.443, ed. by Stadler, vol. 1, p. 158, vv. 17–27: 'dicit Loxus, quod quibus iuguli conclusi sunt et compressi ad humeros et ad pectus, tarditatem indicant sensuum et animi stoliditatem, et manus talis hominis nec arti vel operi alicui quod manibus cum studio fit, esse ydoneas. Iuguli autem qui longe ab humeris separati sunt et largum ac latum sinum faciunt, imbecillitatem animi declarant. Medietas autem inter haec erit optima tam ad virtutem quam ad prudentiam. Pectus autem latum et bene dispositum, cuius latae sunt singulae partes, optimum est: quia hoc est naturale homini. Tenue autem pectus et invalidum et imbecillem significat animum. Pectus autem quod multis carnibus congestum est, indociles indicat et ignaros'. See Schipperges, 'Das medizinische Denken bei Albertus Magnus', p. 285.

46 Anon., *An anima racionalis sit mortalis*, MS CP 439 fol. 15$^{rb}$. Cf. Albertus Magnus, *De homine*, ed. by Anzulewicz and Söder, p. 375, vv. 21–24; Albertus Magnus, *De animalibus*, XIV.2.1, ed. by Stadler, vol. 2, p. 962, vv. 24–31, p. 963, vv. 21–31; Albertus Magnus, *Physica*, VII.1.9, ed. by Hossfeld, p. 535, vv. 18–25; Albertus Magnus, *Liber de natura et origine animae*, II.2, ed. by Geyer, p. 20, vv. 72–78.

With regard to these, Albert does not restrict his consideration to geographical location and ecology, but also considers cosmological place and the influence of cosmological powers on the sublunary realm. The study of these factors takes into account the climatic and meteorological conditions proper to a geographical location and the impact of the seasons and food.

Albert thus considers human beings within a complex network of causes and effects — one in constant transformation, involving the epigenesis of the human organism, its structure, growth, and degeneration. He works from the assumption that this process already begins before birth and ends in physical degeneration and death. He thus conceives of the more proximate endogenous causes and phenomena as being embedded within a much more broadly constituted succession of natural processes that are interconnected and simultaneous — as links in a causal chain. Albert touches on these issues within his theology of creation in his early works, and later in Book 2 of his *Sentences* commentary, when discussing the effects of the heavenly bodies upon human beings — the dispositions, behaviour, actions, and decisions that are conditioned by the complexions of the human body — and the general question of astral determinism as the result of particular constellations of stars and their influence, especially at the moment of birth.[47] A broad spectrum of such themes is addressed in several of Albert's writings in natural philosophy, especially in *De natura loci, De causis proprietatum elementorum, De caelo et mundo, Meteora*, and, from a physiological point of view, in the two commentaries on *De animalibus*.[48]

In this section, I sketch Albert's proposed accounts of the exogenous influence upon the psychophysiological constitution of human beings, which in large measure determines their differing aptitudes for science. I draw upon selected texts from Albert's *De natura loci* and his longer *De animalibus* commentary.

Albert's concept of place and his conception of the properties of place, as well as of the influence of place upon what it contains, are Aristotelian in origin.[49] These views are inspired by Aristotle insofar as Albert adapts his *mesotes* doctrine 'topographically' for the differentiation, explanation, and qualitative evaluation of climates.[50] He identifies geographical location as an active principle of procreation, with a role comparable to the father's, and deduces the properties of geographical location from its relation to cosmological place. Furthermore, he identifies the relation of place to place's content by analogy to the relation of form to matter. In that analogy, the role of form is attributed to all-encompassing cosmological place, from which proceeds the active principle for the formation

---

47 For more on this, see Anzulewicz, 'Der Einfluss der Gestirne'; Anzulewicz, '*Fatum*'.

48 See Cadden, 'Albertus Magnus' Universal Physiology'; Mayer, 'Die Personallehre in der Naturphilosophie von Albertus Magnus', pp. 238–39.

49 See Anzulewicz, 'Perspektive und Raumvorstellung'; Anzulewicz, 'Zwischen Spekulation und Erfahrung'; Tracey, 'Albert the Great on Possible Intellect'.

50 Aristotle, *NE* II.6, 1106b–1107a. See Hursthouse, 'Central Doctrine of the Mean', pp. 97–98; Ross, *Aristotle*, pp. 194–97.

of everything. Albert sets out these foundational assumptions at the beginning of *De natura loci*, with reference to his earlier discussions of the concept and properties of place in his Aristotle commentaries *Physica* and *De caelo et mundo*.[51] He supplements the properties of place with the dynamic principle of change and conservation, which represents place as opposed to place's content.[52]

Place is composed from a mixture of elements and is formed through the activity of the heavenly sphere as the first procreator. Place's general properties and place's effects upon a human being (as upon on all objects in space) configure themselves differently depending on the human being's distance from the heavenly sphere (*ex distantia ab orbe*), which is the overarching place that includes the place of all creatures, as well as on his condition and his relation to the planetary orbits (*ex situ et respectu loci ad vias planetarum*).[53] The forming of place by place's contents occurs through the instrumental causality of fire, which ascends on all sides of the round shell of the lunar sphere and exercises a constitutive and ordering influence upon the matter of the elements in the sublunary realm. With his richly detailed discussion of the mechanisms that generate the spheres of the four elements (fire, air, water, and earth) and their primary qualities (warm, cold, moist, dry), Albert elucidates the natural regularity of the coming into existence, in a manner conditioned by place, of the elements and their properties, as well as the formation of the properties of their mixtures and of the bodies composed from them, including the human body. In this way, the primary qualities of place shape the individual physical constitution of a human being and influence his mental disposition.[54]

---

51 Albertus Magnus, *De natura loci*, I.1, ed. by Hossfeld, p. 1, vv. 9–20: 'De natura locorum, quae provenit ex habitudine loci ad caelum, tractaturi primo facimus mentionem de his quae in *Physicis* determinata sunt; in his enim probatum est, quod locus est generationis principium activum quemadmodum pater. Cuius causa est, quod omne locatum se habet ad locum suum quemadmodum materia ad formam. Et quia superiora ad inferiora se habent sicut formae ad materias suas, sicut in *Caelo et mundo* diximus, oportet, quod superiora semper loca sint inferiorum, et ideo principium formationis inferiorum ex superioribus influitur eis sicut ex principiis activis'. Cf. Albertus Magnus, *Physica*, IV.1.1–15, ed. by Hossfeld, pp. 201–30, esp. IV.1.10, pp. 219–21; Albertus Magnus, *De caelo et mundo*, I.3.9, ed. by Hossfeld, p. 73, v. 49–p. 75, v. 63; ibid., II.1.5, p. 118, vv. 65–66; Anzulewicz, 'Zwischen Spekulation und Erfahrung', p. 75.

52 Albertus Magnus, *De natura loci*, I.1, ed. by Hossfeld, p. 1, vv. 21–24 and 27–28: 'transmutatio omnis contenti est per continens; locus autem continens est; et ideo ex ipso principium est transmutationis eius quod continet. [...] locum habere virtutem activam in locatum non est ambiguum'.

53 On this point and what follows, see ibid., I.3, ed. by Hossfeld, p. 4, vv. 26–41; Anzulewicz, 'Zwischen Spekulation und Erfahrung', pp. 78 ff.

54 Albertus Magnus, *De natura loci*, I.3, ed. by Hossfeld, p. 4, vv. 42 sqq.; ibid., I.5, p. 8, vv. 43–56 and 67–71: 'Si quis autem particulariter velit cognoscere omnes naturas et proprietates particularium locorum in aqua et aëre et terra, sciet, quod non est punctus in eis, qui non habeat specialem proprietatem ex virtute stellarum mediam habitationem commixtorum respicientium; ad quodlibet enim punctum habitationis animalium et plantarum et lapidum variatur circulus horizontis, et ad variationem circuli horizontis totus respectus caeli ad medium habitationis variatur. Qua de causa variantur naturae et proprietates et mores et actus et species eorum quae in eodem loco sensibili videntur generari, in tantum, quod et geminis seminibus et animalibus brutis et hominibus ex hoc

Human beings are also subject to secondary conditions of place, that is, to more proximate ecological influences produced by the landscape, the vicinity of mountains, sea, forests, and swamps, and the climate and weather. These factors affect one's general bodily composition, especially health, well-being, and lifespan, but also one's personality traits, habits, mental condition, and *eo ipso* one's aptitude for science.[55] In his second treatise in *De natura loci* and in several passages of *De animalibus*, Albert analyses the extent to which place, as determined in its properties by geographical length and breadth and as constituted by the configuration of heavenly bodies, shapes material reality as composed of the elements earth, water, air, and fire.[56] He also addresses how place shapes human bodies in unique ways and how place's secondary conditions affect human beings. In what follows, I present briefly only the most important aspects of his view regarding the influence of climates and heavenly bodies, the landscape and weather, and nutrition on the psychophysiological constitution and disposition of human beings.

Albert distinguishes seven fundamental types of climate zones covering the inhabitable regions of earth. Because they are defined on the basis of average climatic conditions, Albert proceeds on the assumption that there have to be two extremes in relation to the mean, and thus that the total number of climates with specifiable locales rises to twenty-one.[57] This climatic variability within inhabitable regions significantly influences the physical properties of all things existing in those places. According to Albert, because the mixtures of elements within bodies are proportional to the variations in climate, there are twenty-one different mixtures.

Albert characterizes the place-conditioned rules governing the natural constitution of the properties in composites in yet another specific respect: every species, he says, is constituted in a way proper only to itself, whereas the constitutional character of all individuals belonging to a species is said to be common. In accordance with the criterion governing species membership, this means that the complexion — that is, the mixture of opposed elements and their interconnection — is one and the same for all human beings. However, if one considers the

diverso respectu proprietates et mores diversi attribuantur. [...] necessario consequitur quodlibet punctum habitationis habere virtutes speciales, quibus informatur id quod locatur in ipso. Et haec causa est, quare nulla rerum generatarum invenitur omnino et per omnia in proprietatibus similis alteri'. Cf. Albertus Magnus, *Mineralia*, II.3.4, ed. by Borgnet, p. 53a.

55 Albertus Magnus, *De natura loci*, I.13, ed. by Hossfeld, p. 21, v. 65–p. 23, v. 10. See Anzulewicz, 'Zwischen Spekulation und Erfahrung', pp. 82–86; Mayer, 'Die Personallehre in der Naturphilosophie von Albertus Magnus', pp. 238–52.

56 Albertus Magnus, *De natura loci*, II.1–4, ed. by Hossfeld, pp. 23–28. Mayer, in 'Die Personallehre in der Naturphilosophie von Albertus Magnus', lists the many passages in *De animalibus* and Albert's other works.

57 Albertus Magnus, *De natura loci*, II.2, ed. by Hossfeld, p. 25, vv. 12–16. On the classification of climates and their characterization, see Al-Farghānī, *Differentie scientie astrorum*, c. 8–9, ed. by Carmody, pp. 13–14.

individual human being as a human being of a particular sort, the mixtures and complexions are as manifold as the number of individuals. Even within an individual, complexions are distinguished in three further respects. Albert accounts for the complexional differentiation within a single individual firstly by reference to an individual who has a qualitatively better complexion, secondly by reference to the dispositions that, in virtue of his composition, he ought to possess, and thirdly by reference to the dispositions that he actually possesses through the interaction of time, place, and other accidental causes. Albert excludes further place-based discussion of this question on the grounds that it belongs to the fields of arithmetic, botany, and zoology.[58]

Albert's analyses of the causes of different properties of geographical locations, and of the formative effect they exercise in correspondence with different climates, contain some intriguing observations concerning the fundamental effects of local climatic conditions on the psychophysiological constitution of human beings and especially on their dianoetic disposition. According to Albert, each of the seven climates is differentiated in a threefold manner, in accordance with the mean and two border zones at each extreme. Each of these climatic segments influences in its own unique way a human being's conception, prenatal development, and growth, the composition of his body's physiology, and the disposition of his soul's faculties. It is taken as a rule of natural philosophy that the constitution of locales, the complexion of composites, and the act of living are more strongly affected in a negative way by cold than by heat, and that in extreme cases coldness renders locales uninhabitable.[59] One practical effect of this is that the inhabitants of hot climate zones seem to be physically weak by nature. The heat causes their pneuma to evaporate, which is why they are said to begin to age very early and be frail already at the age of thirty; physiological and mental weaknesses emerge, as can be observed among the Ethiopians, inhabitants of the second tropical climate zone. Albert explains their black skin colour, lightweight

58 Albertus Magnus, *De natura loci*, II.2, ed. by Hossfeld, p. 25, vv. 17–40: 'Ut autem hoc melius intelligatur, dicemus duas esse cuiuslibet rei compositae ex contrariis commixtiones, unam quidem, quae est secundum proportionem speciei suae, et alteram, quam habet ex loco, in quo est perfecta eius generatio. Illam vero quae est ad speciem, habet communem cum omnibus suae speciei individuis, sicut dicimus unam esse mixtionem contrariorum in homine et aliam in asino et aliam in leone et sic de aliis. Et haec quidem in specie considerata una est. Sed si consideretur secundum esse individui designati in specie, tam multiplex est, quam est multus numerus individuorum. Est autem et in uno individuo multiplex, variata in communi tribus modis, secundum quod proportionatur individuum ad individuum, quod est melioris compositionis in specie sua, et secundum quod comparatur individuum ad dispositiones, in quibus debet esse secundum propriam compositionem, et secundum quod comparatur individuum ad dispositiones, in quibus est secundum tempus et locum et cetera accidentia, quae impressiones suas relinquunt in ipso. De his autem compositionibus singillatim tractare pertinet ad scientiam libri *De numeris* et ad scientiam libri *De vegetabilibus* et ad scientiam libri *De animalibus*'.

59 Ibid., II.2, p. 25, vv. 68 sqq.; ibid., II.3, p. 26, vv. 44 sqq. Cf. Albertus Magnus, *Meteora*, III.1.19, ed. by Hossfeld, p. 119, v. 44–p. 120, v. 56.

bodies, and 'simple-mindedness' as a consequence of the extreme heat and the evaporation of their pneuma.[60]

By contrast, Albert attests that inhabitants of the first tropical climate zone, along the equator, possess a strength of sense perception and a skill in invention whose cause lies in the agitating warmth and the sharpness of spirits. To emphasize his argument, Albert notes that outstanding philosophers, mathematicians, and magicians have owed their special abilities to the geographical, regional, and climatic conditions. Their skills are generated by strong radiation emanating at a perpendicular angle from the heavenly bodies.[61]

For those residing in the sixth and seventh climate zones, Albert claims that there are no initial conditions conducive to the physiological development and functioning of cognitive capacities and operations. The coldness in these climate regions results in a condensation of liquids in the organs, such that the physiological processes of sense perception are slowed down and the reception of sense impressions is obstructed. All the inhabitants of these two climate zones — who include Daciens, Goths, Slavs, and Parthians — are distinguished by their mental dullness and loutishness. When they do not overcome their climatically influenced lethargy through vigorous exercise, they show little interest in judicature, education and science, or art and artisanship. If they manage to free themselves from the cold-induced numbness, however, they reach a much better condition that can last for quite some time.[62]

Albert ascribes entirely different dispositions to the peoples of southern regions, more precisely of the fourth and the neighbouring fifth climate zone, than he does to those of the northern climate zones. The climatic conditions in the southern regions, as well as in locations in other climate zones with corresponding levels of warmth and moisture,[63] influence in a most beneficial manner the

---

60 Albertus Magnus, *De natura loci*, II.3, ed. by Hossfeld, p. 27, vv. 2–4: 'Aethiopes nigerrimi leves quidem sunt corpore et fatui mente propter defectum et evaporationem spiritus animalis'. For indications of how Albert is using Avicenna's teachings on the climes in the *Canon*, see Avicenna, *Liber canonis*, I.2.2.1.11, ed. Venetiis, fol. 32r. See Anzulewicz, 'Zwischen Spekulation und Erfahrung', p. 85; Mayer, 'Die Personallehre in der Naturphilosophie von Albertus Magnus', pp. 243–44.

61 Albertus Magnus, *De natura loci*, II.3, ed. by Hossfeld, p. 26, vv. 85–92: 'In operationibus autem animalibus, qui sub aequinoctiali sunt, vigent propter subtilitatem spirituum et plus in inveniendo propter calidum movens et acumen spirituum eorum. Cuius signum est, quia praecipui in philosophia in India fuerunt et praecipue in mathematicis et magicis propter fortitudinem stellarum super climata illa super quae perpendiculares radios proiciunt planetae'. Cf. Albertus Magnus, *De animalibus*, XII.1.2.20, ed. by Stadler, vol. 1, p. 805, v. 38.

62 Albertus Magnus, *De natura loci*, II.3, ed. by Hossfeld, p. 27, vv. 12–26 and 52–57: 'Operationes autem eorum animales non vigent propter spissitudinem, quippe umor eorum est piger et spissus nec oboedit motui et receptioni formarum animalium. Sunt igitur tales hebetes et stolidi, nisi hoc sit ex studii exercitio. Sed quando moventur, diu durant et efficiuntur multo meliores post exercitium'. See Mayer, 'Die Personallehre in der Naturphilosophie von Albertus Magnus', p. 244.

63 See Albertus Magnus, *De natura loci*, II.4, ed. by Hossfeld, p. 28, vv. 19–21: 'Et habitantes in locis calidis et umidis sunt sicut homines quarti climatis, inquantum sibi proportionantur'. See Mayer, 'Die Personallehre in der Naturphilosophie von Albertus Magnus', p. 244.

physiological processes and cognitive capacities of the human beings residing in them. For this reason, inhabitants of these regions foster the sciences and arts, and their noble demeanour and conduct, good morals, and healthy constitution are crowned with longevity.[64]

Albert's depiction of the properties of climate zones and their formative effects on the physical, physiological, and faculty-psychological constitution of their inhabitants shows that, on his view, the harsh climatic conditions of the northern regions generate a robustness of body, yet a certain sluggishness of mind. The local properties of the southern climates, especially the warmth and the mild weather conditions, have contrary effects. For their inhabitants, a more delicate bodily composition is natural, as is a style of thinking and a way of living marked by lightness and subtlety.

Albert counts seasons and nutrition among the additional exogenous factors affecting the physiological foundations of thought and the structure and function of the brain. For Albert, the impact of seasons on a human being differs depending on the kind of temperament he possesses. The melancholic individual can be more keen-witted in spring than in other seasons, since this is when the pneuma receiving the sensory forms exhibits the highest possible level of agility on account of the season's warmth and humidity. The choleric individual attains the best-tempered form of his physiological constitution — that is, of his pneuma and complexion, which provide the foundation for cognitive processes — during the cold season of winter. The phlegmatic enjoys the best periods for demanding kinds of thinking in summer, the sanguine person in autumn. Conversely, when measured by the cognitive performance of a particular temperament type, summer is the least favourable time for the choleric individual, winter for the phlegmatic, spring for the sanguine, and autumn for the melancholic. As the foregoing discussion showed, Albert regards the intermediate state — the equilibrium of complexion, which for him amounts methodologically to an application to physiology of the Aristotelian doctrine of the mean — as the optimal precondition for scientific aptitude in an individual of any temperament type.[65]

64 Albertus Magnus, *De natura loci*, II.3, ed. by Hossfeld, p. 27, vv. 57–70: 'Huius autem signum est, quod communitas populi meridionalis semper studet circa leges et studia liberalia et artes, de quibus non multum curat populus Dacus et Sclavorum. Propter quod iam quartum clima et vicinum sibi quintum laudabilia sunt, quae media sunt inter istas excellentias, habentia laudabiles utriusque gentis proprietates medias, secundum quod unicuique facile indagare, qui scit medium constitui ab extremis; est enim aetas istorum longa et operationes tam naturales quam animales laudabilissimae et mores boni et studia laudabilia, nisi ex consuetudine inducantur ad prava'. See Mayer, 'Die Personallehre in der Naturphilosophie von Albertus Magnus', pp. 245–46.

65 Albertus Magnus, *De sex principiis*, IV.6, ed. by Meyer, p. 43, v. 39–p. 44, v. 4: 'Anima enim, ut dicitur in *Libro de causis*, est in horizonte aeternitatis et temporis et ideo in actionibus et passionibus suis attingit tempus et sic etiam convenit ei quando. Hoc autem patet in animae parte speculativa quae tamen magis separata videtur esse a corpore quam pars motiva quae corpore utitur in motu. Secundum partem enim speculativam videmus quod *quidam* acutius *speculantur in vere* sicut melancholici, eo quod tunc spiritus qui deferunt formas speculationis subtiliantur et mobiles fiunt

According to Albert, the somatic, physiological foundations for cognitive processes (more precisely, the brain's size and the quality of its composition) and thus the capacity to think are also influenced, in ways that differ by age, by the quality and amount of food that the individual consumes. During childhood, the continuous flow of nutrients into the body's upper parts and the vapours rising into the head impair prudent and logical thinking. As I showed in the conclusion of the third part of this article, for Albert, such turbulence of the pneuma does not occur in adults, and especially not in the elderly. Even in adults, however, overeating not only leads to a temporary impairment of mental capacities, but may also trigger idiokinetic changes in the brain and the nervous system, owing to heightened absorption of incompatible materials. In this way, it may cause a functional insufficiency of the intellect and all the powers of sensation involved in the cognitive process.[66]

## Concluding Remarks

For Albert, the natural foundation for the condition and function of all psychic powers lies in the psychosomatic composition of human beings as produced by particular endogenous and exogenous causes, as well as in the physiological constitution of the brain, nervous system, organs, and pneuma and in the humoural complexion of the organism and of its individual parts. Those psychic powers include thinking itself, along with imagination and memory — two powers of spe-

---

ex calido et humectantur ex humido, ut facilius formas recipiant. E contra *quidam in hieme* acutius speculantur et illi sunt cholerici quibus calor cholerae, quando multiplicatur, commiscit operationes, et siccitas formarum impedit bonam receptionem. Sed in hieme vincente frigore temperatur calor operationes commiscens ita, ut non commisceat sed bene moveat, et humore hiemis inducente phlegma temperatur siccitas ad bonam susceptionem et bonam retentionem ex naturali siccitate cholerae. *Quidam* autem *in aestate* acutius contemplantur sicut phlegmatici, quorum frigus calido excellente temperatur, ut bene moveat spiritus, et humor multus temperatur sicco, ut et bene recipiat ex naturali humido et bene retineat ex temporali sicco. Quidam autem in autumno acutius speculantur ut sanguinei, quibus calorem sanguinis temperat moderata autumni frigiditas; humorem autem sanguinis exuberantem ad temperantiam restringit moderata autumni siccitas'. See Mayer, 'Die Personallehre in der Naturphilosophie von Albertus Magnus', pp. 251–52; Dewender, 'Albertus Magnus über Imagination und Krankheit', p. 200.

66 Albertus Magnus, *De animalibus*, XIV.2.1.29, ed. by Stadler, vol. 2, p. 963, vv. 21–31: 'Fluxus igitur talis nutrimenti efficit quod omnia animalia minoris sunt intellectus quam homo: et pueri etiam in quibus nutrimenti motus et spirituum non resedit, minoris sunt intellectus senibus presbiteris sicut diximus in septimo nostrorum *Physicorum*. Quandoque enim pondus nutrimenti multiplicatur in superiori diminuet intellectum et operationes virium animalium et causa huius est quia tunc multum extraneum admiscetur cerebro, quod sicut organum est principium operationum animalium: et admixto illo efficitur organum grossum et corporale non subtilium spirituum, et erit tunc parvi et pauci motus in operationibus et motibus animae sensibilis et intellectualis'. See Mayer, 'Die Personallehre in der Naturphilosophie von Albertus Magnus', pp. 227 and 254; Dewender, 'Albertus Magnus über Imagination und Krankheit', p. 200.

cial relevance to cognitive capacity and scientific aptitude.[67] While the psychical and bodily constitution and predisposition of human beings may differ greatly between individuals, in Albert's view it remains an expression of the teleological law inherent in human nature, which he interprets as a necessity that operates through internal and external factors *ex suppositione finis*.[68] The rational soul, as the highest natural form, shapes and forms the body, its organs and limbs, to be its instrument. Since, on Albert's view, the human being qua human is his intellect alone and the *telos* of intellect naturally consists in noetic perfection, it is on the basis of this *telos* that the human being's specific psychosomatic and physiological constitution is set, and to this *telos* that it is instrumentally proportioned.[69] As can be seen from my preceding remarks, the constitutional form that comes to concrete expression under the influence of exogenous factors (from which its function can be inferred) stands in a reciprocal relation to that final and necessary formation in respect of function.[70] The complexity and profundity of Albert's psychophysiological and anatomical account of the human being as both natural creature and intellect, as well as his account of the endogenous and exogenous factors that determine an individual's scientific aptitude, contrast markedly with the simple and fragmentary character of the anthropological, metaphysically based 'naturalism' that has been attributed to Alberich of Reims prior to Albert's day, during Albert's lifetime to Roger Bacon, and after Albert to John of Tytyngsale, Henry of Ghent, and John of Jandun.[71]

Albert developed the anthropological views I have outlined on a theoretical foundation furnished by the natural sciences available at his time, in particular on zoology and medicine, with physiology and anatomy as their theoretical disciplines. He was a step ahead of the scholars of his time in these subjects insofar as he integrated medicine into the Aristotelian life sciences, in the process making natural philosophy the foundational science for medicine.[72] By combining Aristotelian natural philosophy with a Galenic medicine that was philosophically influenced by Plato and Aristotle in their pure form, and in the Perso-Arabic form developed most notably by Avicenna, Albert closes the gap in science theory between the medicine of the Salerno school, which until then consisted

---

67 See Albertus Magnus, *De animalibus*, VIII.6.1.231–34, ed. by Stadler, vol. 1, p. 668, v. 22–p. 670, v. 26. See Mayer, 'Die Personallehre in der Naturphilosophie von Albertus Magnus', p. 235.

68 On this point and what follows, see Albertus Magnus, *De animalibus*, XI.1.3.43–50, ed. by Stadler, vol. 1, p. 776, v. 36–p. 779, v. 32; Albertus Magnus, *Liber de natura et origine animae*, I.5, ed. by Geyer, p. 13, vv. 64–81; ibid., I.6, p. 14, v. 49–p. 15, v. 35, esp. p. 15, vv. 16–27; Wallace, 'Albertus Magnus on Suppositional Necessity'; Wallace, 'Scientific Methodology of St Albert the Great'.

69 Albertus Magnus, *De quattuor coaequaevis* (*Summa de creaturis* I), IV.61.4, ed. by Borgnet, pp. 655b–656a; Albertus Magnus, *De animalibus*, I.1.1.2, ed. by Stadler, vol. 1, p. 2, vv. 5–6; Schipperges, 'Das medizinische Denken bei Albertus Magnus', p. 284.

70 See Mayer, 'Die Personallehre in der Naturphilosophie von Albertus Magnus', p. 207.

71 See König-Pralong, '*Omnes homines natura scire desiderant*', pp. 132–33.

72 See Krause, 'Grenzen der Philosophie'; Anzulewicz, 'Albertus Magnus über das Verhältnis von Medizin und Naturphilosophie'.

mainly of Graeco-Arabic and Jewish curative arts and pharmacy, and medicine's theoretical foundation as a natural philosophy and natural science of Aristotelian character.[73] Albert's knowledge of Greek medicine, especially of *Galenica*, and his appropriation of the medical and pharmaceutical expertise of the Perso-Arabic authors acquainted with Galen's legacy, has not yet been adequately studied.[74] Nevertheless, Albert had demonstrable acquaintance with it, as is clear from his use of Galenic texts, especially in his two commentaries on *De animalibus*, and from his use of Avicenna's *Canon*.

My overview may have clarified some of the main lines of Albert's attempt to synthesize the views of Aristotle, Galen, and Avicenna on central medical questions such as those of physiology and pneumology, the theory of humours and temperament theory, and the connection of medicine with the philosophy of Plato and Aristotle. The making of an enormously comprehensive inventory of zoological and medical knowledge, the critical examination of that inventory, and the transmission of contradictory viewpoints such as those of Aristotle and Galen — together with the repudiation of errors and the construction of sophisticated solutions — are all achievements that exerted a lasting influence on the study of medicine and the development of medicine as a science. Notwithstanding some ways in which it may be criticized from the standpoint of modern medicine, Albert's natural-philosophical and medical investigations of human beings remain eminently valuable for the history of science.[75]

---

73 See Albertus Magnus, *De sensu et sensato*, I.1.1, ed. by Donati, p. 20, vv. 38–57; Albertus Magnus, *De anima*, I.1.7, ed. by Stroick, p. 14, vv. 73–85. Also Jacquart and Paravicini Bagliani, *La Scuola Medica Salernitana*; Jacquart, '§ 52. Die Medizin als Wissenschaftsdisziplin'.

74 Although Heinrich Schipperges offers basic information concerning the assimilation of Galen and Perso-Arabic medicine in the Latin West, Albert plays an altogether marginal role in his study. Schipperges, *Die Assimilation der arabischen Medizin*.

75 Heinrich Balss criticizes Albert's views concerning the role of the heart and brain in the process of living — a subject on which clinical opinion remains divided to this day. His criticism seems unwarranted when he concludes: 'Considered as a whole, then, Albert's physiology has nothing original about it; indeed, its views of the heart as the chief seat of the animal functions mean that it is even a step backward compared with Galen'. Balss, *Albertus Magnus als Zoologe*, p. 57 [49]. Albert seeks to find a middle ground between extreme positions by ascribing simultaneously to the heart and brain elementary, mutually complementary life-functions; see Albertus Magnus, *De homine*, ed. by Anzulewicz and Söder, p. 337, v. 42–p. 338, v. 6 (where further quotations and references can also be found). Even as he defends him, Albert offers no shortage of criticisms of Galen, and in so doing exhibits his own independence and arrives at more sophisticated views. See Albertus Magnus, *De animalibus*, III.1.1–6, ed. by Stadler, vol. 1, pp. 277–305; Jacquart, '§ 52. Die Medizin als Wissenschaftsdisziplin', p. 1609. Balss maintains elsewhere that, on account of Albert's observations, he is to be regarded as 'a true biologist in the present-day sense', even though he is not so independent as a theoretician; Balss, *Albertus Magnus als Zoologe*, pp. 140–41 [132–33]. Paul Kopp, too, appraises Albert's psychopathological perspectives critically: Kopp, 'Psychiatrisches bei Albertus Magnus', pp. 59–60. By contrast, Schipperges finds evidence in Albert of a '*summa medicinae*', in which physiology is conceived as the natural history of the human being. Schipperges, 'Das medizinische Denken bei Albertus Magnus', p. 293; Schipperges, 'Eine *summa medicinae* bei

## Works Cited

### *Primary Sources*

Albertus Magnus, *Analytica posteriora*, ed. by Auguste Borgnet, Editio Parisiensis, 2 (Paris: Vivès, 1890), pp. 233–524

———, *De anima*, ed. by Clemens Stroick, Editio Coloniensis, 7/1 (Münster: Aschendorff, 1968)

———, *De animalibus libri XXVI. Nach der Cölner Urschrift*, ed. by Hermann Stadler, 2 vols, Beiträge zur Geschichte der Philosophie des Mittelalters, 15–16 (Münster: Aschendorff, 1916–20)

———, *De caelo et mundo*, ed. by Paul Hossfeld, Editio Coloniensis, 5/1 (Münster: Aschendorff, 1971)

———, *De generatione et corruptione*, ed. by Paul Hossfeld, Editio Coloniensis, 5/2 (Münster: Aschendorff, 1980), pp. 109–213

———, *De homine*, ed. by Henryk Anzulewicz and Joachim R. Söder, Editio Coloniensis, 27/2 (Münster: Aschendorff, 2008)

———, *De intellectu et intelligibili*, ed. by Auguste Borgnet, Editio Parisiensis, 9 (Paris: Vivès, 1890), pp. 477–525

———, *De natura loci*, ed. by Paul Hossfeld, Editio Coloniensis, 5/2 (Münster: Aschendorff 1980), pp. 1–44

———, *De quattuor coaequaevis* (*Summa de creaturis, prima pars*), ed. by Auguste Borgnet, Editio Parisiensis, 34 (Paris: Vivès, 1895), pp. 307–761

———, *De sensu et sensato*, ed. by Silvia Donati, Editio Coloniensis, 7/2a (Münster: Aschendorff, 2017), pp. 1–112

———, *De sex principiis*, ed. by Ruth Meyer, Editio Coloniensis, 1/2 (Münster: Aschendorff, 2006), pp. 1–80

———, *Liber de natura et origine animae*, ed. by Bernhard Geyer, Editio Coloniensis, 12 (Münster: Aschendorff, 1955), pp. 1–44

———, *Metaphysica, libri I–V*, ed. by Bernhard Geyer, Editio Coloniensis, 16/1 (Munster: Aschendorff, 1960)

———, *Meteora*, ed. by Paul Hossfeld, Editio Coloniensis, 6/1 (Münster: Aschendorff, 2003)

———, *Mineralia*, ed. by Auguste Borgnet, Editio Parisiensis, 5 (Paris: Vivès, 1890), pp. 1–103

---

Albertus Magnus'. Nancy G. Siraisi, Joan Cadden, and Thomas Dewender recognize Albert's medical learning and comprehensive knowledge of physiology. Siraisi, 'The Medical Learning of Albertus Magnus'; Cadden, 'Albertus Magnus' Universal Physiology'; Dewender, 'Albertus Magnus über Imagination und Krankheit'.

———, *Physica, libri I–IV*, ed. by Paul Hossfeld, Editio Coloniensis, 4/1 (Münster: Aschendorff, 1987)

———, *Physica, libri V–VIII*, ed. by Paul Hossfeld, Editio Coloniensis, 4/2 (Münster: Aschendorff, 1993)

———, *Quaestiones super De animalibus*, ed. by Ephrem Filthaut, Editio Coloniensis, 12 (Münster: Aschendorff, 1955), pp. 77–309

Al-Farābī, *Al-Farabi's Commentary and Short Treatise on Aristotle's 'De Interpretatione'*, trans. by F. W. Zimmermann (London: The British Academy, 1981)

Al-Farghānī, *Differentie scientie astrorum*, ed. by Francis J. Carmody (Berkeley: University of California Press, 1943)

Algazel [al-Ghazālī], *Metaphysics: A Mediaeval Translation*, ed. by Joseph T. Muckle, St Michael's Mediaeval Studies (Toronto: The Institute of Mediaeval Studies, 1933)

Anselmus Cantuariensis, *Cur Deus homo*, ed. by Franz S. Schmitt, S. Anselmi Cantuariensis Archiepiscopi Opera Omnia, 2 (Seckau: Thomas Nelson, 1938–61; repr. Stuttgart-Bad Cannstatt: frommann-holzboog, 1968), pp. 37–133

Aristotle, *Aristoteles Graece, Ex recensione Immanuelis Bekkeri*, ed. by Academia Regia Borussica, 2 vols (Berlin: Reimer, 1831)

———, *De animalibus: Michael Scot's Arabic-Latin Translation. Part Three, Books XV–XIX: Generation of Animals*, ed. by Aafke M. I. van Oppenraaij, Aristoteles Semitico-Latinus, 5 (Leiden: Brill, 1992)

———, *De generatione et corruptione, translatio vetus*, ed. by Joanna Judycka, Aristoteles Latinus, 9/1 (Leiden: Brill, 1986)

Averroes, *Commentarium magnum in Aristotelis De anima libros*, ed. by F. Stuart Crawford, Corpus Commentariorum Averrois in Aristotelem, 6/1 (Cambridge, MA: The Mediaeval Academy of America, 1953)

Avicenna, *De animalibus* (Venice, 1508; repr. Frankfurt am Main: Minerva, 1961)

———, *Liber canonis* (Venice, 1507; repr. Hildesheim: Olms, 2003)

———, *Liber de anima seu Sextus de naturalibus, IV–V. Édition critique de la traduction latine médiévale*, ed. by Simone Van Riet, introduced by Gérard Verbeke (Leuven: Peeters, 1968)

———, *Liber primus naturalium. Tractatus primus de causis et principiis naturalium. Édition critique de la traduction latine médiévale*, ed. by Simone Van Riet, introduced by Gérard Verbeke (Leuven: Peeters, 1992)

## Secondary Works

Anzulewicz, Henryk, 'Albertus Magnus über das Verhältnis von Medizin und Naturphilosophie', in *Medicina antiqua, mediaevalis et moderna: Historia – filozofia – religia*, vol. 3, ed. by Beata Wojciechowska, Sylwia Konarska-Zimnick, and Lucyna Kostuch (Kielce: Jan Kochanowski University Press, 2023), pp. 97–118

———, 'Der Einfluss der Gestirne auf die sublunare Welt und die menschliche Willensfreiheit nach Albertus Magnus', in *Actes de la V$^{ème}$ Conference Annuelle de la SEAC, Gdańsk 1997*, ed. by Arnold Le Beuf and Mariusz S. Ziółkowski (Warsaw: Institute of Archaeology, Warsaw University, 1999), pp. 263–77

———, '*Fatum*: Das Phänomen des Schicksals und die Freiheit des Menschen nach Albertus Magnus', in *Nach der Verurteilung von 1277: Philosophie und Theologie an der Universität von Paris im letzten Viertel des 13. Jahrhunderts. Studien und Texte*, ed. by Jan A. Aertsen, Kent Emery, Jr., and Andreas Speer (Berlin: De Gruyter, 2001), pp. 507–34

———, 'Perspektive und Raumvorstellung in den Frühwerken des Albertus Magnus', in *Raum und Raumvorstellungen im Mittelalter*, ed. by Jan A. Aertsen and Andreas Speer (Berlin: De Gruyter, 1998), pp. 249–86

———, 'Vermögenspsychologische Grundlagen kognitiver Leistung des Intellektes nach Albertus Magnus', *Acta Mediaevalia*, 22 (2009), 95–116

———, 'Zwischen Spekulation und Erfahrung: Alberts des Großen Begriff vom Raum', in *Représentations et conceptions de l'espace dans la culture médiévale*, ed. by Tiziana Suarez-Nani and Martin Rohde (Berlin: De Gruyter, 2011), pp. 67–87

———, and Philipp A. C. Anzulewicz, 'Einleitung', in Albert der Große, *Über das Gewissen und den praktischen Intellekt: Eine Textauswahl aus 'De homine', den 'Quaestiones' und 'De anima'. Lateinisch-Deutsch*, ed. and trans. by Henryk Anzulewicz and Philipp A. C. Anzulewicz (Freiburg: Herder, 2019), pp. 9–109

Balss, Heinrich, *Albertus Magnus als Biologe* (Stuttgart: Wissenschaftliche Verlagsgesellschaft, 1947)

———, *Albertus Magnus als Zoologe* (Munich: Verlag der Münchner Drucke, 1928)

Barbado, F. M., 'La physionomie, le tempérament et le caractère, d'après Albert le Grand et la science moderne', *Revue Thomiste*, 36 (1931), 314–51

Bono, James J., 'Medical Spirits and the Medieval Language of Life', *Traditio*, 40 (1984), 91–130

Cadden, Joan, 'Albertus Magnus' Universal Physiology: The Example of Nutrition', in *Albertus Magnus and the Sciences: Commemorative Essays 1980*, ed. by James A. Weisheipl (Toronto: Pontifical Institute of Mediaeval Studies, 1980), pp. 321–39

de Libera, Alain, 'Épicurisme, stoïcisme, péripatétisme: L'histoire de la philosophie vue par les Latins (XII$^{e}$–XIII$^{e}$ siècle)', in *Perspectives arabes et médiévales sur la tradition scientifique et philosophique grecque*, ed. by Ahmad Hasnawi, Abdelali Elamrani-Jamal, and Maroun Aouad (Leuven: Peeters, 1997), pp. 343–64

Dewender, Thomas, 'Albertus Magnus über Imagination und Krankheit', in *Gesundheit im Spiegel der Disziplinen, Epochen, Kulturen*, ed. by Dietrich Grönemeyer and others (Tübingen: Niemeyer, 2008), pp. 187–203

Evans, Elizabeth C., *Physiognomics in the Ancient World* (Philadelphia, PA: American Philosophical Society, 1969)

Hödl, Ludwig, '*Opus naturae est opus intelligentiae*: Ein neuplatonisches Axiom im aristotelischen Verständnis des Albertus Magnus', in *Averroismus im Mittelalter und in der Renaissance*, ed. by Friedrich Niewöhner and Loris Sturlese (Zurich: Spur, 1994), pp. 132–48

Hursthouse, Rosalind, 'The Central Doctrine of the Mean', in *The Blackwell Guide to Aristotle's Nicomachean Ethics*, ed. by Richard Kraut (Oxford: Blackwell, 2006), pp. 96–115

Jacquart, Danielle, '§ 52. Die Medizin als Wissenschaftsdisziplin und ihre Themen', in *Die Philosophie des Mittelalters, Bd. 4: 13 Jahrhundert*, ed. by Alexander Brungs, Vilem Mudroch, and Peter Schulthess (Basel: Schwabe, 2017), pp. 1595–1612

———, and Agostino Paravicini Bagliani, eds, *La Scuola Medica Salernitana: Gli autori e i testi* (Florence: Edizioni del Galluzzo, 2007)

König-Pralong, Catherine, '*Omnes homines natura scire desiderant*: Anthropologie philosophique et distinction sociale', in *The Pleasure of Knowledge / Il piacere della conoscenza*, ed. by Pasquale Porro and Loris Sturlese (Turnhout: Brepols, 2015), pp. 121–38

Kopp, Paul, 'Psychiatrisches bei Albertus Magnus', *Zeitschrift für die gesamte Neurologie und Psychiatrie*, 147 (1933), 50–60

Krause, Katja, 'Grenzen der Philosophie: Alberts des Großen Kommentar zu *De animalibus* und die Medizin', *Documenti e studi sulla tradizione filosofica medievale*, 30 (2019), 265–93

Lennox, James G., 'Aristotle on the Biological Roots of Virtue: The Natural History of Natural Virtue', in *Biology and the Foundation of Ethics*, ed. by Jane Maienschein and Michael Ruse (Cambridge: Cambridge University Press, 1999), pp. 10–31

———, *Aristotle's Philosophy of Biology: Studies in the Origins of Life Science* (Cambridge: Cambridge University Press, 2001)

Maienschein, Jane, 'Epigenesis and Preformationism', in *The Stanford Encyclopedia of Philosophy* (Spring 2017 Edition), ed. by Edward N. Zalta, https://plato.stanford.edu/archives/spr2017/entries/epigenesis/

Marechal, Patricia, 'Galen's Constitutive Materialism', *Ancient Philosophy*, 39 (2019), 191–209

Mayer, Claudius Franz, 'Die Personallehre in der Naturphilosophie von Albertus Magnus: Ein Beitrag zur Geschichte des Konstitutionsbegriffs', *Kyklos: Jahrbuch des Instituts für Geschichte der Medizin an der Universität Leipzig*, 2 (1929), 191–257

Meroni, Michele, 'The Doctrine of Spirit, Divinatory Dreams and Prophecy: Analysis of the Arabic Influence in Albert the Great's *Parva Naturalia*' (unpublished doctoral thesis, University of Milan / LMU Munich, 2021)

Misener, Geneva, 'Loxus, Physician and Physiognomist', *Classical Philology*, 18 (1923), 1–22

Phillipson, Traci, *Aquinas, Averroes, and the Human Will* (Milwaukee, WI: e-publications@Marquette, 2017), https://libus.csd.mu.edu/search/a?SEARCH=Phillipson,+Traci&searchscope=1

Popović, Mladen, *Reading the Human Body: Physiognomics and Astrology in the Dead Sea Scrolls and Hellenistic-Early Roman Period Judaism* (Leiden: Brill, 2007)

Rose, Valentin, 'Die Physiognomia des Apuleius nach Polemon mit zusätzen aus Eudoxus und Aristoteles', in *Anecdota Graeca et Graecolatina: Mitteilungen aus Handschriften zur Geschichte der griechischen Wissenschaft*, ed. by Valentin Rose, Part 1 (Berlin: Duemmler, 1864), pp. 61–169

Ross, David, *Aristotle* (London: Methuen, 1968)

Schipperges, Heinrich, 'Das medizinische Denken bei Albertus Magnus', in *Albertus Magnus, Doctor universalis: 1280/1980*, ed. by Gerbert Meyer and Albert Zimmermann (Mainz: Grünewald, 1980), pp. 279–94

———, *Die Assimilation der arabischen Medizin durch das lateinische Mittelalter*, Sudhoffs Archiv. Beihefte, 3 (Wiesbaden: Steiner, 1964)

———, 'Eine *summa medicinae* bei Albertus Magnus', *Jahres- und Tagungsberichte der Görresgesellschaft 1981* (1981), 5–24

Schönberger, Rolf, 'Rationale Spontaneität: Die Theorie des Willens bei Albertus Magnus', in *Albertus Magnus: Zum Gedenken nach 800 Jahren. Neue Zugänge, Aspekte und Perspektiven*, ed. by Walter Senner and others (Berlin: Akademie Verlag, 2001), pp. 221–34

Siraisi, Nancy G., 'The Medical Learning of Albertus Magnus', in *Albertus Magnus and the Sciences: Commemorative Essays 1980*, ed. by James A. Weisheipl (Toronto: Pontifical Institute of Mediaeval Studies, 1980), pp. 379–404

Stubbe, Hannes, *Albertus Magnus: Der erste Kölner und mitteleuropäische Psychologe*, 2nd edition (Aachen: Shaker, 2016)

Taylor, Richard C., '*Cogitatio, Cogitativus* and *Cogitare*: Remarks on the Cogitative Power in Averroes', in *L'élaboration du vocabulaire philosophiques au Moyen Âge*, ed. by Jacqueline Hamesse and Carlos Steel (Turnhout: Brepols, 2000), pp. 111–46

Thomann, Johannes, 'Studien zum *Speculum physionomie* des Michele de Savonarola' (unpublished doctoral thesis, Universität Zürich, 1997)

Tracey, Martin J., 'Albert the Great on Possible Intellect as *locus intelligibilium*', in *Raum und Raumvorstellungen im Mittelalter*, ed. by Jan A. Aertsen and Andreas Speer (Berlin: De Gruyter, 1998), pp. 287–303

Wallace, William A., 'Albertus Magnus on Suppositional Necessity in the Natural Sciences', in *Albertus Magnus and the Sciences: Commemorative Essays 1980*, ed. by James A. Weisheipl (Toronto: Pontifical Institute of Mediaeval Studies, 1980), pp. 103–28

———, 'The Scientific Methodology of St Albert the Great', in *Albertus Magnus, Doctor universalis: 1280/1980*, ed. by Gerbert Meyer and Albert Zimmermann (Mainz: Grünewald, 1980), pp. 385–407

Weisheipl, James A., 'The Axiom "Opus naturae est opus intelligentiae" and Its Origins', in *Albertus Magnus, Doctor universalis: 1280/1980*, ed. by Gerbert Meyer and Albert Zimmermann (Mainz: Grünewald, 1980), pp. 441–63

KATJA KRAUSE

# Chapter 11. Source Mining

## *Arabic Natural Philosophy and* experientia *in Albert the Great's Scientific Practices*

The exact ways in which Albert followed the lead of his Arabic sources as he adopted and reworked the natural philosophical insights they offered for each particular doctrine has been an ongoing concern in the literature. In most cases, the question has been approached from the perspective of a reception history, asking predominantly whether and how far Albert remained truthful to the original teachings contained in his sources.[1] In this chapter, I wish to complement that approach with a different one that underscores Albert's own expressed purposes and the ways he pursues them in the specifics of his natural philosophy. Keeping in mind Albert's own presentation of evidence, definitions, and explanations — his 'intellectual practices', to cite Lorraine Daston,[2] or 'scientific practices' to use a more common term — my aim is to investigate when Albert incorporated

1 There are many very fine studies on Albert's appropriation of Arabic thought, with a particular emphasis on the *Physics* and *Metaphysics*. Among the most important recent ones are Bertolacci, '"Subtilias speculando"'; Bertolacci, 'Le citazioni implicite testuali'; Bertolacci, 'The Reception of Avicenna's *Philosophia prima*'; Bertolacci, 'Albert the Great'; Bertolacci, 'A New Phase of the Reception of Aristotle'; Bertolacci, 'Albert's Use of Avicenna and Islamic Philosophy'; Bertolacci, 'Avicenna's and Averroes's Interpretations'; Bertolacci, '"Averroes ubique Avicennam persequitur"'; Burger, 'Albertus Magnus'; Caminada, 'A Latin Translation?'; Donati, 'Is Celestial Motion a Natural Motion?'; Endress, *Der arabische Aristoteles*; Hasse, 'The Early Albert Magnus'; Hasse, 'Der mutmaßliche arabische Einfluss'; Hasse, 'Avicenna's "Giver of Forms"'; Lizzini, 'Flusso, preparazione appropriata e *inchoatio formae*'; López-Farjeat, 'Albert the Great'; Müller, 'Der Einfluss der arabischen Intellektspekulation'; Schwartz, 'Celestial Motion'; Schwartz, 'Divine Space'; Tellkamp, 'Why Does Albert the Great Criticize Averroes?'; Wéber, 'Un thème de la philosophie arabe'.

2 I draw on Lorraine Daston's distinction between 'intellectual practices', which she identifies as including 'the presentation of evidence and arguments', from 'cognitive practices', which indicate 'a

**Katja Krause** (kkrause@mpiwg-berlin.mpg.de) is professor of the history of science at the Technical University Berlin and leads a research group at the Max Planck Institute for the History of Science.

*Albert the Great and his Arabic Sources*, ed. by Katja Krause and Richard C. Taylor, Philosophy in the Abrahamic Traditions of the Middle Ages, 5 (Turnhout: Brepols, 2024), pp. 311–334
BREPOLS PUBLISHERS 10.1484/M.PATMA-EB.5.136491

certain insights from Arabic natural philosophers, how he reworked them, and why he mined these sources to establish the contours of his own natural scientific enterprise and defend its truth.[3]

By turning from the general meaning of Arabic authorities in Albert's comprehensive science (*scientia*) to the meaning of one particular aspect these sources gave him, that of experience (*experientia, experimentum*), we do not much narrow down the multiplicity of purposes he had in mind. Usually uninhibited by the original intentions of the Arabic sources, Albert's mining of particular experiences from them followed its own epistemic concerns.

To show what these concerns were, and how they differed from those of the original sources, I focus on two cases drawn from Albert's vast natural philosophical corpus. In the first, I show how he 'transhistoricizes' empirical evidence contained in Avicenna's medical *Canon* — by which I mean that he focuses on its epistemic rather than its authoritative value, as will become clear below. In the second, I show how he establishes his mature doctrine of taste in reliance on Averroes's principle of form-matter relations as the best possible explanation of the particular taste of saltiness.

Although this choice of cases is highly selective, and although it neither draws on his own direct experience nor includes experience used to verify or falsify a given scientific theory — as we would perhaps expect its epistemic relevance to be from a post-Scientific Revolution perspective — I nonetheless wish to show that Albert's scientific practices encompassed cases of mining his sources that went far beyond those particular epistemic concerns, which became fixed much later in history.

My aim here is twofold. First, I suggest that Albert's references to experience relied upon an epistemic value utterly different from those familiar to us — one that concerned more the hearer of the science than the scientific object or

---

learned (and learnèd) habitus, which has bodily, mental, and ethical components'. Daston, 'Taking Note(s)', p. 446.

3 See, for instance, Albertus Magnus, *De animalibus*, IX.2.3, ed. by Stadler, vol. 1, p. 714, vv. 18–20: 'In omnibus autem inductis non intendimus, nisi quod ratio dicti Galieni non est sufficiens, sed de ipsa positione eius nichil omnino diximus, an vera sit vel falsa per rationem probantem'; ibid., XI.1.1, vol. 1, p. 761, vv. 1–7: 'Omnibus hiis diversitatibus animalium habitis oportet modo aliud ordiri principium circa causas inveniendas eorum quae diximus. Differentias autem substantiales animalium supra posuimus et differentias membrorum et partium omnium substantiales exsequuti sumus: et insuper posuimus differentias eorum quae accidunt eis tam communiter quam proprie: et oportet utrorumque istorum, prout possumus, invenire causas naturales et veras'; Albertus Magnus, *De anima*, I.1.2, ed. by Stroick, p. 4, v. 54–p. 5, v. 8: 'Utilitas *autem* eius praecipua est, quod *ad omnium* scibilium *veritatem cognoscendam maxime proficit*, et praecipue *ad* notitiam veritatis rerum *naturalium*, cum ipsa sit *principium* formale et essentiale *animalium*, et non nisi per notitiam animae poterunt cognosci corpora animatorum. In corporibus autem animatorum et commixtio est simplicium et ipsa simplicia, et sic ulterius scientia animae proficit ad notitiam omnium corporum naturalium. Ad veritatem autem omnium proficit praeter modum, quem diximus superius, quo videlicet apud se habet lumen, quod est omnis veritatis examen, tribus modis, non simul, sed divisis. Quorum unus est, quia proficit ad naturalem veritatem, eo quod ipsa pars nobilissima est scientiae naturalis'.

approach under scrutiny. Second, I wish my focus on Albert's 'mined experience' to show that his natural science followed an epistemic purpose going beyond the purposes thus far identified by the literature, namely truthfulness to the original, amassing encyclopaedic knowledge of as many sources as possible, avoiding the danger of the double-truth, or responding to his Latin interlocutors.[4] This epistemic purpose was one that Albert himself identified as crucial to his overarching science and that is very familiar to scholarship: the combination of truth with certainty and epistemic comprehensiveness to the extent that these can be achieved through the intellectual practices of defining and explaining.

## Transhistoricizing Empirical Evidence

Probably during the second decade of the 1200s, Michael Scot concluded his Latin translation of Aristotle's *De animalibus* from the Arabic language. Like his Arabic template, the translation comprised three of Aristotle's works, *Historia animalium* (Books I–X), *De partibus animalium* (Books XI–XIV), and *De generatione animalium* (Books XV–XIX). Before Albert's long commentary on the *De animalibus* and his subsequent *Quaestiones super De animalibus*, two other Latin thinkers wrote commentaries on Michael Scot's translation: Peter of Spain (Petrus Hispanus *medicus*), whose commentary, composed around 1240, is still extant in two manuscripts, and Roger Bacon, whose commentary, possibly composed somewhat earlier, is lost to us.[5] Following Peter of Spain's lead, but in his own comprehensive and innovative ways, Albert incorporated large amounts of Avicenna's *Canon* into his *scientia de animalibus*, particularly as regards human anatomy and physiology.[6]

Avicenna's *Canon* is a medical work in five volumes, originally composed in Arabic and translated into Latin in the twelfth century by Gerard of Cremona. It was utilized by thirteenth- and fourteenth-century Latin physicians predominantly for the ideas in Book I, where — among many other anatomical matters

---

4 On these concerns, see, for instance, Donati, 'Alberts des Großen Konzept der *scientiae naturales*'; Tracey in this volume.

5 On the history of Peter of Spain's commentary, see Navarro Sanchez, *Peter of Spain*. Important studies on insights of the *De animalibus* tradition in the thirteenth century include Pouchet, *Histoire des sciences naturelles au Moyen Âge*; Zaunick, 'Albertus Magnus'; Wingate, *Mediaeval Latin Versions*; Pelster, 'Die beiden ersten Kapitel der Erklärung Alberts des Großen'; Gerhardt, 'Zoologie médiévale'; Hünemörder, 'Die Zoologie des Albertus Magnus'; Asúa, 'Organization of Discourse on Animals'; Asúa, 'El Comentario de Pedro Hispano'; Asúa, 'Peter of Spain'; Dold, 'What is Zoology About?'.

6 On Albert's medical learning, see especially Kopp, 'Psychiatrisches bei Albertus Magnus'; Killermann, 'Die somatische Anthropologie'; Shaw, 'Scientific Empiricism in the Middle Ages'; Schipperges, 'Das medizinische Denken'; Schipperges, 'Eine *summa medicinae*'; Demaitre and Travil, 'Human Embryology and Development'; Siraisi, 'Medical Learning of Albertus Magnus'; Asúa, 'Organization of Discourse on Animals'; Asúa, 'Albert the Great'; Theiss, *Die Wahrnehmungspsychologie und Sinnespsychologie*.

— Avicenna discusses the question of sensation in teeth, showing that teeth form the exception to the general Galenic anatomical rule that 'no bone [...] has sensation':[7]

> For Galen said [*dixit*] that experience [*experimentum*] has shown us that they [i.e., teeth] have sensation, about which nature was very careful and produced it with a power that originates in the brain, so that, for this reason, they may also discern between hot and cold.[8]

Presenting Galen's testimony of his experience, probably derived from his *De ossibus ad tirones*,[9] Avicenna emphasizes the great spatio-temporal distance between the two thinkers. Using the past tense, *dixit Galenus*, he locates Galen's experience in a deep history that took place centuries before it was retold in the *Canon*. Avicenna thus creates three separate historical moments, dividing Galen's *experimentum* from his report of his *experimentum* and both of these from Avicenna's own testimony on Galen's report. But the temporal gaps he thus creates are also epistemic gaps. Avicenna provides no report of testing Galen's experience, no account of his own sensation in his teeth in everyday circumstances, no case of his patients' toothache. Instead, Avicenna summons Galen's authoritative *experimentum* to validate the scientific conclusion that teeth are the exception to the rule. In keeping with his own definition of medicine as *scientia*, on a par with philosophy, Avicenna's intellectual practice thus privileges the authoritative value over the evidentiary value of *experimentum*, and incorporates it into the intellectual activity of argument for a conclusion.[10] In short, Galen's *experimentum* is inscribed into Avicenna's *Canon* with an authoritative value.

In contrast, when Albert extracted these insights from Avicenna's *Canon* and incorporated them into his *De animalibus* commentary, he transformed the authoritative value of *experimentum* into an evidentiary value. In his account, Albert no longer emphasizes the spatial, temporal, and epistemic gaps as Avicenna did, but stresses instead the epistemic warrant that *experimentum* supplies for the rational conclusion:

> No bone apart from teeth, as Galen and Avicenna say [*dicunt*], has sensation. For concerning teeth, they say [*dicunt*] that experience shows [*experimentum demonstrat*] that teeth have sensation. And this is decreed by the sagacity of nature, for it has supplied them with sensation together with a sensory power,

---

7 Avicenna, *Liber canonis*, I.1.5, ed. Venetiis, fol. 10ra: 'Nullum preterea ossium ullo modo sentit preter dentes'.

8 Ibid.: 'Galenus enim dixit, quod experimentum nobis demonstrauit eos [sc. dentes] sensum habere: de quo natura sollicita fuit: et fecit ipsum cum uirtute que a cerebro prouenit, idcirco ut ipsi etiam inter calidum et frigidum discernant'.

9 For Galen's report of his own experience, see Claudius Galenus, *De ossibus ad tirones*, ed. by Kühn, vol. 2, p. 754, vv. 13–15: 'Participes vero sunt nervorum mollium, qui a cerebro, dentes soli e reliquis ossibus; unde et soli manifeste sentiunt'.

10 On this distinction, see the Introduction to this volume.

which descends from the brain, so that they may also discern between hot and cold.[11]

Despite striking similarities in content, Albert's report appears to move away from Avicenna's emphasis on experience as evidence that has been reported and towards experience as evidence that warrants or secures the truth.[12] That reading is supported by his use of the present tense (*dicunt*), his conflation of the two *experimenta* of Galen and Avicenna, and his separation between reporting on the subjects of the *experimentum* (*dicunt*) and on its evidentiary function. Rather than emphasizing an authoritative inheritance, Albert puts Galen's and Avicenna's experience at the service of empirically verifying the conclusion that teeth are the exception to the rule. In this way, he privileges the evidentiary value of experience over its authoritative value (regardless of the fact that it is testimonial experience only), and integrates it into the predominant intellectual practices of defining and explaining in his works. Likewise, Albert grants 'the sagacity of nature' the ontological warrant of truth, giving nature an intrinsic authority: an authority of final causality, a wise decree, for the sake of physiological function. This, too, confirms that Albert shifts the epistemic weight away from the transmitted *inheritance* of this piece of knowledge to a natural scientific *explanation* of it.

For present-day tastes, granting evidentiary status to the *experimentum* of Galen and Avicenna to the extent that Albert does seems like an inequitable handling of the evidence and a potential distortion of the conclusion. But Albert was not prey to our epistemic fears. His epistemology was an optimistic one, culminating in a deeply held conviction that 'all activities that arise from nature [*a natura*] are uniform in all things that possess this nature'.[13]

Here, let me stress once again that for Albert, this principle applies not only to the scientific object under investigation — the sagacity of nature for sensation in teeth — but also, and most importantly, to the scientist who is pursuing the investigation. Encompassing natural activities and actions carried out to realize the human 'desire to know by nature', as Albert puts it, elaborating on Aristotle's *Metaphysics*, sense perceptions and the ability to build truthful universals follow 'the nature of the species'.[14] Albert believed this uniformity of nature was true for

11 Albertus Magnus, *De animalibus*, I.2.6, ed. by Stadler, vol. 1, p. 71, vv. 1–6: 'Nullum autem ossium, ut dicunt Galienus et Avicenna, praeter dentes habet sensum. De dentibus enim dicunt, quod experimentum demonstrat, dentes habere sensum: et hoc sagacitate naturae factum est: fecit enim eis sensum cum virtute sensus quae a cerebro descendit, ut ipsi etiam inter calidum et frigidum discernant'.

12 On premodern experience more generally, see Krause with Auxent and Weil, 'Making Sense of Nature in the Premodern World', and the literature quoted there.

13 Albertus Magnus, *Super Ethica*, I.2, ed. by Kübel, p. 12, vv. 32–33: 'Operationes omnes quae sunt a natura, sunt uniformes in omnibus habentibus naturam'.

14 Albertus Magnus, *Metaphysica*, I.1.5, ed. by Geyer, p. 7, vv. 36–40: 'omnes homines natura scire desiderant. Cum enim hoc desiderium sit omnium quorum in specie determinata est natura una, erit hoc desiderium naturale et naturam speciei consequens'.

all individuals (that is, all philosophically trained ones) across space and time.[15] His principled belief in the uniformity of natural being and activities explains to us why Albert found himself able to integrate Avicenna's reference to Galen's *experientia* into his *scientia de animalibus* in the way that he did. His reworking reveals a potential to participate in Galen's act of experiencing, as testified in Avicenna, not experientially but in a conceptual way that transcends linguistic, spatial, and temporal boundaries — precisely because Albert took concepts derived from experience to be, in most cases, universally true as well. The grounds for this sharing lay not in a testing or repetition of experience of the kind demanded by later epistemic ideals, but in the firm belief in the correct and shared workings of the human soul, which ensured a continued and universally human epistemology even for the experiencing of particulars.[16]

Albert's inheritance of Avicenna's testimony to Galen's *experimentum* was not a simple assumption or adoption. It was, rather, a matter of reworking epistemic values, of shifting from authoritative to evidentiary values, grounded in a firm belief that the activities of the soul, at least those of the sensitive and vegetative parts, are universally shared across humanity and have a common goal: they assist the perfection of the human intellect. This reworking was no accident. Albert's intellectual motivation, his scientific program, leaves no doubt that *experimentum* in conjunction with *ratio* leads to specific definitions of animals, the backbone of his *scientia perfecta de animalibus*:

> From all that has been put forth, it is evident that whoever wishes to narrate and convey through teaching what they have cognized by reason and seen by experience of the natures of animals must be in possession of definitions known per se, by which the intention of the one speaking about the natures of the animals is guided according to those definitions. For these definitions are the means to prove everything else that is sought in the natures [of animals], and through them, it must be judged whether what is said to belong to animals by common or proper accidents (1) might be true with certainty, if it can be demonstrated, or (2) might be close to or approach the truth, if it is gathered from probable things, since knowledge by demonstration cannot be had of

---

15 This is not to say that, for Albert, humans cannot engage in diverse activities on an individual level. It is also not to say that they do not have different epistemic dispositions, potencies, or capabilities. In Albert's eyes, all these arise on the physiological level of complexion — an ontological explanation of individual differences in material features. But this applied neither to the level of specific characteristics, nor to the scope of different human activities or actions. Albert's concentration on human nature *as* human enabled him to trust that any activity, be it external sense perception or experience in its technical sense, is uniform, at least for the most part, across the species. Albert supposed his own sensation in his teeth, therefore, to be no different from the sensation that Galen and Avicenna had. On these individual conditions, see Anzulewicz, 'Psychophysiology, Natural Spaces and Climata'; Cadden, 'Albertus Magnus' Universal Physiology'.

16 See also Lorraine Daston's observations on Aristotle in her thought-provoking article on the epistemic fear of error in the early modern and Enlightenment periods. Daston, 'Scientific Error', esp. her discussion of Aristotle on p. 5.

> all things. But concerning some things it is necessary to conjecture, and we believe that the things that do not oppose the natures of animals belong to them with probability.[17]

Albert's reworking of epistemic values, and its scientific goals in specific definitional knowledge of all animals (which equals the truth about them), was an epistemic constant across his incorporations of Avicenna's empirical evidence in his *De animalibus* and in his other works, where he referenced this evidence explicitly.[18]

The specific practice of reworking authoritative experience into experience as warrant cannot, however, be generalized across his oeuvre. Quite different uses of experience — such as the notion that a given teaching results in the best possible explanation of experiential knowledge — appear in Albert's treatments on the five external senses. Ultimately, though, they are grounded in the same overarching epistemic value.

---

17 Albertus Magnus, *De animalibus*, XI.1.1, ed. by Stadler, vol. 1, p. 763, vv. 6–18: 'Ex omnibus igitur inductis manifestum est, quod quicumque vult docendo narrare et tradere quod per rationem cognovit et quod per experimentum vidit de naturis animalium, debet habere diffinitiones notas per se, per quas dirigatur intentio loquentis de naturis animalium secundum illas diffinitiones, quia ipsae sunt medium ad probandum omne aliud quod quaeritur in naturis, et per eas debet iudicari utrum hoc quod dicitur inesse animalibus de accidentibus communibus aut propriis, sit certitudinaliter verum si demonstrari potest: aut sit circa vel prope verum, si ex probabilibus colligitur, quoniam non in omnibus haberi potest per demonstrationem scientia, sed in quibusdam coniecturare oportet, et quae non repugnant animalium naturis, probabiliter haec eis credimus inesse'.

18 For instance, Albertus Magnus, *De homine*, ed. by Anzulewicz and Söder, p. 206, vv. 32–39: 'Praeterea, quaecumque duo sic se habent quod unum percipitur sine altero, illorum unum non est aliud; motus et sonus sic se habent; ergo unum illorum non est aliud. Prima patet per se. Secunda probatur per experimentum, quia multotiens percipimus sonum non percipientes motum. Cum igitur "idem non possit simul sciri et ignorari", ut dicit Avicenna, verum est, quoniam sonus non erit motus aëris'; ibid., p. 269, vv. 17–48; Albertus Magnus, *De anima*, II.3.6, ed. by Stroick, p. 106, vv. 5–9: 'Et ideo ventus non aufert vel affert colores, sed bene obtundit auferendo sonos in parte et non in toto; odores autem et affert et aufert in toto, sicut dicit Avicenna et veritas per experimenta attestatur'; ibid., II.3.29, p. 140, v. 63–67: 'Et ideo si forte Galenus et Avicenna experimentis probant amarum universaliter operari calorem in homine, non erit per hoc probatum, quod amarum in se et simpliciter sit calidum vel cuilibet sit calidum'; Albertus Magnus, *De animalibus*, VI.1.1, ed. by Stadler, vol. 1, p. 444, vv. 6–16: 'Dicit autem Aristoteles, quod ova longa acuti capitis producunt mares avium, rotunda vero et habentia in loco acuti anguli rotunditatem producunt feminas. Et hoc est falsum omnino et vitium fuit ex scriptura perversa, et non ex dictis philosophi: propter quod dicit Avicenna, quod ex rotundis et brevibus ovis producuntur mares et galli: ex longis autem et acutis ovis producuntur gallinae: et hoc concordat cum experientia, quam nos in ovis experti sumus, et cum ratione, quoniam perfectio virtutis in ovo masculino aequaliter ambit et continet extrema: sed eiusdem imperfectio in feminino causa est, quare materia diffluit longius a centro'; ibid., VII.1.6, p. 522, vv. 27–43: 'Similiter autem dicit Avicenna, quod iam expertus est in terra sua, quod aves quaedam aquaticae veniunt in vere ad Mare Mortuum, quod salsius et calidius est alio mari quod Magnum vocatur, et hoc mare aput Arabes vocatur Ihemene, et abinde recedunt ad lacus, qui Demore vocantur: et deinde pertranseunt Nilum et vadunt ad lacum dictum Decaurisme: et quaedam etiam earum perveniunt ad lacum dictum de Trabestem: et quaedam perveniunt ad lacus alios, qui sunt in locis illis'.

## Doctrine of Taste as the Best Explanation of Particular Sensations

The correct workings of the soul across humanity (or at least among philosophers), their power to ensure a universally shared epistemology, required doctrinal reflection. Albert chose to set out that reflection mainly in his commentary on Aristotle's *De anima*, but not only there. Before a solid tradition of commentaries on Aristotle's *De anima* was established, Albert favoured other locations, rooted in the *summa* tradition of the twelfth century. His *Summa de creaturis*, comprising the two autonomous books *De quattuor coaequaevis* and *De homine*, attests to this practice. Indeed, it is in his *De homine* that Albert gives his first coherent and sovereign theoretical account of humans in their nature and activities; in his *De anima*, *Parva naturalia*, and *De natura et origine animae*, he complements, perfects, and reworks many of these early teachings.[19]

These general considerations also apply to Albert's particular doctrines, and his teaching of the sense of taste in its relation to and demarcation from the sense of touch is no exception. In *De homine*, Albert describes the sense of taste as falling under what seem to be two strictly separate, merely conceptual articulations: taste is divided into a sense of alimentation (*sensus alimenti*) and a sense of judgment of flavours (*iudicium saporum*). In its first meaning, as a sense of alimentation, Albert takes taste to be a part of the sense of touch,[20] whereas in the second, as a sense of judgment, he takes it to be distinct from touch.[21] Taste shares with touch its generic object of sensation (the four qualities of hot, cold, wet, and dry), its medium (the tangible, watery, tasteless moisture), and its modality of activity (direct contact between object and subject of sensation), and it overlaps as regards organic location (the tongue and palate, both in the mouth).[22] Yet taste does not

---

19 See also Anzulewicz, 'Die Denkstruktur des Albertus Magnus'; Anzulewicz, '*Memoria* und *reminiscentia*'; Anzulewicz, '*Solus homo est nexus Dei et mundi*'; Anzulewicz, 'Hervorgang – Verwirklichung – Rückkehr'; Anzulewicz, 'Zum anthropologischen Verständnis'; Anzulewicz and Rigo, '*Reductio ad esse divinum*'.

20 Albertus Magnus, *De homine*, ed. by Anzulewicz and Söder, p. 239, vv. 42–45: 'Dicendum quod gustus accipitur duobus modis, scilicet secundum quod est sensus alimenti, ut in multis locis dicit Philosophus, et sic gustus est quidam tactus quadruplici ratione'.

21 Ibid., p. 240, vv. 13–18: 'Accipitur etiam gustus secundum quod est iudicium saporum, et secundum hoc gustus nullo modo tactus est, ut probatum est in obiectione. Et quia per obiectum quod est diffiniens sensum sic distinguitur a tactu, ideo ponitur gustus unus quinque sensuum e diverso a tactu divisus'.

22 Ibid., p. 239, v. 42–p. 240, v. 12: 'Dicendum quod gustus accipitur duobus modis, scilicet secundum quod est sensus alimenti, ut in multis locis dicit Philosophus, et sic gustus est quidam tactus quadruplici ratione. Quarum prima sumpta est ex parte obiecti, quod est alimentum. Cum enim alimentum non nutriat nisi per substantiam, sicut dicitur in primo De generatione et corruptione, oportet quod substantia alimenti tangat id quod nutritur. Cum vero non nutriat nos nisi id ex quo sumus, oportet quod ipsum sit commixtum ex calido, frigido, humido et sicco, sicut et nos commixti sumus, sicut dicitur in secundo De generatione et corruptione. Cum ergo sic tangat per substantiam sine medio extrinseco et immutet per calidum, frigidum, humidum et siccum, quae sunt qualitates tangibiles, patet quod ex parte obiecti gustus quidam tactus est. Secunda ratio sumitur ex parte

share with touch the targeted activity of judging the 'object that defines the sense', of discriminating between the five different flavours of sweet, fat, sour, bitter, and salty.[23]

Framing these early reflections on taste as sense of alimentation and judgment of flavours, Albert's articulations remained solely within the horizon of the Aristotelian corpus. Albert explicitly anchored his considerations in philosophical principles and conclusions derived from Aristotle's *Ethica, De anima, De sensu et sensato, De animalibus, De generatione et corruptione*, and *Physica*. First and foremost, he promotes Aristotle's passing statement in the *Ethica* that 'taste is the judgment of flavour' to the distinguishing criterion of taste from touch; no other criterion served this purpose.[24] In his solution, Albert adduces Aristotle's repeated identification of taste as 'a certain touch' (*quidam tactus*) in the *De anima* and *De sensu et sensato*, though only as regards taste as a sense of alimentation. Stipulating the first reason for this partial identification of taste and touch, Albert relies on two principles from Aristotle's *De generatione et corruptione*: 'alimentation only nourishes through a substance', and taste requires the four elements of hot, cold, wet, and dry alike to be present in our body and in the alimentation. As the third reason for the partial identification, Albert employs Aristotle's insight, in the *Physica*, that 'the termini of these are identical in kind' to explain that the tongue, as one terminus, and the tasted object, as another, must physically touch one another in order to produce the sensation of taste.

The wide range of these borrowings from Aristotle's works was doubtlessly intended as much to display the young Albert's erudition (well before the Stagirite's corpus was officially read at Paris in 1255) as to demonstrate his remarkable

---

materiae, quae est medium gustabilium; hoc enim est humidum aqueum quod est insipidum, et cum hoc sit unum tangibilium, erit gustus secundum hoc quidam tactus. Tertia ratio sumitur ex parte modi gustandi: Non enim gustatur aliquid nisi habendo ultimum linguae coniunctum cum ultimo rei gustatae. Cum ergo illa se tangant, ut dicitur in V Physicorum, "quorum ultima sunt simul", patet quod ex modo gustandi gustus quidam tactus est. Quarta ratio est, quia gustus est tactus in quibusdam membris, sicut in lingua et palato; tactus autem est in omnibus membris corporis; et ideo dicitur quidam tactus quasi tactus particularis'.

23 Ibid., p. 243, vv. 40–49: 'Generatio vero saporis in specie ut a causa efficiente est, secundum quod a calido sufficienter digerente humidum generatur sapor dulcis; a calido autem digerente et subtiliante aqueum humidum in humidum aëreum generatur sapor pinguis; a calido vero non digerente sed adurente secundum aliquem modum, si est quidem cum humido, generatur acidum; et si est cum sicco, generatur amarum vel salsum, si minus adurat. Et secundum hanc generationem loquitur Aristoteles in libro De plantis'. Albert's teaching of the sense of taste covers many more aspects, but these are not our concern here. The five flavours are inspired by pseudo-Aristotle's *De plantis* and not by Avicenna's *Liber de anima* or *Canon*. See also Panarelli, 'Scientific Tasting'.

24 Aristotle, *Ethica Nicomachea*, III.12, 1118a26–1118b1, ed. Gauthier, trans. lincolniensis, p. 197, v. 28–p. 198, v. 4: 'Videntur utique et gustu in parum vel nihil uti. Gustus enim, est iudicium saporum; quod faciunt qui vina probant, et pulmenta condiunt. Non multum autem gaudent hiis, vel non, intemperati, sed usu, qui fit omnis per tactum et in cibis et potibus et venereis dictis. Propter quod et oravit Philoxenus Erixius, pulmentivorax existens, guttur ipsius longius gruis, fieri; ut delectatus, tactu'. See also Albertus Magnus, *Super Ethica*, III.13, ed. by Kübel, p. 208, vv. 13–39; Albertus Magnus, *Ethicorum libri X*, III.3.4, ed. by Borgnet, pp. 258a–259a.

mastery of the particular material. Only half of his borrowings come from contexts in which Aristotle focused on the subject matter of taste, and all of them are brought together to form a well-supported account of the proximity of the two senses on the basis of their shared relational and material properties.

None of Albert's early borrowings make explicit reference to direct or indirect experiences. Nor do they engage with ideas derived from parallel discussions on taste by Arabic-speaking scholars, despite the availability of these authors in Latin and Albert's familiarity with their works at the time.[25] Albert's initial demarcation of taste from touch pursues the goal of structuring explanations by way of relational, anatomical, and physiological criteria, and he mines Aristotle's corpus accordingly. By the time Albert composed his commentary on Aristotle's *De anima*, this simple and merely conceptual distinction of taste as a sense of alimentation and as judgment of flavours was no longer acceptable to him.

Albert's urge to seek a more careful theoretical demarcation of taste from touch, one that could account for alimentation as the special property of taste alone, emerged in his commentary on Aristotle's *De anima*, composed between 1254 and 1257. Considering the criteria by which to distinguish taste from the remaining four senses, he initially classified the object of taste as 'something tastable among the number of things that can be touched',[26] and subsequently turned to Aristotle's well-known distinction between internal and external media. Media played a significant role in both Aristotle's and Albert's theories of sensation, helping to explain how sense objects interact with the environment when they affect the senses. External media, such as air or water, were important in explaining how sense objects become capable of being seen, heard, or smelled. In contrast, the internal medium of 'the moist body of saliva in the mouth and on the tongue', as Albert calls the medium of taste, was important to establish how sense objects become tastable.[27] The definition of the medium of taste as a 'moist body' alone had the disadvantage of suggesting that taste is reducible to touch. For Albert, its material nature as a medium, as opposed to a spiritual or intentional nature, and its moist quality, as opposed to a dry quality, both implied that 'what is tastable is also tangible'.[28] Yet this reference to the moist quality prefigured Albert's theoretical solution to the difficulty, one that turned on a formal difference between taste and touch:

25 In *De homine*, Albert references Avicenna's *Liber de anima* as far as I have been able to establish 345 times and Averroes's *Long Commentary on the De anima* fifty-five times, counting the references that Anzulewicz and Söder list. See Albertus Magnus, *De homine*, ed. by Anzulewicz and Söder, pp. 609–10.

26 Albertus Magnus, *De anima*, II.3.27, ed. by Stroick, p. 137, vv. 85–86: '*gustabile quiddam est* de numero *tangibilium*'.

27 Ibid., p. 138, vv. 2–5: 'Est autem hoc corpus umor salivalis in ore et lingua. Cum enim gustabile sit tangibile, sicut *tactus non* potest *esse* per medium extrinsecum, ita neque gustus'.

28 Ibid., vv. 5–30.

> the moisture in the tastable object is material only, and the flavour in [the object] is the form acting on taste. And therefore, taste is not a part of touch, but a certain species of sensation, just as touch is. [...] moisture is the proper matter of flavour in which it [i.e., flavour] is diffused, and this is in accordance with its material being.[29]

The substructure of form-matter composition aided Albert in differentiating flavour, the active agent, from moisture, its material carrier, in their causality on the sense of taste. The application of this structure to the theme can already be found in Averroes's *Long Commentary on the De anima*: 'the body in which flavour exists is not tastable except insofar as that flavour exists in in a moisture whose relation to this flavour is as [the relation of] matter to form'.[30] But while Averroes anticipated Albert's commitment to a form-matter relationship between flavour and moisture in a tastable body, he did not draw the same conclusion as Albert. Albert suggested independently that only the material part in taste is tangible, whereas the formal part properly distinguishes taste from touch. In this way, he also departed from his earlier demarcation line, the conceptual distinction between taste as a sense of alimentation and as a sense of judgment of flavour, in favour of a realist distinction grounded in the object of taste itself and coupled to taste as a sense of alimentation.[31]

The new realist demarcation inspired by Averroes's insight was nonetheless capable of embracing Albert's previous identification of taste as a sense of judgment of flavour. As the active form of moisture and as that which is subject to change, Albert now suggested, this flavour of alimentation is likewise subject to the judgment of taste:

> But if someone might have wondered how, then, taste is distinguished from touch, it should be answered as before: that flavour in moisture touches according to actuality, yet inasmuch as it is moist, moisture nonetheless does not change taste inasmuch as it is taste, but rather, [it changes] the flavoured inasmuch as it is flavoured. Because of this, taste, in that it is taste, passes

29 Ibid., vv. 9–12 and 19–20: 'umidum in gustabili materiale tantum est, et sapor in eo est forma agens in gustum; et ideo non est pars tactus gustus, sed species quaedam sensus sicut et tactus [...] umidum est propria materia saporis, in qua diffunditur, et est secundum esse materiale ipsius'.

30 Averroes, *Commentarium magnum in Aristotelis De anima libros*, II.101, ed. by Crawford, p. 285, vv. 38–41 (trans. by Taylor and Druart, p. 220, considerably emended): 'Corpus enim in quo existit sapor non est gustabile nisi secundum quod ille sapor existit in eo in humore cuius proportio ad illum saporem est sicut materie ad formam'.

31 Equally important to Albert was the fit between the active form of flavour and the specific matter of moisture, a fit which qualified moisture as the only material carrier of flavour and thus distinguished it from all other material carriers of touchable nature. Flavour could be received by the sense of taste in or with the medium of moisture, and not from the medium, as Albert envisioned it for the more spiritual or intentional media of air and water that enabled sight, hearing, and smell.

> judgment upon flavour, and in this way is distinguished from touch, and is not a certain part of touch.[32]

The judgment that taste exerts over flavour as the active formal constituent of the tastable body here acquires an immediate connection to taste as alimentation. The two contexts which stood side by side in Albert's earlier *De homine* are now connected through the form-matter relation that Albert borrowed from Averroes. Together, they account for the demarcation of taste from touch.[33] Judgment of flavour remains a decisive aspect of distinguishing between taste and touch, but it does so within one encompassing teaching rather than two separate doctrinal aspects.

This embracive understanding of a realist distinction, writes Albert as a self-corrective, is also what 'the three authorities, Aristotle, Averroes, and Avicenna, agree upon'.[34] It seems to pose no problem that neither judgment of flavour, nor flavour as formal aspect of a tastable object, nor the combination of the two occurs in this form in the three authorities. At stake in Albert's final summary of his new realist demarcation is how explanation relates to experience, or in this case, how well the explanation just sketched accounts for the experience of any given flavour:

> And in this way, by saying so, it should be clear that taste has no extrinsic medium, but rather, just as colour is visible and properly acts on sight, so flavour is tastable and acts on taste per se. But it does not act and perfect the sense of flavour by taste without actual humidity, as we have said, just as it can be experienced [*experiri*] when something salty acts on taste. For it does not act without moisture because saltiness is well liquefied, and, touched by moisture, it is dissolved and liquefies on the tongue, and, mixed with that moisture in a corporeal way, it acts on taste, and not otherwise.[35]

---

32 Albertus Magnus, *De anima*, II.3.27, ed. by Stroick, p. 138, vv. 58–65: 'Si autem aliquis quaesiverit, qualiter ergo gustus dividitur a tactu, dicendum sicut prius, quoniam sapor in umido secundum actum tangit, tamen umidum, inquantum umidum, gustum, inquantum gustus est, non immutat, sed potius sapidum, inquantum est sapidum. Propter quod gustus, in eo quod gustus, iudicium saporis est et sic a tactu discernitur et non est pars quaedam tactus'.

33 Thomas Aquinas found a different way, possibly inspired by Avicenna's *Liber de anima*, II.4, where he repeatedly refers to *humor* (rather than the qualities) and its *commixtio* with the tongue to distinguish taste from touch. Thomas Aquinas, *Sentencia libri De anima*, II.21.4, ed. Leonina, p. 155, vv. 49–59: 'alio modo quantum ad obiectum; et sic oportet dicere quod, sicut se habet obiectum gustus ad obiectum tactus, ita se habet sensus gustus ad sensum tactus; manifestum est autem quod sapor, qui est obiectum gustus, non est aliqua de qualitatibus simplicium corporum ex quibus animal constituitur que sunt propria obiecta sensus tactus, set causatur ab eis et fundatur in aliqua earum sicut in materia, scilicet in humido; unde manifestum est quod gustus non est idem quod sensus tactus, set quodam modo radicatur in eo'.

34 Albertus Magnus, *De anima*, II.3.27, ed. by Stroick, p. 138, vv. 65–67: 'Et in hac sententia tres auctores concordant Aristoteles et Averroes et Avicenna'.

35 Ibid., p. 138, vv. 68–76: 'Et *sic* dicendo patet, quod gustus *nullum medium est* extrinsecum, *sed* tamen *sicut color* est *visibilis* et proprie agit in visum, *sic sapor* est *gustabilis et* per se agit in gustum, sed *non*

Albert introduces the experience of the action of the salty quality (saltiness) on taste as a pertinent example within the theoretical context of flavour as a formal agent that is tastable in itself if in moisture and moisture as a material medium that activates the formal agency of flavour. Saltiness, this application of the general teaching to the particular experience reveals, is the formal agent in moisture, which is the material medium. Saltiness acts on taste and, as such, elicits the judgment of taste, whereas moisture is just its carrier.

How saltiness is experienced as salty, and how this particular judgment is passed by taste upon the experience of saltiness, are the two epistemic moments that Albert can now explain together, and thus much better, by way of his new form-matter doctrine. Saltiness is not simply experienced as salty and judged to be such as salty alone, but is experienced as an active form dissolved in a material moisture, and on that basis is judged by way of its elemental commixture. Earlier in Albert's *De homine*, the judgment of taste upon any salty flavour had no explanatory roots in the formality of the object: this theoretical aspect was not bound up with taste as a sense of alimentation and its whole explanatory apparatus.

Albert's explicit appeal to *experiri* in this passage of his *De anima* — to the experiential value of the taste of salt — therefore carries tremendous epistemic weight. The example of saltiness as a transhistorical experience, that is, experience with evidentiary rather than authoritative value, formulated much in the manner of the example of sensation in teeth, enabled Albert to apply the general teaching to the particular experience. It helped him to establish and validate his new-found teaching as the best possible or most plausible explanation of the experience of saltiness. Whether this best possible explanation was grasped universally by Albert and extended to other particular flavours, and how exactly he reached the insight that this is the best possible explanation, remains, to the best of my knowledge, hidden in his own thoughts and never put down on parchment.

None of these facets — the experiential value of saltiness, the matching of the general to the particular, establishing and validating the new teaching in application, and that doctrine's ontological value for demarcating taste from touch — was on the radar of Albert's Greek and Arabic sources. Avicenna, unlike Aristotle and Averroes, did not even include saltiness among the eight specific flavours in his *Liber de anima* II.4, though he added a reference to it in his *Canon*.[36] Aristotle, along with Averroes (who remained particularly close to Aristotle's template), focused on the requirement of a moist inclination in the different

agit et *perficit sensum saporis* in gustu *sine umiditate* actuali, sicut diximus; *sicut* enim experiri potest, cum *salsum* agit in gustum; *illud enim* non agit sine umiditate, quia salsum *est bene liquidum* et tactum umido dissolvitur *et liquefacit linguam* et commixtum corporaliter illi umido agit in gustum, et non aliter'.

36 Avicenna, *Liber canonis*, II.1.3, ed. Venetiis, fol. 83va: 'Sapores autem sunt octo quos ipsi dicunt qui sunt vere sapores post insipidum. Et sunt dulcedo amaritudo acuitas salsedo acetositas ponticitas stipticitas et unctuositas'.

flavours. For both, the example of saltiness simply served as a prime example of such inclinations. They write:

> [Aristoteles Latinus:] In this way, however, there is no medium. But in whatever way colour is visible, in the same way, flavour is tastable. And nothing receives the sensation of flavour without moisture, but moisture in it is either in actuality or in potency. Take something salty, for instance: for it dissolves swiftly and with [moisture], it dissolves on the tongue.
>
> [Averroes Latinus:] That is, things only receive the sensation of flavour, which is called taste, if flavour is in moisture and if moisture is imbued with flavour either in act or in potency. Take something salty, for instance, which is in proximate potency to moisture, because it is dissolved swiftly and it dissolves moistures on the tongue. And therefore, nature has provided saliva in the mouth and she has provided glands in humans for gathering this moisture, so that dry things may be tasted by its facilitation. This is why we say that flavour is flavour in actuality only in a body that is moist in actuality.[37]

The case of saltiness, then, is yet another indication that Albert trusted his authorities, though only insofar as he could mine their insights to establish the truth on flavour that he himself advocated, improve its explanatory value in distinguishing it formally from touch, and emphasize the usefulness of experience as a transhistorical fact. This does not mean he followed his sources to the letter or adopted the epistemic value that they accorded to the example of saltiness. Rather, it means that his reference to the experience of saltiness, whoever its experiential subject may be, could pertinently explain why Albert's doctrine distinguished taste from touch on formal rather than material grounds, and thus on grounds theoretically superior to those of Aristotle, Avicenna, and Averroes together. For us, this example of saltiness might seem like an excellent candidate for a theory of verification. But it seems that Albert did not see its true or ultimate epistemic value in such an objective goal.

---

37 Aristotle, *De anima*, 422a15–422a19, as quoted in *Averroes, Commentarium magnum in Aristotelis De anima libros*, ed. by Crawford, p. 286, vv. 1–7 (trans. by Taylor and Druart, p. 221, considerably emended): 'Secundum autem hunc modum non est medium; sed quemadmodum color est visibile, sic sapor est gustabile. Et nichil recipit sensum saporis absque humiditate, sed in eo est in actu aut potentia humiditas; v. g. salsum; est enim velocis dissolutionis, et cum hoc dissolvit linguam'; Averroes, *Commentarium magnum in Aristotelis De anima libros*, II.102, ed. by Crawford, p. 286, vv. 14–22 (trans. by Taylor and Druart, p. 221, considerably emended): 'Idest, et nichil recipit sensum saporis, qui dicitur gustus, nisi sapor sit in humore, et humor est in saporoso aut in actu aut in potentia, v. g. salsum, quod est humidum in potentia propinqua, cum velociter dissolvitur et dissolvit humores qui sunt in lingua. Et ideo preparavit Natura salivam in ore, et preparavit brancos in homine ad congregandum istam humiditatem, ut ea mediante gustaretur sicca. Unde dicimus quod sapor non est sapor in actu nisi in corpore humido in actu'.

## Mining the Sources for Cognitive Experience

Albert continually weighed the options of thought available to him at any given place and time, and on that basis decided the scientific truth of each matter. His criteria for determining that truth concerning any specific natural scientific teaching derived not from human authoritative or theological parameters, but from his own scientific and anthropological standards.[38] Authority was, as I have shown, instrumental to truth-making, but reworked in its arrangement and design so as to fit the scientific goal. The scientific and anthropological criteria that anchored Albert's truth-making in the specifics of natural scientific teachings were derived from elsewhere. Space does not permit a comprehensive discussion of this point, but there are two clues that attest to Albert's procedures.

In his initial considerations on sensation in his *De anima*, Albert composed one of his famous digressions, a self-standing insertion to fill scientific lacunae in Aristotle's work. Albert's digression on Book II focuses on the grades and modes of abstraction in *scientia*. He presents this with a clear cognitive purpose, one whose content is itself dependent on the science of the soul that he is in the midst of explicating:

> Before we speak of the sensible things one by one, it is necessary for us to speak of the sensible thing in general, because, as we have said, according to reason, objects are prior to acts and acts to powers. And because in the natural sciences [*in physicis*], theorizing of the common things is also prior with regard to us — since in these [sciences] common things are confused in the singulars and are prior with regard to us — we must speak, first of all, of the sensible thing in common. But for an easier understanding of those things of which we shall speak, we shall provide a brief chapter on the mode of apprehension that all apprehensive powers have. For this will be very useful for an easier knowledge of all that follows.[39]

In this introductory passage to a whole book section, as in many others of its kind, Albert unhesitatingly turns epistemic principles to cognitive purposes. The Aristotelian insight that human knowledge of common things is confused but prior with respect to us as human knowers found its application in the order of study for all future natural philosophers under his wing, an order that always proceeded from the common thing to its specific definition. Yet this movement could only occur once the different modes of abstraction available to humans

---

38 Krause and Anzulewicz, 'Albert the Great's *Interpretatio*'.

39 Albertus Magnus, *De anima*, II.3.4, ed. by Stroick, p. 101, vv. 50–61: 'Antequam nos loquamur de sensibilibus singulariter, oportet nos loqui de sensibili generaliter, quia, sicut diximus, obiecta sunt priora actibus et actus potentiis secundum rationem. Et quia de communibus etiam quoad nos prior est speculatio in physicis, eo quod in illis communia confusa sunt in singularibus et priora quoad nos: debemus primo loqui de sensibili in communi. Sed ad faciliorem intellectum eorum quae dicturi sumus, faciemus capitulum breve de modo apprehensionis potentiarum apprehensivarum omnium; hoc enim perutile erit ad omnium sequentium notitiam faciliorem'.

were known to the students of the science of the soul and potentially experienced, practised, and exercised by them in their acquisition and mastery of *scientia*.

For Albert, then, sensation, imagination, estimation, cogitation, and understanding were not just distinct grades of apprehension, but grades that he assigned to order his science of the soul: it was 'by these grades of abstraction and separation [that] the powers of apprehension will be distinguished below'.[40] Principles determined in the science of the soul found their immediate application in its explanations; their truth, in other words, was validated not just in the theories of the science, but in its practices, in its very formulation and articulation through the cognitive processes he had just described. The habit of *scientia*, the practices of hearing, seeing, commenting, studying, analysing, weighing, and reworking ideas and arguments enabled Albert to shape and determine these scientific truths. Conversely, these practices were themselves informed by the doctrines he established; they entered into a feedback loop of a cognitive *scientia*.

It was these scientific and anthropological criteria of truth that, as principles and grounding standards, determined Albert's choices regarding his use of sources, their usefulness, their *locus*, their shape, and their meaning in Albert's work. But they also fixed these sources within Albert's overarching scientific teleology. His practice of science in general, but also in its particulars, was governed by his goals of comprehensiveness, expressed in specific definitional knowledge of all things natural,[41] and pursued for the sake of leading the listener's intellect to its *telos* of perfect knowledge and completion:

> We must investigate the natures of any given sentient being and know that there is a certain noble natural and divine cause in all of them, because none of them was brought into being naturally in vain or without purpose. Rather, whatever, however many, and however much they proceed from the work of nature, they will only be because of that which is the end. And everything that was, is, and will be, was not, is not, and will not be except because of something that is its completion, and because of this, it has a place among things natural, and a wondrous and noble rank. If, therefore, someone should hold the opinion that the cognition of some of these things is base, they had better blame themselves, because their affective cognition is base and corrupt, because they themselves do not take into account the things out of which the human being is composed without the deformity of affection, as when they

40 Ibid., p. 102, vv. 25–27: 'Secundum autem hos gradus abstractionis sive separationis distinguentur inferius vires apprehensivae'.

41 Albertus Magnus, *De animalibus*, XI.1.3, ed. by Stadler, vol. 1, p. 773, v. 34–p. 774, v. 2: 'naturalis debet diffiniendo in scientia animalium dicere et docere de dispositione animae et partium eius quanto magis poterit, quia anima principium est animalium, sicut in libro primo de Anima diximus: et debet narrare assignando dispositionem cuiuslibet animae et dispositionem cuiuslibet modi in partibus animae et diffinire, quid sit animal, et ostendere utrum sit anima pars animalis aut non: et deinde narrare debet accidentia quae accidunt animali et substantiae animae, quae est talis aut talis'.

> cogitate flesh, bone, blood, vein, and similar things. For it is these accidents of their souls that are base, but not the cognition itself.[42]

Explaining sensation in teeth, and doing so by appealing to experience in its evidentiary value without attention to a historical subject that guaranteed its authority, added one universal truth to the comprehensive scientific knowledge of sentient beings, and consequently to the listener's intellectual growth, the ultimate epistemic value that Albert always kept in mind. The study of particular and ignoble matters — such as the body of sentient beings, its members, and their causes — was worth everyone's time, because of the noble and divine teleology inscribed both in these matters and, equally, in the knowing subjects.

Albert's intellectual practices of mining his Arabic natural philosophical sources were thus pursued with the *telos* of gaining the truth, which he saw expressed in definitional knowledge and wished to acquire for the sake of intellectual perfection. The two cases I have presented, the transhistorization of empirical evidence and independent teaching as the best possible explanation of the particular experience of saltiness, were subject to just these epistemic ultimates.

To conclude, Albert's efforts to mine his sources were rooted in the conviction of a shared nature, shared activities, and shared teleology among all capable humans, but particularly among those who had already embarked on the study of the *scientiae naturales* and of the corpus of his philosophy as a whole. They each held *within* them the truth of science, which came about through scientific practices that explicitly included what I have explored here: a kind of 'cognitive empiricism'.[43] Surprisingly for the modern reader, this scientific truth amounted to a universal truth.

All this is reason enough to say that Albert the Great challenges our conventional understanding of empiricism and science. I have tried to show that in order to understand his *experientia*, we must adopt a much broader, more inclusive perspective that values experience as integral to a trained human cognition, which

---

42 Ibid., XI.2.3, p. 794, vv. 6–20: 'debemus inquirere naturas cuiuslibet animalis et scire quod in omnibus animalibus quaedam est causa naturalis nobilis et divina, eo quod nullum omnino naturalium fuit naturatum casualiter aut otiose sive frusta, sed quaecumque et quotcumque et quantumcumque procedunt de opere naturae, non erunt nisi propter hoc quod est finis: et omne quod fuit, est et erit, non fuit neque est neque erit, nisi propter aliquid quod est complementum et propter hoc habet locum in naturalibus et ordinem mirabilem et nobilem. Si ergo aliquis opinetur cognitionem aliquorum ignobilem esse, culpet seipsum potius, eo quod sua affectiva cognitio ignobilis est et vitiosa, eo quod ipse non concipit res ex quibus homo componitur sine turpitudine affectus, sicut quando cogitat carnem et os et sanguinem et venam et hiis similia: accidentia enim animae suae sunt vilia, et non cognitio ipsa'.

43 As this chapter will already have made clear, I do not follow the narrow definition of cognitive empiricism found in Dawes, 'Ancient and Medieval Empiricism', but apply a much broader understanding of scientific practices in which Albert also labelled certain kinds of cognition as empirical.

in turn is expressed in his scientific practices of defining and explaining. Such a perspective not only enriches our understanding of premodern scientific practices, but also invites us to reconsider the epistemic value we place on different forms of scientific experience.

## Works Cited

### *Primary Sources*

Albertus Magnus, *Commentarii in II Sententiarum*, ed. by Auguste Borgnet, Editio Parisiensis, 27 (Paris: Vivès, 1893)

———, *De anima*, ed. by Clemens Stroick, Editio Coloniensis, 7/1 (Münster: Aschendorff, 1968)

———, *De animalibus libri XXVI. Nach der Cölner Urschrift*, ed. by Hermann Stadler, 2 vols, Beiträge zur Geschichte der Philosophie des Mittelalters, 15–16 (Münster: Aschendorff, 1916–20)

———, *De homine*, ed. by Henryk Anzulewicz and Joachim R. Söder, Editio Coloniensis, 27/2 (Münster: Aschendorff, 2008)

———, *De somno et vigilia*, ed. by Auguste Borgnet, Editio Parisiensis, 9 (Paris: Vivès, 1890)

———, *Ethicorum lib. X*, ed. by Auguste Borgnet, Editio Parisiensis, 7 (Paris: Vivès, 1891)

———, *Metaphysica, libri I–V*, ed. by Bernhard Geyer, Editio Coloniensis, 16/1 (Münster: Aschendorff, 1960)

———, *Physica, libri I–IV*, ed. by Paul Hossfeld, Editio Coloniensis, 4/1 (Münster: Aschendorff, 1987)

———, *Quaestiones super libri De animalibus*, ed. by Ephrem Filthaut, Editio Coloniensis, 12 (Münster: Aschendorff, 1955)

———, *Super Ethica, libri I–V*, ed. by Wilhelm Kübel, Editio Coloniensis, 14/1 (Münster: Aschendorff, 1968–72)

———, *Super I librum Sententiarum, distinctiones 1–3*, ed. by Maria Burger, Editio Coloniensis, 29/1 (Münster: Aschendorff, 2015)

Aristotle, *Ethica Nicomachea*, ed. by René Antoine Gauthier, Aristoteles Latinus, 26/1–3 (Leiden: Brill, 1974)

Averroes, *Commentarium magnum in Aristotelis De anima libros*, ed. by F. Stuart Crawford, Corpus Commentariorum Averrois in Aristotelem, 6/1 (Cambridge, MA: The Mediaeval Academy of America, 1953)

———, *Long Commentary on the 'De Anima' of Aristotle*, trans. with an introduction and notes by Richard C. Taylor, with Thérèse-Anne Druart (New Haven, CT: Yale University Press, 2009)

Avicenna, *Liber canonis* (Venice, 1507; repr. Hildesheim: Olms, 2003)

Claudius Galenus, *Claudii Galeni opera omnia*, ed. by Carl Gottlieb Kühn, 2 vols (Leipzig: Knobloch, 1821)

Thomas Aquinas, *Sentencia libri De anima*, ed. by Commissio Leonina, Sancti Thomae de Aquino Opera omnia, 45/1 (Rome: Commissio Leonina, 1984)

### Secondary Works

Anzulewicz, Henryk, 'Die Denkstruktur des Albertus Magnus: Ihre Dekodierung und ihre Relevanz für die Begrifflichkeit und Terminologie', in *L'élaboration du vocabulaire philosophique au Moyen Âge*, ed. by Jacqueline Hamesse and Carlos Steel (Turnhout: Brepols, 2000), pp. 369–96

———, 'Die Emanationslehre des Albertus Magnus: Genese, Gestalt und Bedeutung', in *Via Alberti: Texte – Quellen – Interpretationen*, ed. by Ludger Honnefelder, Hannes Möhle, and Susana Bullido del Barrio (Münster: Aschendorff, 2009), pp. 219–42

———, 'Hervorgang – Verwirklichung – Rückkehr: Eine neuplatonische Struktur im Denken Alberts des Großen und Dietrichs von Freiberg', in *Die Gedankenwelt Dietrichs von Freiberg im Kontext seiner Zeitgenossen*, ed. by Karl-Hermann Kandler, Burkhard Mojsisch, and Norman Pohl (Freiberg: Technische Universität Bergakademie Freiberg, 2013), pp. 229–44

———, '*Memoria* und *reminiscentia* bei Albertus Magnus', in *La mémoire du temps au Moyen Âge*, ed. by Agostino Paravicini Bagliani (Florence: SISMEL, Edizioni del Galluzzo, 2005), pp. 163–200

———, 'Psychophysiology, Natural Spaces and Climata: Albert the Great on the Natural Preconditions of Epistemic Abilities in Humans', forthcoming

———, '*Solus homo est nexus Dei et mundi*: Albertus Magnus über den Menschen', in *Multifariam: Homenaje a los profesores Annelies Meis, Antonio Bentué y Sergio Silva*, ed. by Samuel Fernández Fernández, Juan Noemi, and Sergio Silva (Santiago de Chile: Pontificia Universidad Católica de Chile, 2010), pp. 321–35

———, 'The Systematic Theology of Albert the Great', in *A Companion to Albert the Great*, ed. by Irven M. Resnick (Leiden: Brill, 2013), pp. 15–67

———, 'Zum anthropologischen Verständnis der *Perfectio* bei Albertus Magnus', *Documenti e studi sulla tradizione filosofica medievale*, 30 (2019), 339–69

———, 'Zwischen Faszination und Ablehnung: Theologie und Philosophie im 13. Jh. in ihrem Verhältnis zueinander', in *What Is 'Theology' in the Middle Ages?*, ed. by Mikołaj Olszewski (Münster: Aschendorff, 2007), pp. 129–56

———, and Caterina Rigo, '*Reductio ad esse divinum*: Zur Vollendung des Menschen nach Albertus Magnus', in *Ende und Vollendung: Eschatologische Perspektiven im Mittelalter*, ed. by Jan Aertsen and Martin Pickavé (Berlin: De Gruyter, 2002), pp. 388–416

Asúa, Miguel de, 'Albert the Great and the *Controversia inter Medicos et Philosophos*', *Proceedings of the PMR Conference*, 19/20 (1994–96), 143–56

———, 'El Comentario de Pedro Hispano sobre el *De animalibus*: Transcripcion de las *Quaestiones* sobre la controversia entre medicos y filosofos', *Patristica et Mediaevalia*, 16 (1995), 45–66

———, 'The Organization of Discourse on Animals in the Thirteenth Century: Peter of Spain, Albert the Great, and the Commentaries on *De animalibus*' (unpublished doctoral dissertation, University of Notre Dame, 1991)

———, 'Peter of Spain, Albert the Great and the *Quaestiones De animalibus*', *Physis: Rivista internazionale di storia della scienza*, 34 (1997), 1–30

Bertolacci, Amos, 'Albert's Use of Avicenna and Islamic Philosophy', in *A Companion to Albert the Great*, ed. by Irven M. Resnick (Leiden: Brill, 2013), pp. 601–11

———, 'Albert the Great and the Preface of Avicenna's *Kitab al-Sifa*', in *Avicenna and His Heritage*, ed. by Jules Janssens and Daniel de Smedt (Leuven: Leuven University Press, 2002), pp. 131–52

———, '"Averroes ubique Avicennam persequitur": Albert the Great's Approach to the Physics of the *Šifā'* in the Light of Averroes' Criticisms', in *The Arabic, Hebrew and Latin Reception of Avicenna's Physics and Cosmology*, ed. by Dag Nikolaus Hasse and Amos Bertolacci (Berlin: De Gruyter, 2018), pp. 397–431

———, 'Avicenna's and Averroes's Interpretations and Their Influence in Albertus Magnus', in *A Companion to the Latin Medieval Commentaries on Aristotle's Metaphysics*, ed. by Fabrizio Amerini and Gabriele Galluzzo (Leiden: Brill, 2014), pp. 95–135

———, 'Le citazioni implicite testuali della *Philosophia prima* di Avicenna nel Commento alla *Metafisica* di Alberto Magno: analisi tipologica', *Documenti e studi sulla tradizione filosofica medievale*, 12 (2001), 179–274

———, 'A New Phase of the Reception of Aristotle in the Latin West: Albertus Magnus and His Use of Arabic Sources in the Commentaries on Aristotle', in *Albertus Magnus und der Ursprung der Universitätsidee: Die Begegnung der Wissenschaftskulturen im 13. Jahrhundert und die Entdeckung des Konzepts der Bildung durch Wissenschaft*, ed. by Ludger Honnefelder (Berlin: Berlin University Press, 2011), pp. 259–76 and 491–500

———, 'The Reception of Avicenna's *Philosophia Prima* in Albert the Great's Commentary on the *Metaphysics*: The Case of the Doctrine of Unity', in *Albertus Magnus: Zum Gedenken nach 800 Jahren*, ed. by Walter Senner (Berlin: Akademie-Verlag, 2001), pp. 67–78

———, '"Subtilius speculando": Le citazioni della *Philosophia prima* di Avicenna nel Commento alla *Metafisica* di Alberto Magno', *Documenti e studi sulla tradizione filosofica medievale*, 9 (1998), 261–339

Burger, Maria, 'Albertus Magnus Theologie als Wissenschaft unter der Herausforderung aristotelisch-arabischer Wissenschaftstheorie', in *Albertus Magnus und der Ursprung der Universitätsidee: Die Begegnung der Wissenschaftskulturen im 13. Jahrhundert und die Entdeckung des Konzepts der Bildung durch Wissenschaft*, ed. by Ludger Honnefelder (Berlin: Berlin University Press, 2011), pp. 97–114

Cadden, Joan, 'Albertus Magnus' Universal Physiology: The Example of Nutrition', in *Albertus Magnus and the Sciences: Commemorative Essays 1980*, ed. by James A. Weisheipl (Toronto: Pontifical Institute of Mediaeval Studies, 1980), pp. 321–39

Caminada, Niccolò, 'A Latin Translation? The Reception of Avicenna in Albert the Great's *De praedicamentis*', *Documenti e studi sulla tradizione filosofica medievale*, 28 (2017), 71–104

Daston, Lorraine, 'Scientific Error and the Ethos of Belief', *Social Research* 72 (2005), 1–28

———, 'Taking Note(s)', *Isis*, 95 (2004), 443–48

Dawes, Gregory W., 'Ancient and Medieval Empiricism', *The Stanford Encyclopedia of Philosophy* (Summer 2023 Edition), ed. by Edward N. Zalta and Uri Nodelman, https://plato.stanford.edu/archives/sum2023/entries/empiricism-ancient-medieval/

Demaitre, Luke, and Anthony A. Travil, 'Human Embryology and Development in the Works of Albertus Magnus', in *Albertus Magnus and the Sciences: Commemorative Essays 1980*, ed. by James A. Weisheipl (Toronto: Pontifical Institute of Mediaeval Studies, 1980), pp. 405–40

Dold, Dominic Nicolas, 'What is Zoology About? The Philosophical Foundations of Albert the Great's Science of Animals' (unpublished doctoral dissertation, Technische Universität Berlin, 2023)

Donati, Silvia, 'Alberts des Großen Konzept der *scientiae naturales*: Zur Konstitution einer peripatetischen Enzyklopädie der Naturwissenschaften', in *Albertus Magnus und der Ursprung der Universitätsidee: Die Begegnung der Wissenschaftskulturen im 13. Jahrhundert und die Entdeckung des Konzepts der Bildung durch Wissenschaft*, ed. by Ludger Honnefelder (Berlin: Berlin University Press, 2011), pp. 354–81

———, 'Is Celestial Motion a Natural Motion? Averroes' Position and Its Reception in the Thirteenth- and Early Fourteenth-Century Commentary Tradition of the Physics', in *Averroes' Natural Philosophy and Its Reception in the Latin West*, ed. by Paul J. J. M. Bakker (Leuven: Leuven University Press, 2015), pp. 89–126

Endress, Gerhard, *Der arabische Aristoteles und sein Leser: Physik und Theologie im Weltbild Alberts des Großen* (Münster: Aschendorff, 2004)

Gerhardt, Mia I., 'Zoologie médiévale: préoccupations et procédés', in *Methoden in Wissenschaft und Kunst des Mittelalters*, ed. by Albert Zimmermann and Rudolf Hoffmann (Berlin: De Gruyter, 1970), pp. 231–48

Hasse, Dag Nikolaus, 'Avicenna's "Giver of Forms" in Latin Philosophy, Especially in the Works of Albertus Magnus', in *The Arabic, Hebrew and Latin Reception of Avicenna's Metaphysics*, ed. by Dag Nikolaus Hasse and Amos Bertolacci (Berlin: De Gruyter, 2012), pp. 225–49

———, 'The Early Albertus Magnus and His Arabic Sources on the Theology of the Soul', *Vivarium*, 46 (2008), 232–52

———, 'Der mutmaßliche arabische Einfluss auf die literarische Form der Universitätsliteratur des 13. Jahrhunderts', in *Albertus Magnus und der Ursprung der Universitätsidee: Die Begegnung der Wissenschaftskulturen im 13. Jahrhundert und die Entdeckung des Konzepts der Bildung durch Wissenschaft*, ed. by Ludger Honnefelder (Berlin: Berlin University Press, 2011), pp. 241–58

Hossfeld, Paul, *Albertus Magnus als Naturphilosoph und Naturwissenschaftler* (Bonn: Albertus-Magnus-Institut, 1983)

Hünemörder, Christian, 'Die Zoologie des Albertus Magnus', in *Albertus Magnus Doctor Universalis 1280/1980*, ed. by Gerbert Meyer and Albert Zimmermann (Mainz: Grünewald, 1980), pp. 235–48

Jacquart, Danielle, 'La place d'Isaac Israeli dans la médecine médiévale', *Vesalius*, 4 (1998), 19–27

Killermann, Sebastian, 'Die somatische Anthropologie bei Albertus Magnus', *Angelicum*, 21 (1944), 224–69

Kopp, Paul, 'Psychiatrisches bei Albertus Magnus', *Zeitschrift für die gesamte Neurologie und Psychiatrie*, 147 (1993), 50–60

Krause, Katja, and Henryk Anzulewicz, 'Albert the Great's *Interpretatio*: Converting Libraries into a Scientific System', in *Premodern Translation: Comparative Approaches to Cross-Cultural Transformations*, ed. by Sonja Brentjes and Alexander Fidora (Turnhout: Brepols, 2021), pp. 89–132

———, with Maria Auxent and Dror Weil, 'Introduction: Making Sense of Nature in the Premodern World', in *Premodern Experience of the Natural World in Translation*, ed. by Katja Krause, Maria Auxent, and Dror Weil (New York: Routledge, 2022), pp. 7–19

Lizzini, Olga, 'Flusso, preparazione appropriata e *inchoatio formae*: brevi osservazioni su Avicenna e Alberto Magno', in *Medioevo e filosofia: Per Alfonso Maierú*, ed. by Massimiliano Lenzi, Cesare A. Musatti, and Luisa Valente (Rome: Viella, 2013), pp. 129–50

López-Farjeat, Luis, 'Albert the Great between Avempace and Averroes on the Knowledge of Seperate Forms', *Proceedings of the American Catholic Philosophical Association*, 86 (2013), 89–102

Müller, Jörn, 'Der Einfluss der arabischen Intellektspekulation auf die Ethik des Albertus Magnus', in *Wissen über Grenzen*, ed. by Andreas Speer and Lydia Wegener (Berlin: De Gruyter, 2006), pp. 545–68

Navarro-Sanchez, Francisca, *Peter of Spain, 'Questiones super libro De Animalibus Aristoteles': Critical Edition with Introduction* (London: Routledge, 2016)

Panarelli, Marilena, 'Scientific Tasting: Flavors in the Investigation of Plants and Medicines from Aristotle to Albert the Great', in *Premodern Experience of the Natural World in Translation*, ed. by Katja Krause, Maria Auxent, and Dror Weil (New York: Routledge, 2022), pp. 74–89

Pelster, F., 'Die beiden ersten Kapitel der Erklärung Alberts des Großen zu *De animalibus* in ihrer ursprünglichen Fassung', *Scholastik*, 10 (1935), 229–40

Pouchet, Félix A., *Histoire des sciences naturelles au Moyen Âge: ou Albert le Grand et son époque considérés comme point de depart de l'école expérimentale* (Paris: J. B. Baillière, 1853)

Schipperges, Heinrich, 'Das medizinische Denken bei Albertus Magnus', in *Albertus Magnus, Doctor universalis: 1280/1980*, ed. by Gerbert Meyer and Albert Zimmermann (Mainz: Grünewald, 1980), pp. 279–94

———, 'Eine *summa medicinae* bei Albertus Magnus', *Jahres- und Tagungsberichte der Görresgesellschaft 1981* (1981), 5–24

Schwartz, Yossef, 'Celestial Motion, Immaterial Causality and the Latin Encounter with Arabic Aristotelian Cosmology', in *Albertus Magnus und der Ursprung der Universitätsidee: Die Begegnung der Wissenschaftskulturen im 13. Jahrhundert und die Entdeckung des Konzepts der Bildung durch Wissenschaft*, ed. by Ludger Honnefelder (Berlin: Berlin University Press, 2011), pp. 277–98

———, 'Divine Space and the Space of the Divine: On the Scholastic Rejection of Arab Cosmology', in *Représentations et conceptions de l'espace dans la culture médiévale / Repräsentationsformen und Konzeptionen des Raums in der Kultur des Mittelalters*, ed. by Tiziana Suarez-Nani and Martin Rohde (Berlin: De Gruyter, 2011), pp. 89–119

Shaw, James Rochester, 'Scientific Empiricism in the Middle Ages: Albertus Magnus on Sexual Anatomy and Physiology', *Clio medica*, 10 (1975), 53–64

Siraisi, Nancy G., 'The Medical Learning of Albertus Magnus', in *Albertus Magnus and the Sciences: Commemorative Essays 1980*, ed. by James A. Weisheipl (Toronto: Pontifical Institute of Mediaeval Studies, 1980), pp. 379–404

Tellkamp, Jörg Alejandro, 'Why Does Albert the Great Criticize Averroes' Theory of the Possible Intellect?', in *Via Alberti: Texte – Quellen – Interpretationen*, ed. by Ludger Honnefelder, Hannes Möhle, and Susana Bullido del Barrio (Münster: Aschendorff, 2009), pp. 61–78

Theiss, Peter, *Die Wahrnehmungspsychologie und Sinnesphysiologie des Albertus Magnus: Ein Modell der Sinnes- und Hirnfunktion aus der Zeit des Mittelalters* (Frankfurt: Lang, 1993)

Wéber, Édouard-Henri, 'Un thème de la philosophie arabe interpreté par Albert le Grand', in *Albertus Magnus: Zum Gedenken nach 800 Jahren*, ed. by Walter Senner (Berlin: Akademie-Verlag, 2001), pp. 79–90

Wingate, S. D., *The Mediaeval Latin Versions of the Aristotelian Scientific Corpus, with Special Reference to the Biological Works* (London: The Courier Press, 1931)

Zaunick, R., 'Albertus Magnus, der Prerenaissance-Zoologe', *Ostdeutsche Naturwart*, 2 (1924), 124–28

AMOS BERTOLACCI

# Chapter 12. Inheritance and Emergence of Transcendentals*

## *Albert the Great between Avicenna and Averroes on First Universals*

Recent studies have drawn attention to the centrality of the doctrine of the primary and most universal concepts ('existent', 'thing', 'one', 'true', etc.) — the so-called 'transcendentals' — in both Arabic and in Latin medieval philosophy,[1] and to the seminal role that discussions of the topic in the Arabic cultural

* This paper is a revised and enlarged version of Bertolacci, 'Albert the Great, *Metaph.* IV, 1, 5', which was presented at the conference 'Universals in the XIII Century', organized by Gabriele Galluzzo at the Scuola Normale Superiore in Pisa on 5–7 September 2011. I am deeply grateful to the organizer, all the participants, and especially the late Prof. Francesco Del Punta for invaluable remarks received on that occasion. My sincere gratitude also goes to Prof. David Twetten, as well as to the editors of the present volume, for their careful reading and insightful comments on the first draft, and to Kate Sturge for her help with the style editing. The essay is part of the research project 'Itineraries of Philosophy and Science from Baghdad to Florence: Albert the Great, his Sources and his Legacies (2023–2025)', funded by the Italian Ministry of University and Research (PRIN 2022, 20225LFCMZ), in the framework of the PNRR M4C2 funded by NextGenerationEU.

1 On transcendentals in Arabic philosophy, see Adamson, 'Before Essence and Existence'; Wisnovsky, *Avicenna's Metaphysics in Context*; Menn, 'Al-Fārābī's *Kitāb al-Ḥurūf*'; Aertsen, 'Avicenna's Doctrine of the Primary Notions'; Bertolacci, '"Necessary" as Primary Concept in Avicenna's Metaphysics'; Koutzarova, *Das Transzendentale bei Ibn Sina*; Bertolacci, 'The Distinction of Essence and Existence'; Wisnovsky, 'Essence and Existence'; Menn, 'Fārābī in the Reception of Avicenna's Metaphysics'; Benevich, *Essentialität und Notwendigkeit*; De Haan, *Necessary Existence and the Doctrine of Being*; Janos, *Avicenna on the Ontology of Pure Quiddity*. For a general account of the various formulations of this doctrine in Latin philosophy, see Aertsen, *Medieval Philosophy as Transcendental Thought*, which is now the fundamental study on the topic. Goris, *Transzendentale Einheit*, addresses primarily the Scotist tradition of the transcendental unity.

**Amos Bertolacci** (amos.bertolacci@imtlucca.it) is professor of the history of medieval philosophy at the IMT School for Advanced Studies Lucca, and has also been professor at the Scuola Normale Superiore di Pisa.

*Albert the Great and his Arabic Sources*, ed. by Katja Krause and Richard C. Taylor, Philosophy in the Abrahamic Traditions of the Middle Ages, 5 (Turnhout: Brepols, 2024), pp. 335–370
BREPOLS PUBLISHERS 10.1484/M.PATMA-EB.5.136492

context played in its development in the Latin one.[2] Although the importance of Albert the Great (d. 1280) in the transmission of the doctrine of transcendentals from Arabic into Latin has been noted,[3] his specific contribution still needs precise assessment. Some scholars have stressed (perhaps even exaggerated) the novelty of his approach;[4] others have viewed his formulations of the issue as historically propaedeutic to later, more developed views.[5] Still lacking is a systematic investigation of his position, especially in his commentary on the *Metaphysics*, where scholarly attention has focused primarily on *ens* as the subject matter of metaphysics, leaving the other transcendentals in the background.[6]

In a pioneering article of 1994, Alain de Libera analysed the Latin reception of Avicenna's (Ibn Sīnā, d. 1037) doctrine of transcendental unity, showing how deeply and extensively Averroes's (Ibn Rushd, d. 1198) criticism of this Avicennian doctrine influenced Latin readers. De Libera convincingly documented the fact that many Latin logicians and metaphysicians of the thirteenth and fourteenth centuries, including Albert the Great, shared Averroes's polemical attitude towards Avicenna, drawing from Averroes the arguments by means of which they portrayed and discarded Avicenna's doctrine of transcendental unity. As de Libera put it, 'les Latins se sont approprié le texte d'Avicenne à travers le prisme averroïste'.[7] Among the various texts he discussed, de Libera pointed to an important passage of Albert's commentary on the *Metaphysics*, namely digression IV.1.5, on which I focus in the present paper.

---

2 On the importance of the Arabic discussion of primary concepts for the genesis of the Latin doctrine of transcendentals, see Craemer-Ruegenberg, '"Ens est quod primum cadit in intellectu"'; de Libera, 'D'Avicenne à Averroès, et retour'; Aertsen, '"Res" as Transcendental'; Aertsen, *Medieval Philosophy as Transcendental Thought*, chap. 2.4; Pini, 'Scotus and Avicenna'; Bertolacci, 'Reading Aristotle with Avicenna'.

3 See Aertsen, 'Albert's Doctrine on the Transcendentals'; Aertsen, *Medieval Philosophy as Transcendental Thought*, pp. 46–49 and 177–207.

4 On the basis of the passage of his commentary on the *Metaphysics* in which he refers to *prima et transcendentia* — Albertus Magnus, *Metaphysica*, I.1.2, ed. by Geyer (hereafter *In Metaph.*), p. 5, vv. 13–14 — Albert is credited with a conception of metaphysics as transcendental science, in anticipation of Duns Scotus's later famous formulation (see, for example, Aertsen, 'Albert's Doctrine on the Transcendentals', p. 618). However, this passage is open to various interpretations. In particular, the expression *prima et transcendentia* in Albert's text is closely connected with the analogous expression *causae omnium et principia* that immediately precedes it (*In Metaph.*, I.1.2, p. 5, vv. 12–13). This close link seems to suggest a 'non-transcendental' sense of *transcendens*, that is, it points at what transcends the physical order in a vertical, hierarchical direction, rather than at what transcends the categorial divisions in a horizontal perspective.

5 In Aertsen, *Medieval Philosophy as Transcendental Thought*, the chapter devoted to Albert (chap. 5, pp. 177–207) bears the title 'Albert the Great: Different Traditions of Thought and the Transcendentals', signalling from the very beginning a certain lack of coherence in Albert's global view of the topic. Previous studies expressly devoted to Albert's doctrine of transcendentals are Kühle, 'Die Lehre Alberts des Grossen'; de Libera, 'D'Avicenne à Averroès, et retour'; Tarabochia Canavero, 'I "sancti" e la dottrina'; Gabbani, 'Le proprietà trascendentali'.

6 See Zimmermann, *Ontologie oder Metaphysik?*, pp. 186–98; Noone, 'Albert on the Subject of Metaphysics' including the bibliography.

7 De Libera, 'D'Avicenne à Averroès, et retour', p. 146.

This text is worthy of consideration in several respects. Firstly, it has the structure of a *quaestio*, with *argumenta contra*, *solutio*, and *responsio ad argumenta* — a peculiarity indicating that Albert's commentary on the *Metaphysics*, as well as his other Aristotelian commentaries, cannot be straightforwardly classified as 'paraphrases' but have a wider stylistic frame, including the *commentum per quaestiones*. Secondly, despite being part of a commentary on the *Metaphysics*, the passage relates to a discussion of transcendental unity performed by a non-metaphysician, namely a *sophista* — a term that *prima facie* refers to some Latin logician of the Faculty of Arts contemporary with Albert, although it may simply mean 'opponent of Aristotle'.[8] Finally, the passage reveals Albert's desire to rescue Avicenna from Averroes's criticism. Albert's defensive attitude towards Avicenna is not unusual. In his early work *De homine* (q. 4, a. 3), for example, on the issue of whether a soul can be the form of simple bodies like the heavens, Albert contends that Avicenna's doctrine of the animation of heavens can be saved — that is, can be made acceptable — by 'doing violence to his words'.[9] The case I am going to discuss is different. There, Albert does not save Avicenna by forcing or deforming his text, or by rejecting Averroes's criticism *in toto*, as he does elsewhere,[10] but by modifying the purport of the criticism put forward by Averroes.

As to the first aspect of digression IV.1.5, its *quaestio* structure, a thorough analysis of Albert's method in the Aristotelian commentaries, with regard to our

8 De Libera, 'D'Avicenne à Averroès, et retour', p. 156, views the reference to the 'sophists' (*sophistae*) in the title of the digression as an indication of Albert's dependence on one or more authors of *sophismata*, on account of the expression 'multi Parisienses non philosophiam, sed sophismata sunt secuti' in Albertus Magnus, *De quindecim problematibus*, probl. 1, ed. by Geyer, p. 34, vv. 55–57, as well of the evidence provided by contemporary *sophismata* literature. However, it appears unlikely that an author of *sophismata* could label himself, or be called by his contemporaries, *sophista*. A less stringent use of *sophista* in this case is possible: the term occurs in a non-technical sense (meaning 'opponent of Aristotle') in, for example, Albertus Magnus, *In Metaph.*, IV.2.6, p. 183, v. 97; IV.3.4, p. 191, v. 77. See also the *sophismata Platonis* against Aristotle and the *elenchi sophistici*, stemming again from Plato's doctrine of ideas, mentioned in ibid., VII.2.1, p. 338, v. 33, and VII.2.4, p. 343, vv. 38 and 50 respectively; *rationes sophisticas* against Aristotle are cited in Albert's commentary on the *Physics* (*Physica*, VIII.1.12, ed. by Hossfeld, vol. 2, p. 572, v. 53: 'rationes sophisticas'). In his commentary on the *Liber de causis* (*De causis et processu universitatis a causa prima*, I.3.4, ed. by Fauser, p. 40, v. 19), Albert regards the *Fons vitae* as a spurious work, falsely ascribed to Avicebron by *quidam sophistarum*. A more technical use of the term *sophista* can be seen in *In Metaph.*, IV.1.2, p. 162, v. 82–p. 163, v. 34; but also in this case, the mention of the *obiecta sophistarum* does not seem to designate a particular instance of *sophismata* literature, but merely a structured and logically organized set of objections. See Albert's commentary on the *Liber sex principiorum*, *De Sex Principiis*, IV.5, ed. by Meyer and Möhle, p. 42, v. 10. On the difference between Albert's digression and the specimen of *sophismata* literature to which de Libera refers, see below, note 63. See also Albert's commentary on the *Categories*, *De praedicamentis*, I.3, ed. by Santos Noya, Steel, and Donati, p. 9, vv. 23 ff.; II.10, p. 41, vv. 23 ff.; II.12, p. 44, vv. 66 ff.

9 Albertus Magnus, *De homine*, q. 4, a. 3, ed. by Anzulewicz and Söder, p. 40, vv. 73–74: 'Ad aliud dicendum quod si volumus salvare Avicennam, tunc faciemus vim in verbo eius'.

10 Bertolacci, '"Averroes ubique Avicennam persequitur"'.

text and other similar *quaestiones* has yet to be carried out.[11] De Libera has already taken sufficiently into account the second peculiarity of the digression: its similarity to and possible connection with contemporary *sophismata* literature. In what follows, I will focus on the third interesting aspect of the digression, the defence of Avicenna, by considering the reasons for Albert's vindication of Avicenna against Averroes and, more generally, his attitude to these two major Arabic metaphysicians.

I proceed by arguing three main points. First, Albert takes the criticism of Avicenna directly from the Latin translation of Averroes, not from an intermediate source. Second, Albert defends Avicenna from Averroes's attack because he arguably detects in Averroes's criticism some lack of internal consistency and of faithfulness to Avicenna's actual thought. Third, Albert rescues Avicenna from Averroes's criticism through a direct and keen acquaintance with the Latin translation of Avicenna's metaphysics, rather than merely through the account of Avicenna's position provided by Averroes or by some previous Latin author.[12]

Accordingly, my exposition consists of three parts. The first describes the context, translates the text, and surveys the content of the passage of the *Long Commentary on the Metaphysics* (*Tafsīr mā baʿda l-ṭabīʿa*) in which Averroes criticizes Avicenna. The second focuses on the main problems that affect Averroes's criticism and the degree to which Albert is aware of them. The third part points to the changes that Albert introduces into the Latin translation of Averroes's text when he quotes it in his own commentary on the *Metaphysics*, and to the passages of Avicenna's *Philosophia prima* — the Latin translation of the metaphysical section, *Ilāhiyyāt* (*Science of Divine Things*), of the *Kitāb al-Shifāʾ* (*Book of the Cure/Healing*), his masterpiece on philosophy — that Albert has probably in mind when he defends Avicenna against Averroes.[13]

---

11 See, for example, Albertus Magnus, *Physica*, II.2.3, ed. by Hossfeld, p. 101, v. 84–104, v. 16. References to Albert can be found in Weijers, *In Search of the Truth*.

12 By contrast, de Libera, 'D'Avicenne à Averroès, et retour', p. 155, contends: 'Rien ne prouve, pourtant, qu'Albert soit remonté à l'original [d'Avicenne] pour répondre à l'interprète fantôme d'Ibn Sīnā baptisé du nom de *sophista*'. My impression is that, in this case, Albert uses the original text of Avicenna as well as that of Averroes (see the remarks below, note 63). I have documented Albert's direct recourse to Averroes's *Long Commentary on the Metaphysics* in Bertolacci, 'Reception of Averroes' Long Commentary'; Bertolacci, 'New Phase of the Reception of Aristotle'. For his equally direct recourse to Avicenna's *Philosophia prima*, see Bertolacci, '"Subtilius speculando"'; Bertolacci, 'Le citazioni implicite testuali'.

13 Avicenna, *Al-Shifāʾ, al-Ilāhiyyāt*, vol. 1, ed. by Qanawatī and Zāyid; *Al-Shifāʾ, al-Ilāhiyyāt*, vol. 2, ed. by Mūsā, Dunyā, and Zāyid (hereafter *Ilāhiyyāt*); Avicenna Latinus, *Liber de Philosophia prima sive Scientia divina, I–IV*, ed. by Van Riet; Avicenna Latinus, *Liber de Philosophia prima sive Scientia divina, V–X*, ed. by Van Riet. In what follows, Avicenna's work will be quoted with reference to pages and lines of the edition of the Arabic text, followed between square brackets by the pages and lines of the edition of the Latin translation. Averroes, *Tafsīr mā baʿd aṭ-ṭabīʿa*, ed. by Bouyges; Averroes Latinus, *Aristotelis Metaphysicorum libri XIIII*. In what follows, I will cite Averroes's work indicating the book of the *Metaphysics* and the section of Averroes's exegesis (e.g., Λ.5 = Book Λ, commentum

The digression I discuss contains Albert's version of one of the most significant criticisms that Averroes addresses to Avicenna. The importance of this critique is attested by its length, its articulated structure, and the variety of topics that Averroes touches upon in an anti-Avicennian vein. After the Latin translations of Averroes's Long Commentaries on the Aristotelian corpus in the first decades of the thirteenth century, Latin thinkers — under the same Aristotelian umbrella and in the context of the same Peripatetic tradition — were faced with two alternative views of the theory and practice of philosophy, both coming from Arabic Peripateticism. In fact, Avicenna and Averroes upheld two different formulations of philosophy, in terms of style (paraphrase vs literal commentary), attitude towards Aristotle (free adaptation vs faithful endorsement), and doctrine (inclusion of non-Aristotelian views vs strict adherence to the Peripatetic tradition). Moreover, Averroes frequently and harshly criticizes Avicenna in his commentaries on Aristotle, although to varying degrees depending on the specific type of exegesis adopted (epitome, paraphrase, literal commentary) and the particular Aristotelian work commented upon. This polemical attitude reaches its climax in the *Long Commentary on the Metaphysics*. Hence, the Latin reception of Avicenna's *Shifāʾ* as a *summa* of Peripatetic philosophy was certainly influenced by its counterpart, the systematic exegesis of Aristotle's works by Averroes. The contrast was particularly sharp in the principles of natural philosophy, psychology, and metaphysics, since Latin thinkers had at their disposal both Avicenna's and Averroes's major accounts of Aristotle's *Physics*, *De anima*, and *Metaphysics* in Latin translation.

In response to this situation, two main reactions in Latin culture can be observed. On the one hand, the idea of a conflict between Avicenna and Averroes pervaded Latin philosophy from the thirteenth century onwards, taking inspiration from and amplifying Averroes's criticisms. The divergence became associated with competing cultural institutions (the Avicennian sympathies of the theologians vs the Averroean allegiance of the masters of Arts) and disciplinary fields (the 'physician' Avicenna vs the 'commentator' Averroes). It assumed religious connotations (the 'pious' Avicenna vs the 'sceptic' Averroes), corroborated by pseudo-epigraphical writings (the ps.-Avicennian *Epistula ad Sanctum Augustinum* vs the ps.-Averroean *Tractatus de tribus impostoribus*); it inspired fictive biographical tales showing the two thinkers in a personal clash; and it found vivid expressions in iconography (the 'prince' or 'king' Avicenna vs the Averroes over whom Thomas Aquinas triumphs).

On the other hand, confronted with the manifest disagreement between Avicenna and Averroes, some Latin thinkers adopted a different strategy, both historically significant and theoretically demanding: they undertook to create a synthesis between the two Arabic masters. That harmonization was an arduous path to follow, since it required a profound understanding of Avicenna's and Averroes's standpoints and an intelligent search for a 'third way' in the interpretation

5); the page number and lines of the Arabic edition (e.g., p. 1420, v. 6–p. 1421, v. 16); between square brackets, the folio and sections of the Juncta edition of the Latin translation (e.g., [fol. 292K–M]).

of the various works of Aristotle that they had reworked or commented upon, in terms of approach, style, and doctrine.

Albert the Great is an illuminating example of this second trend. He was certainly aware of the distance separating Avicenna from Averroes, and in his first Aristotelian commentaries (especially those on *Physica* and *De caelo*) indulges in the topos of their antinomy. In his more mature commentaries, however, his attitude evolves; there, rather than insisting on the differences between the two Arabic masters, he tries to establish a consensus among them. The commentary on the *Metaphysics* shows this tendency with particular clarity, and the digression I consider in this paper is a compelling specimen of Albert's mature approach to the issue.

## Averroes's Criticism of Avicenna

The doctrine of transcendentals is the metaphysical doctrine of Avicenna's that Averroes criticizes most harshly in his *Long Commentary on the Metaphysics*. Criticisms of this doctrine are recurrent, lengthy, and disdainful. The text I examine in this section is a prime example of this attitude, being the first, and one of the most extensive, criticisms of the topic in Averroes's *Long Commentary on the Metaphysics*.[14] In it, Averroes draws out several points of dissent, engages in an extended discussion, and refers to Avicenna with expressions of amazement and scorn ('Ibn Sīnā made a serious mistake [...]. What is surprising about this man [...] This man does not distinguish [...]. Several things made this man go astray'). The criticism we are concerned with is deservedly famous, although it has hitherto received only cursory analysis.[15]

The text occurs in the third section of Averroes's exegesis of Book Γ of the *Metaphysics*. In it, Averroes explains *Metaph.* Γ.2, 1003b22–1004a1, a passage whose translation from the Arabic runs as follows:

> Text 1: Arabic translation of *Metaph.* Γ.2, 1003b22–1004a1
> [A: 1003b22–32] Since 'one' and 'being' [*huwiyya*] are a single thing and have a single nature, each one of them follows the other, as principle and cause follow each other. This does not happen because a single definition signifies

14 The other criticisms of Avicenna's doctrine of transcendentals occur in Averroes's commentary on books Γ, Δ, and I of the *Metaphysics*: Γ.3, p. 315, vv. 3–9 [fol. 67G]; Δ.14, p. 557, vv. 16–19 [om.]; Δ.14, p. 558, v. 17–p. 559, v. 14 [fol. 117C–D]); I.5, p. 1267, v. 15–p. 1268, v. 3 [fol. 255B]; I.8, p. 1279, v. 12–p. 1280, v. 11 [fol. 257E–G]; I.8, p. 1282, vv. 8–12 [fol. 257K]. An overview of all of the criticisms of Avicenna in Averroes's *Long Commentary on the Metaphysics* is available in Bertolacci, 'From Athens to Buḫārā'; Bertolacci, 'Avicenna's and Averroes's Interpretations'.

15 See Forest, *La structure métaphysique*, p. 41; Gilson, *L'être et l'essence*, p. 67; O'Shaugnessy, 'St Thomas's Changing Estimate', pp. 252–53; al-Ahwani, 'Being and Substance'; Fakhry, 'Notes on Essence and Existence'; Rashed, *Essentialisme*, pp. 255–56. Related criticisms of Avicenna's doctrine of transcendentals in Averroes's *Long Commentary on the Metaphysics* have been analysed by Menn, 'Fārābī in the Reception of Avicenna's Metaphysics', pp. 62–64.

> both, although it makes no difference as to their relationship if we believe something of this kind. For, if someone says 'a man one', or 'a man is', or 'a man this', he signifies a single thing, and he does not signify different things by repeating them. It is well known that the expression that says 'man is' or 'man one' does not signify different things, since there is no distinction between saying 'man is' and [saying] 'man neither in generation nor in corruption'. The same happens also with the statement regarding 'one'. It is well known that what is added in these [statements] signifies a single thing, and that 'one' does not signify something other and different from 'being'.
>
> [B: 1003b32–33] We also say that the substance of each thing is one not accidentally. Therefore, we say that the substance of every thing is being.
>
> [C: 1003b33–1004a1] It is well known that the forms of 'one' are as many as the forms of 'being', and [that] to a single science belongs the absolute investigation of these forms and the knowledge of what they are. I mean: to a single science belongs the investigation of 'congruent', 'similar', the other things resembling these, etc. In sum, all the contraries refer to this first science.[16]

In this passage, Aristotle holds: (A) that 'being' and 'one' are the same thing and a unique nature, and that neither signifies something different from what the other signifies; (B) that the substance of everything is essentially 'being' and 'one'; (C) that the species of 'being' are as numerous as the species of 'one' and that their study belongs to the same science, namely, metaphysics.[17]

---

16 Averroes, *Tafsīr mā baʿd aṭ-Ṭabīʿa*, ed. by Bouyges, vol. 1, p. 310, v. 2–p. 311, v. 4. The Arabic-Latin translation of this passage in the *Metaphysica nova* that was available to Albert reads as follows in the Juncta printing: '[A] Unum autem et ens, cum sint idem et habeant eandem naturam, consecutio utriusque ad alterum est sicut consecutio principii et causae unius ad alterum, non quia eadem definitio significat utrumque. Nulla autem differentia est inter ea, etsi existimantes fuerimus tali existimatione. Sermo enim dicentis "homo unus" aut "homo est" aut "homo iste" idem significat, et non diversa significat apud iterationem. Manifestum est enim quod sermo dicens "homo iste" et "homo unus" et "homo est" non significat diversa, cum non sit differentia inter dicere "homo iste" et "homo neque in generatione neque in corruptione". Et similiter est etiam de uno. Manifestum est igitur quod additio in istis significat idem et non significat unum aliud ab ente. [B] Et etiam substantia cuiuslibet est una non modo accidentali. Et ideo dicimus quod substantia cuiuslibet unius communis est esse eius. [C] Manifestum est igitur quod formae unius sunt secundum numerum formarum entis et unius scientiae est consideratio similiter de istis formis, scilicet quod unius scientiae est consideratio de convenienti et simili et de aliis rebus similibus. Et universaliter omnia contraria attribuuntur huic primae scientiae' (Averroes Latinus, *Aristotelis Metaphysicorum libri XIIII*, fol. 66G–K, with punctuation changed). The critical edition in preparation by Dag Nikolaus Hasse and Andreas Büttner provides a slightly different text, which does not, however, substantially diverge from that printed in the Juncta edition.

17 A thorough account of the doctrine of this passage, its various possible interpretations, and the scholarly discussions thereupon can be found in Castelli, *Problems and Paradigms of Unity*, pp. 51–55. Averroes holds the second interpretation of lines 1003b32–33 mentioned by Castelli ('the relation of one and being to essences as non accidental', p. 54, n. 8).

In his commentary on this passage of Aristotle, elaborating on all three points, Averroes criticizes Avicenna's position concerning points A and B, excluding from his criticism point C. According to Averroes, Avicenna proposed a view of the mutual relationship between 'being' and 'one' (issue A), and of the relationship between 'being' and 'one' and essence (issue B), that is decidedly different from Aristotle's, and therefore wrong. According to Averroes, Avicenna holds that 'being' and 'one' do not signify one and the same thing (issue A), and that they are not identical to the thing's substance or essence, but rather superadded and accidental to it (issue B).

One should notice that in Averroes's account of Avicenna, Avicenna's position on issue A — the identity or difference of 'being' and 'one' — is adduced as the reason for his position on issue B, their essential or accidental status. Since for Avicenna (*apud Averroem*) 'being' and 'one' do not signify one and the same thing (issue A), they cannot be essential attributes (issue B). The rationale behind the causal relationship of A with respect to B that Averroes posits seems to be that if someone takes 'being' and 'one' to be distinct from one another, that person is forced to endorse their accidentality, because if they were essential attributes, they would necessarily signify one and the same thing: the essence. I will discuss this feature of Averroes's report of Avicenna in detail in the next part of this paper.

Averroes's criticism of Avicenna is reported in Table 1 together with the *loci paralleli* in Avicenna. It consists of three main parts, each of which can be further subdivided. In the first part, Averroes expounds Avicenna's incorrect thesis, underscoring the gravity of its error. In the second, he declares Avicenna's main argument invalid. In the third, he points out the doctrinal roots of Avicenna's error.

In part 1, Averroes posits what he regards as the error of Avicenna: the consideration of 'existent' (the most usual equivalent of 'being' in Arabic philosophy) and 'one' as non-essential features, more precisely as distinct attributes superadded to the essence of things (1.1). In the section that immediately follows, 1.2, Averroes stresses the gravity of this mistake, adding some interesting considerations on the theological background of Avicenna's metaphysics that cannot be addressed in detail here.[18]

In part 2, Averroes ascribes to Avicenna an argument that, in his opinion, functions as the proof of Avicenna's thesis in 1.1. Averroes's intent in this part is to show that this argument is invalid and the reasons why it is invalid. The argument in question acts as a *reductio ad absurdum*, of which Averroes reports only the main part: if 'existent' and 'one' did not signify attributes superadded to the essence — contrary to what Avicenna holds — then they would signify the same notion or item (the Arabic term *maʿnan* occurring here can express both ideas), namely the essence itself; but in that case a proposition such as 'the existent is one' would be a tautology, which is not the case (2.1). Implicitly, the

18 The theological underpinnings of the discussion may explain Averroes's use of the theologically loaded term 'attribute' (*ṣifa*) in section 1.1.

next step in the argument is that the premise leading to the false conclusion just reached — namely the premise that posits that 'existent' and 'one' do not signify attributes superadded to the essence — is false and its contrary — that 'existent' and 'one' do signify attributes superadded to the essence — is true, as Avicenna wishes.

Table 1. Averroes's criticism of Avicenna together with the *loci paralleli* in Avicenna.

| AVERROES, *LONG COMMENTARY ON THE METAPHYSICS*, Γ.3, ED. BY BOUYGES, P. 313, V. 6–P. 314, V. 11 | AVICENNA, *AL-SHIFĀ'*, *AL-ILĀHIYYĀT*, VOL. 1, ED. BY QANAWATĪ AND ZĀYID; VOL. 2, ED. BY MŪSĀ, DUNYĀ, AND ZĀYID |
|---|---|
| [1.1] Ibn Sīnā made a great mistake in this regard, since he believed that 'one' [*wāḥid*] and 'existent' [*mawjūd*] signify attributes that are added [*ṣifāt zā'ida*] to the thing's essence [*dhāt*]. | III.2, p. 103, v. 9 [p. 114, vv. 19–20]: neither of them [i.e., 'one' and 'existent'] signifies the substance of any thing<br>V.1, p. 196, v. 13 [p. 229, v. 37]: unity is an attribute [*ṣifa*] that is joined [*taqtarinu*] with horseness, so that horseness, with this attribute, is one |
| | V.1, p. 198, v. 6 [p. 230, v. 68]: [to be one or many] is like something that is consequent from outside [*yalḥaqu min khārij*] to humanity (cf. V.1, p. 198, v. 3 [p. 230, v. 64]; p. 198, v. 8 [p. 230, vv. 71–72]) |
| | VIII.4, p. 347, v. 9 [p. 402, vv. 45–46]: existence occurs from outside [*ya'riḍu min khārij*] to the quiddities of things other than God |
| [1.2] What is surprising about this man is how he made this mistake despite having heard [the teaching of] the Ash'arite theologians, whose theology he mixed in his divine science. [...] | |
| [2.1] This man argues for his doctrine by saying that, if 'one' and 'existent' signified a single notion/item [*ma'nan*], | cf. VII.1, p. 303, vv. 9–10 [p. 349, vv. 15–17]: If the concept [*mafhūm*] of 'one' were [...] the concept of 'existent', |
| | [...] in every way [*min kulli jiha*] [...] |
| the statement 'existent is one' would be a futility [*hadhr*], like the statement 'existent is existent' and 'one is one'. | then 'many' — qua 'many' — would not be 'existent', as it is not 'one'.<br>cf. I.5, p. 31, v. 10–p. 32, v. 2 [p. 35, v. 62–p. 36, v. 79] |
| [2.2.1] But this [absurdity] would necessarily follow only if someone contended that saying of one and the same thing [*shay'*], 'it is existent' and '[it is] one' signifies a single | |

| **AVERROES, *LONG COMMENTARY ON THE METAPHYSICS*, Γ.3, ED. BY BOUYGES, P. 313, V. 6–P. 314, V. 11** | **AVICENNA, *AL-SHIFĀʾ*, *AL-ILĀHIYYĀT*, VOL. 1, ED. BY QANAWATĪ AND ZĀYID; VOL. 2, ED. BY MŪSĀ, DUNYĀ, AND ZĀYID** |
|---|---|
| notion/item [*maʿnan*] according to a single way [*jiha*] and to a single mode [*naḥw*], | |
| [2.2.2] whereas we have only said that these two [terms] signify a single essence [*dhāt*] in different modes [*anḥāʾ muḥtalifa*], | |
| [2.2.3] not different attributes [*ṣifāt mukhtalifa*] added to it [i.e., to a single essence]. | |
| [2.3] According to this man, therefore, there is no distinction between the expressions that signify different modes of a single essence [*dhāt*], without signifying notions/items added to it, and the expressions that signify attributes added to a single essence, namely other [*mughāyira*] than it in actuality. | |
| [3.1] [Several] things made this man go astray. One of them is that he found that the name 'one' belongs to the derived names [*asmāʾ mushtaqqa*], | III.3, p. 110, vv. 2–3 [p. 122, vv. 67–69]: the predicate [i.e., 'one'] [...] derives its name [*mushtaqq al-ism*] from the name of a simple item, i.e., from the item 'unity' |
| and [that] these names signify an accident [*ʿaraḍ*] and a substance. | |
| [3.2] Another reason is that he believes that the name 'one' signifies a notion/item [*maʿnan*] in the thing, [namely] 'lacking division', | III.2, p. 97, vv. 4–5 [p. 107, vv. 77–79]: 'One' is said equivocally of items sharing the fact of lacking any division in actuality, insofar as each of them is what it is |
| and that this notion/item is different from the notion/item that is the [thing's] nature. | III.3, pp. 106, vv. 12–13 [p. 117, vv. 83–85]: unity does not enter into the determination of the quiddity of any substance [...] |
| [3.3.1] Another reason is that he believes that this 'one' said of all the categories is the 'one' that is the principle of number. | III.1, p. 95, vv. 16–17 [p. 107, vv. 67–69]: 'one' has a tight relation with 'existent' [...] 'one' is a principle, in a way, of quantity |
| But number is an accident [*ʿaraḍ*]. | III.3, p. 110, v. 4 [p. 122, vv. 70–71]: number [...] is an accident [*ʿaraḍ*] |
| Therefore he was convinced that the name 'one' signifies an accident [*ʿaraḍ*] of existents. | III.3, p. 106, v. 15 [p. 117, v. 87]: unity is the notion that is the accident [*ʿaraḍ*]; p. 109, v. 10 [p. 121, vv. 51–52]: the essence of unity is an accidental [*ʿaraḍī*] notion; p. 110, vv. 3–4 [p. 122, vv. 69–70]: that simple item [i.e., unity] is an accident; [...] unity is an accident |
| [3.3.2] But the 'one' that is the principle of number is only one of the existents of which | |

| **Averroes, *Long Commentary on the Metaphysics*, Γ.3, ed. by Bouyges, p. 313, v. 6–p. 314, v. 11** | **Avicenna, *Al-Shifāʾ*, *al-Ilāhiyyāt*, vol. 1, ed. by Qanawatī and Zāyid; vol. 2, ed. by Mūsā, Dunyā, and Zāyid** |
|---|---|
| the name 'one' is said, although it is the worthiest of them to be [said 'one'], as you will learn in the ninth treatise of this book. | |

After expounding Avicenna's argument, Averroes shows that it is based on an incorrect deduction. According to Averroes, the aforementioned counterintuitive conclusion (that the proposition 'the existent is one' is a tautology) follows, properly speaking, not from the premise leading to the absurd conclusion in Avicenna's argument (namely 'existent' and 'one' signify the same notion or item, with no further specification), but from a premise positing that 'existent' and 'one' signify the same notion or item *according to a single way and to a single mode* (2.2.1). Averroes argues that, once the premise leading to it is fully articulated, the absurd conclusion of the argument is harmless with respect to Aristotle's position, since Aristotle, as Averroes interprets him ('we have [...] said'), holds the opposite of the premise at stake ('existent' and 'one' signify the same essence according to different modes, 2.2.2). In 2.2.3, Averroes remarks that Aristotle's and his own thesis is different from the thesis that Avicenna intends to corroborate by means of this argument ('existent' and 'one' signify distinct attributes added to the essence).

In 2.3, Averroes concludes this part of the criticism by maintaining that Avicenna's defective formulation of the premise, leading to the absurd conclusion in his argument, shows that Avicenna missed the fundamental distinction capable of discriminating between his own position and a position like the one advocated by Averroes in the footsteps of Aristotle: namely, a distinction between expressions that signify different modes of an essence (that is, Averroes's position with regard to 'existent' and 'one') and expressions that signify attributes added to the essence (that is, Avicenna's own position with regard to 'existent' and 'one'). The implicit assumption of Averroes's discourse is that Avicenna manifestly lacks an indispensable theoretical tool to deal with such intricate metaphysical topics as the present one (a critique of Avicenna that Averroes also formulates in other cases).

The third part of Averroes's text contains three arguments that he considers to be the remote causes of Avicenna's error in 1.1. All three indicate the non-essential character of unity, arguing either that 'one' is an accident of the essence (3.1, 3.3) or that it is different from the essence (3.2). The exposition of the last of these arguments (3.3.1) is followed by a criticism (3.3.2).

## A Puzzling Criticism

The main tenets of Averroes's report can be summarized as follows (the points made implicitly by Averroes are added in square brackets):

Outline 1: Summary of parts 1–3
1.1 (B) Avicenna's thesis ($t_{IS}$): 'one' and 'existent' signify [distinct] attributes added to the essence
2.1 (A) Avicenna's argument: if ($\neg t_{IS}$) 'one' and 'existent' signified a single notion/item, then the proposition 'the existent is one' would be a tautology [therefore 'one' and 'existent' do not signify the essence, if the essence is meant as the single notion/item in question]
2.2.1 (A) Avicenna's argument emended by Averroes: if ($\neg t_{IR}$) 'one' and 'existent' signified a single notion/item *according to a single way and to a single mode*, then the proposition 'the existent is one' would be a tautology
2.2.2 (B) The emended argument is harmless with respect to Averroes's thesis ($t_{IR}$), according to which 'existent' and 'one' signify a single essence according to different ways and modes
2.2.3 (B) Averroes's thesis ($t_{IR}$) is different from Avicenna's thesis ($t_{IS}$)
2.3 (B) Avicenna is unaware of the distinction between ($t_{IR}$) and ($t_{IS}$), namely between expressions that signify different modes of a single essence vs expressions that signify attributes added to a single essence
3 ($B_1$) The remote cause of Avicenna's error: 'one' signifies an accident (3.1; 3.3.1); it signifies a notion/item different from the essence (3.2)

Averroes's report of Avicenna's position is puzzling in various ways. First of all, it consists of a discontinuous series of distinct sections dealing with different issues and topics, which Averroes assembles from several Avicennian *loci*, rather than from a single text by Avicenna, and integrates with his own views. Moreover, the transitions between the three main parts and their distinct sections show some logical inconsistencies. In particular, the two issues A and B that Averroes causally connects in his report of Avicenna's position appear, in principle, logically independent: one can argue that 'being' and 'one' are identical to one another or different from one another (issue A), regardless of their being essential or accidental features (issue B).[19] Finally, in a few notable instances, Averroes appears to be seriously distorting Avicenna's point of view, either by selecting arbitrarily some of Avicenna's different statements on a given issue, or by reporting the assertions he selects in a form substantially different from Avicenna's original

---

19 One can easily imagine two things, such as 'being' and 'one' in the present case, as essential and distinct from one another, e.g., 'animal' and 'rational' with respect to 'man', or as accidental and identical to one another, e.g., 'unmarried' and 'wifeless' with respect to 'man', whereas Averroes seems to suppose that they are either essential and identical, or accidental and distinct.

one.[20] I will now analyse the three types of problems just mentioned, with regard to (i) the articulation, (ii) the cogency, and (iii) the congruity of Averroes's text with Avicenna's actual position.

Regarding problem (i), the most remarkable aspect of Averroes's criticism is that parts 1 and 3 deal with issue B, namely the relationship between 'existent', 'one', and essence (more precisely, part 3 deals with a specific instance of this issue, as we will see), whereas part 2 deals with both issue A (the reciprocal relation of 'existent' and 'one', without any explicit mention of essence) and issue B. As we can see from the outline, the initial treatment of issue B in section 1.1 is superseded by the discussion of issue A in sections 2.1 and 2.2.1. Issue B surfaces again, in connection with issue A, in sections 2.2.2, 2.2.3, and 2.3, where Averroes speaks significantly of 'a single essence'.[21] In part 3, only issue B is taken into account, although with a narrower scope (part 3 regards only the relationship of 'one', to the exclusion of 'being', with the essence) and different philosophical concepts (the idea of accidentality replaces that of superaddition to essence). To distinguish it from issue B, I therefore label it $B_1$.

Due to this variation of the specific topics dealt with and the fluctuating presence of the consideration of essence, it is not immediately clear how section 1.1, which regards squarely issue B of Aristotle, relates to the subsequent sections 2.1 and 2.2.1, which are supposed to ground section 1.1 but, differently from 1.1, *prima facie* concern expressly only issue A: here, the question is whether or not 'one' and 'existent' signify the same notion or item, regardless of whether the signified notion or item is the essence or something else, and Avicenna is said to offer a negative answer to that question.

Averroes tries his best to provide a coherent account of Avicenna's position. But he does so by a series of terminological shifts that, though surely smoothing the transitions between issue B and issue A (and vice versa), do not eliminate all cleavages. A first shift of this kind emerges in the transition from section 2.2.1 to section 2.2.2. In 2.2.1, Averroes sets apart three elements in the predication of 'existent' and 'one': the 'thing' (*shay'*) of which they are predicated, the 'notion (or: item)' (*ma'nan*) that they signify, and the 'way' (*jiha*) or 'mode' (*naḥw*) by means of which they signify this notion. But in 2.2.2, he replaces the second of these three elements — the neutral term 'notion/item' (*ma'nan*) — with a much stronger term, namely 'essence' (*dhāt*), thus surreptitiously passing from the current issue A to the initial issue B.[22] Conducive to the same result of bridging issue A with issue B is the shift in the meaning of the adjectives 'different'

20 The same tendency to distortion surfaces in other criticisms of Avicenna in Averroes's *Long Commentary on the Metaphysics*: see O'Shaugnessy, 'St Thomas's Changing Estimate', pp. 253–55; Bertolacci, 'Averroes against Avicenna'.

21 In these sections, the adjective 'single' (*wāḥida*) is reminiscent of the previous mention of 'a single notion/item' (*ma'nan*), namely of issue A, in section 2.1; however, the change in the noun, i.e., the reintroduction of consideration of the essence raises issue B anew.

22 In this light, we can guess that the occurrence of the key term 'notion/item' (*ma'nan*) in 2.1, too, is meant by Averroes in the meaning of 'essence'.

(*mukhtalif*) and 'other' (*mughāyir*) in section 2.2.2 and section 2.3 respectively. In 2.2.2, 'different' expresses the idea that 'existent' and 'one' signify features that are distinct *from one another*, whereas 'other' in 2.3 expresses the idea that 'existent' and 'one' signify features that are distinct *from the essence*. Thus, it seems that Averroes is trying to connect parts 1 and 3 with part 2 as coherently as possible by means of an ambiguous use of terminology, helped by the fact that the two main terms he uses to signify the 'notion/item' and the 'essence' (*maʿnan* and *dhāt*) have a wide range of meanings and are constitutively multivocal.

On point (ii), even if we accept these terminological oscillations aimed at easing the interplay between different issues, the thesis that Averroes ascribes to Avicenna in section 2.1 is inconclusive with respect to the doctrine he attributes to Avicenna in section 1.1. In 1.1, Avicenna contends that both 'existent' and 'one' are features added to the essence, therefore extrinsic to the essence and hence non-essential. Section 1.1 is therefore meant to establish that *neither* 'existent' *nor* 'one' signifies the essence. But from the fact that 'existent' and 'one' do not signify the same notion/item in section 2.1, a much weaker thesis follows: even if we assume that the notion/item in question is the essence — thus switching from the present issue A to the original issue B — the contention in 2.1 entails that *either* 'existent' *or* 'one' does not signify the essence, and therefore that *either* 'existent' *or* 'one' is a non-essential feature. In other words, according to section 2.1 only one among 'existent' and 'one' is a non-essential feature, whereas section 1.1 aims to establish that both are non-essential features. This being the case, section 2.1 — as it is formulated, and regardless of the logical weakness that Averroes detects in Avicenna's alleged argument — is far from being an 'argument' for section 1.1, contrary to what Averroes contends.

Other incongruences affect part 3. This part is allegedly intended to explain the remote causes of Avicenna's position in section 1.1; however, it conveys a thesis that in one way is weaker, and in another way stronger, than the doctrine actually ascribed to Avicenna in 1.1. On the one hand, part 1.1 regards the relationship of both 'one' and 'existent' with the essence, whereas part 3 concerns the relationship only of 'one' with the essence, to the exclusion of 'existent'. On the other, in part 1.1 'one', like 'existent', is portrayed as an attribute superadded to a thing's essence; in part 3, by contrast, it assumes — much more pointedly — the status of an 'accident' (*ʿaraḍ*) of essence. Averroes is certainly entitled to ascribe to Avicenna the doctrine of the accidentality of unity, as we will see. But part 3, being presented as an explanation of section 1.1, suggests that for Avicenna 'existent' is also an accident in the same sense as 'one' is. A parallelism of that kind looks much less warranted, as the following exposition will document. Moreover, and paradoxically, it is not immediately clear how part 3, if taken together with section 2.1, supports section 1.1 rather than invalidating it. In part 3, Avicenna contends that 'one' signifies a non-essential feature, or an accident, of the essence. In part 2.1, he holds that 'existent' and 'one' signify different items. This being the case, it would seem that if 'one' signifies an accident of the essence, 'existent' does not signify an accident of the essence as well; but if 'existent' does not signify an accident of the essence, it has arguably good chances of signifying the essence, contrary to what part 1.1 contends.[23]

Finally, regarding point (iii): The fullest expression of Avicenna's view of the mutual relationship of 'existent' and 'one' and of their relationship with essence can be found in the *Ilāhiyyāt* of the *Shifā'*, a work with which Averroes was surely acquainted and from which he mainly drew his knowledge of Avicenna's philosophy.[24] In this work, Avicenna offers a variety of statements on the issue.[25] Section 1.1 can be compared with some of these statements of Avicenna's, with the following differences. First, by calling 'one' and 'existent' 'attributes' (*ṣifāt*), Averroes selects a substantive rarely used by Avicenna.[26] Second, the idea of externality conveyed by the participle 'added' (*zā'ida*) has no verbatim correspondence in Avicenna, although this participle can be compared with the phrase 'from outside' (*min khārij*) that Avicenna uses adverbially, mostly in the case of 'one',[27] but also in the case of 'existent'.[28] Here, Averroes disregards the most frequent Arabic root used by Avicenna to express the relationship of 'existent', 'one', and essence, from the beginning until the end of the *Ilāhiyyāt*, namely the root *l-z-m*, which conveys the idea of inseparable concomitance (lit.: 'clinging') more than that of externality.[29] Significantly, this is the only root used by Avicenna

---

23 Part 3 of Averroes's criticism becomes compatible with and explanatory of the doctrine of part 1.1 only if we assume that 'existent' and 'one' *do* signify the same type of item, i.e., an accident *large loquendo*, as part 1.1 contends, but *do not* signify the same token of this item: since they do not signify the same specific accident, they comply with the requirement of not signifying the same item imposed by part 2.1 on 'existent' and 'one'. But this precision remains entirely implicit in Averroes's text. Not even the corrections that Averroes deems necessary to make part 2.1 conclusive — namely, to assume that 'existent' and 'one' signify the same item in different ways — seem sufficient to solve the impasse.

24 On Averroes's knowledge of Avicenna's *Kitāb al-Shifā'*, see Bertolacci, '"Incepit quasi a se"'.

25 A wide sample of these statements is analysed in Bertolacci, 'Reception of Avicenna', pp. 256–59.

26 It is used only once, at the singular, for 'one', in Avicenna, *Ilāhiyyāt*, V.1, p. 196, v. 13 [p. 229, v. 37] (see Table 1).

27 See, for instance, ibid., V.1, p. 198, v. 6 [p. 230, v. 68], in Table 1.

28 See ibid., VIII.4, p. 347, v. 9 [p. 402, vv. 45–46], in Table 1. See also ibid., V.1, p. 201, v. 15 [p. 234, v. 46]: '[to be one or many] is a concomitant from outside (*lāzim min khārij*) of animal'. In V.1, p. 198, v. 3 [p. 230, v. 64], the adverb *min khārij*, used by Avicenna to describe the relationship of 'one' and 'many' with the 'entity' or essence (*huwiyya*) of man, is not attested by all manuscripts (it is omitted, for instance, in MSS Oxford, Bodleian Library, Pococke 125, Oxford, Bodleian Library, Pococke 110, London, British Library, Oriental and India Office Collections, Or. 7500, and by the Latin translation). Wisnovsky, 'Essence and Existence', p. 28 and n. 5, records one occurrence of the participles *zā'id* and *khārij* in Avicenna's *Ta'līqāt* (*Annotations*) (IV.32, ed. by al-'Ubaydī, p. 164, vv. 18–ult.: 'The existence of each category is extrinsic, *khārij*, to its quiddity and superadded, *zā'id*, to it; whereas the quiddity of the Necessary of Existence is its "thatness"; <and its thatness is not> superadded to [its] quiddity'). In the same context, Wisnovsky points out the doubts still surrounding the composition and Avicenna's authorship of this work.

29 Avicenna, *Ilāhiyyāt*, III.3, pp. 106, vv. 12–13 [p. 117, vv. 83–85]: 'unity does not enter into the determination of the quiddity of any substance, but it is an entity that is a concomitant [*lāzim*] of substance' (cf. III.3, p. 109, v. 10 [p. 121, vv. 51–52]; V.1, p. 201, v. 14 [p. 234, v. 44]); VI.5, p. 292, vv. 2–3 [p. 336, vv. 85–87]: 'There is a distinction [*farq*] between "thing" and "existent"

to describe the mutual relationship of essence and existence when he speaks *ex professo* about it in the *locus classicus* of chapter I.5.[30] Third, Averroes equates the cases of 'one' and 'existent' in his report of Avicenna's position, taking the former as his main reference point. This procedure can be justified by the various statements in which Avicenna ascribes an equal status to the two concepts in terms of their relationship with essence,[31] although nowhere does Avicenna speak jointly of 'one' and 'existent' as notions superadded to the essence. For all these reasons, it is hard to maintain that Averroes's report in section 1.1 faithfully mirrors Avicenna's standpoint: although the idea that unity is superadded to essence has a solid textual basis in Avicenna, and although 'Avicenna's ontology could doubtless be interpreted as implying the thesis that existence is superadded to a thing's quiddity', as the history of *falsafa* attests,[32] Avicenna looks to convey a view of existence and essence in which these two items are, primarily, two inseparable and mutually linked concomitants, the accent falling on their connection rather than their separation.[33]

Whereas textual evidence supporting section 1.1 can be found in Avicenna, with the provisos noted above, the case of part 2 is very different, since the correspondence with Avicenna there is fragmentary and incomplete. Section 2.1 is a *reductio ad absurdum*, made of a premise and a consequence, with the conclusion left unexpressed. Since the premise of the *reductio* is 'if "one" and "existent" signify a single notion/item', the unexpressed conclusion should be that 'one' and 'existent' do not signify a single notion/item. Of this elliptical *reductio ad absurdum*, only the premise has a rough correspondence in Avicenna: it vaguely resembles the premise of a *reductio ad absurdum* that we find in a passage of *Ilāhiyyāt* VII.1 (p. 303, vv. 9–10 [p. 349, vv. 15–17]). But the consequence in

(although "thing" isn't but an "existent"), as there is a distinction [*farq*] between an entity [*amr*] and its inseparable concomitant [*lāzim*]' (cf. VIII.4, p. 346, v. 15–p. 347, v. 2 [p. 401, vv. 33–36]).

30 Ibid., I.5, p. 32, v. 3 [p. 36, v. 81]: 'the notion of "existent" always accompanies it [i.e. the notion of "thing", which signifies the essence] inseparably [*yalzamuhū*] it'; p. 34, vv. 9–10 [p. 39, vv. 37–39]: 'Now you have understood in what [the concept of] "thing" differs from the concept of "existent" and of "supervening", even though ["thing" and "existent"] accompany inseparably each other [*mutalāzimani*]' (cf. VI.5, p. 292, v. 3 [p. 336, v. 87]; VIII.4, pp. 347, 2 [pp. 401, 36]). Other notions that Avicenna uses in the *Ilāhiyyāt* to express the relation of existence and unity with essence are 'supervenience' (verb *dakhala ʿalā*), and — as we have seen — 'joining' (verb *iqtarana*), 'consequence' (verb *laḥiqa*), and 'accidental occurrence' (verb *ʿaraḍa*). Within the discussions of the relationship of essence and existence, the verbs *dakhala ʿalā* (I.7, p. 45, vv. 10–11 [p. 52, vv. 94–95]) and *ʿaraḍa* (VIII.4, p. 346, v. 13 [om.]) and are always used in conjunction with *lazima*. The verb *laḥiqa* is semantically close to *lazima*.

31 See, for example, ibid., III.2, p. 103, v. 9 [p. 114, vv. 19–20], in Table 1.

32 Wisnovsky, 'Essence and Existence', p. 29. At p. 42, n. 43, Wisnovsky points to Bahmanyar's (d. *c.* 1066) adoption of the Avicennian idea that existence and unity relate to the essence 'from outside' (*min khārij*). Wisnovsky also documents that the view of existence as superadded to essence is attributed by al-Suhrawardī (d. 1191) to the followers of the Peripatetics, and recurs in Fakhraddīn al-Rāzī (d. 1210).

33 See Bertolacci, 'Distinction of Essence and Existence', in which I have also argued that for Avicenna, 'existent' has both conceptual and extensional priority over 'thing' and the essence.

Avicenna's original text is different: in *Ilāhiyyāt* VII.1, from the assumption (regarded by Avicenna as wrong) that 'existent' and 'one' have the same concept, the false consequence follows that 'many' is not 'existent', as it is not, strictly speaking, 'one'. In Averroes's report, on the other hand, from the assumption that 'one' and 'existent' signify the same notion or item, it follows that a statement like 'existent is one' is non-informative and similar to a tautology. The actual consequence of the *reductio ad absurdum* in 2.1 remotely echoes another passage of the *Ilāhiyyāt* (I.5, p. 31, v. 10–p. 32, v. 2). There, from the assumption (taken by Avicenna as right) that 'essence' and 'thing' convey similar meanings, the correct consequence follows that a statement like 'the essence is a thing' is non-informative.[34] It is not too far-fetched to maintain that Averroes is somehow conflating these two distinct texts of Avicenna and that this reading results in a misreport of both.

More importantly, neither the imperfection of Avicenna's argument that Averroes underscores in section 2.2.1, nor the ignorance of the fundamental distinction that he imputes to Avicenna in section 2.3, is supported by any explicit text of Avicenna's. On the contrary, Avicenna's actual statements seem to invalidate both points. In the same passage of *Ilāhiyyāt* VII.1 on which Averroes models his report of Avicenna's argument in 2.1, Avicenna makes it clear (p. 303, vv. 9–10) that the *reductio ad absurdum* he proposes is valid only if the concepts of 'existent' and 'one' are the same 'in every way' (*min kulli jiha*), using the same term 'way' (*jiha*) that Averroes, too, employs in 2.2.1. Thus, the distinction of the 'concept' (*mafhūm*) of 'existent' and 'one' and their 'way' of predication in Avicenna's text does not turn out to be dissimilar from the distinction of 'notion/item' and 'way' that Averroes introduces in his emendation of Avicenna's argument.[35] This being the case, it seems difficult to accuse Avicenna, as Averroes does in 2.3, of neglecting the distinction between the expressions that signify different modes

---

34 In *Ilāhiyyāt*, I.5, p. 31, v. 10–p. 32, v. 2 [p. 35, v. 62–p. 36, v. 79], Avicenna supports the distinction of essence and existence by pointing to the fact that the sentence 'the essence so-and-so is existent' is informative, which attests that 'essence' and 'existent' are not synonymous and are therefore conceptually distinct. To corroborate *e converso* this point, he shows that when two terms are identical or synonymous, a sentence in which the one is subject and the other predicate is non-informative. As an example of 'useless redundancy of speech' (*ḥashw min al-kalām ghayr mufīd*, p. 31, vv. 13–14), he mentions the non-informative tautologies 'the essence so-and-so is an essence so-and-so' and 'the essence so-and-so is an essence'. Immediately afterwards (p. 31, vv. 14–17), as an example of 'speech that does not inform about what is not [yet] known', he provides the two non-informative non-tautological sentences: 'the essence so-and-so is a thing' and 'the essence is a thing': despite being non-tautological in so far as the subject is different from the predicate, these two sentences are nonetheless non-informative due to the synonymous relation of 'essence' and 'thing'. In the passage in question, Averroes seems to apply this same kind of reasoning to 'existent' and 'one', and to have in mind the non-informative tautological sentences 'existent is existent' and 'one is one' and the non-informative non-tautological sentence 'existent is one'. However, none of the statements reported by Averroes is mentioned by Avicenna in this passage of I.5.

35 The preceding lines of *Ilāhiyyāt*, VII.1, are: 'everything that is said "existent" in one respect can be said "one" in [another] respect' (p. 303, v. 7 [p. 349, vv. 10–12]). The different 'respect' (*iʿtibār*) by means of which 'existent' and 'one' are predicated of things looks equivalent to the term 'concept' in the passage just recalled.

of an essence and the expressions that signify attributes added to an essence. In a passage like *Ilāhiyyāt* VII.1, Avicenna appears to be quite aware that 'existent' and 'one', regardless of their relation with essence, are not only associated with different concepts, but also predicated in different ways. In other words, Averroes does not seem justified in denouncing the absence in Avicenna's ontology of a theory of the modes of signification, at least as far as 'existent' and 'one' are concerned.

As to part 3, Averroes is certainly entitled to ascribe to Avicenna the doctrine of the accidentality of unity, since Avicenna often speaks of unity (and of number) as an 'accident' (*ʿaraḍ*), due to the intimate connection of unity with the accidental category of quantity and despite the doctrinal tensions that this teaching introduces into his metaphysical system.[36] But part 3, coming after and being closely linked with the previous two parts, suggests that, for Avicenna, 'existent' is also an accident in the same sense as 'one' is. This suggested implication looks unwarranted, however: in the few cases in which Avicenna portrays existence as an accident of essence,[37] he appears to have in mind a logical notion of accident, namely the fact that existence is not part of a thing's essence, rather than a metaphysical notion, namely existence as an adventitious and unstable component of an existing thing.[38]

Among the three parts of our text, part 2 is obviously crucial in so far as it is the most problematic. On the one hand, it deals comprehensively with different issues (issue A, the mutual relationship of 'existent' and 'one', in sections 2.1 and 2.2.1; issue B, the relationship of 'existent' and 'one' with essence, in sections 2.2.2, 2.2.3, and 2.3). On the other, it is perplexing for several reasons. It is incongruous with the preceding part 1.1, which deals only with issue B. It misreports Avicenna's thought, ascribing to him in sections 2.1 and 2.3 arguments

36 Pickavé, 'On the Latin Reception', p. 344, remarks that Averroes's ascription to Avicenna of the accidentality of unity is incompatible with Avicenna's doctrine of individuation by means of non-accidental features (since individuality is a kind of unity, if unity is accidental, also individuality must be so). In my opinion, the incongruence that Pickavé signals has underpinnings in Avicenna's own thought, and does not totally depend on Averroes's report of it.

37 This happens in a single chapter of the work (VIII.4), in two consecutive passages (VIII.4, p. 346, v. 13 [om.]; p. 347, v. 9 [p. 402, vv. 45–46]) in which Avicenna employs first the participle *ʿāriḍ* and then the verb *ʿaraḍa* to portray the relationship of existence ('that-ness', *anniyya*) and essence ('quiddity', *māhiyya*). The first of these two passages, however, is omitted by many Arabic testimonies and by the Latin translation. See Bertolacci, 'God's Existence and Essence'. On the second passage, see Table 1.

38 This is confirmed by Avicenna's joint use of the roots *ʿ-r-ḍ* and *l-z-m* in these passages. For terms stemming from the root *l-z-m* in these contexts, see VIII.4, p. 346, v. 13 [om.]; p. 347, v. 2 [p. 401, v. 36]. More generally, also independently of the relationship of essence and existence, Avicenna often uses terms stemming from the root *ʿ-r-ḍ* in conjunction with terms stemming from the root *l-z-m* (see III.3, p. 109, v. 10 [p. 121, vv. 51–52]; V.1, p. 201, v. 9 [p. 233, v. 38]; V.1, p. 203, vv. 12–14 [p. 235, vv. 86–90]). The term 'accidental' (*ʿaraḍī*), instead of 'accident', that Avicenna uses in one notable case also for unity (III.3, p. 109, v. 10 [p. 121, vv. 51–52]; see Table 1) may suggest that the same idea is also lurking behind Avicenna's conception of the relationship of 'one' and essence, despite his many statements maintaining that unity possesses the status of simple accident.

or errors in which Avicenna actually does not engage. In so far as it contends that, for Avicenna, 'existent' and 'one' do not signify the same item, it *prima facie* prevents part 3 — which argues that for Avicenna 'one' is an accident — from fully supporting part 1.1, which argues that for Avicenna *both* 'existent' *and* 'one' signify an accident.

Averroes's criticism of Avicenna is a resolute disavowal of what Averroes asserts to be Avicenna's doctrine of the transcendentals 'existent' and 'one'. Attacking what is arguably the fundamental metaphysical doctrine of Avicenna, in Averroes's intention this criticism indicates that the entire metaphysics of Avicenna is flawed. Not by chance, the criticism is placed emphatically at the beginning of what Averroes regards as the expository part of Aristotle's *Metaphysics* (namely Book Γ), after the preliminary and previous dialectical books, in order to reassess the Stagirite's original thought against Avicenna's erroneous innovations and deformations.[39]

Albert does not share the same polemical attitude. On the contrary, he builds upon Averroes's text an *excusatio* of Avicenna and a harmonization of the views of the two Arabic philosophers. To this end, he makes part 2 the cornerstone of his citation of Averroes's passage, aware of the key role that this part plays in Averroes's account of Avicenna and arguably also of the problems that it raises. There are good reasons to believe that Albert makes this part of Averroes's text pivotal in his own quotation of the Commentator because it is the only part of Averroes's criticism in which issue A is taken into account: Albert knows by direct acquaintance with Avicenna's *Philosophia prima* that on issue A, despite Averroes's accusations, Avicenna's position is fundamentally congruent with Averroes's standpoint.

## Albert's Solution: Between Averroes and Avicenna

Table 2 displays digression IV.1.5 of Albert's commentary, and compares it with its main sources in Averroes, Avicenna, and the *Liber de causis*.[40] Terms or expressions that are identical in Albert and his sources are reported in bold; further points that are similar, though not identical, in terminology or doctrine are underlined. The most significant additions or changes introduced by Albert vis-à-vis Averroes are indicated by italics.

---

39 A shorter criticism of Avicenna on a related topic is added by Averroes later in the same section of the *Long Commentary on the Metaphysics*: it is a refutation of Avicenna's view of unity as a non-essential feature (Γ.3, p. 315, vv. 3–9 [fol. 67G]). Although related to the criticism considered here, this reference to Avicenna constitutes an independent criticism (see note 14 above), and is not quoted by Albert in the digression IV.1.5.

40 At the beginning of his commentary on Aristotle's *Physics*, Albert explains the purpose of digressions. Digressions are those chapters of his Aristotelian commentaries in which Albert does not analyse Aristotle's text, but either resolves a doubt or fills a doctrinal gap concerning a text previously commented upon (*Physica*, I.1.1, ed. by Hossfeld, p. 1, vv. 27–30).

The digression is appended to the preceding chapter (IV.1.4), in which Albert explains *Metaph.* Γ.2, 1003b22–36, the same passage commented upon by Averroes in the section of his commentary where he places the criticism of Avicenna just analysed (see above, Text 1). In that passage, according to Albert, Aristotle holds, in short, that 'being' and 'one' are the same thing and a unique nature ('ergo ens et unum sunt idem sive una et eadem natura') since they follow each other, although they bear different names.[41] In other words, Albert sees Aristotle's text as dealing primarily with issue A, and issues B and C as ancillary to issue A.[42] The digression under consideration, accordingly, concerns issue A, as is clear from its title and introduction, and aims to defend the correct view of issue A against its proposed denial ('solutionem rationum sophistarum inductarum ad hoc quod ens et unum non sint natura una et eadem'; 'an unum et ens consequuntur se ad invicem sicut unam et eandem rem et naturam significantia'). The other two issues (B and C), and in particular issue B, are intentionally left outside the scope of the digression. This is a fundamental strategic move on Albert's part, for it is on issue A that Albert will be able to construe a consensus between Avicenna and Averroes.

The digression is formally structured as a *quaestio*. After stating in the introduction the topic to be discussed, Albert reports seven arguments attributed to Avicenna (Contra 1–7), by means of which Avicenna allegedly intended to prove that 'being' and 'one' do *not* signify the same nature. Afterwards, in a sort of *responsio*, Albert opposes his personal opinion to these arguments, according to which Aristotle is right in positing that 'being' and 'one' signify the same nature. Finally, Albert refutes each of the arguments attributed to Avicenna (Ad Contra 1–7). The digression ends with a short conclusion restating the main result of the previous chapter.

41 Albertus Magnus, *In Metaph.*, IV.1.4, p. 166, vv. 57–58.

42 That the substance of everything is essentially 'being' and 'one' (*Metaph.* Γ.2, 1003b32–33) is, according to Albert, part of the proof of the main thesis announced in 1003b22–32 (see *In Metaph.*, IV.1.4, p. 166, vv. 40–58). Albert regards Aristotle's further statement, that the species of 'being' are as numerous as the species of 'one' (1003b33–36), as a corollary of the main thesis (*In Metaph.*, IV.1.4, p. 166, vv. 59–66). Issue B is only obliquely hinted at in Albert's formulation of Aristotle's main thesis (*In Metaph.*, IV.1.4, p. 165, vv. 38–39).

Table 2. Albert the Great, *Metaphysica*, digression IV.1.5: Conspectus of Sources.

| **Albert, *In Metaph.*, IV.1.5, p. 166, v. 67–p. 167, v. 72** | **Sources** |
|---|---|
| [Titulus] Et est digressio declarans solutionem rationum sophistarum inductarum ad hoc quod ens et unum non sint natura una et eadem | |
| [Introductio] Dubitabit autem aliquis de inductis, an unum et ens consequuntur se ad invicem sicut unam et eandem rem et naturam significantia. | |
| | Averroes, *Long Commentary on the Metaphysics* Γ.3, Lat. trans. as in *Aristotelis Opera cum Averrois Commentariis*, ed. Venetiis 1562, vol. 8, pp. 67B–E |
| | [1.1] 67B: Avicenna autem peccavit multum in hoc, quod existimavit, quod unum et ens significant dispositiones additas essentiae rei. [1.2] Et mirum est de isto homine, quomodo erravit tali errore [...] |
| [Contra 1] Obicit enim contra hoc Avicenna dicens, **quod si unum et ens significant** eandem naturam, **tunc** ista nomina, unum et ens, sunt synonyma, et est **nugatio**, *quando unum alteri additur*, cum dicitur 'unum ens'. | [2.1] 67C: Et iste homo ratiocinatur ad suam opinionem, dicendo **quod, si unum et ens significant** idem, **tunc** dicere ens est unum esset **nugatio**, quasi dicere unum est unum, aut ens est ens. [...] |
| [Contra 2] Amplius, cum dicitur 'unum ens', haec duo nomina non[43] iunguntur sibi per appositionem, sicut cum dicitur 'animal homo', quia unum non determinat alterum. Videtur igitur, quod unum iungatur enti per denominationem et informationem; hoc enim videtur ex hoc quod numerum et suppositum trahit ab ente sicut denominans a denominato et adiectivum a substantivo. Omne autem denominativum formam quandam aliam ponit super denominatum. *Unum ergo dicit aliquam formam enti additam, cum dicitur 'unum ens'*. | [3.1] 67D: Et fecerunt errare illum hominem res, quarum quaedam est, quia innuit hoc nomen unum de genere nominum denominativorum, et ista nomina significant accidens, et substantiam. |
| [Contra 3] Amplius, **unum** dicit indivisionem, quam non dicit ens, et cum dicitur 'unum ens', indivisionem ponit unum super ens; *addit igitur aliquid enti.* | [3.2] 67D: Et etiam, quia existimavit, quod hoc nomen **unum** significat intentionem in re carente divisibilitate, et quod illa intentio est alia ab intentione, quae est natura illius rei. |
| [Contra 4] Amplius, **unum principium est** numeri. Sicut igitur punctus est naturae continui, licet non sit continuum, ita unum est | [3.3.1] 67D–E: Et etiam, quia existimavit, quod **unum** dictum de omnibus praedicamentis, est illud unum, quod **est** |

| **Albert, *In Metaph.*, IV.1.5, p. 166, v. 67–p. 167, v. 72** | **Sources** |
|---|---|
| naturae numeri, licet non sit numerus; est igitur **unum accidens**. *Cum igitur dicitur 'unum ens', addit unum quoddam accidens super ens.* | **principium** numerorum. Numerus autem est accidens. Unde opinatus fuit iste, quod hoc nomen **unum** significat **accidens** in entibus; […] |
| [Contra 5] Adhuc, […] ens solum est creatum; unum autem est per informationem, quia suum intellectum ponit circa ens praesuppositum; est enim unum ens indivisum; *ergo aliquid addit super ens.* | Cf. *Liber de causis*, IV.37, p. 142, vv. 37–38 (prima rerum creatarum est esse et non est ante ipsum creatum aliud); XVII (XVIII). 148, p. 174, vv. 57–61 (vita autem prima dat eis quae sunt sub ea vitam non per modum creationis immo per modum formae. et similiter intelligentia non dat eis quae sunt sub ea de scientia et reliquis rebus nisi per modum formae); XXXI (XXXII).219, p. 202, vv. 12–13 (omnis unitas post unum verum est acquisita) |
| [Contra 6] Amplius, omne dividens aliquid addit super divisum; unum autem cum multo sibi opposito dividit ens; *ergo addit aliquid enti.* | Cf. Avicenna, *Liber de Philosophia prima*, I.2, p. 13, vv. 16–17 [p. 13, vv. 42–43]: Et ex his quaedam sunt ei quasi accidentia propria, sicut unum et multum. |
| [Contra 7] Amplius, si ens et unum sunt penitus una et eadem natura, quidquid opponitur uni, opponitur et alteri; multum autem opponitur uni; ergo opponitur et enti, quod falsum est; *ergo ens et unum non sunt penitus una natura et eadem.* | Cf. Avicenna, *Liber de Philosophia prima*, VII.1, p. 303, vv. 9–10 [p. 349, vv. 15–17]: Si enim id quod intelligitur de uno omnino [*min kulli jiha*] esset id quod intelligitur per ens, tunc multum, secundum quod est multum, non esset ens sicut non est unum. |
| Haec et similia inducit Avicenna pro se, quando contradicit Aristoteli in supra inductis rationibus. | |
| [Responsio] Quia autem superius inductae rationes [sc. rationes Aristotelis] sunt irrefragabiles, revertemur dicentes, quod ens et unum sunt una et eadem natura […] | |
| [Ad Contra 1] Modus igitur diversus importatus per ens et unum facit, quod nomina non sunt synonyma nec est nugatio, quando sibi iunguntur, nec per appositionem iunguntur sibi. | [2.2.1] 67C: Et hoc non sequeretur, nisi diceremus, quod dicere de aliquo quod est ens et unum, quod significant eandem intentionem et eodem modo.<br>Cf. [2.2.2] 67C: Nos autem diximus, quod significant eandem essentiam, sed modis diversis, non dispositiones diversas essentiae additas. |
| [Ad Contra 2] Et licet unum ponat modum suum, quem importat circa ens sicut circa suppositum suum, tamen modus ille non est alicuius formae alterius ab ente, sed modus | |

| **Albert, *In Metaph.*, IV.1.5, p. 166, v. 67–p. 167, v. 72** | **Sources** |
|---|---|
| negationis, qui sufficit grammatico. Et ideo non est denominativum, sed modum habens denominativi. | |
| Et hoc forte attendit Avicenna, cum dixit esse denominativum. | |
| [Ad Contra 3] Sic igitur licet indivisionem addat super ens et quoad hoc praesupponat ens, hoc tamen non est aliquam formam addere, sed potius modum, qui ex negatione resultat. | |
| [Ad Contra 4] Quod autem dicitur, quod **unum est principium** numeri, dupliciter accipi potest propter aequivocationem principii […] Et hoc modo duplex est unitas […] | [3.3.2] 67E: et non intellexit, quod **unum**, quod **est principium** numerorum, est ex entibus, de quibus dicitur hoc nomen unum, licet sit magis dignum hoc […] |
| [Ad Contra 5] Ex dictis autem patet, qualiter unum sit factum per informationem et ens per creationem et qualiter unum consequitur ens. | |
| [Ad Contra 6] Et ideo dividit ipsum et modum quendam addit ei, | |
| [Ad Contra 7] gratia cuius opponitur multitudini, cui non opponitur ens. Et sic patet omnium praeinductorum solutio. | |
| [Excusatio] Et facile est per haec quae hic dicta sunt, excusare dicta Avicennae, quia pro certo, si quis subtiliter dicta sua respiciat, dicere intendit hoc quod hic dictum est. | Cf. Avicenna, *Liber de Philosophia prima*, I.4, p. 26, vv. 17–18 [p. 30, v. 59]; III.2, p. 103, vv. 7–8 [p. 114, vv. 17–19]; VII.1, p. 303, vv. 6–9 [p. 349, vv. 9–15] (see below, Texts 2–4) |
| [Conclusio] Ex omnibus autem inductis hoc accipiendum est, quod ens et unum unam dicunt naturam, et ideo species unius sunt species entis. […] | |

This digression is remarkable in many ways. Although he is not named in the title, Avicenna is its main focus, since, according to Albert, it is he who casts doubt on Aristotle's doctrine by disagreeing with it (see Contra 1). He is the only author who is referred to by name, being mentioned four times throughout the digression, which thus includes almost one sixth of the twenty-six occurrences of the name 'Avicenna' in Albert's commentary on the *Metaphysics*. Averroes, by

43 I read *non* with manuscript P: *non* is omitted in the edition.

contrast, despite being the main source of Albert's digression, is never mentioned by name. Albert's emphasis on Avicenna does not seem coincidental: it looks as though he wants to attract the reader's attention, signalling that something important is at stake regarding this Arabic master. What Albert does, in fact, is worth considering. In the first part of the digression, at the beginning and the end of the exposition of Avicenna's arguments, Albert introduces Avicenna as an adversary of Aristotle.[44] But contrary to expectation, in the third part of the digression, after refuting the arguments previously attributed to Avicenna, Albert does not emphasize Avicenna's error, but instead excuses Avicenna's arguments, showing the similarity between Avicenna's doctrine and the true Aristotelian position.[45] This ambivalent attitude, both anti-Avicennian and pro-Avicennian in one and the same text, is quite striking.

The twofold tenor of the digression has a double explanation. On the one hand (a), Albert reports its main source (the passage of Averroes's *Long Commentary on Metaph.* Γ discussed above) selectively and in a modified form, in a way that is quite lenient towards Avicenna's actual position; on the other (b), Albert has independent access to Avicenna's text, on the basis of which he is able to evaluate whether and to what extent Averroes's report of Avicenna's position is faithful or distorting.

(a) Albert takes the first four arguments of Avicenna (Contra 1–4), as well as the basic elements of the answer to them (Ad Contra 1–4), from parts 2–3 of Averroes's text. The sequence of the arguments is exactly the same in Averroes and Albert, and the general structure of the two texts is largely similar.[46] Albert himself constructs the subsequent three arguments of Avicenna (Contra 5–7) along the lines of the first four, drawing freely on Avicenna's *Philosophia prima*,[47] as well as from some propositions of the *Liber de causis*.[48]

---

44 'Obicit enim contra hoc Avicenna dicens, quod' (beginning of Contra 1); 'Haec et similia inducit Avicenna pro se, quando contradicit Aristoteli in supra inductis rationibus' (end of Contra 7).

45 At the end of Ad Contra 2, the refutation of the second argument attributed to Avicenna closes as follows: 'Et hoc forte attendit Avicenna, cum dixit [*sc.* unum] esse denominativum'. Likewise, after Ad Contra 7, at the end of the refutation of all the arguments attributed to Avicenna, the excusatio appears to be an apology for and total rehabilitation of Avicenna's doctrine: 'Et sic patet omnium praeinductorum solutio. Et facile est per haec quae hic dicta sunt, excusare dicta Avicennae, quia pro certo, si quis subtiliter dicta sua respiciat, dicere intendit hoc quod hic dictum est'.

46 Albert does not reproduce sections 2.2, 2.3, and 3.3.2 in the first part of the digression immediately after 2.1 and 3.3.1, as in Averroes, but uses 2.2 and 3.3.2 in the answer to the single arguments in the third part of the digression. The close correspondence between the parts of Averroes's text and the arguments attributed to Avicenna by Albert proves that Albert drew upon Averroes's text while writing the digression.

47 Despite the presence of the expression 'unum et idem' in Avicenna's *Philosophia prima*, VII.1, ed. by Van Riet, p. 303, v. 8 [p. 349, v. 13], Albert's expression 'una et eadem natura' in Contra 7, p. 167, vv. 10–11 and 14, comes from 'idem et una natura' in the *Translatio media* of *Metaph.* 1003b22, an expression that Albert uses also in *In Metaph.*, IV.1.4, p. 166, vv. 57–58.

48 Latin text in *Liber de causis*, ed. by Pattin. On the connection that Albert sees between the theological part of Avicenna's metaphysics (treatises VIII–X.3) and the content of the *Liber de causis*, see

However, Albert substantially modifies the content of Averroes's text, in three main ways. First, Albert completely omits part 1, as well as all the sections of part 2 (2.2.2, 2.2.3, and 2.3) that — like part 1 — deal with issue B, that is to say, with the relationship of 'existent' and 'one' with essence. He therefore quotes only section 2.1 of part 2 in its original place, and takes inspiration from section 2.2.1 in the Ad Contra 1 for the idea that 'existent' and 'one' are predicated of the same thing in different ways (*modus diversus*).[49] Second, he consequently shifts the balance of Averroes's report towards section 2.1 — Avicenna's view of the mutual relationship of 'existent' and 'one' (issue A) — as the initial and main element of Avicenna's position. Third, he rephrases part 3 (issue $B_1$) so as to bring it into agreement with section 2.1 (issue A) rather than leaving it congruent, within the limits seen above, with the omitted part 1 (issue B), as it is in Averroes.

The first change, the total exclusion of the sections of Averroes's report of Avicenna dealing with issue B, is, of course, especially important.[50] As we have seen, these sections are the only passages of Averroes's text in which Avicenna's doctrine of the relationship of essence and existence is attacked. Thus, by omitting them, Albert excludes Avicenna's distinction of essence and existence from the scope of his own criticism of Avicenna in digression IV.1.5. This might be a further instance of Albert's defence of Avicenna in the digression, this time silent or implicit,[51] worth being considered in the analysis of Albert's attitude to Avicenna's view on essence and existence.[52]

The second change is a consequence of the first. Because of the omission of part 1 of Averroes's text, section 2.1 comes to the forefront of Albert's report of Averroes's criticism of Avicenna. Albert quotes this section faithfully, almost verbatim. In it, Albert, like Averroes, deals with issue A of Aristotle, namely the mutual relationship between 'being' and 'one', a point that Albert stresses by adding to Averroes's text the formula *quando unum alteri additur* (in italics in Table 2).

As a third and final change, in the other part of Averroes's criticism that Albert quotes, namely part 3, the arguments that in Averroes's text support Avicenna's view that 'one' is added to essence (issue $B_1$) are changed by Albert in order to

---

Bertolacci, '"Subtilius speculando"', pp. 327–36. On his reception of the *Liber de causis*, see Krause and Anzulewicz, 'From Content to Method'.

49 The same idea is also present in section 2.2.2 of Averroes's criticism (issue B). Albert might have considered also this section, although he diverts the idea supposedly taken from it from issue B to issue A.

50 A similar emphasis on issue A rather than on issue B can be seen, in ways different from Albert's, in Roger Bacon and in the sophisma 'Tantum unum est' (see below, note 63).

51 Likewise, in the corresponding passage of his commentary on the *Metaphysics*, Albert omits the criticism in which Averroes attacks Avicenna's doctrine that 'existent' and 'one' signify non-essential features of things (*Long Commentary on the Metaphysics* I.8, p. 1279, v. 12–p. 1280, v. 11 [fol. 257E–G]).

52 See Vargas, 'Albert on Being and Beings', p. 646. Other useful hints can be found in the other parts of the section 'Albert the Great on Metaphysics', ed. by Carasquillo, Twetten, and Tremblay.

support part 2.1 and issue A. In Albert's version (Contra 2–4), these arguments are rephrased to corroborate the view that 'one' is an addition to 'being': the quotation of each of these arguments ends with formulae, absent in Averroes, that stress the additional character of 'one' with respect to 'being' (in italics in Table 2).[53] Something similar happens with the subsequent three arguments (Contra 5–7), added from the *Liber de causis* and Avicenna's *Philosophia prima*.[54]

Albert's *modus operandi* in the present case serves a double purpose. First, with regard to Averroes, by omitting some passages of Averroes's criticism of Avicenna and changing the content of others Albert recasts in a coherent setting the multifarious attack directed by the Commentator against Avicenna's doctrine of 'existent' and 'one'. Second, with regard to Avicenna, by focusing on part 2.1 of Averroes's text and on issue A, Albert drives the reader away from an element of Avicenna's metaphysics genuinely at variance with Aristotle's and Averroes's views, namely Avicenna's account of issue B, and directs attention instead to a doctrine — Avicenna's treatment of issue A — that is compatible with Aristotle's and Averroes's standpoint. By thus recasting the entire discussion under the umbrella of issue A, Albert neutralizes Averroes's criticism with respect to Avicenna's true position; at the same time, he makes Avicenna's true position excusable vis-à-vis Averroes's attack, which does not affect Avicenna's authentic standpoint, but only Averroes's own (mis)representation of it. In fact, Avicenna does not uphold the account of issue A that Averroes ascribes to him, and, as we have seen, advocates a view of it that is not contrary to Aristotle's and Averroes's.

(b) Significantly, the last argument that Albert ascribes to Avicenna in the first part of the digression (Contra 7) is taken directly from the passage of Avicenna's *Philosophia prima* (VII.1, p. 303, vv. 9–10 [p. 349, vv. 15–17]) that Averroes misreports in section 2.1 of his Long Commentary. Albert, in contrast, reports faithfully this passage by Avicenna, which he was evidently able to access independently of Averroes. We can therefore assume that Albert knew this passage first-hand, that he was able to evaluate the inaccuracy of Averroes's report of it, and possibly that he could even glimpse the presence in Avicenna's work of a theory of the different ways in which 'existent' and 'one' signify things.[55]

Likewise, when Albert excuses Avicenna in the final part of the digression, he very probably has in mind a series of passages of Avicenna's *Philosophia prima*

---

53 'Unum ergo dicit aliquam formam enti additam, cum dicitur "unum ens"' (Contra 2), etc.

54 See 'ergo [*sc.* unum] aliquid addit super ens' in Contra 5, and 'ergo addit aliquid enti' in Contra 6. Contra 7 ends with 'ergo ens et unum non sunt penitus una natura et eadem', which still regards issue A.

55 The idea of a *modus diversus importatus per ens et unum* is no doubt the leitmotif of Albert's reply to the arguments attributed to Avicenna in the last part of the digression, starting with Ad Contra 1: in proposing this idea, Albert is certainly beholden to Averroes's own view (section 2.2.2 of Table 1). It looks likely, however, that the final *excusatio* of Avicenna also reflects Albert's awareness of the presence of this same idea in Avicenna. Although the *Philosophia prima* renders the crucial expression 'in every way' (*min kulli jiha*) in the Arabic text of *Ilāhiyyāt*, VII.1 rather vaguely as *omnino*, Albert had at his disposal other texts of Avicenna's work on the same point (see Texts 2–4).

where Avicenna repudiates the view according to which 'one' is subordinated to 'existent' and asserts their equality. These passages may be laid out as follows.

Texts 2–4: Avicenna, *Philosophia prima*

[2] I.4, p. 26, vv. 17–18 [p. 30, v. 59]: [...] unum parificatur ad esse.

[3] III.2, p. 103, vv. 7–8 [p. 114, vv. 17–19]: Unum autem parificatur ad esse, quia unum dicitur de unoquoque praedicamentorum, sicut ens, sed intellectus [*mafhūm*] eorum [...] diversus est.

[4] VII.1, p. 303, vv. 6–9 [p. 349, vv. 9–15]: Scias autem quod unum et ens iam parificantur in praedicatione sui de rebus [*ashyā'*], ita quod, de quocumque dixeris quod est ens uno respectu [*bi-ʿtibār*], illud potest esse unum alio respectu [*bi-ʿtibār*]. Nam quicquid est, unum est, et ideo fortasse putatur quia id quod intelligitur [*mafhūm*] de utroque sit unum et idem, sed non est ita: sunt autem unum subiecto [*bi-l-mawḍūʿ*], scilicet quia, in quocumque est hoc, est et illud.

The *parificatio* of 'one' and 'existent' stated in these passages is crucial to Avicenna's way of reshaping the structure of Aristotle's *Metaphysics* in the *Ilāhiyyāt*. It is the basis of Avicenna's framing of metaphysics as a science that deals, at the same time and at equivalent levels, with both 'existent' and 'one', being epistemologically both an ontology and a henology.[56] These texts were in all likelihood very familiar to Albert.[57] In particular, he must have been acquainted with the longest and most informative of them (Text 4), since this text immediately precedes the passage of Avicenna's *Philosophia prima* that Albert reports in Contra 7. In these texts of the *Philosophia prima*, Avicenna denies that 'one' adds something real to 'existent'. According to Avicenna, 'existent' and 'one' are coextensional and bear two totally distinct concepts, along the lines of the conceptual distinction also admitted by Aristotle, Averroes, and Albert.

On the basis of the evidence that the *Philosophia prima* gives him, Albert takes Avicenna's conception of the mutual relationship of 'existent' and 'one' to be analogous to the doctrine of Aristotle in *Metaph.* Γ.2, endorsed also by Averroes and by Albert himself in their commentaries on *Metaphysics*. Consequently, Albert can excuse Avicenna from Averroes's attack in the last part of the digression.[58]

---

56 The relevance of these texts is discussed in Bertolacci, 'The Structure of Metaphysical Science'; Bertolacci, *Reception of Aristotle's 'Metaphysics'*, chap. 6.

57 This does not mean, of course, that Albert endorses every single point of Avicenna's position. In *In Metaph.*, IV.1.4, p. 165, vv. 38–39, for example, he seems to reject that *ens* and *unum* are simply the same according to subject, contrary to what Avicenna's Text 4 asserts.

58 Avicenna says that 'existent' and 'one' are predicated of the same set of things, or the same subjects, according to a different concept (*conceptus, id quod intelligitur*; Ar. *mafhūm*) or respect (*respectus*; Ar. *iʿtibār*; see Texts 3–4). Besides the conceptual distinction, he also takes into account, albeit obliquely, the presence of a different 'way' (Ar. *jiha*) of signification (VII.1, p. 303, vv. 9–10 [p. 349, vv. 15–17]; see Table 2). It is not clear whether the terms 'concept' and 'respect' in these texts are synonymous, or whether the latter term is closer in meaning to 'way'. What is sufficiently clear is that Albert considers

To sum up: On the relationship between 'being', 'one', and essence (issue B), Averroes criticizes an aspect of Avicenna's philosophy that can be regarded as non-Aristotelian or anti-Aristotelian, since Avicenna contends, contrary to Aristotle, that 'existent' and 'one' are distinct from essence (the former is distinct but inseparably connected with essence; the latter is said to be an accident). Aristotle, by contrast, in the passage of *Metaph.* Γ.2 (1003b32–33) commented upon by Averroes and Albert (Text 1 above), affirms that the substance or essence of a thing is 'one' and 'being' not accidentally, that is, essentially.[59] Albert arguably sides with Averroes against Avicenna on issue B.[60] On the mutual relationship of 'being' and 'one' (issue A), however, Averroes's criticism of Avicenna appears pointless to Albert, since Albert knows that Avicenna holds that 'existent' and 'one' signify the same thing in different ways, and that 'one' adds nothing real to 'existent'; in other words, Albert is aware that Avicenna agrees with Aristotle and Averroes in regarding 'being' as different from 'one' not in reality, but only conceptually. Sure of Avicenna's real position, and by shifting the target of Averroes's criticism of Avicenna from issue B to issue A, Albert paves the way for his apology for Avicenna in the final part of the digression.[61] By excusing Avicenna, as well as by avoiding any mention of Averroes in the digression, Albert portrays the contrast between Averroes and Avicenna much less harshly than Averroes does in his *Long Commentary on the Metaphysics*. In his own commentary on the *Metaphysics*, Albert never mentions Avicenna again as holder of a doctrine of transcendentals criticized by Averroes.

---

the difference in concept, respect, or way that Avicenna affirms between 'existent' and 'one' to be remarkably similar to the difference in the way of signifying that Averroes accuses him of neglecting.

59 I do not take into account here whether a distinction of essence and existence is envisaged by Aristotle himself in other loci of the *Corpus*, as in the famous distinction of the questions 'what it is' and 'if it is' in the *Posterior Analytics*, or in the polarity between the universality of essence and the individuality of existence in the *Metaphysics* (the notorious issue of whether Aristotle regards the essence as individual or universal in the *Metaphysics* is fiercely debated). Castelli, *Problems and Paradigms of Unity*, contends that in *Metaph.* Γ.2, 1003b32–33, 'the basic idea is that the essence of each being is one and a certain being primitively and not by accident' (p. 66; see also pp. 208 and 266).

60 The criticism of Avicenna in Averroes's *Long Commentary on the Metaphysics* (Γ.3, p. 315, vv. 3–9 [fol. 67G]) that follows the one discussed here concerns Avicenna's doctrine of the extrinsic relationship of unity to essence. Its purport is summarized by Albert in the chapter preceding the digression (*In Metaph.*, IV.1.4, p. 166, vv. 40–53); Albert cites this criticism silently, however, without any reference to either Avicenna or Averroes. In this case, Albert seems to endorse Averroes's critical stance without openly reproaching Avicenna.

61 The *excusatio* of Avicenna at the end may be one of the reasons why the title of the digression does not ascribe the error in question to Avicenna, but generally to some sophists (*sophistae*). Likewise, when Albert subsequently refers to the present digression (*In Metaph.*, X.1.5, p. 437, vv. 33–34), he replaces the four explicit mentions of Avicenna here by a single and more vague reference to *quidam*. The occasion of this retrospective reference is Albert's report of another criticism by Averroes against Avicenna's doctrine of transcendentals (*Long Commentary on the Metaphysics*, *Aristotelis Metaphysicorum libri XIIII*, I.5, p. 1267, v. 15–p. 1268, v. 3 [fol. 255B]).

## Conclusion

De Libera rightly remarks that Albert's defence of Avicenna in our digression is due to an intention 'de rectifier une lecture étroite ou incorrecte de la lettre du texte avicennien'.[62] In this paper, I have argued that the 'reading of Avicenna' against which Albert reacts is the one proposed by Averroes in the *Long Commentary on the Metaphysics*. Averroes's interpretation of Avicenna is 'narrow or incorrect' because Averroes's account is neither coherent, insofar as he ascribes to Avicenna contrasting doctrines, nor well grounded, insofar as he presents as Avicennian a doctrine that Avicenna in fact does not uphold. Albert seems to be somehow aware of these shortcomings. He 'rectifies' Averroes's account of Avicenna's position by excusing Avicenna for the thesis that Averroes erroneously ascribes to him.

It seems sufficiently clear that Albert builds this digression directly upon Averroes's *Long Commentary on the Metaphysics* and integrates it with recourse to Avicenna's *Philosophia prima*, two works whose Latin translations he reads, in this as in other cases, first-hand and without mediation.[63] In fact, the present digression is the only case of a quotation of Avicenna in Albert's commentary on the *Metaphysics*, which is partially taken from another source (that is, Averroes), and not directly from the Latin translations of Avicenna's works.[64]

Although exceptional in many ways, the present digression can be taken as emblematic of Albert's more general attitude towards Arabic metaphysics in his commentary on the *Metaphysics*. In other instances of controversy between Averroes and Avicenna over metaphysical issues as well, Albert frequently seeks a harmonization that can minimize the points of dissent and reconcile, as far as possible, the contrasting positions of his two Arabic sources. More visibly in the digression I have discussed than in the rest of the commentary, Albert strives to smooth out the incompatibility between those positions. In all these regards, his aim is to rework Averroes's and Avicenna's metaphysical writings in order to create a unified and coherent system of Arabic Peripatetic metaphysics that can serve as a non-controversial tool for an insightful interpretation of Aristotle.[65]

---

62 De Libera, 'D'Avicenne à Averroès, et retour', p. 155.

63 It would be difficult to explain otherwise either the changes that he introduces into Averroes's criticism or the final *excusatio* of Avicenna. The joint presence of a criticism of Avicenna and of a defence of him in the digression is very likely the fruit of Albert's direct recourse to the Latin translations of Averroes's and Avicenna's texts, rather than a borrowing from an intermediate source. In this context, one may notice that the four *rationes* ascribed to Avicenna in the anonymous sophisma 'Tantum unum est' (MS Paris, BNF, Lat. 16135; see de Libera, 'D'Avicenne à Averroès, et retour', pp. 156–57) are only partially similar to the ones proposed by Albert as Contra 1–4. The same holds true of four arguments that 'one' and 'being' are not the same and do not signify the same item in Roger Bacon (de Libera, 'D'Avicenne à Averroès, et retour', pp. 150–51). In Bacon, moreover, these arguments are not ascribed to Avicenna, but remain anonymous.

64 See Bertolacci, '"Subtilius speculando"', pp. 297–300.

65 See Mulchahey, *'First the Bow Is Bent in Study…'*.

As shown by the cases of Porphyry and al-Kindī with respect to Plato and Aristotle, the need for philosophical consistency is felt especially urgently in periods of crisis and transformation, involving changes in the milieu within which philosophy is practised in a given culture or the introduction of the discipline into a foreign culture. Albert did something analogous with respect to Avicenna and Averroes in a further step in the history of philosophy. The thirteenth century was a crucial period of this kind, as the 'new' Aristotle entered Latin culture for the first time, through and together with Arabic *falsafa*, triggering the resistance of traditional Latin philosophy to a foreign world view that was rooted in a pagan master, Aristotle, and intimately linked to a 'heretic' religion, Islam. Albert seems to be perfectly aware that his endorsement of Arabic philosophy creates an unbridgeable gap between his own interpretation of Aristotle and that of previous and contemporary Latin philosophers, who were still unaware of — or consciously hostile to — Arabic sources. In the specific case of the *Metaphysics*, he reacted to such reactionary tendencies by striving for philosophical congruence between Avicenna and Averroes, as the two main Arabic interpreters of Aristotle's work. Thus, the philosophical enterprise for which Albert is famous is possibly not only 'to make Aristotle intelligible to Latin readers', but also to make Arabic philosophy, especially metaphysics, acceptable to Latin culture.

Albert's digression is revealing in another respect as well: it marks the transition from a first phase of Albert's attitude towards Arabic philosophy, in which Avicenna is still an established philosophical authority to be defended against the *novitas* of Averroes, to a second phase, in which Averroes has gained the status of the most authoritative commentator on Aristotle. The shift reverberates in the institutional contexts of the time, where, on the one hand, Avicenna's philosophy was the essential element of the theologians' aspiration of integrating philosophy into theology, and, on the other, Averroes's interpretation of the Stagirite was the quintessence of the Arts masters' aim of making philosophy an independent discipline. The digression analysed in this paper partakes in both phases. It retains traces of the first phase insofar as it contains the only explicit apology for Avicenna against Averroes to be found in Albert's commentary on the *Metaphysics*, as opposed to the numerous such apologies in Albert's previous commentaries on Aristotle. It reflects the second phase insofar as Averroes's commentary emerges as a true 'companion' to Aristotle's *Metaphysics*, providing not only a full-fledged understanding of Aristotle's text, but also a glimpse of Avicenna's teachings on key metaphysical doctrines by means of his criticisms.[66]

66 I have documented how Albert's defence of Avicenna against Averroes's attacks changes throughout his commentaries on Aristotle in Bertolacci, '"Averroes ubique Avicennam persequitur"'.

## Works Cited

### *Primary Sources*


Albertus Magnus, *De causis et processu universitatis a causa prima*, ed. by Winfried Fauser, Editio Coloniensis, 17/2 (Münster: Aschendorff, 1993)

———, *De homine*, ed. by Henryk Anzulewicz and Joachim R. Söder, Editio Coloniensis, 27/2 (Münster: Aschendorff, 2008)

———, *De praedicamentis*, ed. by Manuel Santos Noya, Carlos Steel, and Silvia Donati, Editio Coloniensis, 1/1B (Münster: Aschendorff, 2014)

———, *De quindecim problematibus*, ed. by Alfons Hufnagel, Bernhard Geyer, Jakob Weisheipl, and Paul Simon, Editio Coloniensis, 17/1 (Münster: Aschendorff, 1975)

———, *De sex principiis*, ed. by Ruth Meyer, Editio Coloniensis, 1/2 (Münster: Aschendorff, 2006), pp. 1–80

———, *Metaphysica, libri I–V*, ed. by Bernhard Geyer, Editio Coloniensis, 16/1 (Münster: Aschendorff, 1960)

———, *Metaphysica, libri VI–XIII*, ed. Bernhard Geyer, Editio Coloniensis, 16/2 (Münster: Aschendorff, 1964)

———, *Physica, libri I–IV*, ed. by Paul Hossfeld, Editio Coloniensis, 4/1 (Münster: Aschendorff, 1987)

———, *Physica, libri V–VIII*, ed. by Paul Hossfeld, Editio Coloniensis, 4/2 (Münster: Aschendorff, 1993)

Averroes, *Tafsīr mā ba'd aṭ-ṭabī'a* [Long Commentary on the *Metaphysics*], ed. by Maurice Bouyges, 4 vols (Beirut: Imprimerie catholique, 1938–48)

Averroes Latinus, *Aristotelis Metaphysicorum libri XIIII: Cum Averrois Cordubensis in eosdem Commentariis* [Long Commentary on the *Metaphysics*], Aristotelis Opera cum Averrois Commentariis, 8 (Venice, 1562; repr. Frankfurt am Main: Minerva, 1962)

Avicenna, *Al-Shifā', al-Ilāhiyyāt*, vol. 1 (books I–V), ed. by Jūrj Sh. Qanawatī and Sa'īd Zāyid, vol. 2 (books VI–X), ed. by Muḥammad Yusuf Mūsā, Sulaymān Dunyā, and Sa'īd Zāyid (Cairo: al-Hay'a al-'āmma li-shu'ūn al-maṭābi' al-amīriyya, 1960)

———, *Ta'līqāt* [*Annotations*], ed. by Ḥasan M. al-'Ubaydī (Baghdad: Dār ash-Shu'ūn ath-Thaqāfīya 'Āfāq 'Arabīya, 2002)

Avicenna Latinus, *Liber de Philosophia prima sive Scientia divina, I–IV*: *Édition critique de la traduction latine médiévale*, ed. by Simone Van Riet, introduced by Gérard Verbeke (Leuven: Peeters, 1977)

———, *Liber de Philosophia prima sive Scientia divina, V–X*: *Édition critique de la traduction latine médiévale*, ed. by Simone Van Riet, introduced by Gérard Verbeke (Leuven: Peeters, 1980)

*Liber de causis*, in Adriaan Pattin, 'Le *Liber de causis*: Édition établie à l'aide de 90 manuscrits avec introduction et notes', *Tijdschrift voor filosophie*, 28 (1966), 90–203

### *Secondary Works*

Adamson, Peter, 'Before Essence and Existence: Al-Kindī's Conception of Being', *Journal of the History of Philosophy*, 40 (2002), 297–312

Aertsen, Jan A., 'Albert's Doctrine on the Transcendentals', in *A Companion to Albert the Great: Theology, Philosophy, and the Sciences*, ed. by Irven M. Resnick (Leiden: Brill, 2013), pp. 611–19

———, 'Avicenna's Doctrine of the Primary Notions and Its Impact on Medieval Philosophy', in *Islamic Thought in the Middle Ages: Studies in Text, Transmission and Translation*, ed. by Anna Akasoy and Wim Raven (Leiden: Brill, 2008), pp. 21–42

———, *Medieval Philosophy as Transcendental Thought: From Philip the Chancellor (ca. 1225) to Francisco Suárez* (Leiden: Brill, 2012)

———, '"Res" as Transcendental: Its Introduction and Significance', in *Le problème des transcendantaux du XIV*e *au XVII*e *siècle*, ed. by Graziella Federici Vescovini (Paris: Vrin, 2002), pp. 139–56

al-Ahwani, Ahmed F., 'Being and Substance in Islamic Philosophy, Ibn Sina versus Ibn Rushd', in *Die Metaphysik im Mittelalter: Ihr Ursprung und ihre Bedeutung: Vorträge des II. Internationalen Kongresses für mittelalterliche Philosophie, Köln 31.8–6.9.1961*, ed. by Paul Wilpert (Berlin: De Gruyter, 1963), pp. 428–36

Benevich, Fedor, *Essentialität und Notwendigkeit: Avicenna und die Aristotelische Tradition* (Leiden: Brill, 2018)

Bertolacci, Amos, 'Albert the Great, *Metaph.* IV, 1, 5: From the *Refutatio* to the *Excusatio* of Avicenna's Theory of Unity', in *Was ist Philosophie im Mittelalter?: Akten des X. Internationalen Kongresses für mittelalterliche Philosophie der S.I.E.P.M. 25. bis 30. August 1997 in Erfurt*, ed. by Jan A. Aertsen and Andreas Speer (Berlin: De Gruyter, 1998), pp. 881–87

———, 'Averroes against Avicenna on Human Spontaneous Generation: The Starting-Point of a Lasting Debate', in *Renaissance Averroism and its Aftermath: Arabic Philosophy in Early Modern Europe*, ed. by Anna Akasoy and Guido Giglioni (Dordrecht: Springer, 2013), pp. 37–54

———, '"Averroes ubique Avicennam persequitur": Albert the Great's Approach to the *Physics* of the *Šifāʾ* in the Light of Averroes' Criticisms', in *The Arabic, Hebrew and Latin Reception of Avicenna's Physics and Cosmology*, ed. by Dag Nikolaus Hasse and Amos Bertolacci (Berlin: De Gruyter, 2018), pp. 397–431

———, 'Avicenna's and Averroes's Interpretations and Their Influence in Albertus Magnus', in *A Companion to the Latin Medieval Commentaries on Aristotle's Metaphysics*, ed. by Fabrizio Amerini and Gabriele Galluzzo (Leiden: Brill, 2014), pp. 95–135

———, 'Le citazioni implicite testuali della *Philosophia prima* di Avicenna nel Commento alla *Metafisica* di Alberto Magno: analisi tipologica', *Documenti e studi sulla tradizione filosofica medievale*, 12 (2001), 179–274

———, 'The Distinction of Essence and Existence in Avicenna's Metaphysics: The Text and Its Context', in *Islamic Philosophy, Science, Culture, and Religion: Studies in Honor of Dimitri Gutas*, ed. by Felicitas Opwis and David C. Reisman (Leiden: Brill, 2012), pp. 257–88

———, 'From Athens to Buḫārā, to Cordoba, to Cologne: On the Transmission of Aristotle's *Metaphysics* in the Arab and Latin Worlds during the Middle Ages', in *Circulation des savoirs autour de la Méditerranée: philosophie et sciences (IX^e–XVII^e siècle)*, edited by Graziella Federici Vescovini and Ahmad Hasnawi (Florence: Cadmo, 2013), pp. 217–33

———, 'God's Existence and Essence: The *Liber de Causis* and School Discussions in the Metaphysics of Avicenna', in *Reading Proclus and the Book of Causes*, vol. 3, ed. by Dragos Calma (Leiden: Brill, 2022), pp. 251–80

———, '"Incepit quasi a se": Averroes on Avicenna's Philosophy in the *Long Commentary on the De anima*', in *Contextualizing Premodern Philosophy: Explorations of the Greek, Hebrew, Arabic, and Latin Traditions*, ed. by Katja Krause, Luis Xavier López-Farjeat, and Nicholas A. Oschman (New York: Routledge, 2023), pp. 408–35

———, '"Necessary" as Primary Concept in Avicenna's Metaphysics', in *Conoscenza e contingenza nella tradizione aristotelica medievale*, ed. by Stefano Perfetti (Pisa: ETS, 2008), pp. 31–50

———, 'A New Phase of the Reception of Aristotle in the Latin West: Albertus Magnus and His Use of Arabic Sources in the Commentaries on Aristotle', in *Albertus Magnus und der Ursprung der Universitätsidee: Die Begegnung der Wissenschaftskulturen im 13. Jahrhundert und die Entdeckung des Konzepts der Bildung durch Wissenschaft*, ed. by Ludger Honnefelder (Berlin: Berlin University Press, 2011), pp. 259–76 and 491–500

———, 'Reading Aristotle with Avicenna: On the Reception of the *Philosophia Prima* in the *Summa Halensis*', in *The Summa Halensis: Sources and Context*, ed. by Lydia Schumacher (Berlin: De Gruyter, 2020), pp. 135–54

———, *The Reception of Aristotle's 'Metaphysics' in Avicenna's 'Kitāb al-Šifā'': A Milestone of Western Metaphysical Thought* (Leiden: Brill, 2006)

———, 'The Reception of Averroes' Long Commentary on the *Metaphysics* in Latin Medieval Philosophy until Albertus Magnus', in *Via Alberti: Texte – Quellen – Interpretationen*, ed. by Ludger Honnefelder, Hannes Möhle, and Susana Bullido del Barrio (Münster: Aschendorff, 2009), pp. 457–80

———, 'The Reception of Avicenna in Latin Medieval Culture', in *Interpreting Avicenna: Critical Essays*, ed. by Peter Adamson (Cambridge: Cambridge University Press, 2013), pp. 242–69

———, 'The Structure of Metaphysical Science in the *Ilāhiyyāt* (*Divine Science*) of Avicenna's *Kitāb al-Šifā'* (*Book of the Cure*)', *Documenti e studi sulla tradizione filosofica medievale*, 13 (2002), 1–69

———, '"Subtilius speculando": Le citazioni della *Philosophia prima* di Avicenna nel Commento alla *Metafisica* di Alberto Magno', *Documenti e studi sulla tradizione filosofica medievale*, 9 (1998), 261–339

Carasquillo, Francisco J. Romero, David Twetten, and Bruno Tremblay, eds, 'Albert the Great on Metaphysics', in *A Companion to Albert the Great: Theology, Philosophy, and the Sciences*, ed. by Irven M. Resnick (Leiden: Brill, 2013), pp. 541–721

Castelli, Laura M., *Problems and Paradigms of Unity: Aristotle's Account of the One* (Sankt Augustin: Academia, 2010)

Craemer-Ruegenberg, Ingrid, '"Ens est quod primum cadit in intellectu": Avicenna und Thomas von Aquin', in *Gottes ist der Orient – Gottes ist der Okzident*, ed. by Udo Tworuschka (Cologne: Böhlau, 1991), pp. 133–42

De Haan, Daniel D., *Necessary Existence and the Doctrine of Being in Avicenna's 'Metaphysics of the Healing'* (Leiden: Brill, 2020)

de Libera, Alain, 'D'Avicenne à Averroès, et retour: Sur les sources arabes de la théorie scolastique de l'un transcendental', *Arabic Sciences and Philosophy*, 4 (1994), 141–79

Fakhry, Majid, 'Notes on Essence and Existence in Averroes and Avicenna', in *Die Metaphysik im Mittelalter: Ihr Ursprung und ihre Bedeutung: Vorträge des II. Internationalen Kongresses für mittelalterliche Philosophie, Köln 31.8–6.9.1961*, ed. by Paul Wilpert (Berlin: De Gruyter, 1963), pp. 614–17

Forest, Aimé, *La structure métaphysique du concret selon Saint Thomas d'Aquin* (Paris: Vrin, [1931] 1956)

Gabbani, Carlo, 'Le proprietà trascendentali dell'essere nel *Super Sententiarum* di Alberto Magno', *Medioevo*, 28 (2003), 97–138

Gilson, Étienne, *L'être et l'essence* (Paris: Vrin, [1948] 1972)

Goris, Wouter, *Transzendentale Einheit* (Leiden: Brill, 2015)

Janos, Damien, *Avicenna on the Ontology of Pure Quiddity* (Berlin: De Gruyter, 2020)

Koutzarova, Tiana, *Das Transzendentale bei Ibn Sina (Avicenna): Zur Metaphysik als Wissenschaft erster Begriffs- und Urteilsprinzipien* (Leiden: Brill, 2009)

Krause, Katja, and Henryk Anzulewicz, 'From Content to Method: The *Liber de causis* in Albert the Great', in *Reading Proclus and the 'Book of Causes'*. Vol. 1: *Western Scholarly Networks and Debates*, ed. by Dragos Calma (Leiden: Brill, 2019), pp. 180–208

Kühle, Heinrich, 'Die Lehre Alberts des Grossen von den Transzendentalien', in *Philosophia perennis: Abhandlungen zu ihrer Vergangenheit und Gegenwart*, ed. by Fritz-Joachim von Rintelen (Regensburg: Habbel, 1930), vol. 1, pp. 131–47

Menn, Stephen, 'Al-Fārābī's *Kitāb al-Ḥurūf* and His Analysis of the Senses of Being', *Arabic Sciences and Philosophy*, 18 (2008), 59–97

———, 'Fārābī in the Reception of Avicenna's Metaphysics: Averroes against Avicenna on Being and Unity', in *The Arabic, Hebrew, and Latin Reception of Avicenna's Metaphysics*, ed. by Dag Nikolaus Hasse and Amos Bertolacci (Berlin: De Gruyter, 2012), pp. 51–96

Mulchahey, Marian Michèle, *'First the Bow is Bent in Study…': Dominican Education before 1350* (Toronto: The Pontifical Institute of Mediaeval Studies, 1998)

Noone, Timothy B., 'Albert on the Subject of Metaphysics', in *A Companion to Albert the Great: Theology, Philosophy, and the Sciences*, ed. by Irven M. Resnick (Leiden: Brill, 2013), pp. 543–53

O'Shaugnessy, Thomas, 'St Thomas's Changing Estimate of Avicenna's Teaching on Existence as an Accident', *The Modern Schoolman*, 36 (1959), 245–60

Pickavé, Martin, 'On the Latin Reception of Avicenna's Theory of Individuation', in *The Arabic, Hebrew, and Latin Reception of Avicenna's Metaphysics*, ed. by Dag Nikolaus Hasse and Amos Bertolacci (Berlin: De Gruyter, 2012), pp. 339–63

Pini, Giorgio, 'Scotus and Avicenna on What It Is to Be a Thing', in *The Arabic, Hebrew, and Latin Reception of Avicenna's Metaphysics*, ed. by Dag Nikolaus Hasse and Amos Bertolacci (Berlin: De Gruyter, 2012), pp. 365–87

Rashed, Marwan, *Essentialisme: Alexandre d'Aphrodise entre logique, physique et cosmologie* (Berlin: De Gruyter, 2007)

Tarabochia Canavero, Alessandra, 'I "sancti" e la dottrina dei trascendentali nel Commento alle *Sentenze* di Alberto Magno', in *Was ist Philosophie im Mittelalter? Akten des X. Internationalen Kongresses für mittelalterliche Philosophie der S.I.E.P.M., 25. bis 30. August 1997 in Erfurt*, ed. by Jan A. Aertsen and Andreas Speer (Berlin: De Gruyter, 1998), pp. 517–21

Vargas, Rosa E., 'Albert on Being and Beings: The Doctrine of *Esse*', in *A Companion to Albert the Great: Theology, Philosophy, and the Sciences*, ed. by Irven M. Resnick (Leiden: Brill, 2013), pp. 627–48

Weijers, Olga, *In Search of the Truth: A History of Disputation Techniques from Antiquity to Early Modern Times* (Turnhout: Brepols, 2013)

Wisnovsky, Robert, *Avicenna's Metaphysics in Context* (Ithaca, NY: Cornell University Press, 2003)

———, 'Essence and Existence in the Eleventh- and Twelfth-Century East (*Mašriq*): A Sketch', in *The Arabic, Hebrew, and Latin Reception of Avicenna's Metaphysics*, ed. by Dag Nikolaus Hasse and Amos Bertolacci (Berlin: De Gruyter, 2012), pp. 27–50

Zimmermann, Albert, *Ontologie oder Metaphysik? Die Diskussion über den Gegenstand der Metaphysik im 13. und 14. Jahrhundert: Texte und Untersuchungen* (Leiden: Peeters, [1965] 1998)

DAVID TWETTEN

# Chapter 13. The Emanation Scheme of Albert the Great and the Questions of Divine Free Will and Mediated Creation*

## Albert between Aquinas and the Arabs

There is no doubt that a major figure standing in the middle between 'Aquinas and "the Arabs"' (as one research project is named) is Albert the Great. But there is plenty of doubt as to how Albert's 'mediation' should be read. A key doctrine is Albert's theory of emanation. Until not long ago, Albert scholars either failed to recognize or ignored the emanation scheme in Albert's emanationism. Part of the reason may be our habit of starting with Aquinas and approaching Albert through Aquinas's cosmology. On the present topic, Aquinas freely uses the language of *emanation*, *procession*, or *influx* for any origination that, properly speaking, involves no change or motion.[1] Thus, he speaks of thoughts as *emanating* from

* My gratitude goes to Adriano Oliva for hosting the conference entitled 'Albert between Aquinas and "the Arabs"', where this paper (without its discussion of mediate creation) was originally presented in 2012; to Henryk Anzulewicz for his penetrating and helpful reactions; to Maria Burger, Ruth Meyer, and Bruno Tremblay for their generous assistance; to Michael Jordan and Jules Janssens for their keen eye; as well as to Isabelle Moulin, Catarina Rigo, Thérèse Bonin, and Jörg Tellkamp for hearing me out.

1 Thomas Aquinas, *De substantiis separatis*, cap. IX, ed. Leonina, p. D58, vv. 184–90: 'In his autem quae fiunt absque mutatione vel motu per simplicem emanationem sive influxum, potest intelligi aliquid esse factum praeter hoc quod quandoque non fuerit; sublata enim mutatione vel motu non invenitur in actione influentis principii prioris et posterioris successio'. Cf. *Quaestiones disputatae de potentia*, q. 10.1c. Thérèse Bonin, 'Emanative Psychology of Albertus', pp. 45, 54 n. 13, and 55 n. 27, has perceptively shown that Aquinas's language regarding the emanation of properties from essence comes from Albert. We find similar language among the university masters: e.g, Bacon [ps.], *Quaestiones super quatuor libros Physicorum*, III.5 ob 2, ed. by Delorme and Steele, p. 140, v. 17: 'omne

**David Twetten** (david.twetten@marquette.edu) teaches in the Philosophy Department of Marquette University, and publishes mainly in ancient and medieval philosophy.

*Albert the Great and his Arabic Sources*, ed. by Katja Krause and Richard C. Taylor, Philosophy in the Abrahamic Traditions of the Middle Ages, 5 (Turnhout: Brepols, 2024), pp. 371–442
BREPOLS PUBLISHERS 10.1484/M.PATMA-EB.5.136493

intellects, or of 'all being' as *emanating* from God in 'creation' (whether or not there was a first moment in time).[2] But Aquinas, unlike Albert, never uses *emanation* of what 'proceeds', *in the genus of substance*, from creatures themselves. In short, Aquinas's language hews close to Pseudo-Dionysius's, for whom creatures emanate causally as to their being from God,[3] but do not proceed as a whole *from subordinate causes*, as for Proclus.[4] So Aquinas's 'soft' emanationism may obscure the comparatively 'hard' version of Albert.

Similarly, the simplicity of Aquinas's cosmology apparently has made it difficult for scholars even to acknowledge the complexity of Albert's. Aquinas rarely affirms any 'higher causes' other than God and angels. Albert agrees with Aquinas's view only in his earliest works on the issue, written perhaps prior to his arrival at Paris circa 1240.[5] For the rest of his career after 1246, as I have shown elsewhere,[6] he affirms Intelligences, not angels, as celestial movers. After the mid-1260s, Albert provisionally affirms, with the Arabic philosophers, both celestial souls and Intelligences. These are demonstrated to exist, he thinks, on good Aristotelian grounds: they are simply the best scientific explanation of the rotation of the invisible spheres that carry the Sun and the planets. Albert himself even takes the 'Arabic cosmology' as ultimately the best reading of Aristotle himself. For Albert, the *Liber de causis*, which affirms a plurality of higher souls

---

accidens emanat a principiis subjecti'; Bonaventure, *Commentaria in quatuor libros Sententiarum*, II.27.1.2c, ed. by Collegium Bonaventurae, p. 657a: 'potentia animae non est ipsa animae essentia, sed potius emanat et procedit ab illa'. On Albert's novelty against the Avicennian background for this language of emanation, see Ehret, 'Flow of Powers'.

2 See Thomas Aquinas, *Summa theologiae*, I.44–45, esp. 44.2, ad 1: 'nunc autem loquimur de rebus secundum emanationem earum ab universali principio essendi'; Aquinas, *Scriptum super Sententiis*, II.1.1.3, ad 4: 'influentia primi agentis, quae est creatio'; Aquinas, *In octo libros 'Physicorum' Aristotelis expositio*, VIII.2.18: '[moventia vel mobilia] ipsum esse non acquisiverunt per mutationem vel motum, sed per emanationem a primo rerum principio: et sic non sequitur quod ante primam mutationem sit aliqua mutatio'.

3 See Pseudo-Dionysius, *De coelesti hierarchia*, IV.1, ed. by Heil and Ritter, p. 177C. As we shall see, there is a very Albertian language in Aquinas, who writes: 'Flumina ista sunt naturales bonitates quas Deus creaturis influit, ut esse, vivere, intelligere, et hujusmodi'; Thomas Aquinas, *Scriptum super Sententiis*, III, prol. In the second part of this paper, I discuss the meaning of 'emanation', indicating that the sense of the term as consistent with divine free will was well established prior to Albert and Aquinas.

4 See esp. Pseudo-Dionysius Latinus, *De divinis nominibus*, V.2, 816C, trans. by Sarracenus; Roques, *L'univers Dionysien*, pp. 78–80.

5 Or, better, Aquinas was satisfied throughout his career with a simplified version of the cosmological system that the Latins inherited from the Arabic philosophers, a version perhaps inspired by that of early Albert and other masters (but, interestingly, not one that Albert held during most of the time Aquinas studied under him: from 1246 on, Albert rejected angels as celestial movers): angels are the proximate causes of the motion of the heavens; they are the equivalent of the philosophers' Intelligences or celestial souls. For Albert's early reversals on this issue, see Twetten, 'Albert the Great's Early Conflations', pp. 29–41. The texts are rehearsed also in Krause and Anzulewicz, 'From Content to Method', pp. 191–201.

6 Twetten, 'Albert the Great, Double Truth, and Celestial Causality'; Twetten, 'Albert's Arguments for the Existence of God'.

and Intelligences (props IV–V), is based on a lost work of Aristotle that completes the train of thought left unfinished in *Metaphysics* Lambda.[7] As Alain de Libera has brilliantly explained, we misread Albert if we forget that his historiographical categories are therefore quite different from our own.[8] What for us (and as Aquinas began to see) is an Arabic adaptation of Proclus's *Elements of Theology*, and hence is 'Neoplatonic' according to our nineteenth-century categories, Albert takes to be the highest fulfilment of Peripateticism.[9]

But even when Albert's 'Arabic cosmology' has been acknowledged, it has been mainly interpreted as contradicting his Christian thought. De Libera puts the matter as follows:

> The admission of Intelligences and celestial souls passes beyond the mere cosmological problem of the animation of the heavens. To accept separate Intelligences is also, for example, to accept the thesis of the mediating role of separate substances in the *processus* of a causal emanation that generates a universe made of realities at once emanating and emanated; it is to deny that God alone creates and [that he creates] without intermediary; in short, it is to deny the very idea of creation in the Christian sense of the term, conforming [instead] with the slogan 'from one comes only one', which the Parisian condemnations of 1277 will meet head on.[10]

Of course, there are various ways of interpreting this situation. On the one hand, the original claim, found in Pierre Duhem, Martin Grabmann, and Bruno Nardi, was that Albert's Aristotelian paraphrases do not report his personal thought, since they contradict Christian belief.[11] On the other hand, the most prominent reading today (correctly) takes the paraphrases to reflect Albert's personal thought and therefore sees Albert as a 'precursor of radical Averroism'.[12] Loris Sturlese, for example, regards Albert as welcoming 'in toto the pagan cosmology', including necessary, eternal emanation and mediate creation, despite its conflict with the

---

7 Albertus Magnus, *De causis*, II.1.1, ed. by Fauser, p. 59, v. 32–p. 60, v. 5 and p. 61, vv. 65–68; ibid., II.5.24, p. 191, vv. 17–23.

8 See especially de Libera, 'Albert le Grand et Thomas d'Aquin'.

9 On the *Liber de causis*, see note 18.

10 De Libera, *Métaphysique et noétique*, p. 63. This statement, taken over literally from the original version of the book, de Libera, *Albert le Grand et la philosophie*, p. 46, is also largely incorporated into de Libera, *Raison et foi*, p. 284.

11 Duhem, *Le système du monde*, vol. 5, pp. 431–32 and 446–47; Grabmann, 'Die Lehre des heiligen Albertus Magnus', p. 302, also p. 294; Nardi, 'La posizione di Alberto Magno', pp. 122–25. Nuanced alternatives are possible, such as that Albert takes seriously a philosophical presentation of materials; but, as in the case of the Neoplatonic emanation he inserts within Aristotelianism, he does not always clarify how they harmonize with each other or with Church teachings. See Kaiser, 'Zur Frage der eigenen Anschauung Alberts', pp. 54, 58–62.

12 Piché, *La Condamnation parisienne de 1277*, pp. 185–86.

Christian faith.[13] As a result, Sturlese's Albert is, after 1250, a philosophical rationalist, a 'second Averroes'.[14]

I present an alternative reading of Albert's cosmology, while insisting that Sturlese raises an important problem. After all, de Libera is correct regarding the conflict, whether intended or not, between the action of the Bishop of Paris in 1277 and Albert's cosmology: I count at least ten of the condemned articles as entirely or partly affirmed in Albert's cosmology, and some ten more as ascribable to Albert with some qualifications.[15] It is probably not accidental, then, that Aquinas goes in a different direction, from 1251 on, from his teacher, though it is a direction that, I believe, cannot be correctly understood without seeing Albert between Aquinas and the Arabs.

## Emanation in the Context of Arabic Philosophy

This paper represents a preliminary effort to identify the emanation scheme presented in the most fully finished cosmological thinking of Albert. Others have recently made ground-breaking contributions to emanation as found in early Albert,[16] but what has not been explained is how cosmic emanation works at the level of substance in causes below God. I introduce Albert's final emanative cosmology by identifying principles within Albert's incredibly dense texts and

13 Sturlese, *Storia della filosofia tedesca*, pp. 94–100.

14 Ibid., pp. 78 and 85. De Libera follows Sturlese in speaking of a 'second Albert' after Albert's radical 'epistemological turn' in 1250, when he articulates 'a vision of the autonomy of philosophical research' 'more daring than that of Boethius [of Dacia]'; de Libera, *Raison et foi*, pp. 166, 265–69, 279, esp. p. 268 n. 18.

15 In cosmology, I count six articles as justifiably Albertian: 61, 71, 92, 95, 189, and 219 (Piché's enumeration); at least four as partially justifiable: 67, 112, 204, and 218; and eleven as potentially justifiable with qualification: 36, 43–44, 62, 64, 66, 73, 82, 94, 212, and 215. We know that Albert's paraphrases, as also Aquinas's commentaries, influenced the Arts masters at Paris, so we need not infer that it is false that the principal target of the Condemnation was the Arts masters and their students, as is insisted upon by Hissette, *Enquête sur les 219 articles condamnés à Paris*. This provisional list leaves out articles on other themes that could potentially be ascribed to Albert, such as the six propositions mentioned by Hissette, 'Albert le Grand et Thomas d'Aquin', pp. 228–29, not to mention the many that could be ascribed to Aquinas. See, for example, art. 65, mentioned by Palazzo, 'Scientific Significance of Fate', p. 59.

16 See especially Anzulewicz, 'Die Emanationslehre des Albertus Magnus'; Krause and Anzulewicz, 'From Content to Method', p. 200. For Schwartz, 'Celestial Motion', pp. 287–89, Albert rejects celestial souls and Intelligences as mediating causes in 1246 and for the rest of his career. See also Schwartz, 'Divine Space', p. 108. It is, of course, correct to emphasize Albert's early criticism of mediate creation, as has been done in the case of the mediate creation of the human soul, a doctrine Albert finds in Gundissalinus; Fidora, 'From Arabic into Latin into Hebrew', pp. 21–28. After this paper was written, I discovered my agreement with Milazzo, 'Commentaire', p. 271, who concludes: 'L'émanation nécessaire et déterminée d'Avicenne fait place à une émanation volontaire et libre chez Albert le Grand'. Some disagreements will emerge in what follows.

reducing those texts to a sort of system.[17] Albert's emanation scheme is set forth only in his *On the Causes and Procession of the Universe from the First Cause* of circa 1267, Book II of which constitutes his paraphrase of the *Liber de causis* (an original Arabic composition from the Circle of al-Kindī, based on rearranged and adapted passages selected from Proclus's *Elements* and from Plotinus's *Enneads*).[18] As I lay out the teaching found there, I appeal to earlier texts, especially from the Dionysian writings of 1248–50. Albert incorporates elements of thought contained there into his scheme.

The hypothesis thus far confirmed by my research is that Albert's emanation scheme is harmonizable with Christian teachings as found in the Church councils up to his time, although it is not explicitly harmonized by Albert with respect to all of its details. I must leave for another paper the stages by which Albert came to develop his emanation scheme by reflecting on the principle *Ab uno non est nisi unum*, which he always regarded as true in its proper sphere. The historical development must be considered before one addresses the complex question of the relation of philosophy and theology in Albert, and therefore the relation of Albert's philosophical and theological works. What emerges from the present, comparatively systematic discussion is that Albert's emanation scheme does not affirm mediate creation, *pace* Sturlese and de Libera. At the same time, I also take up another major obstacle to the emanation scheme's being harmonizable with the theological notion of creation: that it appears to be necessitarian. In addition to conflicting with divine free will, necessitarianism would entail the eternity of the world, as is evident from Article 58 condemned in 1277:

> 58 (34). Quod Deus est causa *necessaria* primae intelligentiae: qua posita ponitur effectus, et sunt simul duratione.

17 For other recent approaches, see Baldner, 'Albertus Magnus on Creation'; Bonin, *Creation as Emanation*; Moulin, 'Éduction et émanation chez Albert le Grand'; Molina, 'Movens-Efficiens-Agens'; Hankey, '*Ab uno simplici non est nisi unum*'. This paper focuses on the Arab or Avicennian emanation scheme, a necessitarian version of which is criticized, as is well known, in Albert's early works. See Albertus Magnus, *Commentarii in II Sententiarum*, 1.10 sc 1; 3.3 sc 4 and ad 5; 14.6c and ad 2; 3.15, ad 4, ed. by Borgnet, pp. 27b–28a, 65a–66a, 265b–266b, 92ab; Anzulewicz, 'Die Emanationslehre des Albertus Magnus', pp. 224 and 227; and note 108 below. Albert, just as Aquinas, adopts the language of emanation for creation even when he appears not to follow the Arabic emanation scheme, as Anzulewicz has stressed in regard to Albert, *Commentarii in II Sententiarum*, d. 1, a. 12c, ed. by Borgnet, p. 34b; Anzulewicz, 'Die Emanationslehre des Albertus Magnus', pp. 236–37. See also Albertus Magnus, *De bono*, I.1.7c, ed. by Kühle, p. 15, vv. 53–59; Albert, *Super I Sententiarum*, 1.1 [q. 6c], ed. by Burger, p. 37, vv. 40–47; Albert, *Commentarii in I Sententiarum*, 35.12.2 ad 2, ed. by Borgnet, p. 201b. Compare also the language of *influere* in Albertus Magnus, *De IV coaequaevis*, q. 7, a. 2c and ad 1, ed. by Borgnet, pp. 401a–402a; and q. 21, a. 1 sc, p. 461a. The language is borrowed from the Latin *Liber de causis*, which uses *influere* and *superfluere* thirty-three times, though not *emanare*.

18 See especially D'Ancona and Taylor, '*Le Liber de causis*'; D'Ancona, 'The *Liber de causis*'.

> 58 (34). That God is the necessary cause of the first Intelligence, [such that] when [the cause] is affirmed, the effect is affirmed, and they are simultaneous in duration.[19]

Albert's main presentation of emanation is found in Tract 4 of his personal writing, *De causis* Book I, and so it is grounded, I observe, upon Tract 3's defence of the free will production of all things by God. Therefore, I begin with a discursus on Albert on divine free will.

But first let us consider a question about terminology: To what does Albert commit himself in espousing *emanation* (or its close synonyms *procession, flux, influx*, etc.)? Whereas for us it calls to mind a set of doctrines associated with the pagan philosophers, Albert — no less than Aquinas — finds the Latin *emanare* used repeatedly in the Christian treatise composed (purportedly) by Dionysius, the colleague of the Apostle Paul, the *De divinis nominibus*. There, it is John Saracen's translation of *ekbluzō*.[20] Granted, a broad view of emanation is opened up for Albert by the frequent use of *fluere, procedere*, and their derivatives in an 'Aristotelian' source so authoritative that it is worth paraphrasing, the *Liber de causis* (for Albert a treatise that Ibn Daud 'ordered' *modo geometrico* — as theorems and their proof — consisting of his comments on propositions drawn from Aristotle, especially in the lost epistle *De principio universi esse*, as well as from al-Fārābī, Avicenna, and al-Ghazālī[21]). Still, it seems odd to us that in the *Physics* paraphrase of circa 1250, Albert can associate emanative language with *will* in both a creationist and a necessitarian sense:

> Bene enim viderunt Peripatetici et ipse Aristoteles, quod absque dubio mundus procedit a prima causa per intellectum et voluntatem, sed dicunt, quod procedit per modum necessitatis [...] nos autem dicimus, quod absolute fluunt a prima causa secundum electionem suae voluntatis.[22]

---

19 Piché, *La Condamnation parisienne de 1277*, pp. 185–86.

20 For the texts on flux in the *Liber de causis* on which Albert bases his conception, see Lizzini, 'Flusso', p. 131. For the distinction between *fluere* and *influere*, and the problem of univocity raised by Albert, see Lizzini, *Fluxus (fayḍ)*, pp. 10–11, 22, 114.

21 Albertus Magnus, *De causis*, II.1.1, ed. by Fauser, p. 59, vv. 9–22 and p. 61, vv. 39–68; see also notes 87 and 134 below, as well as Alarcón, 'S. Alberto Magno'; de Libera, 'Albert le Grand et le platonisme', pp. 92–95. For the ascription to Ibn Daud, see also Albert, *De caelo et mundo*, I.3.8, ed. by Hossfeld, p. 73, vv. 30–31. As D'Ancona explains in 'Nota sulla traduzione latina del *Libro di Aristotele*', there is evidence in Arabic of a work ascribed to Aristotle corresponding to the title *Liber de causis*, a title that accurately reflects the status of the *Theology of Aristotle* among philosophers. Albert appears to be drawing inferences from (misleading) information in manuscripts, some of which we possess; Saffrey, 'Introduction', p. xxiii. See also D'Ancona, 'Un "quindicesimo libro"'; Anzulewicz, 'Einleitung', pp. xxix–xxxiv.

22 Albertus Magnus, *Physica*, VIII.1.15, ed. by Hossfeld, p. 579, vv. 55–59 and p. 580, vv. 48–50, a chapter entitled 'Et est digressio declarans, qualiter orbis et caelestia corpora procedunt a causa prima'. In this passage, as throughout much of the tract, he follows the discussion of Maimonides' *Guide* closely, adding that neither of these positions can be proved except with probable arguments. For Maimonides on the necessitarian will of Aristotle's First Cause, see note 69 below.

> The Peripatetics and Aristotle himself have seen well that, without doubt, the world *proceeds* from the First Cause through intellect and will, but they say that it *proceeds* by way of necessity [...]. We say, however, that [the heavens] *flow* absolutely [i.e., without mediation] from the First Cause according to the *choice* of his will.

Yet we find other Parisian masters who use emanation language in a way consistent with classical theism.[23] Is this a monstrosity of the Latin thirteenth century?

Even in the context of Arabic thought, contemporary with the 'authorship' of the *Liber de causis*, *emanation* can be used as consistent with divine free creation. Peter Adamson has helpfully distinguished four elements in our standard Plotinian notion of 'emanationism':

> (a) the world is caused eternally by God (or the One);
> (b) the world is caused necessarily by God (or the One);
> (c) the cosmos consists of primary *archai*: God, Intellect, and Soul; and
> (d) God's immediate causality is limited to a first effect, which becomes the immediate cause of the second effect, and so on.[24]

As Adamson goes on to show in the pages that follow, al-Kindī affirms a true emanationism even though he denies (a) and (b): namely, eternal and necessary creation. The same can be said not only of Maimonides,[25] but also of Albert. Albert's Arabic 'emanationism' and emanation scheme — or better, so as to avoid necessitarian connotations, his 'derivationism' — found only in the last of his philosophical works, consists in a highly nuanced and developed account of (d) that God immediately causes only one. That account begins, in Tract 3 of *De causis* 1, with the denial, just as in al-Kindī, that the universe proceeds necessarily and eternally from God. Nonetheless, the denial is not as forthright as one might expect, giving de Libera and Sturlese some grounds, it would appear, for questioning Albert's commitment to divine freedom.

---

23 See, for example, William of Auxerre, *Summa aurea*, I.12.4.8, ad 1, p. 355, vv. 30–34 [Appendix XLII]: 'Sed omnes res praeter peccatum emanant a divina providentia. [...] Ergo nullus licite potest velle voluntate rationis contrarium alicuius quod scit emanare vel emanasse a divina providentia'. We also find awareness of necessary emanation among the masters, as in Robert Kilwardby, *Quaestiones in librum secundum Sententiarum*, q. 3 sc, ed. by Leibold, p. 14, vv. 69–79.

24 Adamson, *The Arabic Plotinus*, p. 185. For the language of emanation, see esp. Armstrong, '"Emanation" in Plotinus'; Trouillard, 'Procession néoplatonicienne'; Dörrie, 'Emanation'; Hasnawi, 'Fayḍ (épanchement, émanation)'; Janssens, 'Creation and Emanation in Ibn Sīnā'; Lizzini, *Fluxus (fayḍ)*, pp. 27–87.

25 Maimonides, *Dalālat al-ḥā'irīn (Le guide des égarés)*, II.4, ed. by Munk, pp. 14b17–15a14; ibid., II.11, pp. 24b1–6; ibid., II.12, pp. 25b21–26a23; ibid., II.21, pp. 47b10–16; ibid., II.22, pp. 48b7–50a15; ibid., III.17, pp. 32b4–13. See Twetten, 'Aristotelian Cosmology and Causality', pp. 386–88. For discussion of the view of Davidson and others that Maimonides may have secretly agreed with Plato's creation account, which presupposes uncreated, eternal matter, see Seeskin, *Maimonides on the Origin of the World*, pp. 172–81.

## Divine Free Will in Albert

### *'De causis et processu' I.3 and an Objection*

At first sight, Albert's account of divine freedom in Book I, Tract 3 looks as if it could pertain equally to a doctrine of necessary emanation like Plotinus's. Albert's dominant theme from the outset of the tract is something also espoused by Plotinus:[26] the First is absolutely free because nothing prior to it can possibly constrain it. As Albert puts it, 'Since the First has absolutely no dependence in relation to any cause, it is agreed that it is free from all necessity'.[27] In fact, Albert opens his discussion by quoting the Latin Aristotle: what is free is precisely what is a *causa sui*.[28] Albert concludes: the first principle is maximally a *causa sui* in acting, and therefore 'most free' in acting.

Despite this apparently Plotinian account of freedom, close inspection of the text reveals Albert's affirmation also that the alternatives 'to act' and 'not to act' are in the divine power. If freedom requires alternatives, this affirmation would seem to be what Albert needs to account for divine freedom. He responds to an objection thus:

> Adhuc autem, per hoc quod dicitur, quod non sit in ipso agere et non agere, nihil probatur. Hoc enim dupliciter dicitur. Non esse enim in aliquo agere et non agere potest esse per obligationem ad unum et impossibilitatem ad alterum. Alio modo potest esse per libertatem ad unum et ad alterum. Sed quia melius est esse unum quam alterum, propter hoc non transponitur de uno in alterum. Sicut in casto est caste agere et non caste et in liberali dare et non dare. Sed quia melius est caste agere et liberaliter dare quam non caste agere et avare retinere, ideo non transponitur castus et liberalis in oppositum suae actionis. Et sic agere et non agere quidem est in primo, sed non potest

26 See Plotinus, *Enneads*, VI.8.13–14.

27 Albertus Magnus, *De causis*, I.3.1, ed. by Fauser, p. 35, vv. 46–48: 'Cum igitur primus ad nullam penitus causam habeat dependentiam, constat, quod ab omni necessitate liber est'.

28 Ibid., p. 35, vv. 10–15, citing Aristotle, *Metaphysics*, I.2, 982b25–26, in James of Venice's translation. For other instances, see Albertus Magnus, *De homine*, ed. by Anzulewicz and Söder, p. 507, vv. 23–24 (ob 2); p. 520, vv. 48–52 (ob 10); p. 522, vv. 17–20 (ad 2); p. 523, vv. 48–50 (ad 5); Albert, *Commentarii in II Sententiarum*, 24.5.1c, ed. by Borgnet, pp. 401b–402a (quoted below at note 64; the phrase here, as sometimes, must be taken in the ablative, corresponding to the original Greek phrase and its Latin translation); ibid., 25.1.3c, ed. by Borgnet, p. 424a (quoted at note 44 below); Albert, *Ethica*, III.1.1, ed. by Borgnet, p. 196a (quoted in note 61 below); Albert, *Summa theologiae*, I.79.1, ed. by Borgnet, p. 839b; ibid., II.16.1, p. 209b; ibid., II.91.2, p. 186a: 'Et ideo, sicut dicit Aristoteles in primo *primae philosophiae*, sicut liberum dicimus hominem qui causa sui est, et sicut liberam dicimus scientiam quam propter seipsam volumus, et causa sui est, et non propter aliud: ita dicimus liberum arbitrium, quod in omnibus operibus et motibus sibi est causa, et non potest agi vel cogi ad aliud'. See also Spiering, '"*Liber est causa sui*"'.

non agere, quia melius est emittere bonitates quam retinere, et minimum inconveniens in primo impossibile est.[29]

> Furthermore, nothing is proved by saying [in the third objection] that in [God] there is not [both] acting and not acting [thereby implying change, according to the objector]. For, this [acting and not acting] is said in two ways. For, (a) the non-being in some 'acting and not acting' *can* be through 'being bound' in relation to one, and through an impossibility in relation to the other. Alternatively, (b) it can be through freedom in relation to one as well as in relation to the other. But because it is better for one to be than the other, therefore there is no transitioning from one into the other; just as in a chaste person, there 'is' acting chastely and not chastely; and in a generous person, [there 'is'] giving and not giving. But because it is better to act chastely and to give generously than to act non-chastely and to withhold greedily, for this reason the chaste and the generous person do not transition into the opposite of their action. In this way, to act and not to act are, indeed, in the First. *But it is not able not to act.* Because it is better to emit than to withhold goodness; and the least unfitting thing is impossible in the First.

Still, the final italicized text indicates a deeper problem than before. The culmination of Albert's point, that alternatives do exist in God's power, is the admission that God can nonetheless perform only one of the alternatives: the one that is better. So, even when Albert recognizes in God a freedom that Plotinus overlooks, he seems to reduce it to something that could equally satisfy the necessitarian's account of 'freedom'.

### *Response: Freedom with and without Alternatives in Albert*

In the following, I argue that it is false that divine freedom in Albert could be necessitarian. Since this is not the place for an exhaustive treatment, I limit myself to the following essential points. 1) There can be no question that Albert, including in works post-1250, ascribes to God freedom (*libertas*; what we often call *free will*), as well as *liberum arbitrium* (free arbitration[30]) and choice (*electio*). 2) Definitions of human freedom among the early theology masters at Paris only occasionally state the need for alternatives. If absence of alternatives makes Albert a determinist, then nearly all of these masters must also be held to deny divine freedom. 3) Albert does not think he needs to appeal to a freedom among alternatives in order to account for *human* freedom, although he does acknowledge it as an important characteristic. 4) In fact, Albert's deepest account of freedom

---

29 Albertus Magnus, *De causis*, I.3.1, ed. by Fauser, p. 36, vv. 20–34 (italics added).

30 Albert insists that *arbitrium* of *liberum arbitrium* must not be confused with the *iudicium* of reason; see note 62 below. The translation *free judgment* is thus excluded, and *free decision* is potentially misleading. I shall leave the term partially untranslated: *free arbitrium*.

is already voluntarist without expressly appealing to alternative possibilities, as his use of John of Damascus reveals. 5) For divine freedom as well, though Albert does affirm freedom among alternatives, his voluntarism makes its explicit mention unnecessary. 6) Still, Albert's robust account of God's free choices raises problems for divine simplicity. How can his *Super De divinis nominibus* justify the Avicennian claim there that for God to act through will is for God to act through his essence?

My project is one that begins by looking for alternative possibilities in Albert's account of divine freedom. In the course of finding them there, we discover an account of the will so voluntarist that, for Albert, they become inessential for articulating freedom.

*Libertas, liberum arbitrium, and electio Are Ascribed to God.* If it is important in general to attend to vocabulary in reading the Scholastics, it is especially so in the case of freedom in Albert, as we shall see. We see him using these three terms of the divine, *libertas, liberum arbitrium*, and *electio*, already in writing *Sentences* I–II, circa 1245–46. Albert insists, for example, that unlike 'things that act through the necessity of nature', God acts 'through free choice of the will' (*per liberam electionem voluntatis*).[31] But the clearest statement for our purposes is found after 1270 in *Summa theologiae* II (assuming that this second book was composed by Albert or at least under his direction):[32]

> [p. 211b: Utrum in Deo sit liberum arbitrium?]
> [p. 212b] Solutio. Dicendum, quod liberum arbitrium est in Deo, et summa libertate liberum, qui ita liber est, quod 1) semper facit omne quod vult, nec 2) indigentia nec necessitate nec debito obligatur ad aliquid [...]
> Ad aliud [2] dicendum, quod in Deo est electio, sed aliter quam in nobis, sicut et in Deo aliter est consilium quam in nobis.
> [p. 213a, Solutio quaestiunculae] [L]iberum arbitrium convenit Deo et creaturae [...] communitate analogiae, quae est secundum prius et posterius: libertas enim per prius est in Deo, et per posterius in creatura.[33]
>
> [p. 211b: Whether in God there is free *arbitrium*?]

31 Albertus Magnus, *Commentarii in II Sententiarum*, 1.10, ad 13, ed. by Borgnet, p. 30a: 'illa agunt per necessitatem naturae: Deus autem non, sed per liberam electionem voluntatis'; see also ad 9, pp. 29b–30a. This tenth article contains perhaps Albert's earliest presentation of the Arabic emanation scheme, of which he gives an extended criticism to this effect: 'hoc totum absurdum sit' (sc. 1 ad 2; p. 28a). This part of the *Sentences* commentary II, as I have shown (without drawing attention to the issue of emanation; see Anzulewicz, 'Die Emanationslehre des Albertus Magnus', pp. 222, 224, 227, 233), marks a turning point, as Albert criticizes himself for having previously accepted the cosmology of the Arabic philosophers; Twetten, 'Albert the Great's Early Conflations', pp. 33–37.

32 For a discussion of authenticity, see Siedler and Simon, 'Prolegomena', pp. v–xvi; Wielockx, 'Zur "Summa Theologiae" des Albertus Magnus'.

33 Albertus Magnus, *Summa theologiae*, II, q. 94.2, ed. by Borgnet, pp. 211b–213a.

> [p. 212b] Response: It should be said that free *arbitrium* is in God — free with the highest freedom — who is so free that he always *does everything that he wills*, and neither by need, necessity nor obligation is he bound to something [...]
> Ad 2: It should be said that choice [*electio*] is in God, but in a way other than in us, just as also counsel is in God in another way than in us.
> [p. 213a, Response to the sub-question]: [F]ree *arbitrium* belongs to God and creature [...] by a community of analogy, which is according to prior and posterior. For, freedom is first in God, secondarily in the creature.

*Definitions of Human Freedom in the Early Parisian Masters Need not State Alternatives.* In order to appreciate Albert's account, it is useful to see it against the background of the discussions of the preceding Parisian masters of theology. William of Auxerre in the *Summa aurea*, observes Odon Lottin, appears to be the first to consider definitions of freedom separately from definitions of free *arbitrium*, as subsequently do Philip the Chancellor and Albert in *De homine*.[34] William rejects two of four definitions of freedom — each apparently devised by him — because they use *flexibilitas* (that is, to the opposites good and evil) and thereby imply an agent who can do evil or good, unlike the blessed or the devil.[35] Philip considers a number of definitions of *libertas* before preferring Anselm's, including three inspired by William's four: 'flexibility to opposite acts' (*ad oppositos actus*), 'habitual liberation from coercion', and 'doing what one wants';[36] but he omits the definition of the *Summa Duacensis* (a work he appears to know) that uses

34 Lottin, 'Libre arbitre et liberté', p. 64 (subsequent parentheses in the text body refer to this magisterial work). Debate on the definition of *liberum arbitrium* was lively in the twelfth century. Since the formula drawn from Augustine, 'the power of doing good or evil', included evil, Anselm (see Lottin, 'Libre arbitre et liberté', pp. 12, 16) introduced an alternative that applied to pre-fallen humans and to Christ: 'the power of preserving rectitude'; Anselm, *De libertate arbitrii*, I, III, and XIII, ed. by Schmitt, p. 207, vv. 11–13; p. 212, vv. 19–20; p. 225, vv. 6–7. Abelard's formula, which was criticized for suggesting that after the fall, one could through *liberum arbitrium* do good without grace, introduced alternatives (as had also an earlier formula apparently contemporary with the school of Anselm of Laon; Lottin, 'Libre arbitre et liberté', p. 18 n. 1): 'faculty of deliberating and determining *what one wants to do*, whether it should be done or whether what one has chosen should not be pursued' (ibid., p. 23).

35 Lottin, 'Libre arbitre et liberté', pp. 64–70. See William of Auxerre, *Summa aurea*, II.10.4, ed. by Ribaillier, p. 283: 'flexibilitas in utramque partem' (flexibility to each of two alternatives, such as heaven or hell); 'flexibilitas in id quod vult'; cf. 'privatio coactionis'. The first of these phrases is drawn from Jerome, *Epistle* XII, as quoted by Peter Lombard, *Sententiae in IV libris distinctae*, II.25.2, ed. by Collegii S. Bonaventurae, p. 462: 'Qualiter in Deo accipitur liberum arbitrium'. Albert uses this phrase in early and late works: Albertus Magnus, *De IV coaequaevis*, q. 31, a. 1 ob 1, ed. by Borgnet, p. 502b; Albertus Magnus, *Summa theologiae*, II.1.1 ob 1b, ed. by Borgnet, p. 223b; ibid., II.94.2, qa. sc 2, p. 212b: 'Si univoce convenit liberum arbitrium Deo et rationali creaturae?'.

36 These last two definitions are at play even in texts of Albert: see Albertus Magnus, *De causis*, I.3.1, ed. by Fauser, p. 35, vv. 46–49 (and note 27 above), also vv. 21–24: 'libertas [...] a coactione' and I.3.3, p. 38, vv. 29–30: 'primum [principium...] potest quidquid vult'; Albertus Magnus, *Summa theologiae*, II.94.2c, ed. by Borgnet, p. 212b.

non-evil alternatives: the free power of acting on or not acting on a counselled desire (*potestas libera faciendi uel non faciendi appetitum consiliatum vel appetitum non consiliatum*).[37] The Dominican master Hugh of St Cher, in *Scriptum super Sententiis* (*c.* 1229–30), returns to William's definitions, among which he defends 'flexibility in relation to what is judged ought to be done'; another is adopted by Hugh's predecessor, Roland of Cremona, in his *Summa*: 'flexibility in relation to what one wants'.[38]

In short, the Parisian masters and their sources do not focus on *libertas* as requiring in its notion alternatives, although such a notion is gradually emerging as one feature of freedom. If absence of alternatives makes Albert a determinist, then most of the Scholastic theologians after Anselm must be held to deny *divine* freedom as well.

*Albert's Account of Human Freedom Contains, but Does not Require, Alternatives.* The six-article sequence of the *De homine*, composed perhaps in Germany even before Albert's Parisian studies, contains Albert's earliest treatment of the express question: What is *libertas*? Albert there shows himself so rooted in the discussion of the Parisian masters that, although he reviews many of the aforementioned definitions, including the three stemming from William of Auxerre,[39] he makes little mention of alternatives.[40] Scholars maintain that the *Sentences* commentary

---

37 Lottin, 'Libre arbitre et liberté', pp. 79–83, 87–88, 90; as Lottin points out (p. 55), we also find a definition through alternatives in the earlier *Glossae super Sententias* of the Parisian master Peter of Capua. See Philip the Chancellor, *Summa de bono*, I, ed. by Wicki, pp. 183–91. For the *Summa Duacensis*, see Principe, *Philip the Chancellor's Theology*, pp. 19–20.

38 Lottin, 'Libre arbitre et liberté', pp. 101–02 and 107.

39 See Lottin, 'Libre arbitre et liberté', p. 120; Siedler, *Intellektualismus und Voluntarismus*, pp. 100–05; Michaud-Quantin, *La psychologie de l'activité*, pp. 206–07, 210, 212, 217–18, 227. For *libertas facendi quod vult*, see Albertus Magnus, *De homine*, ed. by Anzulewicz and Söder, p. 506, vv. 32–34; p. 507, vv. 12–14 and 25–27; p. 511, vv. 53–54; p. 513, vv. 33–34; p. 520, vv. 15–30; p. 521, vv. 22–40; p. 526, vv. 18–30. For *flexibilitas ad oppositos actus*, see ibid., p. 519, v. 45–p. 520, v. 3; p. 520, vv. 15–21 and 61–72; p. 521, vv. 1–3. For *libertas a coactione*, see ibid., p. 510, vv. 1–3; p. 511, vv. 53–59; p. 514, vv. 60–61; p. 520, vv. 8–14; p. 522, vv. 36–37; p. 523, vv. 1–11, 19–22, 45–53.

40 But see Albertus Magnus, *De homine*, ed. by Anzulewicz and Söder, p. 513, vv. 55–56: 'libertas autem volendi vel non volendi voluntatis est'. Although normally *flexibilitas*, as elsewhere, implies the opposites good and evil, Albert's last use of the term in the *De homine* (ibid., p. 523, vv. 36–38) seems to imply mere neutral alternatives (a sense that we shall see employed in the *Sentences* commentary at notes 44 and 45 below): 'Ad aliud [4] dicendum quod liberum velle attribuitur libertati primae [i.e., libertati naturae v. gratiae] propter flexibilitatem voluntatis, quae non magis inclinatur ad unum quam ad alterum'. If this is correct, then perhaps we see neutral alternatives also here: 'Dicimus quod libertas arbitrii multipliciter determinatur. Si enim determinatur ad actus, in quos potest per se ex natura sui, tunc est flexibilitas ad oppositos actus'; ibid., p. 520, vv. 58–61. Cf. also the formula *freedom of desire*, versus freedom of *arbitrating*: 'Libertas etiam appetitus consistit in hoc quod facultatem habet inclinandi se in iudicatum, vel declinandi ab ipso'; ibid., p. 508, vv. 14–16 (where it is said that non-rational animals lack all three free powers, *arbitrium*, *iudicium*, and *appetitus*). Accordingly, Miteva, 'I Want to Break Free', pp. 11–17, highlights the importance of alternatives for free will in Albert.

contains a significant rethinking of *liberum arbitrium*.[41] In the first major treatment of the topic there, Albert responds to the question: Do angels have free *arbitrium*? He refers to William's three definitions (indicated with Roman numerals), but focuses on 'doing what one wants' interpreted in terms of alternatives, acting or refraining from acting (segment [a]):

> Dicendum, quod liberum arbitrium convenit Angelis per naturam: sed libertas ejus non est flexibilitas in bonum et in malum: quia flexibilitas in malum, ut dicit Anselmus, nec est libertas, nec pars libertatis: sed liberum est in faciendo quod vult, scilicet, quod non obligatur ad actum unum, sicut naturalis virtus agens, et quod non impellitur ad necessitatem actus quin possit abstinere si vult: et haec libertas semper remanet apud Angelum quocumque se vertat. Quod autem non inclinatur ad malum quando confirmatur in bono, hoc non est ex coactione, sed ex immobilitate spontanea in bono quod est confirmatio.[42]
>
> > It should be said that free *arbitrium* belongs to angels by nature. But [an angel's] freedom is not its (i) *flexibility* toward good and evil; because flexibility toward evil, as Anselm says, is neither freedom nor a part of freedom. But *the free* consists in (ii) *doing what one wants*; that is, [a][it is] what is not bound to one act, as is a natural power in acting, and what is not impelled to the necessity of an act, but is able to abstain if it wants. And, this freedom always remains for an angel wherever it turns itself. The fact that it is not inclined to evil when it is confirmed in the good is not *from coercion* (iii), but from the unshakeableness, of its own accord, in the good, which is 'confirmation' [in heaven].

Accordingly, when Albert turns to the express treatment of free *arbitrium* in d. 25, he introduces Augustine's definition of the *will* using alternatives:[43]

41 Lottin, 'Libre arbitre et liberté', pp. 125–26; Drouin, 'Le libre arbitre', pp. 106, 116–20; Siedler, *Intellektualismus und Voluntarismus*, pp. 73, 78–79, 101–02; Michaud-Quantin, *La psychologie de l'activité*, pp. 205–06, 208–10, 214–16, 221–22; McCluskey, 'Worthy Constraints', pp. 507, 513–14, 531–32 (who also aptly criticizes Lottin). See also Schönberger, 'Rationale Spontaneität'; Hoffmann, 'Voluntariness, Choice, and Will'; Spiering, 'An Innovative Approach to *Liberum Arbitrium*'.

42 Albertus Magnus, *Commentarii in II Sententiarum*, 3.7c, ed. by Borgnet, p. 72b.

43 See the similar definition in Albertus Magnus, *Summa theologiae*, II.99.1, ed. by Borgnet, p. 233a, which, nevertheless, makes no appeal to alternatives: 'motus animi liber sui, qui nullo cogente impellitur ad ea quae facit, et adipiscitur quae adipiscitur, et non agitur ad aliquid, sed agit ex propria libertate'. Compare also the formula associated with choice in Albertus Magnus, *Commentarii in II Sententiarum*, 24.5, qa 1c, ed. by Borgnet, p. 401b: 'suum [liberi arbitrii] enim est velle arbitrium (arbitratum?) sibi, et hoc dicit [Johannes Damascenus] eligere, duobus propositis, hoc est [*illi* for *est*], praeoptare et praeeligere: et proponere duo est rationis, *praeoptare autem alterum* est voluntatis: libertatem [libera?] autem vocat a voluntate' (emphasis added; parentheses contain the emendations of Pfortzen, based on the quotation of John of Damascus; see Siedler, *Intellektualismus und Voluntarismus*, p. 74).

[N]ulla potentia in actu omnino causa sui est nisi voluntas. [...] Quod autem causa sui est, liberum est, ut dicit Philosophus: et ideo dicitur communiter, quod voluntatis causa nulla quaerenda est: et ideo diffinitur infra distinctione sequenti, quod 'voluntas est motus animi, cogente nullo, ad aliquid adipiscendum, vel non admittendum'.[44]

> [N]o potency in act is entirely a cause of itself unless it is the will. [...] What is a cause of itself, moreover, is free, as the Philosopher says [Aristoteles Latinus, *Metaph.* I.2, 982b25–26]. For this reason, it is commonly said that no cause of the will should be sought. And, for this reason it is defined below in a subsequent distinction [d. 35]: 'the will is a motion of the mind, without anything forcing [it], *toward attaining something or toward not admitting it*'.

A similar definition is found in Book III, and it amounts to as clear a statement of alternatives as can be found:[45]

[N]on est diffinitio libertatis arbitrii, posse velle bonum, vel malum: [...] sed libertas ejus consistit in hoc quod possit velle hoc, et non velle hoc, et posse velle diversum ab hoc.[46]

> [T]he definition of *freedom of arbitrium* is not: *being able to will good or evil*; [...] but freedom of [*arbitrium*] consists in the fact that willing this

44 Albertus Magnus, *Commentarii in II Sententiarum*, 25.1.3c, ed. by Borgnet, p. 424a. The definition is almost literally that of Augustine, *De duabus animabus contra Manichaeos*, X.14, ed. by Zycha, p. 68, vv. 23–25. One might argue that the *or* is conjunctive; see also Albertus Magnus, *De homine*, ed. by Anzulewicz and Söder, p. 508, vv. 14–16, quoted in note 40 above; Albertus Magnus, *De anima*, III.4.10, ed. by Stroick, p. 242, vv. 9–11: 'Quia si homo arbiter suus non esset de agendo vel non agendo et si liber non esset, non puniret vel praemiaret actus eius legislator'.

45 See also Albertus Magnus, *Peri hermeneias*, II.2.5, ed. by Borgnet, p. 449a: 'quae habent potestates naturales a forma naturali [...] non habent potestates animales sive rationales quae sunt animae rationalis, quae maxime libera est et ex materia non obligatur ad agendum hoc vel illud, sed potest agere hoc vel illud, et etiam oppositum. Et potest agere hoc et potest non agere hoc per voluntatis libertatem. Quaecumque ergo non habent tales potestates, sed naturales, cum natura non sit nisi ad unum, non possunt nisi ad unum: et hoc necessario faciunt'.

46 Albertus Magnus, *Commentarii in III Sententiarum*, 18.2, ad 1, ed. by Borgnet, p. 315b. The entire passage is important: 'Alia est confirmatio liberi arbitrii in eo quod est simpliciter ultimum, et confirmatur in illo per adhaesionem, sicut confirmati in patria: et haec non tollit voluntatem faciendi quod vult, sed ponit necessitatem non coactionis in eo, sed immutabilitatis in fine illo. [...] Tertia est confirmatio potentiae ad actum hunc per necessitatem naturae, ita quod oportet hunc operari, et non posset non operari hunc, nec posset alium operari: et haec confirmatio tollit naturam totam liberi arbitrii: quia omnis potentia rationalis est /b/ ad opposita: et nullus in patria est confirmatus, ita quod ipsum semper oporteat idem numero velle, et non posset hoc non velle: et hujus oppositum ponitur per naturam liberi arbitrii in omnibus, scilicet posse hoc, et posse cessare ab hoc, et posse aliud ab hoc. Et ideo patet ex hoc quod non est diffinitio libertatis arbitrii, posse velle bonum, vel malum: sed potius accidit ei posse velle malum ex defectu: sed libertas ejus consistit in hoc quod possit velle hoc, et non velle hoc, et posse velle diversum ab hoc. In hoc enim cognoscitur liberum, quod non agit per necessitatem naturae, et hoc modo flexibile habuit Christus liberum arbitrium, et hoc sufficit merito cum statu viatoris: et ideo non procedit objectio prima'.

> and not willing this [given thing] are possible, as also is being able to will something different from this.

Notice in the passage quoted the expression Albert sometimes uses: *libertas arbitrii*, 'freedom of arbitration', as if there is another kind of freedom. What freedom is found outside of *liberum arbitrium*? Before turning to that question, we should have before our eyes a very strong text on freedom between alternatives from the oldest of the Aristotle paraphrases:

> Dependet autem scientia huius problematis a scientia alterius, quod est, utrum per necessitatem naturae fluunt entia causata a prima causa vel per electionem voluntatis, et illud sciri non potest nisi per primam philosophiam. Sed quia alteram partem omnes supponunt Peripatetici et Aristoteles probat, quod per voluntatem et scientiam producit prima causa res, ideo hoc hic supponimus accipientes cum hoc propositionem in praecedentibus probatam, quod omne quod agit per voluntatem et scientiam, liberum est ad agendum, si vult, et ad dimittendum actionem, si vult; et similiter liberum est ad agendum hoc modo vel illo, quocumque modo voluerit, dummodo voluntas eius et scientia sufficienter sit causa ad causandum, et hoc quidem iam in praehabitis ex verbis Aristotelis habetur.[47]

> The understanding of this problem [as to whether God or time precedes the world temporally] depends on the understanding of another, [namely, as to] whether beings caused by the First Cause flow [from it] through the necessity of nature or through choice of the will — and that cannot be understood except through first philosophy. But because all of the Peripatetics suppose, and Aristotle proves, the second part [of the disjunction], [namely,] that the First Cause produces things through will and understanding, therefore we suppose this here, accepting therewith the proposition proved in what came before, [namely,] that everything that acts through will and understanding is *free to act, if it wants, and to omit acting, if it wants; similarly, it is free to act in this way or that, in whatever way it has willed*, since its will and understanding are sufficiently the cause of causing — and this, indeed, is already maintained in what has been previously established from the words of Aristotle.

*The Source of the Voluntarism Prior to liberum arbitrium in Albert's Act Theory.* It is best to think of Albert's act theory as starting with the Parisian masters' account of *liberum arbitrium*, and overlaying upon it the stages of the human act as identified in the appetitive psychology of the Latin John of Damascus.[48] In

---

47 Albertus Magnus, *Physica*, VIII.1.13, ed. by Hossfeld, p. 575, vv. 24–39.

48 See esp. Lottin, 'La psychologie de l'acte humain', pp. 399–414; Drouin, 'Le libre arbitre', pp. 94–102, 106–09, 118; Michaud-Quantin, *La psychologie de l'activité*, p. 138. According to Lottin, 'La psychologie de l'acte humain', pp. 423–24, Albert is the first among the Latins to engage in the serious assimilation of Damascene's act theory. But Damascene's text had been regularly used at the

the Latin Damascene, *voluntas*, which belongs to humans alone among material things, translates two Greek terms. I distinguish the two senses by appending the Greek term to the Latin translation. *Voluntas-[thelēsis]*, in Damascene, can name the mere power of willing (*simplex virtus volendi*; par. 15) or the rational desire (*appetitus rationalis*) — which is natural and vital (as is the heartbeat) — for natural goods such as being, living, and motions of intellect and sense (par. 8).[49] By contrast, the *voluntas-[boulēsis]* names a rational desire for ends that, whether in our power or not, typically leads to further deliberation about means, and hence to inquiry, judgment, proposals, choice, impulse, and use (par. 9–11).[50] In Damascene's Latinized Greek, both *thelisis* and *bulisis* take place 'freely in *arbitrium*' (*libere arbitrio*; *autexousiōs*), as do the acts that follow them (par. 12). Indeed, 'everything rational is free in *arbitrium*', and therefore also angels (II.17.3), God (III.18.6 and III.33.2),[51] and humans, who are made in the image of God (III.14.9).[52]

Over the course of a career, whether in *De homine*, the *Sentences* commentary, or the *Summa theologiae*, Albert adopts John's account of the stages of the human act,[53] and he comes to associate the object of *voluntas-[boulēsis]* with what is a

---

University of Paris in commenting on *Nicomachean Ethics* 1–3 — a phenomenon aided by the shared terminology resulting from the single translator, Burgundio. See Zavattero, 'Voluntas est duplex', pp. 65–66.

49 John of Damascus, *De fide orthodoxa*, XXXVI (II.22), par. 8, 12, ed. by Buytaert, pp. 135–36, 140–41. See Frede, 'John of Damascus on Human Action'; Adelmann, 'Theory of Will in John Damascene'; Zavattero, 'Voluntas est duplex'. For the Greek background to John's account, see Gauthier, 'Saint Maxime le Confesseur'.

50 Albertus Magnus, *De IV coaequaevis*, q. 25, a. 1c, ed. by Borgnet, pp. 487a–488a, follows John's language closely in ascribing both wills to angels. Notice Albert's early characterization of *boulēsis*: 'vult ea quae determinantur ex ratione, quoniam sunt bona, et non tantum ex natura'. According to Zavattero, 'La βούλησις nella psicologia dell'agire', pp. 144–46, Albert appears to be the only theologian in the first half of the thirteenth century to understand John's account of *boulēsis*, and he corrects the Parisian Arts masters, who take *voluntas-*[*boulēsis*] to concern means to ends. On the other hand, it is not clear that early Albert takes John's many acts of the will as acts of reason alone as opposed to will, thereby transforming John from a voluntarist into an intellectualist (ibid. pp. 148, 150).

51 John of Damascus, *De fide orthodoxa*, LXII (III.18), par. 6, ed. by Buytaert, p. 254, vv. 69–71: 'Autexusios autem (id est libere) volebat et divina et humani voluntate. Omni enim rationali naturae omnino innata est autexusios (id est libera arbitrio) voluntas'.

52 Burgundio regularly (mis)translates the adjective *autexousios* as *liberum arbitrio*. Notice that the Latin John of Damascus expands the adjective as equivalent to the denial of being forced, in the case of both the human and the divine (citing Maximus the Confessor, as Buytaert notes): 'Sed naturalem quidem voluntatem dicentes, *non coactam* hanc dicemus, sed autexusios (id est liberam arbitrio); si enim rationalis, omnino autexusios (id est libera arbitrio). "Non solum enim divina et increabilis natura *nihil coactum* habet, sed neque intellectualis et creabilis; hoc autem manifestum. Natura enim ens bonus Deus, et natura conditor et natura Deus, *non necessitate* haec est. *Quis enim est qui necessitatem inducit?*"' John of Damascus, *De fide orthodoxa*, LVIII (III.14), par. 19, ed. by Buytaert, p. 222, vv. 147–53 (emphasis added).

53 In addition to texts in the following note, see Albertus Magnus, *De homine*, ed. by Anzulewicz and Söder, p. 493, vv. 4–13 and 30–34; p. 505, v. 53–p. 506, v. 11; p. 515, vv. 45–49; p. 516, vv. 14–22;

matter of counsel and deliberation.[54] Accordingly, Albert locates *liberum arbitrium* as prior to choice (*electio*), the proper act of *liberum arbitrium*, and as after the intellect's counsel and deliberation.[55] *Liberum arbitrium* is a power distinct from, and dependent upon, both the intellect and the will (*voluntas-[thelēsis]* as a power).[56] The power of the will (*voluntas-[thelēsis]*), for Albert, has a freedom

---

p. 517, vv. 30–33; p. 518, vv. 50–53 (cf. p. 296, vv. 20–23; p. 488, vv. 20–26; p. 505, vv. 47–49; p. 512, vv. 6–24). For the *Sentences* commentary, see: 'Et hoc patet ex Damasceno qui dicit, quod primum inquirit, deinde disponit, tertio ordinat, deinde dijudicat et sententiat, deinde eligit, deinde vult, deinde impetum facit ad opus'; Albertus Magnus, *Commentarii in II Sententiarum*, 24.5.1, ad 3, ed. by Borgnet, p. 402b (in addition to ibid., 25.1 ob 3 and ad 3, pp. 423a and 424ab; Albert, *Commentarii in III Sententiarum*, 35.8 ob 1, ed. by Borgnet, p. 652b). For *Summa theologiae* II, see Albertus Magnus, *Summa theologiae*, II.93.2, ed. by Borgnet, p. 203a; ibid., II.97.1c, ad 1 and qa 1c, p. 222ab, esp. ad 1: 'Unde ex ratione est, quod [liberum arbitrium] arbitratur, quod judicat, quod disponit, quod sententiat, quod eligit: ex voluntate enim est, quod vult, et appetit, et impetum facit, et agit, et utitur'; ibid., II.99.1 ob 2–4 and 7, ad 2–5 and 7, pp. 232a–234b (cf. ibid., II.91.2 ob 4, p. 185b; ibid., II.93.1 ob 2, p. 200a; ibid., II.16.2 ob 2, p. 211a). See also Albertus Magnus, *Ethica*, VI.1.6, ed. by Borgnet, pp. 403b–404a.

54 Albertus Magnus, *De homine*, ed. by Anzulewicz and Söder, p. 489, vv. 13–42 (see also p. 488, vv. 20–48; p. 490, vv. 63–68; p. 491, vv. 14–43), ascribes to John also a third will, called the 'rational' or 'deliberative' will, thereby earning praise from Lottin for separating out the deliberative role (Lottin, 'La psychologie de l'acte humain', p. 404): 'Albert contredisait tous ses contemporains qui identifiaient la boulesis à la volonté délibérée'. By contrast, in *Summa theologiae*, I.79.1.1 prol., ed. by Borgnet, p. 838a, and *Summa theologiae*, II.99.1, ad 2 and 7, ed. by Borgnet, pp. 233a–234b, Albert collapses *boulēsis* and the 'rational' or 'consultative' or 'deliberative' desire into one, although he makes clear in the aforementioned ad 2, p. 233b, that the will per se regards the end, and that it is called *boulēsis* only because it regards deliberable as opposed to naturally willed goods. Cf. note 50 above.

55 In the *De homine*, ed. by Anzulewicz and Söder, p. 513, vv. 39–44 (sol.), Albert appeals to the following phrase of John of Damascus to link *liberum arbitrium* with choice (*electio*): 'electio autem est duobus praeiacentibus eligere, et optare hoc prae altero'; John of Damascus, *De fide orthodoxa*, XXXVI (III.22), par. 11, ed. by Buytaert, p. 137. See also *De homine*, ed. by Anzulewicz and Söder, p. 505, vv. 10–18 (prol.); p. 506, vv. 3–5 (sol.); p. 512, vv. 19–24 (ob 27); p. 517, vv. 42–48 (ob 3). For McCluskey ('Worthy Constraints', pp. 507, 513–14, 531–33), whereas the *De homine* regards *liberum arbitrium* as having both an intellectual and voluntative role, subsequent works take it as purely voluntary. Cf. Lottin, 'Libre arbitre et liberté', p. 126 (referring to Albert, *Commentarii in II Sententiarum*, 24.5.1c, ed. by Borgnet, p. 402a, quoted below at note 64): 'Dans la *Summa de homine*, [...] l'arbitrage est l'acte essentiel du libre arbitre. Dans le "Commentaire" cette explication a disparu et fait place à l'explication devenue classique depuis Philippe'. The assessments of the results vary. Whereas McCluskey draws out the advantages of the early view, Lottin praises the development ('Libre arbitre et liberté', p. 126): 'Aussi bien, la liberté du choix n'est-elle plus présentée comme supérieure à celle de la volonté. La volonté est libre, écrit-il maintenant, parce qu'elle peut ne pas suivre le décret de la raison; et la liberté du libre arbitre n'est autre que cette liberté de la volonté'. By contrast, Michaud-Quantin (*La psychologie de l'activité*, pp. 209–10 and 221–22) takes the radical rupture in Albert's thought to be his ascription of *causa sui* to *liberum arbitrium* rather than to the will in the *Summa theologiae* (quoted in note 28 above).

56 Early Albert speaks of acts of *voluntas* that precede *liberum arbitrium*: Albertus Magnus, *De homine*, ed. by Anzulewicz and Söder, p. 489, vv. 13–42 (sol., ad 1–3) and p. 491, vv. 14–43 (sol.; *bulisis* and *thelisis*); p. 505, vv. 56–62 (sol.; the *voluntas* is only of an end); p. 513, vv. 55–56; p. 514, vv. 3–21 (sol.). Many, after Lottin, see Albert as rejecting this view by omitting a prior act of the will in the account of *Commentarii in II Sententiarum*, 24.5.1c, ed. by Borgnet, p. 401b–402a (quoted at note 64 below); Lottin, 'Libre arbitre et liberté', pp. 125–26, followed by Malik, 'Albert der Große

that belongs to it by nature. So, also, the *liberum arbitrium* has a freedom (*libertas arbitrii; libertas electionis*) that derives from the will,[57] and ultimately from the rational soul as flowing from God.[58] In the passages just cited, the key notion for understanding the freedom of the will that is prior to *liberum arbitrium* is *causa sui*. The will, says Albert, is the active and motive part of the soul by which it can freely do what it wants and by which it is made 'master of its own actions'.[59] For this reason, he concludes, will is found maximally in God, next in angels, and finally also in humans.

It would be a mistake to conclude that there are two kinds of human freedom for Albert, one with and another without alternatives. Instead, we need to appreciate how strongly voluntarist Albert's account of the will is,[60] as becomes especially clear after 1246 (which is not to deny the crucial role of cognition in all appetite[61]). The key note of this voluntarism is the claim that the will can always

---

und das Willensproblem', p. 394; see also Michaud-Quantin, *La psychologie de l'activité*, p. 208. As a result, Albert is taken to transition from a four-step to a three-step model of human action; Drouin, 'Le libre arbitre', pp. 108–09, 111–12, 117–20; McCluskey, 'Worthy Constraints', pp. 506–07 and 513–14. Nevertheless, Albert's omission could be incidental, since prior acts of will seem to be quite in keeping with his adoption of *thelēsis* and *boulēsis* among the stages of the human. Cf. also Albertus Magnus, *Ethica*, VI.1.6, ed. by Borgnet, p. 404a; *Summa theologiae*, II.97.1, ad 1, ed. by Borgnet, p. 222a, quoted in note 53 above. In short, rather than focus on a few key passages, I see continuity on this point throughout Albert's works based on his adoption of John of Damascus's model.

57 Albertus Magnus, *Commentarii in II Sententiarum*, 24.5.1c, ed. by Borgnet, pp. 401b–402a (quoted below at note 64).

58 Albertus Magnus, *Summa theologiae*, I.79.1.1c, ed. by Borgnet, pp. 839b–840a.

59 See also Albertus Magnus, *Commentarii in II Sententiarum*, 25.1.3c, ed. by Borgnet, p. 424a (quoted above at note 44).

60 For helpful analysis of definitions of voluntarism, intellectualism, and their implications, as well as how early Albert is more intellectualist than Philip the Chancellor, see McCluskey, 'Human Action and Human Freedom', pp. 5–8, 26–39, 72, 154–57, 217–30, 257–58, 262. On Philip, see McCluskey, 'Roots of Ethical Voluntarism'. For Albert's voluntarism, see also Siedler, *Intellektualismus und Voluntarismus*, p. 57, in addition to the scholarship cited in note 55 above.

61 See esp. Albertus Magnus, *Ethica*, III.1.13, ed. by Borgnet, pp. 210b–211a (emphasis added): 'Principium autem secundum veram rationem principii accipitur, scilicet quod sic principium est, quod a nullo principiatum, et ex seipso principium, sicut est voluntas non coacta vi, nec metu territa: et hoc principium in ipso non localiter tantum, sed secundum liberae facultatis potestatem, sicut dicimus in nobis esse, quorum nos sumus *causa ex nobis ipsis*, et quae libere possumus facere et dimittere. *Quia autem appetitus non videt sicut oculus, ideo duce indiget: nec movetur nisi ductu demonstrati voliti: et ideo participans qualiter rationem, demonstrationem accipit voliti secundum quod volitum sub circumstantiis quae ipsum volitum efficiunt demonstratum*: et sic oportet quod voluntarium sciat singularia'. Compare also ibid., III.1.1, p. 196a: 'Liber autem est, ut dicit Philosophus, qui causa sui est [...]. Hoc autem modo nullum animalium brutorum suum est, eo quod omnia opera eorum acta sunt a natura. Hoc etiam modo *intellectus et ratio contemplativa non sua sunt* in veritate contemplatorum: *obligatione enim syllogistica coguntur ad assensum. Quod vero primum causa libertatis est, voluntas est*: eo quod hoc et a naturae obligatione liberum est, et ab obligatione argumentorum voluntarium. Ergo esse hominem et liberum facit hominem, et omnem operationem ejus facit liberam, ita quod *nulla causa suarum operationum est, nisi quia sibi voluntariae sunt*'; see also Schönberger, 'Rationale Spontaneität', pp. 230–31.

reject what the intellect judges.[62] Albert introduces this note in explaining the will as a *causa sui*:

> Et ut hoc intelligatur, sciendum quod sicut dicit Philosophus, 'Liber est qui / 402A/ est causa sui, et non causa alterius: et quod est obligatum alteri, non est liberum'. Unde potentiae affixae organis sunt obligatae illis, non potentes in actum ultra naturam receptibilitatis organi [...] Altera obligatio alteri est ab objecto, quando convincit potentiam de consensu in ipsum, sicut omnes potentiae apprehensivae animae rationalis, ut intellectus, et ratio, et hujusmodi, convincuntur rationibus ad consentiendum in verum: et ideo non sunt causa sui, sed alterius. Sed voluntas neutro modo est obligata: quia non est affixa potentia organo, nec etiam necessario consentit: quia ratione dictante hoc esse faciendum vel non, adhuc se habet ad quod voluerit,[63] et potest contrarium velle, quam quod ex ratione dijudicatum est: et ideo voluntas libera est, et electio liberi arbitrii ab illa parte habet eamdem libertatem.[64]

---

62 It is easy to exaggerate the novelty of Albert's strong phraseology in his *Sentences* commentary and subsequent works (see note 55 above). *Pace* Lottin, this key note appears to be entailed already in *De homine* by the following. (1) 'Free *arbitrium* does not follow the judgment of reason of necessity'; Albertus Magnus, *De homine*, ed. by Anzulewicz and Söder, p. 516, vv. 28–29 (a. 2, ad 28, summarizing the sol.); so also p. 523, vv. 7–10. But (2), again *pace* Lottin, the will, given its properties, is already said to be the first ground of the freedom of free *arbitrium*; ibid. p. 522, vv. 1–4 (emphasis added): 'Dicendum quod libertas primo est in voluntate, ut dicunt Sancti. Cum enim liberum arbitrium sit potentia consequens ad rationem et voluntatem, accipit ab ipsis *ea quae habet in seipso*'. (3) It follows that will itself does not of necessity follow the judgment of reason, as is stated at ibid., p. 492, vv. 17–18; cf. p. 508, vv. 14–16, quoted in note 40 above. Furthermore, the first natural freedom is said to be a *causa sui*; ibid., p. 523, vv. 48–50 (ad 5). *Pace* Michaud-Quantin, the same claim is also already made of *liberum arbitrium*. *Free* in *free arbitrium*, says Albert, 'conveys that *causa sui* is in what follows reason and will'; ibid., p. 522, vv. 17–20 (ad 3). Finally, *pace* McCluskey, Albert explains in the same place that *arbitrium* in *liberum arbitrium* 'does not express the *judgment* of reason [...], but *what is inclinative*, of its own accord, in relation to the decree of reason *or* to the appetite of the will' (arbitrium non dicit iudicium rationis, [...] sed spontaneum inclinativum ad decretum rationis vel appetitum voluntatis); ibid., p. 522, vv. 17–20 (ad 3), based on p. 513, v. 55–p. 514, v. 9 (a. 2 sol.).

63 Albert in the *Summa theologiae* appears to have in mind here the purported definition of Peter the Apostle, in Pseudo-Clement's *Itinerarium*: 'arbitrii potestas est sensus animae habens virtutem, qua *se possit ad quos velit* actus inclinare'; see Albertus Magnus, *De homine*, ed. by Anzulewicz and Söder, p. 512, vv. 43–47, 72 and p. 516, vv. 30–33.

64 Albertus Magnus, *Commentarii in II Sententiarum*, 25.5.1c, ed. by Borgnet, pp. 401b–402a. For the immediately preceding context, see note 43 above. Elsewhere, Albert clarifies that, although *causa sui* refers to an efficient cause in the case of the will, the notion is compatible with the will's having a final cause: 'dicendum, quod quando dicit Anselmus, quod voluntas nullam habuit causam, intelligit hoc de voluntate comparata ad volentem, non ad volitum: comparata enim ad volentem, sibi causa est in omnibus, quia principium actionis suae est in ipsa. Sed comparata ad volitum, oportet quod habeat causam id quod faciat nuntium de volendo: et illud potest esse rectum, et non rectum, ut dictum est'; *Summa theologiae*, II.21.3, ad sc 2, ed. by Borgnet, p. 264b; see also ibid., I.79.2.1, part. 1, ed. by Borgnet, p. 844b.

> And, so that this may be understood, it should be known that, as the Philosopher says, 'The free person is for the sake of himself or herself, and not for the sake of another'; and what is bound to another is not free. Hence, powers affixed to organs are bound to them, not capable of an act beyond the nature of the organ's receptibility [...]. A second *binding to another* is from the object: when it convinces the power as to [its] consent to it; just as all of the apprehensive powers of the rational soul, such as intellect, reason, etc., are convinced by reasoning to consent to what is true. Therefore, they are not for their own sake, but for the sake of another. But in neither [of these two ways] is the will bound. For, it is not a power affixed to an organ, and also it does not consent of necessity. For, with reason stating that this should or should not be done, [the will] still has [the same] relation to what [it] has willed, and it can will the contrary of what has been judged by reason. For this reason, the will is free, and the choice of free *arbitrium* has, on that [same ground], the same liberty.

We mistake Albert's thought if, after the manner of Aquinas or Scotus, we start by assuming that the first acts of the will are necessary, then associate freedom with posterior acts that involve alternatives. Under this assumption, Albert's explanation of divine freedom in his *De causis* looks thin when it employs, apparently, nothing more than the indeterminism of the First as *causa sui*. But now we see the degree of voluntarism that is, for Albert, involved in freedom at its deepest root: in the power of will prior to the power of *liberum arbitrium*. As a *causa sui*, the will is able to accept or reject what the intellect proposes. Far from requiring explicit alternatives to account for freedom, Albert's account of *causa sui* implies that the will is its own motor and that alternatives to what reason presents are already built into it.

*God's Freedom also Admits of Alternatives, though Albert's Voluntarism Makes Their Expression Seem Unnecessary*. We have just seen Albert's inference that freedom exists maximally in God, then in angels and humans. In fact, in the same place, Albert insists that there is a sense of freedom that belongs to each, not equivocally, but *per prius et posterius*, by analogy: to God first, who is the cause and master of his actions; because God's substance and will differ only 'according to the mode of signifying.'[65]

What are the notes of freedom that God possesses? First, it is already obvious, in the same text and from our discussion above, that Albert has in mind God's will as *causa sui*. Second, in at two places, he affirms alternatives in God's will, just as in humans':

> [L]iberum ex natura non obligatur nec ex habitu ad hoc vel ad illud: et haec est libertas inseparabilis /430A/ in Deo, et Angelo, et homine: nihil enim

65 Albertus Magnus, *Summa theologiae*, I.79.12c, ad 4, ed. by Borgnet, p. 841a.

agunt quin possint illud non agere, et quin possint aliud agere: et hoc notatum est supra, et probatum est in III Sententiarum.[66]

> What is free is not bound, whether by nature or by habit, to this or that. This is the liberty that is inseparable in God, angel, and human. For, they do nothing without being able not to do that, and without being able to do [something] different. This was noted above and was proved in *Sentences* III.

Finally, for Albert, God has both *thelēsis* and *boulēsis* — the central point in Albert's only express question: In God, what is the reality of the will (*quid voluntas in Deo secundum rem*)?

> [E]rgo dicendum, quod utroque modo voluntas est in Deo. *Thelēsis* enim est in Deo: quia vult se et ad divinitatem pertinentia, scilicet quod omnipotens et simplex sit et hujusmodi: *Boulēsis* autem, quia vult etiam ea quae ex consilio, hoc est, ex definitione aeterna determinavit, sicut dicitur, Isaiae, xlvi, 10: *Consilium meum stabit, et omnis voluntas mea fiet.* Dictum est enim in praehabitis *de providentia*, qualiter consilium cadit in Deum.[67]

> It should therefore be said that will is in God in both ways. For, *thelēsis* is in God because he wills himself and what pertains to divinity, namely, that he is omnipotent, simple, etc. But *boulēsis* [is in God] because he wills also what [comes] out of [his] counsel, namely, [what] he has determined out of eternal decree [*definitio*]. Thus, it is said (Isaiah 16:10): 'My council will stand, and my entire will shall be done'. In fact, in previous treatments on providence [q. 67], it has been stated how counsel befalls God.

Albert goes on to spell out God's different will acts (ad 3). In some cases, the object of his will is God himself or 'what is eternally in God': 'to understand, to be wise, to exist, to know', etc. In other cases, 'God wills that there exist created goods, or the goods of the saints, or some other things exterior [to him]'.

We see what a robust understanding Albert has of the divine will and divine freedom. This raises objections, of course, which Albert treats in the aforementioned question. Does not this broad notion of will, analogous to the human will, compromise divine simplicity?

*To Act per essentiam is To Act per libertatem: Divine Simplicity and Precontainment of Conceptual Distinctions.* Perhaps Albert's most important discussion of divine freedom arises in answer to his question, when commenting on Pseudo-Dionysius's

66 Albertus Magnus, *Commentarii in II Sententiarum*, 2.25.4c, ed. by Borgnet, pp. 429b–430a. The second text is from *Summa theologiae*, I.63.3.2, ad 3, p. 651b: 'non sequitur, quod posita causa ponatur effectus. Hoc enim tripliciter potest impediri, scilicet [1] si causa libera sit et dominus sui actus: tunc enim quamvis immobilis sit, et immutabilis, et aeterna, potest agere et non agere pro libertate sua. Et talis causa est propositum divinae voluntatis'.

67 Albertus Magnus, *Summa theologiae*, I, 79.1.1 ad 1, ed. by Borgnet, p. 840a.

*De divinis nominibus*, chapter 4: 'Does God act through his essence?'[68] Albert constantly answers Yes, and understands Dionysius as affirming the same there. The objections point out that what acts through its essence apparently 1) acts by the necessity of nature, 2) acts eternally, and 3) causes only one effect, since God's essence is one simple being (*esse*) (p. 117, vv. 36–60). It seems to follow for a believer, then, that God acts through will and not through his essence. Albert comes to Dionysius's defence, appealing expressly to the necessary demonstrations found in Avicenna (vv. 71–73). If God's will were really distinct from his essence, God would be a composite and would not be the first being (p. 118, vv. 1–4); instead, part of God would be the First. Also, a supervenient will would be an act perfecting a potency, and so would be more divine than God (vv. 5–8).[69] Therefore, God acts through his essence just as a light source illumines, says Avicenna (vv. 9–12). Yet God's essence is both perfect and simple (vv. 14–20). As perfect, anything that has nobility, such as life, wisdom, will, and so on, is in the divine essence 'according to the truth of these *rationes*'; whereas insofar as the divine essence has the highest simplicity, each of these must be the same as the other in reality (*secundum rem*).[70] Now, as the Philosopher concludes in *Metaphysics* Lambda, God's substance *is* his being, action, and intellection. Therefore, in acting through his essence, God is also acting through intellect and will.[71] His acting through the order of his wisdom and through the freedom of his will are the same as acting through his essence — though not, clarifies Albert, according to the necessity of his essence, namely, according as that necessity entails being constrained to act (vv. 20–30).

But how, then, does one explain the multiplicity of God's effects? Albert answers that the diverse processions are all reduced to a principle that is one 'in

68 Albertus Magnus, *Super Dionysium De divinis nominibus*, IV.8, ed. by Simon, p. 117, vv. 22–23. The phrases *per ipsum esse* and *per ipsam essentiam* in Pseudo-Dionysius Latinus, trans. by Sarracenus, translate αὐτῷ τῷ εἶναι and αὐτῇ τῇ ὑπάρξει in Pseudo-Dionysius, *De divinis nominibus* 4.1 (693B), ed. by Suchla.

69 Cf. Avicenna, *Metaphysics of 'The Healing'*, IX.1, trans. by Marmura, pp. 302–04. The same teaching, of a first agent that emanates necessarily by will (and not as the Sun necessarily illumines or casts a shadow) is ascribed to Aristotle by Maimonides, as Albert could read: Maimonides, *Dux seu Director dubitantium aut perplexorum*, II.21 (20), ed. by Giustiniani, p. 52v, v. 52–p. 53r, v. 6.

70 Aquinas, who as a student assistant worked with Albert on the Dionysian paraphrases, could find in this passage the outline of his *Summa theologiae*, I.3–4 as a framework for discussing the divine names.

71 Similarly, to the question in Albertus Magnus, *De IV coaequaevis*, q. 1, a. 5c, ed. by Borgnet, p. 314a, whether creation is a work of nature or of will, Albert had answered: both. Insofar as will is divided against nature, it is a work of will; but insofar as God's will is identical in reality with his nature, it is also an act of nature (I am indebted to Henryk Anzulewicz for this reminder). In another parallel passage, Albert speaks, perhaps after the manner of Avicenna's 'necessary being through another', of an effect's occurring necessarily because of the caused necessity in the effect rather than in the cause. Albert, *Super Dionysium De divinis nominibus*, I.41, ed. by Simon, p. 23, vv. 51–60. Janssens, in 'Creation and Emanation in Ibn Sīnā', pp. 457–59, brings out the fact that Avicenna departs from the *Theology of Aristotle*, for which emanation is from the divine being alone, not through intellect and will.

reality', but multiple 'in the *rationes*' or notions that belong to its attributes (ad 1, vv. 35–39). Similarly, the multiple radii of a circle all derive from one and the same centre — which one, simple centre point has a distinct conceptual relation to each point on the circumference.

Still, will the effects of the divine essence not be eternal? God's internal action or *velle* is eternal, acknowledges Albert. If effects are not eternal, this is because of their own deficiency or because of the order of divine wisdom, which precontains all effects as conceptually distinct within it.[72] So God has an eternal will, but multiple effects can emerge in time.

The principal discussion of divine freedom in the *Summa theologiae* raises an objection that can help us here. If the predicate *freely creates the angel Gabriel* (my example) is said of God's essence, will not that action be necessary?[73] In response, Albert distinguishes two ways in which predicates name the divine essence: absolutely, as in *God wills to understand*; and together with the connotation of an effect. Predicates in the second case name the divine essence but connote creatures subject to the divine will. As absolute predicates name the same reality but differ by mode of signifying, so diverse modes of signifying prevent one from inferring from *wills all good things* that God *is* all good things.

Albert's adoption of a cause that acts through its essence, appropriating for himself language that in Proclus and his heritage is necessitarian, is a good example of how far he is willing to go in drawing the philosophers' insights into the tent of his faith-based wisdom — inspired by the example of Dionysius.[74] What I show next is that he goes much farther than has been imagined. So far, I have underlined Albert's notion of an eternal divine will that eternally and freely says 'yes, no, or later' to a plurality of creatures precontained within and conceptually distinct from the divine knowledge. With this robust conception of divine freedom in the background, Albert goes on to articulate, with Avicenna, a law-like emanation or derivation of these creatures from the One.

## Emanation or Derivation as a Law-like System in Albert

### *'Creating' ex nihilo versus 'Informing' or 'Inflowing' Form*

Albert's doctrine of emanation or derivation must be seen against the background of his account of creation *ex nihilo*. The account is quite Avicennian, but the language is also drawn from Boethius's *quod est / esse* distinction. Let us review it

72 Albertus Magnus, *Super Dionysium De divinis nominibus*, IV.9, ad 6, ed. by Simon, p. 118, vv. 64–74; see also IV.70, p. 180, vv. 19–37; Albertus Magnus, *Physica*, VIII.1.4, ed. by Hossfeld, p. 557, vv. 32–48.

73 Albertus Magnus, *Summa theologiae*, I.79.1.1, ob 3 and ad 3, ed. by Borgnet, pp. 838b and 840b.

74 For the necessitarian background in Porphyry and Proclus of 'causes that act through their being or essence', see Chase, 'Discussions on the Eternity of the World'.

briefly. For Albert, 'being simply taken' (*esse simplex* or *simpliciter*) is the simplest conception that any intellect forms of a thing, signifying a thing's first formal feature, its 'being'.[75] But *esse* is an abstract term signifying a form, and not yet a 'thing that is', *id quod est*. Tract 1 of the *De causis* mounts a proof of the existence of a first cause in which *quod est* and *esse*, a thing and its being, are identical in reality. All other things have an *esse* that is distinct from 'that which they are'.[76] *Esse* as a simple form can be possessed by more than one instance, and it is multiplied by being received by *id quod est*.[77] Since such things do not have *esse* of themselves (as already in their *id quod est* or as caused by their own *quod est*), no such thing has *esse* unless it receives it from something that has it of itself, the necessary being. Thus, all *esse* proceeds from God and from God alone.

Albert goes further, however, and concludes that God *immediately* causes *only esse*.

Two sentences from one passage in the *De causis* paraphrase explain what Albert has in mind:

> 1. 'Since each of the things that are subsequent [to *esse*] presupposes in its concept what precedes it, each is produced, not *ex nihilo*, but out of something [*ex aliquo*], in which there is something inchoate of its *esse*'.
> 2. 'Therefore, nothing subsequent [to *esse*] comes to be through creation'.[78]

---

75 In early Albert, *esse* signifies essence, but in late Albert it can also signify existence; see esp. Ducharme, '*Esse* chez saint Albert le Grand'; Vargas, 'Albert on Being and Beings'. For *esse simplex*, see Albertus Magnus, *De causis*, II.1.17, ed. by Fauser, p. 81, vv. 19–27, quoted in note 78 below.

76 The teaching is apparently first found already in Albertus Magnus, *De IV coaequaevis*, q. 21, a. 1, ad sc 1–2, ed. by Borgnet, p. 465a, where Albert blends Boethius and the *Liber de causis*. See also especially Albertus Magnus, *Super I Sententiarum*, 2.1 [q. 2 ad 3], ed. by Burger, p. 46, vv. 38–42; 2.4 [q. 3c], p. 55, vv. 56–67.

77 An early version of this doctrine is found in Albertus Magnus, *Super Dionysium De divinis nominibus*, V.32 sol., ed. by Simon, p. 322, vv. 24–30. See also Albertus Magnus, *De IV coaequaevis*, q. 21, a. 1c and ad 3, ed. by Borgnet, pp. 463b–464a; *Super I Sententiarum*, 2.4 [q. 3 ad sc 4], ed. by Burger, p. 56, vv. 24–66.

78 Albertus Magnus, *De causis*, II.1.17, ed. by Fauser, p. 81, vv. 38–42. The entire passage is important, within which I underscore the translated text. Ibid., vv. 19–47 (on *Liber de causis*, IV, ed. by Fauser, p. 88, vv. 63–68; lemmata italicized, as identified in the Cologne edition): '*Esse* enim *simplex* mentis conceptus *est* ad nihil formatus vel determinatus, quo quaelibet res esse dicitur, cum de ipsa quaeritur per quaestionem, an sit. Propter quod in superioribus libri praecedentis ostensum est, quod quaestio, an est, non nisi per causam primam determinari potest. *Esse* enim, quod dicto modo *simplex* conceptus *est* et informis et in quo sicut in ultimo stat resolutio, non nisi causae primae creatum esse potest. Hoc enim non educitur ex aliquo, in quo formalis incohatio sit ipsius, sicut vivere educitur ex esse et sentire ex vivere et rationale ex sensibili. Omne enim rationale sensibile est, et omne sensibile vivens, et omne vivens ens, sed non convertitur. Et quia esse virtutem suam influit super omnia sequentia, propter hoc sicut esse actus est entium, ita "vivere viventibus est esse, et sentire est esse sentientibus, et ratiocinari est esse rationalibus", ut dicit Aristoteles. Et hanc virtutem, quod scilicet quodlibet istorum sit esse eorum quorum est, sequentia non possunt habere nisi a primo, quod est esse. Quodlibet enim sequentium cum supponat in intellectu suo praecedens se, non ex nihilo, sed ex aliquo producitur, in quo est incohatio sui esse. Nihil ergo sequentium potest fieri per creationem. Sequens enim se habet ad praecedens ut informans ipsum et determinans. Productio igitur istorum

Of course, Albert's point is not that nothing but mere *esse* is caused by God. Everything that is and that is other than God is caused to be by God, but only *esse* is caused *ex nihilo*.[79] All other forms presuppose *esse* and add to *esse*. All other forms, then, are mediately caused. Albert reserves one term for this kind of post-creative causation: *informatio* or 'informing'. Secondary causes are responsible for the flow of form, for *informatio*, but not for the emanation of being (*esse*). In the *Super De divinis nominibus*, Albert puts the whole matter succinctly as follows:

> Dicendum, quod esse simpliciter secundum naturam et rationem est prius omnibus aliis; est enim prima conceptio intellectus et in quo intellectus resolvens ultimo stat. Ipsum etiam solum per creationem producitur non praesupposito alio, omnia autem alia per informationem, scilicet supra ens praeexistens, ut dicit Commentator in *Libro de causis*. Illud autem est primum procedens ab alio, quod non procedit supposito quodam; et ita relinquetur, quod inter omnes processiones divinas esse sit primum.[80]
>
> > Being [*esse*] taken simply is prior to all other things in nature and in notion. For, it is the first conception of the intellect, and in it the intellect comes ultimately to a standstill in [the operation of] resolution [i.e., in analysis into principles]. Also, [*esse*] alone is produced through creation without anything else presupposed, whereas all other things [are produced] by 'being informed' [*per informationem*], namely, as upon a

---

non per creationem, sed per informationem est. Relinquitur igitur, quod *esse sit primum* et *creatum et* quod *alia* causata *non creata sint* et quod nullum causatorum *prius* esse possit quam esse'. For what is apparently Albert's first use of 'the first, simple concept' to explain creation using the *Liber de causis*, see Albertus Magnus, *Super I Sententiarum*, 2.4 [q. 3c and ad 1–2], ed. by Burger, p. 55, v. 56–p. 56, v. 15.

79 For the affirmation in Albert's *De causis* that the first in all things, *ens*, is created *ex nihilo*, see below at note 98; and in the early works, at note 95. Hence, I cannot agree with the conclusion that in this work, causing the *esse* of all things sounds like, but is not, creating *ex nihilo*; or that for the Philosopher, the First Cause does not cause everything that has *esse*, in Albert's view; see Baldner, 'Albertus Magnus on Creation', pp. 67–68. Nonetheless, there is a problem as to whether Albert proves, or thinks he proves, that *id quod est* is caused *ex nihilo*. It seems that Albert could and should argue that *id quod est* is included in *ens*, although it is other than *esse*, but then how is only one thing immediately caused, namely, *esse*? Nonetheless, it may be the case that for Albert here, the philosophers prove that *esse* is caused (alone) *ex nihilo*, that all things that have *esse* are, as such, caused *ex nihilo*, but that *id quod est* (alone) is presupposed and is not caused *ex nihilo*. If so, then Baldner (p. 71) has reason to interpret the *De causis* in light of the later *Summa theologiae*, I.53, to the effect that the philosophers were unable to know, and therefore never affirmed, that something comes to be out of 'pure nothing' (*ex puro nihilo*). I cannot take up this question here.

80 Albertus Magnus, *Super Dionysium De divinis nominibus*, V.20 sol., ed. by Simon, p. 314, vv. 4–13. Albert sees the doctrine that everything besides *esse* results from *informatio* in *Liber de causis* IV; see Albertus Magnus, *De causis*, II.1.23 (lemma from *Liber de causis*, IV, ed. by Fauser, p. 88, v. 65), ed. by Fauser, p. 87, vv. 60–63: 'Patet etiam, quod *non est* aliquod *post primam causam latius* ipso *nec* aliquod *creatum prius ipso*, quia omnia sequentia ipsum formando, non creando, in esse deducta sunt'. I have argued that a similar creationist transformation of the thought of the *Liber de causis* is found in al-Kindī, *On the True, First, Perfect Agent*; Twetten, 'Aristotelian Cosmology and Causality', pp. 354–57.

> preexisting being, as the Commentator in the *Liber de causis* says. But that which proceeds without anything presupposed is the first thing that proceeds from another. And so, it remains that *esse* is the first among all of the divine processions.

Before we turn to this flow of form, notice that Albert's 'derivationism' fits the principle *ab uno non nisi unum*: 'from one comes only one'. Albert finds the principle in Aristotle's *De generatione et corruptione*; when properly understood, as speaking of the 'order of nature' and not of time, 'all philosophers preceding us have supposed [it]', he says, and only one person has ever denied it: Ibn Gabirol (from whom Albert distances himself sharply).[81] Some theologians have denied the principle by misunderstanding it, explains Albert, but it can be clarified through Dionysius.[82] Albert's fullest expression of the principle states: 'from one agent [acting] in respect to [its] *form* [*ab uno agente secundum formam*], which remains in the same condition, there is *immediately* only one thing in respect to *[that] form*'.[83] The added qualification 'in respect to form and remaining in the same condition' takes advantage of the claim that many conceptually distinct intelligibilities are contained in the divine, some of which are prior to others. Notice what follows: from the one First agent acting in respect to its first form as such comes, according to the order of nature, only one: the 'first form', being.[84]

---

81 Albertus Magnus, *De causis*, I.4.8, ed. by Fauser, p. 55, vv. 72–80, esp. vv. 72–76: 'Supponentes autem propositionem, quam omnes ante nos philosophi supposuerunt, scilicet quod ab uno simplici immediate non est nisi unum secundum naturae ordinem'. See also Albertus Magnus, *Physica*, VIII.1.13, ed. by Hossfeld, p. 576, vv. 44–48; and Gerard of Cremona's translation in Aristoteles Latinus, *De generatione et corruptione*, II.10, 336a27, ed. by Judycka, p. 74, vv. 16–17: 'idem enim et similiter habens semper idem innatum est facere', which is a rather literal rendering of the Greek: 'τὸ γὰρ αὐτὸ καὶ ὡσαύτως ἔχον ἀεὶ τὸ αὐτὸ πέφυκε ποιεῖν'.

82 Albertus Magnus, *De causis*, I.1.10, ed. by Fauser, p. 22, vv. 1–16.

83 Ibid., II.2.7 (on *Liber de causis*, VI [VII]), ed. by Fauser, p. 100, vv. 50–52: 'Hoc tamen constat, quod ab uno agente secundum formam et eodem modo se habente immediate non est nisi unum secundum formam'. See also the objector's formulation in Albertus Magnus, *Super Dionysium De divinis nominibus*, II.42 ob 2, ed. by Simon, p. 71, vv. 26–29 (emphasis added): 'omnia quae habent aeque immediatam processionem ab uno *secundum formam*, sunt unum *secundum formam*, quia idem *eodem modo* non producit nisi unum'. The phrase in the first quotation 'eodem modo se habente' reflects Aristoteles Latinus, quoted in note 81. But the reduplicative expression is found also in Avicenna Latinus, *Liber de philosophia prima*, IX.4, ed. by Van Riet, p. 481, vv. 50–51: 'ex uno, secundum quod est unum [*min ḥaytha hūa wāhid*], non est nisi unum'. Notice that the *ex uno unum* principle, with this qualification, can be made consistent with another principle that Albert accepts. In rational and volitional agents, as for Maimonides, *Guide*, II.22, many can come from one since such agents can consider many *rationes*; see Albertus Magnus, *Super Dionysium De caelesti hierarchia*, I, q. 3c, ed. by Simon, p. 293; *Summa theologiae*, II.3.3.1.1 ad 1, ed. by Borgnet, p. 26; also Siedler and Simon, 'Prolegomena', pp. xiv–xv.

84 For Albert, it is absurd to say that the *ab uno unum* principle applies only to what acts through essence, not to what acts through will, according to the argument at Albertus Magnus, *De causis*, II.4.14 (on *Liber de causis*, XXII [XXIII]), ed. by Fauser, p. 167, vv. 20–29: 'nec potest intelligi, quod ab omnimode uno per se sint diversa aequali processione. Si enim dicatur, quod hoc est verum in per essentiam agentibus et non in his quae agunt per voluntatem, hoc absurdum est. In primo enim

*Background on This Twofold Causality: Informing vs Creating.* If one looks back over Albert's works, one can see that he discovers the aforementioned doctrine of 'creation of *esse* alone' vs 'information' in two passages from 'Aristotle's' *Liber de causis*,[85] where one reads, first:

1. 'The first of created things is *esse* [being], and nothing else is created before it' ('Prima rerum creatarum est esse et non est ante ipsum creatum aliud'; *Liber de causis*, IV, p. 88, v. 63).

Albert cites this opening proposition of *Liber de causis* IV with approval in his *De quattuor coaequaevis*, but it is explained and adopted into his reasoning in Book I of his *Sentences* commentary.[86] The second passage is from the 'commentary' in *Liber de causis* XVII (XVIII):[87]

---

idem est voluntas quod essentia. Et sicut primum invariabile est secundum essentiam, ita invariabile est secundum voluntatem. Sicut ergo adhuc sequitur, quod ab uno non sit nisi unum, sic a voluntate, quae nullo modo diversificatur secundum volita, non est nisi unum'. Without diversification in the will, even it causes only one.

85 The earliest explicit ascription of the work to 'Aristoteles' is in Albertus Magnus, *De resurrectione*, IV.2.2 ob 4 and ad sc 5, ed. by Kübel, p. 341, vv. 38–41; p. 342, vv. 50–52 (quoting *Liber de causis*, XVII [XVIII]). The work is ascribed to 'the Philosopher' at Albertus Magnus, *De natura boni*, II.3.2.2, ed. Filthaut, p. 76.16–18; *De sacramentis*, I.5 sc 3, ed. by Ohlmeyer, p. 12, vv. 55–57, and V.1.3.2, ad 6, p. 60, vv. 27–31; in addition to *De resurrectione*, II.2, ad 5, ed. by Kübel, p. 260, vv. 9–10. In the last work attributed to Albert, Book I refers to the 'Philosopher's' *Liber de causis*, whereas three times the work is ascribed to 'Aristoteles' in Book II: Albertus Magnus, *Summa theologiae*, II.4.2.1c, ed. by Borgnet, p. 82b (citing the opening proposition of *Liber de causis*, XIX [XX] and XX [XXI], ed. by Fauser); ibid., II.76.1.2c, p. 67a; ibid., II.106.5c, p. 281b: 'potest responderi ad argumentum factum in contrarium per illam propositionem Aristotelis in libro de *Causis*, quod "omnis virtus congregata et unita est fortior quam divisa"', citing the opening proposition of *Liber de causis*, XVI [XVII]. Similarly, in 1271, Albert cites 'Aristotle in the *Liber de causis*' (as well as in the 'epistle' *De principio universi esse*); Albertus Magnus, *Problemata determinata*, I, ed. by Weisheipl, p. 47, v. 9 (referring to the opening proposition of *Liber de causis*, XIX [XX], ed. by Fauser). For evidence that the original context of the composition of the *Liber de causis* involved the Aristotelianism of the Circle of al-Kindī, see Taylor, 'Contextualizing the *Kalām fī maḥḍ al-khair*'.

86 Albertus Magnus, *De IV coaequaevis*, q. 2, a. 1, ob and ad 2, ed. by Borgnet, pp. 320a, 321a; Albert, *Super I Sententiarum*, 2.4 [q. 3c and ad 1–2], ed. by Burger, p. 55, v. 56–p. 56, v. 15; Albert, *Commentarii in I Sententiarum*, 46.13c, ed. by Borgnet, p. 447b. See also Albert, *Commentarii in I Sententiarum*, 22.1.1c, ed. by Borgnet, p. 568b.

87 For reference to this passage as a comment by a commentator, see the texts at notes 80, 89, and 95, and Albertus Magnus, *Super I Sententiarum*, 2.4 [q. 4 ob 4], ed. by Burger, p. 57, vv. 14–18; Albert, *Commentarii in IV Sententiarum*, 10.10c, ed. by Borgnet, p. 262b; Albert, *Super Dionysium De divinis nominibus*, II.3 ob 1, ed. by Simon, p. 45, vv. 25–26; ibid., III.2 ob 2, p. 101, vv. 39–45; ibid., IV.4 ob 2, p. 115, vv. 29–34; ibid., IV.57 ob 4, p. 165, vv. 15–20; Albert, *Summa de mirabili scientia Dei*, I.23.1.1 ob 7, ed. by Siedler, p. 170, vv. 38–40. From the earliest works, Albert ascribes to the *Liber de causis* the genre of propositions plus commentary. See Albertus Magnus, *De incarnatione*, VI.1, ad 1, ed. by Backes, p. 220, vv. 24–26; *De resurrectione*, IV.2.2 ob 4, ed. by Kübel, p. 341, vv. 38–39; *De IV coaequaevis*, q. 1, a. 3 ob 4, ed. by Borgnet, p. 312b; q. 2, a. 1, p. 321a; q. 4, a. 1, qa. 1, ad 1, p. 360a; q. 21, a. 1 ob 7, p. 462a; q. 24, a. 2, p. 479a; *De homine*, 'Utrum animae rationales immediate creentur a deo, an ab intelligentiis angelicis', ob and ad 3–6, ed. by Anzulewicz and Söder, p. 75, vv. 46–60 and p. 76, vv. 61–62; p. 390, vv. 30–33; p. 471, vv. 66–70. Almost invariably a passage cited as a comment

2. 'The First Being is "at rest" and is the cause of causes. If it gives being [*ens*] to all things, it gives it to them *through the mode of creation*. But the First Life gives life to those things under it, not through the mode of creation, but *through the mode of form*. Similarly, Intelligence gives knowledge [*scientia*], et cetera, to those things under it only *through the mode of form*'.[88]

Albert summarizes this doctrine of the 'Commentator of the *Liber de causis*' as follows:

> Et commentum ibidem: Ens est per creationem, bonum autem per informationem; et vocat per creationem esse id quod effluens a primo non praeponit sibi aliquid in quo fundetur, per informationem autem id quod non effluit nisi supposito alio in quo fiat.[89]
>
> The comment on the same place [says]: 'a being' [*ens*] is *through creation*, but 'the good' is *through information*. He calls 'being [*esse*] through creation' that which, in outflowing from the First, presupposes for itself nothing on which it is grounded; however, [he calls] 'through information' that which outflows only with something else presupposed in which it comes to be.

In one of his earliest works, Albert already adopts the teaching of this second passage, when responding to an objection that appealed to *Liber de causis* XVII (XVIII). Albert distinguishes between two effusions of 'goodnesses' flowing from the First. Either they

> sive effluant per modum creationis ut ens, quod nihil supponit in creatura, sive effluant per modum informationis, scilicet quod aliquid supponit in creatura ut vita, quae supponit ens.[90]

---

is not a chapter's opening proposition. At *De IV coaequaevis*, q. 3, a. 1 sc 3, ed. by Borgnet, p. 339a (second redaction), Avicenna is named as a Commentator in the *sed contra*, whereas elsewhere the Philosopher is named; q. 38, a. 1, ad 1, p. 552a. The objections in *De homine* now name al-Fārābī as a Commentator, at Albertus Magnus, *De homine*, 'Utrum intellectus agens sit intelligentia separata vel non', ob 12, ed. by Anzulewicz and Söder, p. 409, vv. 42–47; now the Philosopher himself, p. 85, vv. 10–12; and it is even acknowledged in a *sed contra* that some have doubted that the work is by Aristotle, p. 584, vv. 26–27. See also notes 21 and 134. For a similar treatment, which I discovered after writing this, see the forthcoming paper by Anzulewicz, 'Der *Liber de causis* als Quelle'.

88 *Liber de causis*, XVII (XVIII), ed. by Fauser, p. 151, vv. 74–77: 'Redeamus autem et dicamus quod ens primum est quietum et est causa causarum, et, si ipsum dat omnibus rebus ens, tunc ipsum dat eis per modum creationis. Vita autem prima dat eis quae sunt sub ea vitam non per modum creationis immo per modum formae. Et similiter intelligentia non dat eis quae sunt sub ea de scientia et reliquis rebus nisi per modum formae'. The Latin *Liber*, like the Arabic original, apparently uses interchangeably concrete and abstract terms, such as *ens* and *esse*. Since Albert usually uses these terms in distinct ways, I translate *ens* as 'a being' in what follows; see also the studies in note 75 above.

89 Albert, *Super I Sententiarum*, 1, cap. 3 [q. 7c], ed. by Burger, p. 37, vv. 42–47.

90 Albertus Magnus, *De resurrectione*, IV.2.2 ad 4, ed. by Kübel, p. 342, vv. 41–44.

> flow out *through the mode of creation*, as [does] 'a being' [*ens*], which presupposes nothing in the creature; or they flow out *through the mode of information*, that is, which presupposes something in the creature — such as life, which presupposes 'a being' [*ens*].

Another early work, *De bono*, develops the doctrine of *Liber de causis* XVII (XVIII) while suggesting how being (*esse*) is contained in the subsequent forms as something that is determined,[91] as well as how it is therefore the *primum creatum* within each creature:

> Si enim consideretur intentio boni et intentio entis, in unoquoque ens erit creatum primum et causa primaria, et bonum erit per informationem in ente et secundum. Intentio enim entis est intentio simplicissimi, quod non est resolvere ad aliquid, quod sit ante ipsum secundum rationem. Bonum autem resolvere est in ens relatum ad finem. Et hoc accipitur ex verbis Dionysii in V capitulo *De divinis nominibus*, ubi loquens de ente dicit sic: 'Ante alias *dei* participationes esse propositum est, et est ipsum secundum se esse senius eo /12/ quod est per se vitam esse, et eo quod est per se sapientiam esse, et eo quod est per se similitudinem divinam esse, et alia, quaecumque existentia *et* participantia ante omnia *illa* esse participant'.[92]
>
> If one considers the concept of 'the good' and the concept of 'a being' [*ens*], 'a being' [*ens*] will be the *first [thing] created* and the primary cause in *each [thing] whatsoever*, and 'the good' will be the second [thing], and [will occur] *through information* [*per informationem*] in 'a being' [*in ente*]. For, the concept of 'a being' [*ens*] is the concept of the simplest [thing], which cannot be resolved *in notion* into anything that is before it. But it is possible to resolve 'the good' into 'a being [*ens*] related to the end'. And, this [teaching] is taken from the words of Dionysius in *On Divine Names* 5, where he speaks of 'a being' [*ens*] as follows: 'Being [*esse*] is established before the other participations of God, and "being [*esse*] per se" is senior to "*being* life per se", to "*being* wisdom per se", and to "*being* a divine likeness per se".[93] Whatever other things exist and participate, [these] participate in being [*esse*] before [participating in] all of those'.

Albert does not here miss the convergence of Dionysius and the *Liber de causis* — which is not, in fact, accidental given the influence of Syriac Christians on the Kindī Circle from which the *Liber de causis* emerged.[94]

---

91 For an echo of this idea in Albert's *De causis*, see below at note 98.

92 Albertus Magnus, *De bono*, I.1.6c, ed. by Kühle, p. 11, v. 79–p. 12, v. 4.

93 Italicized English terms indicate where Albert's text differs from Dionysius Latinus, *De divinis nominibus*, V.5 (820A), trans. by Sarracenus, p. 337, col. 1–4.

94 See especially D'Ancona, *Recherches sur le 'Liber de causis'*.

The whole doctrine of 'creation of *esse* alone' versus 'information', which Albert ascribes to the Philosopher based on these two passages, is summed up well in Book I of his *Sentences* commentary:

> [S]ic patet per Philosophum in libro *Causarum*, qui dicit, quod prima rerum creatarum est esse vel essentia, et non est ante ipsum creatura alia: ergo non est in quo fundetur /448/ essentia, sicut id circa quod fit, ut forma subjecti alicujus: et ideo etiam secundum ordinem naturae non potest essentia in ratione essentiae esse nisi creata de nihilo. Omnia autem alia (ut dicit Commentator ibidem) fiunt per informationem circa essentiam: et dat exemplum de bono, quod secundum ordinem naturae praemittit sibi intellectum essentiae in qua est: ergo patet, quod essentia dicit intellectum suum super nihil fundatum, et in omnibus priorem.[95]
>
> > This is clear through the Philosopher in the *Book on the Causes,* who says that the first of created things is being [*esse*] or essence [i.e., 'beingness'], and there is no other creature prior to it. Therefore, there does not exist [that] in which essence is grounded, as though [it were] that upon which [essence] comes to be as the form of some subject. For this reason also it is not possible in the order of nature for 'essence under the notion of essence' to exist unless it is created from nothing. As the Commentator [in the *Book on the Causes*] says in the same place, however, all other things come to be *through information* upon essence [or 'beingness']. He gives the example of 'the good', which presupposes for itself, in the order of nature, the notion of essence [or 'beingness'], in which it is. Therefore, it is clear that essence ['beingness'] expresses as its notion 'founded on nothing, and first within all [things]'.

As is by now clear, the foundation on which Albert bases his emanation scheme in *De causis* has from the outset of his career been adopted and ascribed to Aristotle and Dionysius. Before turning to that scheme, which was novel circa 1267, let us consider the strategies for describing the flow of forms that Albert introduces — inspired especially by Isaac Israeli and Jewish Neoplatonism — to supplement 'Aristotle' and Dionysius.

---

95 Albertus Magnus, *Commentarii in I Sententiarum*, 46.13c, ed. by Borgnet, pp. 447b–448a. For other early references to the distinction between creation and 'information', in addition to those in note 87 above, see Albertus Magnus, *Super Dionysium De caelesti hierarchia*, II [q. 10 ob 3], ed. by Simon and Kübel, p. 29, vv. 11–15; *Super Dionysium De ecclestica hierarchia*, II, ed. by Burger, p. 28, vv. 19–21; *Super Dionysium De divinis nominibus*, III.2 ob 2, ed. by Simon p. 101, vv. 42–45. One also finds five references in the *Metaphysica*, one in Albert's *De causis* (in addition to those cited in notes 78 and 98), and eight in the *Summa theologiae*, including Book I, q. 23, where Albert adopts it in his own name: Albertus Magnus, *Summa de mirabili scientia Dei*, I.23.1.1, ad 6, ed. by Siedler, p. 124, vv. 50–54. Nevertheless, in the same Book I (ibid., q. 28c, ob and ad 3, p. 213, vv. 21–34 and p. 214, vv. 1–53) he rejects 'information' in the account of transcendentals 'being', 'one', 'good', and 'true', opting instead for a solution in terms of *rationes* or *modi essendi* that appears similar to that, for example, of Aquinas's *De veritate*, I.1c.

### *Two Descriptive Strategies for a Procession of Forms*

*Concept-Procession.* Albert rejects as *pessimus* the error of those who, given the principle 'from one only one', say that all things are one, and who therefore identify the entire diffusion of the First in all things with the *esse* of all things.[96] He admits that it is difficult to see what causes any diminution (and diversity) in the effect of the First Cause,[97] and he often resorts instead to two different strategies for describing the emergence of forms. Sometimes Albert appeals to wider concepts, as in a Porphyrian tree, which potentially contain within themselves specific or determinate forms that, in turn, actualize the generic forms. I call this 'Concept-Procession'. Albert's stock example, found also in the *Liber de causis*, is the Neoplatonic triad of 'being, living, and understanding'. Just as Aristotle says that the sensate soul stands to the intellectual as a triangle to a quadrilateral, so being stands to living, and being and living together stand to understanding. The key text also rehearses the systematic creationist account that we have already seen:

> Primum enim in omnibus est ens, quod quia nihil ante se supponit secundum intellectum, necesse est, quod ex nihilo sit. Et ideo in omnibus in quibus est, necesse est ipsum fieri per creationem. Per creationem enim fit, quod ex nihilo fit. Vita autem ante se supponit ens secundum naturam et intellectum et ex esse producitur sicut determinatum ex confuso. Unde vita non dicit simplicem esse conceptum, sed dicit esse formatum ad aliquid. Vita igitur per creationem fieri non potest, quia fit ex aliquo. Relinquitur igitur, quod fiat per informationem. Similiter autem est de intellectivo et scitivo. Hoc enim supponit ante se et vivere et esse. Propter quod per creationem fieri non potest. Supponit enim aliquid ante se, in quo est in potentia, sicut tetragonum est in trigono. Producitur enim tetragonum ex trigono duobus sibi in hypothenusa coniunctis trigonis orthogonis. Unde intellectivum et scitivum ex esse producitur per informationem, ex vita autem per determinationem ad vitae speciem intellectualem.[98]

---

96 Albertus Magnus, *De causis*, I.4.5, ed. by Fauser, p. 49, vv. 14–15. Cf. ibid., I.4.3, p. 45, v. 23–p. 46, v. 30, esp. p. 46, vv. 7–11, and the pantheistic position ascribed to Hermes Trismegistus, Asclepius, etc. (Albert often has in mind the views of David of Dinant and Amaury of Bène).

97 Ibid., p. 48, vv. 82–83.

98 Albertus Magnus, *De causis*, II.3.13, ed. by Fauser, p. 150, vv. 44–63 (on *Liber de causis* XVI [XVII]). What is *per informationem*, then, refers to things insofar as they have a form that presupposes the form *esse*. What *is per determinationem*, by contrast, refers to things insofar as they have a form that presupposes a form other than *esse*. Informing gives content to what has the first form of being; determining gives specificity to what already has content. See also at note 162 below. For subsequent evidence of Concept-Procession, see Albertus Magnus, *Summa theologiae*, II.3.3.1.4 ad 2, ed. by Borgnet, p. 29b. For recognition of Albert's link of Porphyrian logic and emanation, see Booth, 'Conciliazioni ontologiche', pp. 69–71.

> The first in all things is 'a being' [*ens*] — which is necessarily *ex nihilo*, since it presupposes nothing prior to itself in understanding. And for this reason, in all things in which it is, it is necessary that it come to be through creation. For, what comes to be *ex nihilo* comes to be through creation. But life presupposes 'a being' as prior to itself in nature and in understanding, and it is produced out of being [*esse*], as [is] the determinate out of the undifferentiated [*confusum*]. Hence, *life* does not express the simple concept of *esse*, but it expresses *esse* that is formed in relation to something. Therefore, life cannot come to be through creation, because it comes to be out of something [*ex aliquo*]. It follows that it comes to be through [a process of] information [*per informationem*]. The case is similar for what has intellect and knowledge. For, this presupposes both 'being' [*esse*] and 'living' prior to itself. Therefore, it cannot come to be through creation. For, it presupposes something prior to itself in which it is potentially, just as a quadrilateral is in a triangle. [...] Hence, what has intellect and knowledge is produced out of *esse* through [a process of] information, whereas [it is produced] out of life through a 'determination' [*per determinationem*] to the intellectual form of life.

Being (*ens* or *esse*), then, repeats Albert, is only created (that is, it is only caused by the First alone without mediation). But living is an 'informing' of being, whereas understanding is an informing of being as well as a determination of living. This strategy of procession highlights how forms in a series from lower to higher are related to each other and how they are integrated into a new unity.

*Light-Procession.* Albert's predominant strategy, expressly inspired by the Jewish Platonist Isaac Israeli, involves the metaphor of light to account for a different series of forms: forms ordered not from more general to more specific, but from higher to lower. I call it 'Light-Procession'. The further from the light source, the more the form is 'diminished' or 'overshadowed'. Thus, corporeality is in the shadow of vegetative soul, says Albert, which is shaded by the sensate soul, which is shaded by Intelligence.[99] Whereas the 'Concept-Procession' highlights formal causality, 'Light-Procession' seems to highlight efficient causality, especially the idea that the higher lights are filtered through all that is below them: their highly intelligible light is diffused as it extends from its source.

---

99 See, e.g., Albertus Magnus, *De causis*, I.4.5, ed. by Fauser, p. 48, vv. 58–79: 'And so it is the case for all things that a subsequent differentia of "a being" is always constituted upon a certain "decline" [*occasum*] or "shadowing" of what is prior, just as the sensible [is] "in the shadow" of the intellectual, the vegetative [is] "in the shadow" of the sensible, and body determined by corporeity alone [is] "in the shadow" of the vegetative, whereas body determined by contraries [is] "in the shadow" of the heaven, which is determined by corporeity alone. [...] And this, indeed, Isaac already said before us in his *Book on Definitions*'. The centrality of of emanation and light metaphysics even in Book 1 of Albert's work is underscored by Beierwaltes, 'Der Kommentar zum "Liber de causis"', pp. 197–98; see also Hedwig, *Sphaera lucis*, pp. 177–81.

Albert's two descriptive strategies give us the background needed in turning to his general account of the 'flow' of form, or 'informing' causality, with which he begins tract 4 of *De causis* Book I. He sharply distinguishes the 'flow' of forms from 'equivocal' and 'univocal' causes.[100] Flux or emanation is neither so purely equivocal that nothing formal is in any way shared between cause and effect, nor so purely univocal that the same form is 'brought about in another subject', which then possesses it in the same way as does its cause.[101] As Albert prefers to put it, to flow or emanate as such expresses no 'transmutation per se' of that to which such intelligible form is communicated. Thus, water, the first image used in Tract 4, preserves the notion of the integrity of the flow in communicating form. If the river bank is changed by what flows through it, it is in any case not changed into water, and whatever change thereby occurs can only be accidental to the rush of water from its source into the deep, while the flow is diffused into lakes, pools, rivulets, and swamps. Albert's other example is the blueprint design that exists in one way in the architect, in another in the master-builder, in a third way in the workers, in a fourth in the house built, and so on.[102] What is crucial is that all flowing involves some 'verticality', some transmission of a form that is only analogously one according to prior and posterior. Of course, what Albert has in mind, as he makes clear, is not a temporal, horizontal flow, but a timeless, simultaneous, vertical outpouring from a primary font.[103] Perhaps two quotations capture this best:

> Supernal being [*esse*] is threefold: the being of the First Cause, the being of the Intelligence, and the being of the 'Noble Soul' [or sphere-soul]. Inferior being, however, is what falls away from [supernal being] through diverse grades of defect. Therefore, Plato said that true being is in the First Cause, but its form is in the Intelligence, and the image of the latter is in the soul; resonances or shadows of the same, however, are in generable and corruptible things. For, *form* 'remains outside' [*foris manet*] forming and terminating —

---

100 See esp. Albertus Magnus, *De causis*, ed. by Fauser, I.4.1. By *equivocal causation* there, Albert apparently means cause and effect that are radically heterogeneous (e.g., Sun and heat), whereas *influx* includes forms that are analogously one according to prior and posterior. See ibid., I.4.6, p. 49, v. 73–p. 50, v. 5, where Albert also holds that the First efficient cause, from which one must say the 'second' *flows*, is *neither* in the same *genus* as the 'second' (and so is a univocal cause), *nor* is it an equivocal cause; thus the 'second' (form) is said to be a 'quasi instrument' of the First.

101 Ibid., I.4.1, ed. by Fauser, p. 42, vv. 38–48 and vv. 59–63: 'Non enim fluit nisi id quod unius formae est in fluente et in eo a quo fit fluxus. Sicut rivus eiusdem formae est cum fonte, a quo fluit, et aqua in utroque eiusdem est speciei et formae. Quod non semper est in causato et causa. Est enim quaedam causa aequivoce causa. Similiter non idem est fluere quod univoce causare. Causa enim et causatum univoca in alio causant aliquando. A fonte autem, a quo fit fluxus, non fluit nisi forma simplex absque eo quod aliquid transmutet in subiecto per motum alterationis vel aliquem alium. [...] Unde cum causa nihil agat nisi in subiecto aliquo existens, fluxus autem de ratione sua nihil dicat nisi processum formae ab ipso simplici formali principio, patet, quod fluere non est idem quod causare'.

102 Ibid., I.4.6, ed. by Fauser, p. 50, vv. 2–11.

103 See esp. ibid., I.4.8, ed. by Fauser, p. 55, vv. 67–72.

> as simple being [*esse simplex*], which is from the First Cause; whereas image [in the mind of the sphere-soul] is what imitates, but it does not imitate what the whole expresses, but rather that which stands apart [*distat*], and it points out the form that it imitates according to a proportion [*analogia*] to it. But resonance and shadow express [the same] only confusedly and according to the 'standing-apart' [*distantia*] of remote similitude.[104]

The *remaining simple goodnesses* that flow from the First Cause, *as* do being [*esse*], *life, the illumination* [*lumen*] of intelligence, and *what is similar to these,* that is, the noble and immaterial [goodnesses] that pertain to the substantial *esse* of things, having exemplary *esse* in the First Cause, *are* formal and ideal causes of all things that have exemplary goodnesses in the First Cause, [goodnesses] *descending from the First Cause* itself. But they descend *first over the first effect,* which *is Intelligence. Then* through the illumination of the Intelligence *they descend over the rest of effects,* both *intelligible* and *corporeal.* But they descend into them *through the mediation of Intelligence.*[105]

### *Albert's Derivation-Scheme*

At this point, my reader will object: You have given us absolutely no mechanism by which a 'flow of forms' *other* than creation has been introduced. You have so far only *described* what the flow looks like if there *were* a mechanism. By approaching the topic in this way, I have followed Albert's order, and I emphasize the importance of the mechanism that he introduces at the end of Tract 4, which is, in effect, an 'emanation scheme' — or, better, a 'derivation-scheme' — borrowed from Arabic philosophy. What is distinctive of such a scheme is the linking of the triad One-Intellect-Soul with the triad within 'Arabic Aristotelianism' that is very evident in al-Fārābī and that has its proximate roots in the transformation of

104 Ibid., II.1.7 (on *Liber de causis*, II), ed. by Fauser, p. 69, vv. 40–53: '[E]sse superius triplex est, scilicet esse causae primae et esse intelligentiae et esse nobilis animae. Esse autem inferius est, quod deficit ab illo diversis defectus gradibus. Propter quod dixit Plato, quod esse verum est in causa prima, forma autem huius in intelligentia, et imago illius in anima, resonantia autem vel umbra eiusdem in generabilibus et corruptibilibus. Forma enim foris manet, esse simplex, quod a prima causa est, formans et terminans. Imago autem quae imitatur, non imitatur autem, quod totum exprimit, sed id quod distat, et secundum analogiam sui monstrat formam, quam imitatur. Resonantia autem et umbra non nisi confuse exprimunt et secundum distantiam similitudinis longinquae'.

105 Ibid., II.3.6, ed. by Fauser, p. 145, vv. 5–15: 'Et *reliquae bonitates simplices* fluentes a causa prima, *sicut* esse, *vita et lumen* intelligentiae *et quae sunt eis similia,* hoc est, ad esse substantiale rerum pertinentes nobiles et immateriales, esse exemplare in prima causa habentes, *sunt causae* formales et ideales *omnium rerum habentium bonitates* exemplares in causa prima *descendentes ab* ipsa *causa prima.* Descendunt autem *primo super causatum primum,* quod *est intelligentia. Deinde* per lumen intelligentiae *descendunt super reliqua causata* tam *intelligibilia* quam *corporea.* Sed descendunt in ea *mediante intelligentia*' (lemmata from *Liber de causis*, XV [XVI], ed. by Fauser, p. 144, vv. 7–9).

late antique pagan Neoplatonism effected by Alexandrian and Syriac Christians: 'God-Intelligences-celestial souls'.[106]

*Stage 1: Procession of the intelligentia of esse.* Albert introduces such a scheme in several places,[107] but Tract 4, Chapter 8 of his personal Book I offers the most thorough presentation (and it is only in the *De causis* that the scheme appears to be adopted by Albert as his own).[108] The first requirement of any such scheme is to explain how the one immediate effect of God is a cosmic Intelligence.[109] For Albert, following Dionysius's *De divinis nominibus* V as well

---

106 For a review of the history, see Twetten, 'Aristotelian Cosmology and Causality'.

107 See, e.g., Albertus Magnus, *De causis*, II.5.3, ed. by Fauser, p. 170, v. 67–p. 171, v. 17. All of Albert's reports of the derivation-scheme prior to his *De causis* use only two 'considerations': Albertus Magnus, *Commentarii in II Sententiarum*, 1.10, sc 1 ob 2, ed. by Borgnet, pp. 27b–28a (ascribed to the 'philosophers'; see note 31 above); ibid., III.3, ad 5, pp. 65b–66; *Super Dionysium De caelesti hierarchia*, I, q. 3c, ed. by Simon and Kübel, p. 10, vv. 11–39 (ascribed to Aristotle and his followers); *Super Dionysium De divinis nominibus*, VII.3 sol., ed. by Simon, p. 339, vv. 9–35; *Physica*, VIII.1.15, ed. by Hossfeld, p. 579, v. 55–p. 580, v. 70 (ascribed to Aristotle and the Peripatetics); *Metaphysica*, I.4.3 (digr.), ed. by Geyer, p. 49, v. 69–p. 50, v. 40 (inspired by Pythagoras); ibid., XI.2.17 (on Λ.7, 1073a14), p. 504, v. 81–p. 505, v. 2; ibid., XI.2.20 (digr.), p. 508, vv. 28–47 and p. 509, vv. 6–19. In the latter, Albert answers two objections and is not otherwise critical: the objection that the flow of intelligences does not extend to matter applies to the Platonists, not to the Peripatetics (p. 508, v. 75–p. 509, v. 17). By the end of Book XI, Albert rejects the creationist reasoning of Maimonides that he adopted in his *Physica*, VIII, and hence he will need another way to defend a non-necessary derivation of creatures from God; ibid., XI.3.7, ed. by Geyer, p. 542, vv. 7–25; see below at notes 167 and 182.

108 Albertus Magnus, *Super Dionysium De caelesti hierarchia*, I q. 3c, ed. by Simon and Kübel, p. 10, vv. 22–26 and vv. 34–36, admits, with Maimonides explicitly (and Averroes), 'from one comes only one', but only for natural things. But, he says, 'Aristotle and his followers thought poorly' in affirming a derivation-scheme, given this principle (vv. 11–21). The scheme is contrary to the faith, he says, and Maimonides criticizes it, arguing that things proceed from God, not by necessity of nature, but through liberality of will (vv. 22–26). See also Albertus Magnus, *Physica*, VIII.1.15, ed. by Hossfeld, p. 579, v. 55–p. 580, v. 70 (criticizing Aristotle's necessitarianism; see above, at note 22). The evidence indicates that Albert in *Super Dionysium De divinis nominibus* is also critical of the philosophers' scheme. He takes the agents in such a scheme, ascribed to Plato, to account for form alone, not matter (so, *Commentarii in II Sententiarum*, I.10, sc. 1, ad 2, ed. by Borgnet, p. 28a; *Super Dionysium De caelesti hierarchia*, I, q. 3c, ed. by Simon and Kübel, p. 10, vv. 26–31), with the result that there is uncreated matter and no divine knowledge of particulars; *Super Dionysium De divinis nominibus*, II.44–45, ed. by Simon, p. 72, v. 35–p. 74, v. 20 (where he associates the scheme with Plato and states preference for the approach of Aristotle, which appeals to divine exemplar *rationes*); also ch. 7.3 sol., p. 339, vv. 9–35, where the philosophers in general are faulted. Nevertheless, in the same work Albert, appealing to Avicenna for his reading of Dionysius, adopts an incipient version of the scheme that makes no mention of cosmology: a triadic procession of the one form, being, according to three *comparationes* or *considerationes*; ch. 5.32 sol., ad 4, p. 322, vv. 24–30, vv. 50–59 (see also note 77 above). For an *interpretation* by Albert of the emanation scheme such that even he in *De causis* I is critical of it, see Albertus Magnus, *De causis*, I.3.5, ed. by Fauser, p. 40, vv. 59–70 (discussed below at note 181).

109 For the assertion that the Intelligence is the *primum creatum*, see, e.g., Albertus Magnus, *De causis*, II.2.19, ed. by Fauser, p. 112, vv. 53–54; ibid., II.4.10, p. 164, vv. 2–4 (with *Liber de causis*, XXII [XXIII]); cf. ibid., II.3.4 (with *Liber de causis*, XV [XVI]), p. 143, vv. 47–48. That the *primum*

as *Liber de causis* IV, the first creaturely procession is being (*esse*). Hence, Albert must explain how the first effect is, as in the Plotinian tradition, not only being, but also Intelligence. The key to Albert's whole theory of the 'flow' is taking every intellect of itself, and therefore that of the First, as an agent or acting intellect.[110] Through its intellect, the First Agent precontains, as was said above at note 72, the conceptually distinct *rationes* of all effects, in virtue of which it is, as Albert puts it, 'universally' active. Accordingly, after Albert reminds us of the first supposition of the derivation-scheme, which we have already seen above:

> 1) from one (in form) comes only one (in form)

he adds a second supposition:

> 2) the First Intellect, which is universally acting (God), 'constitutes' things by 'emitting' intelligences (with a lower-case 'i').[111]

I say 'lower-case i' because, as we shall see, Albert expressly distinguishes the (lower-case) 'intelligence' that is identical to the *primum creatum* (*esse simplex*) from the first of the ten celestial (upper-case) 'Intelligences'. The First Intellect (God), says supposition (2) — who is above the ten Intelligences — emits intelligibilities (*intelligentiae*, lower case). As Thérèse Bonin has shown, Albert means by *intelligentia* here the concept of, or the *intellectus* of, *esse*.[112]

I would add several points, however. First, unlike Bonin, I emphasize that this 'intelligence' (lower-case) is (a) an ontological form that is (b) other than God (although it is not yet conceived *as existing* in *id quod est*, indicates Albert).[113] In English, 'intelligibility' conveys better than 'intelligence' or 'concept' the realist character of the form *esse*. In fact, there are many forms or intelligibilities (*intelligentiae*, lower case) that flow from God as First Intellect, and not only *esse*, or the

---

*causatum* is Intelligence, see ibid., II.2.7, p. 100, vv. 36–37; ibid., II.3.4, p. 143, vv. 3–5; ibid., II.3.6, p. 145, vv. 11–12 (with *Liber de causis*, XV [XVI]). Albert most frequently says that *esse* as *primum creatum* is intelligence (in the lower case sense) or is in Intelligence (for Albert's specification of the two senses of 'intelligentia', see ibid., II.1.21, p. 85, vv. 72–88).

110 See note 120 below. For *agent intellect* as a name of God as well as of creatures, see, for example, notes 111 and 116.

111 Albertus Magnus, *De causis*, I.4.8, ed. by Fauser, p. 55, vv. 72–80, and ibid., vv. 80–84: 'We suppose also that the universally acting intellect acts and constitutes things only by actively understanding and by emitting intelligences. When it understands in this way, it by itself constitutes the thing at which the illumination of its intellect terminates' ('Supponimus etiam, quod intellectus universaliter agens non agit et constituit res nisi active intelligendo et intelligentias emittendo. Et dum hoc modo intelligit, seipso rem constituit, ad quam lumen sui intellectus terminatur').

112 Bonin, *Creation as Emanation*, pp. 87–90. Albert himself uses the term *conceptus* and links *intelligentia simplex* to Aristotle's *intelligentia indivisibilium* (Aristotle, *De anima*, in James of Venice's translation), the intellection of simples rather than of propositions; Albertus Magnus, *De causis*, II.1.19, ed. by Fauser, p. 83, vv. 58–59, vv. 69–75. For Albert's development of this thought, see the text quoted in note 117 below. For a strongly 'conceptualist' reading of Albert here, see Porro, '*Prima rerum creatarum est esse*', pp. 62–64; Porro, 'University of Paris', pp. 283–84.

113 Albertus Magnus, *De causis*, I.4.8, ed. by Fauser, p. 55, v. 88.

first effect. It is significant that Albert connects these intelligibilities to the divine agent intellect that radiates them.[114] This illuminating intellect is the source of the 'flow' of forms. These forms are received in what receives them, in a supposit other than God: in *id quod est*. Thus, Albert's way of setting up the derivation-scheme allows him to connect it with the Boethian structure of all beings, a structure that he adopts in his early writings, the *De quattuor coaequaevis* and the *Sentences* commentary, and that he inherits from the early thirteenth-century Parisian masters: all things are composed of form or essence (*esse*) and *id quod est*. Again, there are many forms besides *esse* proper, such as goodness, oneness, substance, and life, but all are other than the supposit that receives them.[115]

*Stage 2: Intelligentia of esse, Once Received, Is Intellect.* How, then, does Albert develop a derivation-scheme from the two aforementioned principles? From God as intellect itself proceeds the first intellectual illumination (*lumen*), the 'intelligence' of *esse*, as Bonin has taught us. Albert later puts the matter as follows, when paraphrasing *Liber de causis* IV:

> When we say '*esse* is simple intelligence', we do not understand that it is the Intelligence that is the intellectual substance multiplied in ten orders, but that it is 'intelligence', that is, a *form* brought forth [*producta*] into being by the illumination of the [divine] agent intellect [...], just as we call 'to be' [*esse*] the 'first intelligence', 'to live' [*vivere*] the 'second intelligence', 'sensible' the 'third intelligence', and so on.[116]

This form *esse*, of course, is received into 'that which is' (*id quod est*). Thus, the form received becomes an intellect or an Intelligence with a capital 'I'.[117] Here is how Albert initially puts it:

---

114 Cf. ibid., I.4.5, ed. by Fauser, p. 48, vv. 45–51: 'We have said that the first font is an intellect universally acting in such a way that there is nothing among the *things understood* in any way that it does not actualize in the manner in which it is an intellect. Furthermore, we call thing understood everything that is, in any way whatever, capable of being understood. For, this cannot be understood unless it is constituted in the illumination of the First Intellect' ('Diximus enim, quod primus fons est intellectus universaliter agens ita quod nihil est de intellectis quocumque modo quod non agat eo modo quo intellectus est. Intellectum autem dicimus omne quod quocumque modo intelligi potest. Hoc enim non posset intelligi, nisi in lumine primi intellectus constitueretur').

115 See, again, the works cited in note 75 above.

116 Albertus Magnus, *De causis*, II.1.19, ed. by Fauser, p. 83, vv. 58–69: '*Quamvis* autem *esse*, quod est *creatum primum, intelligentia simplex sit, tamen* propter habitudines et potentias, quas habet, *compositum est ex finito et infinito*. Quando autem dicimus esse intelligentiam simplicem, non intelligimus, quod sit intelligentia, quae substantia intellectualis est in decem ordines multiplicata, sicut in anteriori libro determinatum est, sed quod est intelligentia, hoc est, forma a lumine intellectus agentis in esse producta et in simplici illo lumine per intentionem accepta, sicut dicimus esse intelligentiam primam et vivere intelligentiam secundam et sensibile intelligentiam tertiam et sic deinceps' (lemmata from *Liber de causis*, IV, ed. by Fauser, p. 88, vv. 63–68). Note especially the form *sensibile*, which is clearly below God.

117 One can see Stages 1 and 2 in germ circa 1251 in Albertus Magnus, *Physica*, VIII.1.15, ed. by Hossfeld, p. 579, vv. 60–66: 'et ideo dicunt [Aristotle and the Peripatetics], quod a prima causa

> Therefore, when the first universally acting intellect understands itself in this way, the *illumination* [*lumen*] of the intellect, which is from it, is the first form and the first substance, [which] in all things holds the form of the one who understands, except in 'that which is from another'.[118]

A previous statement in Chapter 5 helps clarify:

> What is proximate [to the First], it is agreed, is from that which is *ex nihilo*. For, according [as it is] 'that which is', it has no principle of its own essence [or beingness]. For, were it to have such a principle, it would have it of itself — which is entirely absurd. It has as the principle of its own *esse* that which is before it. Therefore, the first illumination falls on it through the fact that *esse* in it is other than 'what is'. And, this, indeed, is an Intelligence.[119]

*Stage 3: Derivations from the Considerations of Intellect.* It is clear from Albert's subsequent descriptions that he means 'upper-case' Intelligence in the last sentence quoted.[120] A caused or created Intelligence reflects on something other than itself, and so is both active and passive. In particular, it reflects on three different things: 1) on its source, 2) on itself and on its own essence, and 3) on itself as *ex nihilo* and in potency. Thus, Albert discovers the origin of the triadic 'considerations' that form the standard Arabic derivation-scheme. It will suffice simply to read Albert's report of the scheme in Tract 4, Chapter 8:

> And, in 'that which is from another' [i.e., in the resulting Intellect], there is found a threefold comparison: namely, (1) [the comparison] to the First Intellect [God], from which it is and by which being [*esse*] belongs to it;

---

procedit intelligentia, quae est secundum naturae ordinem causatum primum; licet enim primum secundum rationem resolutionis sit esse, tamen in his quae sunt, simplicius et potentius est intelligentia sive substantia intellectualis, quae quidem, quia intelligit se et suus intellectus reflectitur super se'.

118 Albertus Magnus, *De causis*, I.4.8, ed. by Fauser, p. 55, vv. 84–88: 'Dum ergo primus intellectus universaliter agens hoc modo intelligit se, lumen intellectus, quod est ab ipso, prima forma est et prima substantia habens formam intelligentis in omnibus praeter hoc quod ab alio est'. Notice that *first form* and *first substance* here, as often, refer to first in the order of things made (or creatures), v. first absolutely, as the *lumen-lux* distinction helps make clear. See ibid., p. 56, vv. 18–19: 'Intelligentia ergo, quae inter factas substantias prima est'.

119 Ibid., I.4.5, p. 48, vv. 51–58: 'Quod autem proximum ab illo est, constat, quod ex nihilo est. Secundum enim "id quod est" nullum habet suae essentiae principium. Si enim tale principium haberet, a seipso haberet. Quod omnino absurdum est. Sui autem esse principium habet id quod ante ipsum est. Primum ergo lumen occumbit in ipso per hoc quod aliud est in ipso esse et "quod est". Et hoc quidem intelligentia est'.

120 See ibid., I.4.5, ed. by Fauser, p. 48, vv. 58–62: 'But an Intelligence is what acts of itself. If it is made a possible [intellect], a recipient of things understood — and of the thing understood — of an agent intellect, it falls away somehow [from intelligence per se]. For, what receives the power of understanding from some other is not of itself an intellect' ('Intelligentia autem de se agens est. Et si efficiatur possibilis et recipiens intellecta et intellectum agentis intellectus, occumbens quodammodo est. Non enim de se intellectus est, quod ab alio quodam intelligendi accipit virtutem').

> (2) [the comparison] to itself as 'that which is' [*id quod est*]; and (3) [the comparison] to this [fact, namely,] that it is 'in potency' insofar as it is *ex nihilo*. [...] /56/ The first Intelligence, therefore, (1) has necessary being [*necesse esse*] only according as it understands itself to be from the First Intellect.[121] But (2) according as it understands itself as 'that which is', the illumination [*lumen*] of the First Intellect falls [*occumbit*] on it, by which [illumination] it *understands* itself to be from the First Intellect. And in this way it is necessary that an inferior is constituted under it — and this is the second substance, which is either soul or what is in the place of soul in the heavens. But (3) according as it understands itself to be *ex nihilo* and to have been 'in potency', it is necessary that that level [*gradus*] of 'substance that is in potency' begin. And, this is matter under the first form, which is the matter of the celestial body, which is called the *primum mobile*.[122]

In this text we see emanate, in descending vertical order, Intelligence, celestial soul, and heavenly body.[123] A major feature of developed derivation-schemes is accounting also for a 'horizontal' causality: ten Intelligences corresponding to ten celestial spheres with ten sphere-souls. This feature ensures the continuation of the vertical emanation (intelligence, celestial soul, heaven) for each of the subsequent nine celestial spheres.[124] And so, Albert continues:

---

121 The use of *necesse esse* (through another) betrays the Avicennian roots. If we think of this as follows (as is, after all, not so un-Avicennian, and certainly not un-Thomistic), it is not shocking that Albert can hold it personally: the first Intelligence, once it exists, exists necessarily under continued divine (conserving) causality. But Albert actually says that the Intelligence is necessary only in the (logical) moment that it turns back to the First — a claim that is apparently not Avicenna's. Also, Albert introduces *necesse esse* only at the level of the first 'comparison'.

122 Albertus Magnus, *De causis*, I.4.8, ed. by Fauser, p. 55, v. 89–p. 56, v. 13: 'Et in hoc quod ab alio est, triplicem habet comparationem, scilicet ad primum intellectum, a quo est et quo sibi est esse; et ad seipsum secundum "id quod est"; et ad hoc quod in potentia est secundum hoc quod ex nihilo est. Antequam enim esset, in potentia erat, quia omne quod ab alio est, factum est et in potentia erat, antequam fieret. Intelligentia ergo prima non habet necesse esse nisi secundum quod intelligit se a primo intellectu esse. Secundum autem quod intelligit seipsam secundum "id quod est", occumbit in ea lumen intellectus primi, quo intelligit se a primo intellectu esse. Et sic necesse est, quod inferior constituatur sub ipsa. Et haec est secunda substantia, quae vel anima dicitur vel id quod in caelis est loco animae. Secundum autem quod intelligit se ex nihilo esse et in potentia fuisse, necesse est, quod incipiat gradus substantiae, quae in potentia est. Et hoc est materia sub prima forma, quae est materia corporis caelestis, quae vocatur mobile primum'.

123 If Albert's derivation-scheme contains the standard triad in the vertical order, nonetheless, his other strategies typically include more than these three vertically ordered substances. He even records the view of 'some others' (*alii*) who identify ten levels, in obvious parallel to the spheres, although the lower members include non-celestial entities, such as purpose (*propositum*), fortune, and chance. See ibid., I.4.6, p. 51, v. 40–p. 52, v. 5.

124 At one point Albert alludes to the fact that, although the third Intelligence is in the same (horizontal) order as the second, since both are intellectual in nature, nevertheless there is even there in the third less purity than in the second, and so on successively to the last Intelligence; ibid., II.2.14, p. 108, vv. 4–7.

> And once [the first Intelligence] understands itself in this way [as having received the overflow of illumination from the intellect of the First], it constitutes, by the same principle [*ratio*], the Intelligence of the second order. This also understands itself according [as it is] 'that which is' [*id quod est*], and in this way it constitutes the proximate mover [of the sphere, the celestial soul]. It also understands itself according as it is in potency, and in this way it constitutes the second mobile [thing] [*secundum mobile*], which is the second heaven. For, in an active intellect, to understand itself is to emit an intellectual illumination [*lumen*] for the constitution of a thing.[125] And in this way is had the second Intelligence, the second mover [or sphere-soul], and the second mobile [thing]. And, once that Intelligence, again, understands itself to be from the First Intellect [God], it necessarily understands itself in the overflowing illumination. And in this way the Intelligence of the third order is constituted. [...] And, in this way it is not difficult to determine the Intelligences, the movers, and the heavens as far as to the 'heaven of the moon'.[126]

Several features of Albert's scheme point to Avicenna as its principal source: 1) it involves a threefold 'comparison' at each of the ten levels (as opposed to the twofold 'comparison' of, for example, al-Fārābī's *Political Treatise*, a work not available in Latin); and 2) it uses 'from one comes only one' and 'necessary being'.[127]

*Albert's Achievement.* When we couple Albert's derivation-scheme with the strategies I mentioned in the previous subsection, Concept-Procession and Light-Procession, we begin to form a picture of the law-like way in which Albert thinks

---

125 Notice that this is Supposition (2) above, generalized for all (active) intellects, which share in the same intelligibility as the divine Active Intellect.

126 Albertus Magnus, *De causis*, I.4.8, ed. by Fauser, p. 56, vv. 38–64 (the beginning of the quotation is underscored, following the helpful context): 'Sic ergo habemus constitutionem primae intelligentiae, quae vocatur intelligentia primi ordinis. Habemus etiam constitutionem proximi motoris primi orbis, quem quidam vocant animam caeli primi. Et secundum quod intelligit se in potentia esse, habemus constitutionem primi orbis sive primi caeli. Cum autem lumen intellectus primi principii fluat in primam intelligentiam et exuberet, constat, quod exuberatio luminis iterum refertur ad primum. Et dum sic intelligit se, per eandem rationem constituit intelligentiam secundi ordinis. Haec etiam intelligit se secundum "id quod est" et sic constituit motorem proximum. Intelligit etiam se, secundum quod in potentia est, et sic constituit mobile secundum, quod est secundum caelum. Intelligere enim se in activo intellectu est lumen intellectuale emittere ad rei constitutionem. Et sic habetur secunda intelligentia et secundus motor et secundum mobile. Et dum illa intelligentia iterum intelligit se esse a primo intellectu, necesse est, quod intelligat se in lumine exuberante. Et hoc modo constituetur intelligentia tertii ordinis. Intelligit etiam se secundum "id quod est" et sic constituetur motor tertii mobilis. Intelligit etiam se, secundum quod in potentia est, et sic constituetur tertium mobile sive tertium caelum. Et hoc modo non est difficile determinare intelligentias et motores et caelos usque ad caelum lunae'.

127 See Avicenna, *Metaphysics of 'The Healing'*, IX.4, ed. by Marmura, p. 330, vv. 6–10 and p. 331, vv. 2–13.

the diversity of form emerges from the First Cause or God. He has even said that form emerges in a 'necessary' way from secondary causes, given the original creative act. God alone creates *esse*, but all other form, although it emanates from the divine active intellect, proceeds through mediating secondary causes: through the Intelligences as active intellects and emanating causes, and through celestial souls as the movers of the heavens. Once the heavens are introduced, it is easy to connect Albert's derivationism with what *we* know to be the purely Aristotelian cosmology, unmixed with Proclus. According to Aristotle, the cause of the emergence of individual form-matter composites involves the Sun and the 'oblique sphere': 'Man and the Sun generate a man'.[128] For Albert, the differentiae 'corporeal', 'living', 'sensate', etc., flow from the divine intellect — through the Intelligences *and* spheres — into the sublunar realm. One text in Tract 4 summarizes the matter thus:

> [I]t must be said that the goodness or the illumination that flows from the First flows from it, for the lighting up of all matter, into the Intelligence; and from the Intelligence of the first order [the goodness or illumination] flows into that [Intelligence] that belongs to the second order, and so on. And, from any Intelligence, [the goodness or illumination flows] into its own orb, and from the last orb into the 'sphere of the active and passive', and from the 'sphere of the active and passive' into the centre of each being [*entis*], in which, just as a 'formative' power, it 'forms' the matter to a species [*materiam format ad speciem*]. For, this is the order that all Peripatetics have affirmed.[129]

This last point is important. Similar statements help us gather an idea of Albert's project: to present the common teaching of the Peripatetics. At the same, he acknowledges that each philosopher 'assigns a very different manner to this "flow and influx"'.[130] So, for this doctrine Albert draws on and reports the views of an even wider variety of authors than is usual for him, including Plato, Hermes

128 This account reaches the Latin Scholastics having been modified and reread by Alexander of Aphrodisias, under Avicennian emanationism, and by Averroes; see Freudenthal, 'Astrologization of the Aristotelian Cosmos', pp. 249–54 and 262–63; Cerami, *Génération et substance*; Twetten, 'Whose Prime Mover Is More (un)Aristotelian', pp. 386–89.

129 Albertus Magnus, *De causis*, I.4.6, ed. by Fauser, p. 50, vv. 12–21: 'Et secundum hunc modum dici oportet, quod bonitas fluens a primo sive lumen ad totius materiae illustrationem a primo fluit in intelligentiam et ab intelligentia primi ordinis fluit in eam quae est ordinis secundi, et sic deinceps, et ab intelligentia qualibet in proprium orbem et ab ultimo orbe in sphaeram activorum et passivorum et a sphaera activorum et passivorum in centrum cuiuslibet entis, in quo sicut virtus formativa materiam format ad speciem. Hic enim est ordo, quem omnes posuerunt Peripatetici'. See esp. Moulin, 'Éduction et émanation chez Albert le Grand'; Moulin and Twetten, 'Causality and Emanation in Albert', pp. 713–21.

130 Albertus Magnus, *De causis*, I.4.3, ed. by Fauser, p. 45, vv. 23–24: 'Modus autem istius fluxus et influxus ab antiquis Peripateticis valde diversus assignatur'.

Trismegistus, Asclepius, Isaac Israeli, and Ibn Gabirol. It still comes as a bit of a shock, nevertheless, when Albert reassures us thus: 'But it is not difficult to reduce the statements of Trismegistus to those of Aristotle'.[131]

### *The Problem of Mediate Creation*

Two contemporary authorities, as we have seen, have ascribed to Albert, in light of his 'turn' after 1250, the consciously heterodox claim that, given Albert's acceptance of 'from one comes only one', God immediately can create *ex nihilo* only one thing — and that to create any further things, God requires other posterior causes, such as at least one cosmic Intelligence. Sébastien Milazzo appears to accept this ascription, but saves Albert from heterodoxy because, for Milazzo's Albert, being created *ex nihilo* is no different from being created from a (preexisting) potency or possibility.[132] It is important to return to this issue now, not for an exhaustive treatment, but to sketch an alternative interpretation clearly indicated in the texts.

*Divine Causality and Equal Omnipresence.* In one of his earliest works, Albert outlines and ascribes to Aristotle, based on 'Aristotle's' *Liber de causis* (see note 85 above), as well as to Dionysius,[133] an account of the immediacy of God's causality in relation to a plurality of effects that is consistent with the derivation-scheme we have watched him ultimately develop in his *De causis* some twenty-five years later. The *De resurrectione* responds with considerable sophistication to an objection that affirms the co-causality of Christ and God in relation to our resurrection:

---

131 Ibid., I.4.6, p. 51, vv. 38–39: 'Dicta vero Trismegisti non difficile est ad dicta Aristotelis reducere'. However, Albert apparently intends this judgment to be narrow in scope, applying to the threefold doctrine he summarizes beginning at p. 50, v. 22 and p. 50, v. 72. See also Porreca, 'Albert the Great and Hermes Trismegistus'.

132 Milazzo, in his 'Commentaire' on Albert's *Traité du flux*, p. 244, writes: 'Chaque moteur et orbe est créé par le *medium* qu'est l'*intelligentia* face à l'intellect universellement agent. Plus encore, ce mode de procession ne semble pas entrer en conflit avec la notion d'une *creatio ex nihilo*, puisque le troisième mode d'être de l'intelligence est, en ce qu'elle est en puissance, *ex nihilo*'. Milazzo thereby offers a distinctive interpretation of *De causis*, I.4.8; see below at notes 122 and 189. Given this reading, all mediate creation is also creation *ex nihilo*, with the result that 'la *creatio mediante intelligentia* est ainsi la clef de voûte permettant d'articuler la *processio* et la *creatio ex nihilo*' (p. 263). Accordingly, Milazzo, 'Introduction', p. lxix, observes: 'Ce ne serait trop dire que cette création à partir de la *possibilitas* de la chose est une création *ex nihilo*: car la *possibilitas* ou la potentia en soi d'une chose, sans Dieu qui l'informe, n'est rien'. As a result, he must ask: 'si le *nihil* n'est que la *possibilitas* ou la *potentia* de la chose n'est-il pas aussi la nature suressentielle de Dieu?' (p. lxix n. 85), and he must take pantheism to be a major potential stumbling block for Albert. See again note 189 below.

133 Earlier, Albert had ascribed the doctrine in question to *both* 'the Philosopher' (Aristotle) and Dionysius; Albertus Magnus, *De sacramentis*, I.3, ad 6, ed. by Ohlmeyer, p. 60, vv. 27–31. In fact, Albert's wording is most literally found in Dionysius Latinus, *De divinis nominibus*, III.1 (680B), trans. Sarracenus, p. 122, col. 4: 'Etenim ipsa [Trinitas] quidem universis adest, non autem omnia ipsi adsunt', a passage referred to also under Dionysius's name at Albertus Magnus, *De resurrectione*, I.1 sc 1b ob 4 and ad 4, ed. by Kübel, p. 240, vv. 25–27, p. 241, vv. 33–40.

> Ad aliud dicimus, quod nihil prohibet duo referri ad causam primam ut immediate et tamen unum esse ab altero. Aliter enim est in causa prima et in causis proximis naturalibus, quia dicit Philosophus, quod prima causa aequaliter adest omni rei, et ita non est mediata alicui. Unde natura et effectus naturae immediate sunt a deo, et tamen unum est causa alterius, licet non eodem modo causalitatis, sed natura est conformis causato suo, causa autem prima 'eminens proprietatibus causati'.[134]
>
> In response to [the fifth objection], we say that nothing prevents two things' being related to the First Cause as immediately [from it], and, nevertheless, one is from the other. The case is not the same for the First Cause and for proximate natural causes. For, the Philosopher says that the First Cause is equally present to every thing,[135] and in this way *it is not mediated* in relation to any. Hence, nature and the effects of nature are immediately from God, and nevertheless one is the cause of the other, although not in the same mode of causality: nature is co-formal with its effect, whereas the First Cause 'surpasses the properties of the effect'.[136]

In this early work, then, as in the subsequent *Sentences* commentary,[137] Albert ascribes to Dionysius and Aristotle a comparable doctrine, what I call 'equal omnipresence' — although the ultimate source is actually Proclus, *Elements* 142. The anonymous author/adaptor of the Arabic *Liber de causis* has made Albert's interpretation possible by reading, in place of Proclus's 'the gods' (οἱ θεοί; Proclus, *Elements* 142, p. 124, vv. 26–27: the gods are in the same way present to all things, but all things are not in the same way present to the gods), 'the First Cause' (*al-ʿilla al-ʾūlā; Liber de causis* XXIII, ed. by Taylor, p. 239, vv. 2–4): the First Cause is equally present to every thing.

---

134 Albertus Magnus, *De resurrectione*, II 2.2 ad 5, ed. by Kübel, p. 260, vv. 6–15.

135 I italicize the opening line of *Liber de causis*, XXIII (XXIV) to which Albert refers (and again ascribes to the 'Philosopher's' *Liber de causis* in Albertus Magnus, *De IV coaequaevis*, q. 1.1, ad 4, ed. by Borgnet, p. 310a). The passage as a whole contains the inspiration for Albert's account of how God creates a plurality, as well as for Albert's derivation-scheme: '*Causa prima existit in rebus omnibus secundum dispositionem unam*, sed res omnes non existent in causa prima secundum dispositionem unam. Quod est quia, quamvis causa prima existat in rebus omnibus, tamen unaquaeque rerum recepit eam secundum modum suae potentiae. Quod est quia ex rebus [1] sunt quae recipiunt causam primam receptione unita, et ex eis sunt [2] quae recipiunt eam receptione multiplicata, et ex eis sunt quae recipiunt eam receptione aeterna, et ex eis sunt quae recipiunt eam receptione temporali, et ex eis sunt quae recipiunt eam receptione spirituali, et ex eis sunt quae recipient eam receptione corporali'; *Liber de causis*, XXIII (XXIV), ed. by Fauser, p. 166, vv. 73–79. Notice that this passage displays the 'geometrical' character of 'proposition' (by Aristotle) plus 'proof of' or 'commentary on' the proposition (by a 'commentator'); see also notes 21 and 87 above.

136 As also at Albertus Magnus, *De resurrectione*, I.1, ad sc 1b ob 3, ed. by Kübel, p. 241, vv. 17–20, Albert paraphrases Gerard of Cremona's heavily adapted translation: Aristoteles Latinus, *De caelo*, I.1 268a15; see Albertus Magnus, *De caelo et mundo*, I.1.1, ed. by Hossfeld p. 3, v. 97.

137 Albertus Magnus, *Commentarii in III Sententiarum*, 2.1, ad 2, ed. by Borgnet, pp. 22b–23a: 'Ad aliud dicendum, quod licet Creator ab omnibus creatis aequaliter distat, non /23/ tamen omnia ab ipso distant aequaliter, sicut dicit Dionysius in libro de *Coelesti hierarchia*, et Aristoteles in libro de *Causis*'.

Absent in *De resurrectione*'s discussion of divine 'equal omnipresence' are crucial features of the derivation-scheme in Albert's *De causis*: the doctrines of *esse* as *primum creatum* or first created, and 'from one comes only one'.[138] One might imagine that these doctrines are inconsistent with the 'equal omnipresence' and immediacy of all effects to God. I show otherwise. We can already begin to see their compatibility by considering the last paragraph of *De homine*, where Albert again introduces divine 'equal omnipresence'. Albert asks about the order of the universe, and he affirms a threefold order, the third of which relates all 'to God as creator' (p. 595, v. 5):

> Tertius ordo universi est ad deum, sicut dicit Dionysius quod deus omnibus aequaliter adest, sed non omnia aequaliter sibi assunt, eo quod non aequaliter suas bonitates percipiunt. Bonitates autem suae sunt esse, vivere et cognitio sensibilium et intelligibilium; et entia quae plura de his percipiunt et nobiliori modo, sunt ei propinquiora; quae vero pauciora et modo minus nobili, sunt remotiora per convenientiam essentiae, ut existentia tantum in ultimo gradu, existentia autem viva cognitiva et immortalia in primo, existentia autem et viva et existentia viva et sentientia et existentia viva et sentientia rationalia mortalia in mediis gradibus, eo quod istae sunt causae entitatis participatae a prima causa, et sunt ordinatae, ita quod esse est prima, quae sola influit super causatum, sed nulla alia sine ipsa; vivere autem est secunda, quae non influit sine prima, sed bene influit sine tertia, et sic est de aliis.[139]
>
> > The third order belongs to the universe in relation to God: as Dionysius says, 'God is equally present to all things, but all are not equally present to him', in that all do not receive his 'goodnesses' equally. His 'goodnesses' are 'to be' [*esse*], to live, and the cognition of sensible and intelligible things. Beings [*entia*] that receive more of these things and in a nobler way are nearer to him; but those [that receive] fewer and in a less noble way are more remote as befits [their] essence. Thus, things that merely exist are in the last grade, whereas existing, living, cognitive, and immortal things [are] in the first [grade]; in the intermediate grades, moreover, are: existing and living things; existing, living, and sentient things; and existing, living, sentient, rational, and mortal things. For, those ['goodnesses'] are *causes* of being-ness [*entitas*] participated[140] from the First Cause; and they are ordered such that *esse* is first, which inflows alone upon an effect, but none

138 Depending on what wording is used, the earliest statement of each principle is in Albert's *De IV coaequaevis*. But see also Albertus Magnus, *De sacramentis*, III.5 ob 1, ed. by Ohlmeyer, p. 32, vv. 24–26: 'Unius causae et eodem modo se habentis, ut dicit Philosophus in II *De generatione et corruptione*, est idem effectus'.

139 Albertus Magnus, *De homine*, II.2.2.3c, ed. by Anzulewicz and Söder, p. 595, vv. 32–49.

140 *Participatae* agrees with both *causae* and *entitatis*. *Entitas* can signify *esse* alone, as in *entitas est prima creatura*, or the *esse* of a determinate form, or the *esse* of matter; see Albertus Magnus, *Commentarii in I Sententiarum*, 46.1c (diffinitio 2), ed. by Borgnet, p. 444a; *De resurrectione*, I.6.2, ad 3, ed. by Kübel, p. 249, vv. 38–43. I take *participatae* with *entitatis* in the sense of *esse*, *vivere*, etc.

> of the others [exist] without it. To live, however, is second, which does not inflow without the first ['goodness'], but does inflow well without the third; and so on in the case of the rest.

Here *esse* is unique among the effects of divine, equally omnipresent causality: *esse* flows first, and none of the other forms can be without it. Only given *esse* can the other forms or entities be.[141] Yet, as is made explicit in *De resurrectione*, 'equal omnipresence' is consistent with the order involved in mediated causation. *Esse*, the first thing created, mediates the causal influx in existing, living things, just as *vivere* further mediates the divine causality of existing, living, sensate things. The divine causality is equally present to all through *esse*, I propose. Because of *esse*, the First Cause is immediately present in the influx of *vivere*, etc.[142] Each of these inflowing forms is present at the substantial level of a thing, its *entitas*: *esse* is first, which mediates a thing's *vivere*, *cognoscere*, and so on: without *esse*, there is no *vivere* or *cognoscere*. So all of the forms in a substance are present at once from the origin of that substance through the simultaneous influx from the First Cause, beginning — in the order of nature, not of time — with *esse*.[143] This is the doctrine early Albert ascribes to an Aristotle who, in fact, has been 'Neoplatonized' thanks to the *Liber de causis* and Dionysius together.

Albert continues to ascribe 'equal omnipresence' to the Philosopher's *Liber de causis* in his question-commentary on the *Divine Names*.[144] This fact helps us understand why Chapter 2 there famously ascribes to Aristotle and Dionysius alike efficient, formal, exemplar, and final causality as belonging to the First.[145] As a result, says Albert, in regard to the procession of things from the First (p. 72, vv. 35–36), the opinion of Aristotle should be preferred to that of Plato as truer and as in no way contrary to the faith, insofar as Aristotle's eductive explanations take into account form and matter rather than form alone. Accordingly, upon arriving at the source text for 'equal omnipresence', Albert writes:

---

141 Cf. *Liber de causis*, XIX (XX), ed. by Fauser, p. 158, vv. 79–80: 'Et bonitas prima non influit bonitates super res omnes nisi per modum unum, quia non est bonitas nisi per suum esse'.

142 Compare also the similar use of *esse*, above at note 92.

143 We see the non-temporal cosmic influx in a parallel text that affirms 'equal omnipresence' and also clarifies the role of participation: 'duplex est processio bonitatis divinae in entia: quaedam enim est boni, secundum quod bonum est diffusivum sui et esse per actum creationis, secundum quem modum vocat ea quae non sunt tamquam ea qua sunt: et sic procedit in omnia bonum, secundum quod dicitur, *Vidit Deus cuncta quae fecerat, et erant valde bona* [Genesis 1]. Et licet ipsum sit aequaliter se habens ad omnia, non tamen aequaliter se habent alia ad ipsum: et ideo dicit Dionysius, quod participatur quod est imparticipabile: in se quidem imparticipabile, quia aequaliter adest: sed in aliis quae non aequaliter sibi adsunt, participatum: quia aliquid recipit secundum esse tantum: aliquid secundum esse, et vitam: aliquid secundum esse, vivere, et sentire: et aliquid secundum esse, vivere, sentire, movere, et intelligere'; Albertus Magnus, *Commentarii in I Sententiarum*, 14.1 (Quid sit temporalis processio?), ad 3, ed. by Borgnet, pp. 390b–391a.

144 Albertus Magnus, *Super Dionysium De divinis nominibus*, V.2, ed. by Simon, p. 303, v. 46.

145 Ibid., II.45 sol., p. 73, v. 41–p. 74, v. 20.

> [C]um dicitur: deus adest omnibus uno modo, omnia autem non uno modo assunt ei, sic intelligendum est, quod ipse non abest ab aliqua re, dans omnibus esse et permanentiam, res vero non recipiunt uno modo ab ipso, sed diversimode, et secundum unam processionem eius et secundum plures; secundum plures quidem, quia quaedam participant tantum esse, quaedam vero cum hoc vivere, quaedam vero cum his etiam sentire, quaedam vero ulterius ratiocinari et quaedam cum his etiam intelligere, participantia de bonitatibus primi, quidquid participabile est a creatura, secundum unam vero diversimode, sicut sapientiam ipsius aliter participant intellectualia et aliter rationalia et secundum obscuram resonantiam sensibilia, sicut etiam sol eodem modo agit in inferiora, quae non eodem modo recipiunt ipsum. Et hoc est etiam, quod dicitur in *Libro de causis*, quod prima causa similiter se habet ad omnes res, sed non omnia eodem modo se habent ad ipsam.[146]
>
> When it is said, 'God is present to all things in one way, but all things are not in one way present to him', it should be understood as follows. He is not absent from anything, as he gives 'to be' [*esse*] and permanence. But things do not receive in one way from him, but in diverse ways [...] just as also the Sun acts in the same way upon things here below, which [things] do not receive [its effect] in the same way. This is also what is said in the *Book on the Causes*: the First Cause is related in a similar way to all things, but all things are not related in the same way to it.

As I have already indicated, Albert's question-commentary on the *Divine Names* and *Physica* paraphrase (of circa and post-1250) affirm as philosophically preferable the simultaneous creation *ex nihilo* of a plurality of effects, even in a first moment of time, following the solution of Maimonides — a solution Albert later rejects at the end of *Metaphysica* XI, circa 1263.[147] As we have seen in the quotations from *De resurrectione* and *De homine*, this position, which involves 'equal omnipresence' and immediacy, is not inconsistent with there being an order among created effects such that one causes the other. The *Divine Names* question-commentary does not emphasize this point, however. Its concern seems

---

146 Albertus Magnus, *Super Dionysium De divinis nominibus*, III.7 sol., ed. by Simon, p. 105, vv. 45–64 (with underscoring for the end of the ellipsis).

147 See notes 22, 83, and 107–08 above. A passage from Tract 1 of the paraphrase of *Physics* 8.1 shows how sharply Albert there, with Maimonides, contrasts 'Aristotle's' necessitating will and emanation scheme through mediating Intelligences (Maimonides' 'Aristotle' in *Guide* 2.20–22 seems to be Avicenna) with the creationism that Albert had previously defended in chapter 13 through particularizing argumentation: 'Et differunt opiniones nostrae in tanto, quod ipsi [Peripatetici et ipse Aristoteles] dicunt caelos fluere a causa prima mediantibus intelligentiis, quae sunt primae in ordine eorum quae sunt, nos autem dicimus, quod absolute fluunt a prima causa secundum electionem suae voluntatis'. Albertus Magnus, *Physica*, VIII.1.15, ed. by Hossfeld, p. 580, vv. 45–50. On Maimonides' particularizing arguments in *Guide* 2.19, see Davidson, *Proofs for Eternity*; Seeskin, *Maimonides on the Origin of the World*. For Albert's early reliance on Maimonides, see Rigo, 'Zur Rezeption des Moses Maimonides'; Di Segni, 'Early Quotations from Maimonides's *Guide*'.

to be to affirm Maimonides' particularist reasoning in defending divine free will in creating all things. It emphasizes that a plurality of effects can come from one simple First Cause in light of a plurality of divine ideas that are merely conceptually distinct from each other and from the one, simple Godhead[148] — a doctrine that Albert associates alike with Dionysius and with 'Aristotle', the author of the propositions and some commentaries in the *Liber de causis.*

*The Rejection of Mediated Creation.* An objection arises: to focus only on *Liber de causis* XXIII (XXIV) and 'equal omnipresence' is to ignore the (in)famous doctrine of *Liber de causis* III, a doctrine that Albert appears to embrace in the derivation-scheme of his later *De causis*: 'causa prima creavit esse animae mediante intelligentia' ('the First Cause created the being [*esse*] of soul by the mediation of Intelligence'). Still, already in *De quattuor coaequaevis* Albert is aware of, and sidesteps, the mediate creation suggested in this text. The *Liber de causis* is an authority and deserves a charitable reading to explain its imprecise language. Albert writes: '*creation* is affirmed in the authority for any causality whatsoever, not for the eduction from nothing into being [*esse*]'.[149] Similarly, in *De homine* Albert explains the Intelligence's 'mediate creation of soul' through its generation of animated bodies by moving the spheres, that is, not through the celestial soul's emanation, but through the coming to be of non-human terrestrial souls.[150]

A significant shift in Albert's thought on the present topic comes not in 1250, but when he paraphrased *Metaphysics* Lambda circa 1263 with the help of Averroes's Long Commentary. He affirms what is, for him, a rather narrow reading of Aristotle's cosmology, asserting with the Peripatetics no more supernal causes than there are motions that need explanation; and he rejects Maimonides' particularist reasoning in philosophy.[151] The former assertion is hardly reconcilable with 'Aristotle's' *Liber de causis,* as will soon become evident when Albert turns to paraphrase it.[152] Furthermore, his rejection of Maimonides requires him to find

148 One finds this doctrine at least as early as Albertus Magnus, *De IV coaequaevis*, q. 21, a. 1, ad 4, ed. by Borgnet, p. 464a: 'dicendum, quod prima causa in creatione est unum ut multa: unum in substantia, et ut multa in ideis, quod per Augustinum patet in libro LXXXIII *Quastionum*'; as well as in *Commentarii in II Sententiarum*, 1.10, ad 9, ed. by Borgnet, p. 30a (cf. ob 9b, p. 26b). Compare Albert's exposition of the image of the center of a circle in *De divinis nominibus*, II.5, p. 644A: 'just as all the radii of a circle meet in one center, so the *rationes* that are diverse in creatures, such as good and beautiful, are united in one, simple God'; Albertus Magnus, *Super Dionysium De caelesti hierarchia,* VII, expos. 14, q. 3c, ed. by Simon and Kübel, p. 114, vv. 78–89.

149 Albertus Magnus, *De IV coaequaevis*, q. 1, a. 3, ad sc 4, ed. by Borgnet, p. 313a: 'Unde creatio ponitur in auctoritate pro causalitate quacumque, et non pro eductione in esse de nihilo'. See also Anzulewicz, 'Die Emanationslehre des Albertus Magnus', p. 222.

150 Albertus Magnus, *De homine*, 'Utrum animae rationales immediate creentur a deo, an ab intelligentiis angelicis' ob and ad 4–6, ed. by Anzulewicz and Söder, p. 75, vv. 47–60 and p. 76, vv. 61–69.

151 For Maimonides, see above notes 83, 107–108, 147 and at note 182 below. For the narrow Aristotelianism, see Twetten, 'Albert's Arguments for the Existence of God', pp. 663–69.

152 Perhaps Albert in his *De causis* suggests a reason for the disharmony with Aristotle's *Metaphysics*: Aristotle in Books XII and XIII of the *Metaphysics*, i.e., in Books M and N, made determinations

an alternate philosophical account of how a plurality of creatures proceeds from God. The account is supplied by the *Liber de causis*, but it is coupled with the derivation-scheme of Book I, Tract 4, adapted from Avicenna.

When Albert turns to Book II's paraphrase, he must, of course, take up *creavit esse [...] mediante intelligentia*. Yet Albert could scarcely be clearer in rejecting mediate creation. First, in clarifying *Liber de causis* IV, he insists that 'all of the ancients' ascribe creation *ex nihilo* to the First Cause:

> Supponitur autem ab omnibus antiquis, quod *esse primum* causatum *est*. Propter quod ab antiquis non tantum causatum, sed *creatum* esse dicitur. Sicut enim saepius dictum est, creare ex nihilo producere est. Quod autem causat non supposito quodam alio, quo /81/ causet, consequenter sequitur, quod causet ex nihilo. Primum autem causat non supposito quodam alio, quo causet. Primum ergo causat ex nihilo. Causatio ergo ipsius creatio est.[153]
>
> > All of the ancients have supposed that being [*esse*] is the first thing caused. For this reason being [*esse*] is called by the ancients, not only 'caused', but also 'created'. For, as has very often been said, to create is to produce out of nothing. But it follows for that which causes without the supposition of anything else with which it causes that it causes out of nothing. But the First causes without the supposition of anything else with which it causes. Therefore, the First causes out of nothing. Therefore, its causation is creation.

Accordingly, upon arriving at the second mention in the *Liber de causis* of the Intelligence's mediation in 'creating' soul, Albert writes:

> Causae primae enim virtus creativa est et praeparativa omnium. Virtus autem intelligentiae, in quantum est intelligentia, non est creativa, sed potius creati formativa et determinativa. Et similiter virtutes omnium sequentium. Nihil enim sequentium virtutem creativam habere potest. Omnia enim sequentia causant causante quodam alio praesupposito in ordine causalitatis. Causant ergo ex aliquo et non ex nihilo. Talis autem modus causalitatis non est creativae virtutis. Adhuc, virtus causalitatis per formationem minor est, quam sit virtus causalitatis per creationem, et est casus ab illa. Virtus ergo creativa non nisi primae causae est.[154]
>
> > The power of the First Cause is creative and 'preparative' of all things. However, the power of an Intelligence as such is not creative; instead, it is *formative* and *determinative* of what has been created. [Something]

---

regarding separate substances only 'according to opinion', whereas the *Liber de causis* determines 'according to the full truth'; Albertus Magnus, *De causis*, II.1.1, ed. by Fauser, p. 59, v. 36–p. 60, v. 5.

153 Albertus Magnus, *De causis*, II.1.17, ed. by Fauser, p. 80, v. 73–p. 81, v. 4 (lemma from *Liber de causis* IV, ed. by Fauser, p. 88, v. 65).

154 Albertus Magnus, *De causis*, II.2.17 (Quod intelligentia omnem suam virtutem recipit a causa prima), ed. by Fauser, p. 110, vv. 6–19 (on *Liber de causis* VIII [IX]).

> similar [is true of] the powers of all subsequent things. Indeed, none of the subsequent things can have creative power. For, all subsequent things cause without the supposition of any other thing causing in [some] order of causality. Therefore, they cause 'out of something', not out of nothing. But such a mode of causality does not belong to the creative power. Furthermore, the power of causality '*through formation*' is less than the power of causality 'through creation', and it is a falling away from that. Therefore, the creative power belongs to the First Cause alone.

Notice especially Albert's use, again, of the distinction between 'information' and 'creation', a distinction that we witnessed first in *De resurrectione*, written more than twenty-five years prior.

Nevertheless, Albert does more than insist that God alone, not an Intelligence, 'creates'. In commenting on *creavit esse [...] mediante intelligentia* of *Liber de causis* III, he explains precisely how we should conceive the mediating role of the Intelligence in the flow of forms. His account uses the aforementioned strategy of Concept-Procession, as follows:[155]

> Et hoc est, quod quidam antiquorum dixerunt, quod *prima causa creat animam mediante intelligentia* et *alatyr*,[156] non quod intelligentia pro medio prima causa utatur, sed quod forma intelligentiae media sit in esse diffinitionis animae, sicut sensibile medium est in esse diffinitionis hominis, cum dicitur vivum sensibile rationale. Et hoc modo terminus in esse animae nobilis est proportio ad alatyr, non quod alatyr sit faciens animam vel constituens, quia mobile non constituit motorem, sed e converso mobile per motorem constituitur. Haec autem omnia in prima causa unitive sunt et simpliciter, sed procedentia ab ipsa diversitatem accipiunt et compositionem, sicut in sequentibus erit manifestum. Tale ergo esse et constitutionem in esse a prima causa accipit anima nobilis. Propter quod in tertio gradu est inter causas primarias. [...] Et sic dixerunt antiqui Peripatetici, quod *prima causa creat animam* nobilem intellectualem *mediante intelligentia*, non quod intelligentia, quae est causa secunda, utatur pro medio, sed quia lumine eiusdem intellectus agentis, quo constituit intelligentiam, constituit et nobilem animam et hominis animam, in quantum nobilis est in esse naturae intellectualis.[157]

> This is what certain of the ancients said: *the First Cause creates the soul by mediation of the Intelligence* and of *alatyr* — not that the First Cause uses the Intelligence as a middle [or means], but that the form of the Intelligence is a middle in the 'being' [*esse*] of the definition of the soul,

155 See notes 92–93 above.

156 One manuscript reads *alachir*, perhaps drawn from *al-ʿaql* or intellect. Albert reads *alatyr* and conjectures, as we shall see in the next quoted text, a cosmological explanation of the term: the celestial sphere according to its size, direction, and velocity.

157 Albertus Magnus, *De causis*, II.1.13, ed. by Fauser, p. 76, vv. 23–39 and p. 76, vv. 79–85 (lemma from *Liber de causis* III, ed. by Fauser, p. 79, v. 77).

> just as 'sensible' is a middle in the being [*esse*] of the definition of 'human' when [human] is called 'living, sensible, rational [thing]'. In this way, the terminus in being of the Noble Soul is the proportion to *alatyr* — not that the *alatyr* makes or constitutes the soul; for, what is mobile does not constitute the mover, but, conversely, the mobile is constituted by the mover. All these things, moreover, exist simply and unitedly in the First Cause, but as they proceed from it they receive diversity and composition. [...] Therefore, the Noble Soul receives [its] sort of being [*esse*] and [its sort of] constitution in being [*esse*] from the First Cause. For this reason, it is [located] in the third degree among the primary causes. [...] In this way, the ancient Peripatetics said that the First Cause creates the noble intellectual soul by the mediation of Intelligence — not that it uses the Intelligence, which is the second cause, as a middle [or means], but because by the illumination of this same [divine] agent intellect by which it constitutes the Intelligence, it constitutes also the Noble Soul — and the human soul inasmuch as it is noble in the being [*esse*] of an intellectual nature.[158]

As Albert emphasizes, then, the First Cause uses the causality of the Intelligence not as an instrument or tool, as it were,[159] but as the (filtered) source of an intermediate form in the definition of the subordinate effect, such as in the soul's definition: that is, as source of a form that is in the middle between the other forms in the soul's definition. I say 'filtered' because the ultimate source of this mediating form is the divine agent intellect, which is full of forms.

In another, similar passage in the same chapter, commenting on *Liber de causis* III, we find brought to fruition the notion of a 'flow' of forms that is neither equivocal nor purely univocal, as Albert had described in Book I, Tract 4: there is a sharing of form between cause and effect without the effect's being univocal with its cause.[160] At the same time, Albert insists that things that are mediately caused thanks to the flow of forms are, like all things, also created by the First Cause. How can such things be both mediately caused and immediately created? They are created with respect to their being or *esse*, explains Albert. Therefore, he begins by rehearsing again what creation is:

---

158 The mediate causality that Intelligence exercises over the intellectuality of the celestial and human intellectual soul alike perhaps helps explain why sometimes in previous Aristotelian paraphrases, Albert apparently ascribes to Intelligence below God the source of the human intellectual soul, as is perceptively seen by Hasse, 'Avicenna's "Giver of Forms"', pp. 238–41 and 245. Albert does not say, of course, that the Intelligence is the creator or cause of *esse* of the human soul. He need merely invoke the distinction between creation and 'information', which, as we have seen, he discovers in 'Aristotle's' *Liber de causis* already in his earliest works.

159 Albert does speak of the 'second' as a 'quasi instrument' of the First, where the 'second' apparently refers to a form; see note 100.

160 See above at note 100, in the discussion of 'Light-Procession'.

Primae autem causae, quae causat non causante quodam alio, proprius actus creare est. Quod enim causat non causante quodam alio ante se, ex nihilo facit omne quod facit. Si autem praesupponeret aliud ante se causans, non ex nihilo faceret, sed id quod iam est, formaret in id quod facit et causat. Actus igitur primae causae proprie creatio est. Esse autem, 'quo res est', primum est, quod ante se nihil praesupponit. Esse igitur in omnibus quae sunt, primae causae proprius effectus est, sicut in superiori libro nos probasse meminimus. Oportet igitur, quod esse animae a prima causa creetur, sicut esse omnium a prima causa creatum est. Anima autem dicit esse formatum ad speciem. Formatio autem ista oportet, quod causam habeat in formatore et agente. Aliter enim formator non esset per se agens et univoce. Sicut dicit Aristoteles in II *Physicorum*, quod 'Polyclitus non est causa statuae nisi per /76/ accidens, sed Polyclitus statuarius causa statuae est per se'. Si ergo anima nobilis formata est in principium motus ad formam intelligentiae, oportet, quod anima nobilis formam participet intelligentiae. Non autem participat formam intelligentiae ut intelligentiae nisi per intellectum vel naturam intellectualem. Forma igitur animae nobilis et esse non deducitur ad perfectam speciem motoris nisi mediante intellectualitate. Propter quod animam, prout principium motus est in corpore nobili, intellectualitate determinari necesse est. Ex hoc tamen non habet, quod anima sit. Anima enim secundum quod anima principium et causa motus est in eo quod movetur a seipso. Non autem principium motus est nisi forma determinante et proportionante ipsam ad mobile. Oportet igitur, quod differentia ultima finiens et determinans esse nobilis animae sit ad mobile circulariter inclinatio et proportio. Caelestis autem circulus apud sapientes Arabum 'alatyr' vocatur. Esse igitur animae est intellectualitate formatum et ad alatyr determinatum. Hoc igitur modo anima nobilis a causa prima procedit, ut in esse nobilis animae constituatur.[161]

> To create is the proper act of the First Cause, which causes without anything else's causing. For, what causes without anything else prior to it causing makes out of nothing everything that it makes. But were it to presuppose another [thing] causing prior to it, it would not make [its effect] *ex nihilo*, but it would 'form' that which already exists into that which it makes and causes. Therefore, the act of the First Cause is properly 'creation'. Being [*esse*] 'by which a thing is', moreover, is what is first, which presupposes nothing prior to itself. Therefore, the being [*esse*] in all things that are is the proper effect of the First Cause [...]. The being [*esse*] of the soul must therefore be created by the First Cause, just as the being [*esse*] of all things is created by the First Cause. What is more, the soul is said to be 'formed to a species'. Yet, that 'formation' must have a cause in 'what forms' and in [what is] an agent. Otherwise 'what forms' would not be a per se and univocal agent. Thus, Aristotle

161 Albertus Magnus, *De causis*, II.1.13, ed. by Fauser, p. 75, v. 56–p. 76, v. 22 (emphasis added).

> says in *Physics* II: 'Polyclitus is only a per accidens cause of the statue, but the sculptor Polyclitus is the per se cause of the statue'. If, therefore, the Noble Soul has been 'formed' into a principle of motion according to the form of the Intelligence, the Noble Soul must participate in the form of the Intelligence. But it only participates in the form of Intelligence as Intelligence through [its] intellect or intellectual nature. Therefore, the form and being [*esse*] of the Noble Soul is only brought into [its] complete species as mover by the mediation of intellectuality. For this reason, soul, according as it is a principle of motion in the Noble [celestial] Body, is necessarily *determined* by intellectuality. Nevertheless, not from this does it have the fact that it is soul. For, soul as such is a principle and cause of motion in what is moved by itself. But it is only a principle of motion by a form determining and proportioning it to the mobile [body]. Therefore, the ultimate differentia that finishes and determines the *esse* of the Noble Soul must be an inclination and a proportion to what is circularly mobile. But the celestial circle is called 'alatyr' by the Arab philosophers. Therefore, the being [*esse*] of the soul is 'formed' by intellectuality and *determined* to the alatyr. In this way the Noble Soul proceeds from the First Cause so that it is constituted into the being [*esse*] of a Noble Soul.

Sometimes we have seen Albert using 'forming' and 'determining' almost as synonyms. But strictly speaking, 'forming' is the result of the 'mediating' cause, Intelligence, whereas 'determining' is subsequent in a definition — in this case arising in relation to a further effect, namely, the heavenly body.[162]

With these texts in mind, we will have no difficulty reading Albert's subsequent paraphrase of *creavit esse [...] mediante intelligentia* in *Liber de causis* III, although our scholars correctly find no clear rejection of mediate creation here. Albert paraphrases as follows:

> *Et* sicut ex praehabitis patet, *has* tres *operationes non efficit anima* ex se secundum 'id quod est' nec ex distantia, qua distat a primo, sed secundum quod *est exemplum virtutis superioris*, cuius virtus a superiori causa sibi influxa est et salvatur in ipsa. Et huius quidem explanatio *est*, quod *prima causa creavit esse animae*; non quidem simplici esse, sed *mediante intelligentia*, quod est esse formatum ad lumen et virtutem intelligentiae; nec tantum mediante illo, sed esse illud produxit ulterius secundum *alatyr*, scilicet ut sit proportionatum secundum influentiam vitae et motus caelesti circulo per modum, qui dictus est.[163]

> *And*, as is clear from what has previously been established, *soul does not effect these* three *operations* ['divine', 'intellectual', and 'animate'] from itself

162 See also note 98 above.

163 Albertus Magnus, *De causis*, II.1.16, ed. by Fauser, p. 80, vv. 8–19 (lemmata from *Liber de causis* III, ed. by Fauser, p. 79, vv. 76–77).

> according to 'that which [it] is' [*id quod est*], nor based on the 'standing apart' [*distantia*] by which it 'stands apart' [*distat*] from the First, but [instead] according as [soul] is *the exemplification of a superior power,* [that is,] of that whose power has been 'inflowed into' it [soul] — and is preserved in it — by a superior cause. Indeed, the explanation of this *is* that *the First Cause created the being* [*esse*] *of soul* — not, in fact, by mere being [*simplici esse*], but *by the mediation of Intelligence* — which [medium] is 'formed being' [*esse formatum*], ['formed'] according to the illumination and power of Intelligence. Nor [does this occur] merely with that [illumination] mediating, but [the First Cause] produced that being [*esse*], further, according to the *alatyr,* that is, so as to be proportioned to the celestial circle, according to the inflowing of life and of motion, in the manner that has been stated.

When Albert admits here that 'the First Cause created the being [*esse*] of soul by the mediation of Intelligence', he refers only to the 'formed being [*esse*]', the 'being intellectual', that belongs to the noble soul 'through the information' of Intelligence. The distinction between causing 'through creation' and causing 'through information' is of crucial importance. To cause *being* alone is not the same as to cause, mediately, 'formed being'.

As we have seen, Albert can include in 'noble soul' the human soul as intellectual. But one final paraphrase makes clear that Albert refers to the mediating role that an Intelligence has, not merely over supernal causes, such as celestial souls, but also over humans, that is, over things herebelow. As a result, the promise of *Liber de causis* I is fulfilled: the First Cause exerts its causality more powerfully in each thing than does any second cause. Albert paraphrases as follows:

> *Et* hoc *quidem exemplificare* possumus per ea quae supra diximus, *esse* scilicet, *vivum et hominem,* sive per esse, vivere et intelligere. In qualibet enim *re* causata *primum est esse,* in quo fundantur omnia sequentia; *deinde* est *vivum,* per quod esse formatur; *postea* est *homo* sive rationale, per quod determinatur esse et vivere ad speciem. [...] /68/ *Non enim figitur* sive fundatur in esse *causatum causae secundae nisi per virtutem* sive influentiam *causae primae.* Et huius quidem causa *est, quia quando secunda causa facit rem* per formam suam, tunc *causa prima, quae est* superior quam causa secunda, per *virtutem* suae influxionis *influit super rem* eandem esse in quo fundatur formatio causae secundae.[164]

> *And* we can, *indeed, exemplify* this through what we have said above, namely, [through] *being* [*esse*], *alive, and human* — or through being [*esse*], living, and understanding. For, in any caused thing, the first [thing] is being [*esse*], on which are founded all subsequent things; *next* is *alive,*

164 Albertus Magnus, *De causis,* II.1.6, ed. by Fauser, p. 67, vv. 46–52 and p. 68, vv. 26–33 (lemma from *Liber de causis* I, ed. by Fauser, p. 67, vv. 78–79 and p. 68, vv. 77–78 [with some modifications]).

> through which being [*esse*] is formed; *afterwards* is *human* or rational, through which [two], being [*esse*] and living [*vivere*] are determined to a species. [...] For, *the effect of the second cause is fashioned* — or founded in being [*esse*]— *only through the power,* or the inflowing, *of the First Cause.* The reason for this *is that when the second cause makes a thing* through its form, *the First Cause, which is* above the second cause, through *the power* of its influx, *inflows over the* same *thing* being [*esse*], on which is founded the 'formation' of the second cause.

In sum, Albert's *De causis* stands in continuity with his long-standing teaching, starting from *De sacramentis,* on the 'equal omnipresence' — and immediacy — of the First Cause to all of its effects. From the outset, this doctrine was understood to be consistent with there being an order among effects such that one also causes another. In the *De causis,* Albert continues to affirm 'equal omnipresence': 'The First, about which we have spoken, penetrates all things because of its hyper-simplicity; and there is nothing to which it is lacking in virtue of its existing always and everywhere'.[165] But now he explains the order among effects, using 'from one comes only one', and 'the first effect is being [*esse*]' (from the *Liber de causis*). Whereas God alone causes *esse,* every subsequent form in a thing's essence and definition is filtered also through the Intelligences, the remote movers of the heavens. Such subsequent forms presuppose *esse,* which is created by God alone, and so they are said by Albert to be caused not through creation, but 'through the information' of the Intelligences. An Intelligence, in so causing, is not so much a tool as a filter through which the form flows, with a certain degree of univocity, from the divine agent intellect, with the result that the form enters — and in this lies mediation — into the middle, or into the midst of a thing's essence and definition. Despite this 'mediation' in causality of forms that are posterior to being or *esse,* it remains, as 'the Philosopher' teaches in the *Liber de causis,* that God is more causally efficacious in each effect than is any second cause. And so, the First Cause can continue to be said to be immediately present to each effect through an ordered causality (except in the case of being or *esse*). Let's distinguish between 'immediately present as a cause' and 'immediately causing'. The First Cause is immediately present to its effects, but causes all forms mediately except being or *esse.*

*Albert's Preference of 'Aristotle's' Emanationism over Maimonides' Creationism.* With *De sacramentis* and 'equal omnipresence', then, I have established a line of continuity in Albert's thought on divine causality, a line that underlies his

---

165 Albertus Magnus, *De causis,* I.4.1, ed. by Fauser, p. 43, vv. 54–56: 'Primum enim, de quo locuti sumus, propter suam nimiam simplicitatem penetrat omnia; et nihil est, cui desit ubique et semper existens'. Cf. ibid., II.2.14, p. 108, vv. 1–2, vv. 8–13: 'prima causa est et operatur in omnibus quae sunt [...]. In omnibus enim generaliter hoc est, quod nulla bonitas fluit nisi per bonitatem divinam, quae in ipsa est; et quod nulla habet esse nisi per bonitatem illam, secundum quod a primo est, sive sit illa bonitas in primo et a primo immediate, sive sit a primo operante in alio'.

philosophizing, first with Maimonides' particularization arguments, then with Avicenna's derivation-scheme. Albert in his *De causis* continues to single out Maimonides' rejection of the derivation-scheme belonging to 'Aristotle and all of those who have followed Aristotle'.[166] As Albert puts it, 'Rabbi Moses alone went against this way [of Aristotle], intending to theologize'.[167] The passage is an important one, forming a prologue, as it were, within the chapter in Book II where Albert begins his paraphrase of the opening lines of *Liber de causis* I. At the outset of the chapter, Albert raises an objection: Given that 'from one comes only one', it would seem that since the First Cause stands in one, same causal relation to all, each of its effects is equally near to it, and one effect is never prior to another. In response, Albert summarizes in a paragraph 'Aristotle's' emanation system and the reason for the ordered flow of things. Then he adds: 'And, this is the *way* that we have already pursued in what precedes'.[168] Thus, this passage, as the Cologne editors indicate, links Book II's discussion of *Liber de causis* I with Albert's own explication of the derivation-scheme in Book I, Tract 4 — and especially in Chapter 8.

The reader may be surprised or dismayed at Albert's preference for 'Aristotle's' emanationism over Maimonides' creationism at this point. Were it not for the express linkage with the personal thought of Book I, one might imagine that Albert could only write thus under the disclaimer that he is merely reporting the Peripatetic position. Albert proceeds to suggest why Maimonides (and Albert himself previously) mistakenly opposed 'Aristotle'. The passage requires interpretation, but perhaps what Albert means is the following. Divine wisdom is unique in that God, by understanding his idea of being, emits the form being; by understanding his idea of good, emits the form goodness, by understanding the idea of life, emits the form life, and so on. By contrast, Maimonides starts with the plurality of things as if that comes first, then asserts that divine wisdom, as a result of knowing that plurality, produces it. For Albert, this account is excessively modelled on human wisdom, which starts from things. In the end, however, Albert admits, 'what Rabbi Moses says converges with what the ancient

---

166 The phrase opens Albert's paragraph summary of 'Aristotle's' emanation system, ibid., II.1.6, ed. by Fauser, p. 66, vv. 66–78, in the chapter entitled 'Quid sit causa ordinis istius, quod una primaria est et altera secundaria': 'Aristoteles enim et omnes qui Aristotelem secuti sunt, dicunt, quod procedens a primo in eo quod procedens est vel fluens, efficitur secundum et inferioris gradus quam primum. Procedit enim in diversitate essentiae eo quod primum indivisibile sit secundum essentiam. Et cum esse habeat a primo, necesse est, quod habeat convenientiam cum primo. Et per esse, quod habet, efficitur causa. Sed per "id quod est" efficitur causa distans a primo. Secundum ergo quod distans est, secundum est. Secundum autem quod causa est, effluens est et processus emittens. Et id quod procedit ab ipso, tertium est eadem ratione distans et differens'.

167 Ibid., II.1.6, p. 66, vv. 92–93: 'Rabbi Moyses autem solus ivit contra hanc viam theologizare volens'. Albert appears to have in mind his previous criticism of Maimonides in his *Metaphysica* XI, as if to say: Maimonides' approach was to introduce into his philosophizing what belongs exclusively to the realm of faith in the miraculous. On Maimonides, see again the passages referred to in note 151.

168 Albertus Magnus, *De causis*, II.1.6, ed. by Fauser, p. 66, vv. 81–82: 'Et hanc viam iam in antehabitis prosecuti sumus'.

Peripatetics said'.[169] For both Albert's Maimonides and Albert's Aristotle, in other words, the divine wisdom, in virtue of conceptually distinct ideas, is the cause of the plurality of things, rather than the reverse. 'Aristotle's' way is to be preferred because it explains the ordered plurality of things in a law-like way as proceeding from the divine wisdom itself.

Another passage regarding the divine will may help us understand the point. As Albert explained in his *De causis*, Book I, Ibn Gabirol affirmed that two things, both form and matter, proceed from God, with neither flowing through the mediation of the other.[170] In order to defend an orderless plurality, Ibn Gabirol had therefore to reject the principle 'from one comes only one' and to affirm the primacy of the divine will. Against the latter affirmation, Albert reacts as follows:

> Adhuc, multipliciter probatum est, quod voluntas ut voluntas accepta universi esse non potest esse primum principium. Voluntas enim ut voluntas diversis disponitur ad volendum diversa. Diversis autem ad agendum diversa disponi primum principium penitus absurdum est. Adhuc, agens per voluntatem ante se habet agens aliud, quod est agens per essentiam simplicem. Agens ergo, quod est principium universi esse, non est agens per voluntatem.[171]
>
> > It has been proved in many ways that the will, taken as will, cannot be the first principle of being in general [*universi esse*]. For, the will *as will* is disposed by diverse things toward willing diverse things. But it is thoroughly absurd that the First Principle be disposed by diverse things toward performing diverse [actions]. Furthermore, an agent [acting] through will has prior to it another agent that acts through [its] simple essence. Therefore, the agent that is the principle of being in general [*universi esse*] is not an 'agent through will'.

This surprising passage helps bring together our discussion of free will and mediated emanation. For Albert, the First Cause is an agent that, as we have seen,[172] acts freely through its essence — which essence is extramentally identical to its will — in (i) creating *ex nihilo* all being (*esse*) and in (ii) emanating the rest of things in a law-like way such that from one conceptually distinct divine idea, one form enforms subsequent beings thanks to the mediation of the Intelligences and the celestial souls that move the heavenly bodies. As a result, Albert, unlike his Maimonides, affirms a certain order in the plurality of effects of the First Cause such that only one form most properly reflects the essence of the First Cause: being (*esse*), which together with mediated forms constitutes a multiplicity of things that are (simultaneously) caused (even *ex nihilo*).

---

169 Ibid., p. 67, vv. 17–19: 'Et ideo, quod dicit Rabbi Moyses, ad idem redit cum eo quod dixerunt antiqui Peripatetici'.

170 Ibid., I.1.6, p. 13, v. 68–p. 14, v. 5; ibid., I.4.8, p. 55, vv. 72–80.

171 Albertus Magnus, *De causis*, I.1.6, ed. by Fauser, p. 14, vv. 6–15.

172 See above at notes 68 and 71.

## Final Objections

In the personal part of the last of the Aristotelian paraphrases, the work that culminates philosophical wisdom, or metaphysics, Albert has changed his mind on his assessment of Peripatetic derivation-schemes. They *can* be harmonized with creation *ex nihilo*, it appears, and he proposes and defends a version that is, in fact, harmonious. One may still doubt, of course, that Albert intends to accept that scheme as his own. This is the question of the role of disclaimers in his paraphrases, and, indeed, of the purpose of his paraphrases on Dionysius, Aristotle, and Plato (projected) in the first place.

This is not the place to take up that question. But elsewhere I have charted the remarkable degree of continuity in the accounts of 'celestial causality' between Albert's philosophical paraphrases and in his personal works.[173] I have also shown the surprising degree to which the disclaimers concern this material.[174] 'Celestial causality' was a major issue for Albert, and one on which he changed his mind radically several times. Now I add to that issue another major element of the 'Peripatetic' cosmology passed on by the Arabic philosophers, the 'derivation-scheme'. There is no question that the disclaimers give Albert freedom to express himself without criticism, whether or not he personally accepts the cosmology he welcomes. On this issue, unlike for celestial causality, we lack evidence outside of the paraphrases on Albert's late personal stance regarding a nuanced scheme. In his late *Summa theologiae*, assuming that the relevant passages can be shown to be authentic (see note 32 above), he returns to his previous habit of criticizing the necessitarian scheme that involves mediated creation. However, the evidence of his increasing personal acceptance and defence of celestial causality over the course of his writings gives us good grounds for supposing his (provisional) acceptance of the late recommended derivation-scheme as well. In fact, we find Albert in *Summa theologiae* II still affirming that the first thing that proceeds from 'the First' or God is a pure, simple Intelligence, despite the fact that other things also proceed from him, 'as is philosophically demonstrated in the *Liber de causis*'; for what proceeds from him first, as Avicenna says, is *esse* alone, in which all the rest that proceed are not distinguished.[175]

At the same time, we may still wonder why Albert in his *De causis* seems so nonchalant in his treatment of the problem of free will, especially since circa 1250–51 he had criticized Aristotle for his necessitarianism and used that as

---

173 Twetten, 'Albert the Great, Double Truth, and Celestial Causality'.

174 Twetten, 'Albert the Great's Early Conflations', pp. 25–26.

175 Albertus Magnus, *Summa theologiae*, II.3.3.1.4 ad 2, ed. by Borgnet, pp. 28b–29a: 'Et hoc modo primum procedens a primo principio, est intelligentia pura et simplex [...] et sic est de aliis procedentibus a primo, sicut philosophice in libro de Causis demonstratur: et hoc modo procedentia a primo principio, sunt multa processibus, et unum in principio effectivo: quod enim immediate a primo procedit, quod facit debere esse in omnibus et omnia, ut dicit Avicenna, hoc est simplex esse, in quo procedentia non distinguuntur'.

a basis for rejecting the Peripatetic emanation scheme.[176] Could this fact be grounds for thinking Albert is merely engaging in a commentatorial exercise? The objection gains force from a loose end that we have not yet tied off. Why does Albert say, in a culminating moment in Tract 3 on freedom, *De causis* I, that God is not able not to act, not able to withdraw from causing created goods (see above, at note 29)? Could these words be grounds for seeing Albert as a closet radical Aristotelian?

The key to answering these questions can be found in Albert's distinction, already made in *De homine*, between *electio* and *eligentia* (the term Albert prefers for Aristotle's *proairesis*, the translation in the *Ethica vetus*). The *electio* of *eligentia*, says Albert, 'follows the rules of reason taken from the nature of what is choice-worthy; and for this reason it is not "free", but is *forced* by right reason' (emphasis added).[177] Albert's teaching is that, whereas the *electio* that is the proper act of free *arbitrium* is free, *proairesis*, since it is the efficient cause of the moral *virtues*, is not 'free'. Freedom is in the will and in free *arbitrium*, but not in reason insofar as it is ruled by logic and is passively receptive of objects from the world. To the extent that *proairesis*, which is by definition good, depends on right reasoning, freedom is not in it. Albert offers a parallel account for the will in general. The will is related to opposites when it lacks a disposition, he says, but 'once disposed through the conception and the determination of the thing willed, it is [directed] to one'.[178] At the same time, Albert also offers this clarification: the thing willed is a voluntary cause, and as such does not remove freedom. Perhaps he would agree to our seeing a distinction here between what is 'radically free' versus 'free though caused voluntarily'.

We begin to see what is behind Albert's claim that God, like the chaste person, cannot but do the good — in God's case, the good of his eternal plan. His will is free considered in itself, but considered as disposed by his eternal plan to do good, it cannot deviate. Albert takes up this issue after arguing in the *Super De divinis nominibus* that God's acting through his essence is his acting through will:

---

176 See notes 22, 47, and 108 above. Albert in his *De causis* recognizes that Aristotle, Avicenna, and Averroes rejected will in the First, by which Albert apparently means his notion of a will that is free. Albertus Magnus, *De causis*, I.3.2, ed. by Fauser, p. 36, vv. 62–65: 'Multi autem Peripateticorum in primo negabant esse voluntatem, Aristoteles scilicet, Theophrastus, Porphyrius, Avicenna et Averroes. Et de hoc quinque inducebant rationes'.

177 Albertus Magnus, *Commentarii in II Sententiarum*, 24.7c, ed. by Borgnet, p. 404b: 'Electio autem eligentiae sequitur regulas rationis sumptas ex natura eligibilis: et ideo illa non est libera, sed cogitur per rationem rectam'.

178 Albertus Magnus, *Commentarii in I Sententiarum*, 40.22, ad 2, ed. by Borgnet, p. 335b (underscoring sets off the quotation from its context): '[D]icendum, quod est causa voluntaria, sed determinata ad volitum: hoc enim non tollit libertatem a voluntate. Unde voluntas indisposita est oppositorum, sed disposita per conceptionem et determinationem voliti, est ad unum, et potest averti ab illo. Sed non sic est in Deo: quia sua voluntas non mutatur a volito in volitum contrarium, eo quod hoc est imperfectionis et defectus, sicut in III Sententiarum determinatum est'.

Unde concedimus eum agere ex ordine suae sapientiae et per libertatem suae voluntatis et tamen secundum suam essentiam, non tamen secundum necessitatem essentiae, secundum quod necessitas importat coactionem ad actum, nisi dicatur necessitas finis propter immobilitatem.[179]

> Hence, we grant that [God] acts from the order of his wisdom both through the freedom of his will, and nevertheless according to his essence — not nevertheless according to the necessity of the essence, according as necessity implies being *forced* to act — *unless it be called the necessity of the end on account of [the end's] immobility.*

The necessity that determines the divine will is the necessity of the end, which requires means. Similarly, Albert speaks of the saints and the blessed in heaven: their 'confirmation' is an adhesion to the good, which 'does not remove the will of doing what one wants, but posits in [one] a necessity, not of coercion, but of immutability in that end'.[180]

Albert does reject one version of the Peripatetic emanation scheme in *De causis* I: a version that holds that God requires secondary causes to effect his end.[181] Albert's burning issue in *De causis* is no longer necessity, but mediate creation and divine reliance on secondary causes. He has a highly voluntaristic account of the will, and he is confident that the Peripatetics have simply missed that notion. Were they to ascribe will to God under this notion, they would immediately see that God is free. Albert held something quite different in his first Aristotelian paraphrase circa 1251: Aristotle believes God has a will, but God acts necessarily; therefore, Aristotle's emanation scheme is false. By 1267, Albert has spent over fifteen years on his Aristotelian paraphrases to reach this high point: the culmination of Aristotle's metaphysics in the Treatise on First Causes. By now, he has become persuaded, apparently contrary to his original expectations, that the Peripatetic project can, for the most part, be made to make sense on the issue of emanation.

At the same time, we know that by circa 1263, at the end of *Metaphysica* XI, Albert had become convinced that his account in the *Physics* paraphrase was mistaken.[182] In the latter, he had followed Maimonides and argued, admittedly with probable, not demonstrative reasoning, that because a plurality follows immediately upon the First Cause, the First must act not by the necessity of nature, but by free will. As I have discussed elsewhere, a famous passage in *Metaphysica* XI.3.7 reveals him criticizing precisely this reasoning, as professed by those who (like

---

179 Albertus Magnus, *Super Dionysium De divinis nominibus*, IV.9 sol., ed. by Simon, p. 118, vv. 25–30.

180 Albertus Magnus, *Commentarii in III Sententiarum*, 18.2, ad 1, ed. by Borgnet, p. 315a: 'haec non tollit voluntatem faciendi quod vult, sed ponit necessitatem non coactionis in eo, sed immutabilitatis in fine illo'.

181 Albertus Magnus, *De causis*, I.3.5, ed. by Fauser, p. 40, vv. 59–70.

182 Albertus Magnus, *Physica*, VIII.1.15, ed. by Hossfeld, p. 579, v. 55–p. 580, v. 70. For Aristotle's necessitarian will and Maimonides' particularist reasoning in reaction, see notes 22, 47, 69, and 147 above.

himself, I argue) have improperly 'poured' philosophy into theology by presenting Maimonides' argument.[183] Thus, at least since circa 1263, he has stopped criticizing the Peripatetic derivation-scheme and has been open to reconciling it *in his philosophical writing* with his personal beliefs. The development in *De causis* I and II fits his distinctive understanding of how a person of faith should approach philosophy. Albert does not, in the familiar image, mix water and wine, pouring philosophy and theology into the same vessel, but he offers a rethinking of the Peripatetic derivation-scheme such that it is philosophically satisfying and harmonious with Christian belief.

I have emphasized the development in Albert's emanationism at least since 1263, at variance with an account that observes mainly continuity.[184] However, it is also important not to overlook the unity and continuity in Albert's emanationism. Therefore, I itemize, in a preliminary way, the points of continuity in Albert's early and late emanationism, so that we do not lose sight of the overarching unity in Albert's distinctive way of relating theology and philosophy, a philosophy deeply inspired by the Latinized Arabic thinkers. Throughout his career Albert agrees: 1) all creatures in their plurality emanate, flow, or proceed from God, who alone creates the *esse* of all things, and who is equally omnipresent to all; 2) the plurality in this emanation is to be explained in virtue of the plurality of divine ideas, and the 'Aristotelian' principle 'from one comes only one' remains true; 3) this emanation is the result of eternal divine free will, which in God is the same as the divine essence, so that creatures can be said to flow from God as a cause acting through its essence, and not through mere will, as if through an occurrent accident; 4) there is a flow of forms, an 'information', from God that is non-creative, but presupposes the creation of *esse*.[185] Late Albert corrects

183 Twetten, 'Albert the Great's Early Conflations', pp. 55–61.

184 See Anzulewicz, 'Die aristotelisch-platonische Synthese Alberts', pp. 296–300; Anzulewicz, 'Die Emanationslehre des Albertus Magnus', esp. p. 237. For the argument that the overarching structure of Albert's metaphysical thought, within which emanation becomes intelligible, is consciously theological and Platonic, not Aristotelian, see Anzulewicz, 'Einleitung', pp. xxix–xxxii; Anzulewicz, 'Albertus Magnus als Vermittler', pp. 64, 67, 71, 74 n. 45, and 86; Anzulewicz, 'Pseudo-Dionysius Areopagita', esp. 269; also Anzulewicz, 'Die Denkstruktur des Albertus Magnus', pp. 381–83. I suggest that the fundamental insight here is on target but is best saved by thinking of the Dionysian paraphrases as providing an overarching structure for the Peripatetic paraphrases, since Albert's understanding of what is Platonic and what is Aristotelian, even of what is theology versus philosophy, is extremely foreign to our own and develops over the course of Albert's writings. What is most obvious is that Albert in his *De causis* regards his derivation scheme as Peripatetic and philosophical (for previous works, see above, notes 107-108). Can we conclude that in previous works he regards his emanationism in general as more Platonic than Aristotelian? Does he understand Dionysius as a Platonist more than as an Aristotelian or a Peripatetic? On Albert's historiographical categories, see de Libera, 'Épicureisme, stoïcisme, péripatétisme'; Anzulewicz, 'Die platonische Tradition bei Albertus'.

185 See the texts cited above beginning at note 86. Late Albert adds a fifth point: (5) created substances, namely, Intelligences and celestial souls, share in this 'informing' causality, as we have seen. Anzulewicz, 'Die Emanationslehre des Albertus Magnus', pp. 223–25, 237, apparently ascribes to Albert as early as *De quattuor coaequaevis* a non-creative, formal emanation through supernal

earlier Albert on the following: it is a misuse of philosophy to think that for God's causality to be free it is necessary that whatever in the universe exists from the beginning must flow all at once, equally, and immediately from the divine intellect and will; instead, all effects subsequent to the first effect, *esse*, which alone is caused without mediation or created, should be conceived philosophically as mediated by supernal created causes, so that all flow or unfold in an ordered, law-like way from the divine free will — whether from eternity or in a first moment in time.

## Conclusion

My goal in this paper has been to identify and systematize Albert's emanation scheme or derivationism, his theory of the 'flow of forms', and to show how it is consistent with his personal thought, especially on the questions of divine free will and mediate creation. It is not difficult to see that, though Albert affirms, with 'all of the Peripatetics', the principle 'from one comes only one', he does so in a way that, *pace* Professor Sturlese, denies mediate creation. True, in the Condemnations of 1277, we find Article 44 (28): 'that from the one First Agent there cannot be a multitude of effects'.[186] But Albert does not hesitate to affirm that the entire multiplicity of effects are from God, even though each one is from only one aspect of God as such. Similarly, according to Article 64 (33), 'the immediate effect from the First must be one only and most similar to the First'.[187] Albert emphasizes that the first effect is first in the order of nature, not in the order of time (or in the order of the miraculous beyond nature), hence it can be simultaneous with other effects that are posterior in the order of nature and mediated in some way by the first effect. Finally, Article 43 (68) asserts that God cannot cause diversity without mediating causes.[188] This is strictly false in Albert's view, since God alone causes at least one diversity, that within the first created

---

substances, namely, a qualified version of the 'Neoplatonic doctrine of the *Liber de causis*'. This may be correct, but the texts cited thus far do not prove it. At *De IV coaequaevis*, q. 1, a. 3 ad sc 4, ed. by Borgnet, p. 313a, Albert reduces the causality of the angel or Intelligence, as suggested in the formula 'causa prima creavit esse animae mediante intelligentia', to the moving of the heavens whereby an angel causes the generation of *terrestrial* souls. Unless point (5) is made explicit, Albert's teaching prior to the Aristotelian paraphrases may be 'emanationist' merely in the comparatively uninteresting sense that *esse* and other forms distinct from it emanate from God alone.

186 Piché, *La Condamnation parisienne de 1277*, p. 94: 'Quod ab uno primo agente non potest esse multitudo effectuum'.

187 Ibid., p. 100: 'Quod effectus inmediatus a primo debet esse unus tantum et simillimus primo'.

188 Ibid., p. 94: 'Quod primum principium non potest esse causa diuersorum factorum hic inferius, nisi mediantibus aliis causis, eo quod nullum transmutans diuersimode transmutat, nisi transmutatum'. The article actually refers to the causation of sublunar generation and corruption, which is only possible through the opposed motion of the oblique sphere according to Aristotle, *De generatione et corruptione*, II.10. For the charge that many other of Albert's readings of the *Liber de causis* have been condemned, see Imbach, 'Notule sur le commentaire du *Liber de causis*', pp. 322–23.

substance, whose *id quod est* receives its *esse* (although Albert *could* say that these come from different *rationes* in God).[189] However, for Albert, every other form of created things beside *esse* is caused by God through the mediation of Intelligences and whatever other primary, supernal causes are above that form. Intellect is full of forms, says the *Liber de causis*, and from God's intellect, through the mediation of Intelligences, celestial souls, and their spheres, God causes the diverse specific forms of all creatures in a cascade of forms. Under the supposition of the divine will, the flow of forms occurs in a law-like way: wherever there is a plurality after the first plurality, it has a mediating cause.

Must Albert hold, then, that God could not have done otherwise? Clearly, for Albert, the position of Avicenna is false: that the universe emanates *necessarily* from the divine intellect and will. The emanation of being, and then the derivation or procession of diverse forms, are all under the free divine eternal will, which from all eternity has said 'yes, no, or later' to 'this' and 'that'. Nonetheless, late Albert finds this derivation, at least at the highest universal level (and after the first effect), to unfold or evolve in a law-like way. All at once, at a point before which (conceptually) there was nothing, there was a 'big bang', an outpouring of being and form. Late Albert takes Aristotle and the Peripatetics to have identified, on good philosophical grounds, the key principles behind this law-like derivation.

189 Perhaps the weakest point of Albert's theory, considered on its own terms, is the absence of a full explanation of the cause of *id quod est*, whether in his cosmology or in his metaphysics, generally. A sign of this weakness is that *id quod est*, even when it is said to be potency (or possibility), can also be taken to be non-being or nothing, as by Milazzo; see notes 79 and 132 above (and at note 122) — but cf. Albertus Magnus, *De causis*, II.1.6, ed. by Fauser, p. 66, vv. 85–87, where *id quod est* must differ from *quod est ex nihilo*. See ibid., I.4.2, p. 44, vv. 37–50: 'Si quaeritur vero, cum dicitur "influere", in quo sit continentia importata per praepositionem, dicendum, quod in possibilitate rei, cui fit influxus. Quae possibilitas rei est ex seipsa. In antehabitis enim iam determinatum est, quod omne id quod de nihilo est, nihil est ex seipso et ex seipso non habet nisi ad esse possibilitatem. Quae possibilitas, cum impletur ab eo quod est causa esse ipsius, continet esse defluxum in ipsam. [...] Iam enim habitum est, quod secundum "id quod ipsum est" nihil habet nisi receptionis et continentiae possibilitatem'. In ibid., I.4.5, p. 49, vv. 36–48, Albert admits that in the first procession, there is no *id quod est* from which it differs from the First; nevertheless, the *id quod est* in this case exists in potency, not in act, prior to existing. We find *id quod est* as possibility in an early as well as in the very last work: Albertus Magnus, *De homine*, ed. by Anzulewicz and Söder, p. 416, vv. 38–42 (sol.) and p. 417, vv. 64–68 (ad 2); *Summa de mirabili scientia Dei*, I.20.2, ad 6, ed. by Siedler, p. 102, vv. 21–30. Other texts suggest an explanation of possibility that avoids a difficulty. Possibility is only said improperly to precede creation, as existing in the power of the creator; Albertus Magnus, *Commentarii in II Sententiarum*, 1.10, ed. by Borgnet, p. 29a; cf. *Summa theologiae*, I.78.1, ed. by Borgnet, pp. 828b–829a. See also Sweeney, '*Esse Primum Creatum* in Albert the Great', pp. 642–44; Bächli-Hinz, *Monotheismus und neuplatonische Philosophie*, pp. 193–94; Baumgarten, 'L'interpretation de la proposition 90', p. 166.

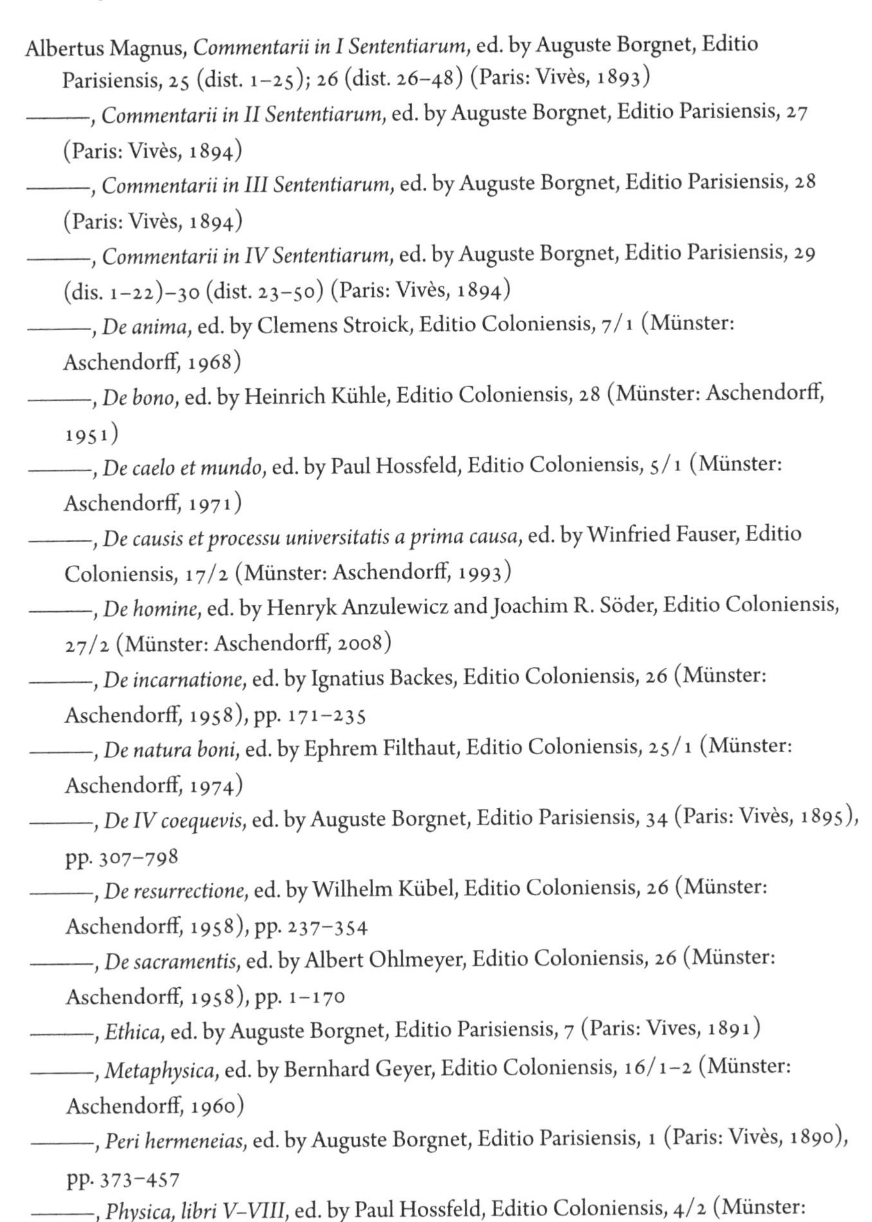

## Works Cited

### *Primary Sources*


Albertus Magnus, *Commentarii in I Sententiarum*, ed. by Auguste Borgnet, Editio Parisiensis, 25 (dist. 1–25); 26 (dist. 26–48) (Paris: Vivès, 1893)

———, *Commentarii in II Sententiarum*, ed. by Auguste Borgnet, Editio Parisiensis, 27 (Paris: Vivès, 1894)

———, *Commentarii in III Sententiarum*, ed. by Auguste Borgnet, Editio Parisiensis, 28 (Paris: Vivès, 1894)

———, *Commentarii in IV Sententiarum*, ed. by Auguste Borgnet, Editio Parisiensis, 29 (dis. 1–22)–30 (dist. 23–50) (Paris: Vivès, 1894)

———, *De anima*, ed. by Clemens Stroick, Editio Coloniensis, 7/1 (Münster: Aschendorff, 1968)

———, *De bono*, ed. by Heinrich Kühle, Editio Coloniensis, 28 (Münster: Aschendorff, 1951)

———, *De caelo et mundo*, ed. by Paul Hossfeld, Editio Coloniensis, 5/1 (Münster: Aschendorff, 1971)

———, *De causis et processu universitatis a prima causa*, ed. by Winfried Fauser, Editio Coloniensis, 17/2 (Münster: Aschendorff, 1993)

———, *De homine*, ed. by Henryk Anzulewicz and Joachim R. Söder, Editio Coloniensis, 27/2 (Münster: Aschendorff, 2008)

———, *De incarnatione*, ed. by Ignatius Backes, Editio Coloniensis, 26 (Münster: Aschendorff, 1958), pp. 171–235

———, *De natura boni*, ed. by Ephrem Filthaut, Editio Coloniensis, 25/1 (Münster: Aschendorff, 1974)

———, *De IV coequevis*, ed. by Auguste Borgnet, Editio Parisiensis, 34 (Paris: Vivès, 1895), pp. 307–798

———, *De resurrectione*, ed. by Wilhelm Kübel, Editio Coloniensis, 26 (Münster: Aschendorff, 1958), pp. 237–354

———, *De sacramentis*, ed. by Albert Ohlmeyer, Editio Coloniensis, 26 (Münster: Aschendorff, 1958), pp. 1–170

———, *Ethica*, ed. by Auguste Borgnet, Editio Parisiensis, 7 (Paris: Vives, 1891)

———, *Metaphysica*, ed. by Bernhard Geyer, Editio Coloniensis, 16/1–2 (Münster: Aschendorff, 1960)

———, *Peri hermeneias*, ed. by Auguste Borgnet, Editio Parisiensis, 1 (Paris: Vivès, 1890), pp. 373–457

———, *Physica, libri V–VIII*, ed. by Paul Hossfeld, Editio Coloniensis, 4/2 (Münster: Aschendorff, 1993)

———, *Problemata determinata*, ed. by James Weisheipl, Editio Coloniensis, 17/1 (Münster: Aschendorff, 1975)
———, *Summa theologiae sive de mirabili scientia Dei, libri I pars I (qq. 1–50A)*, ed. by Dionys Siedler, Editio Coloniensis, 34/1 (Münster: Aschendorff, 1978)
———, *Summa theologiae, pars prima*, ed. by Auguste Borgnet, Editio Parisiensis, 31 (Paris: Vivès, 1895)
———, *Summa theologiae, pars secunda* (qq. I–LXVII), ed. by Auguste Borgnet, Editio Parisiensis, 32 (Paris: Vivès, 1895)
———, *Summa theologiae, pars secunda* (qq. LXVIII–CXLI), ed. by Auguste Borgnet, Editio Parisiensis, 33 (Paris: Vivès, 1895)
———, *Super Dionysium De caelesti hierarchia*, ed. by Paul Simon and Wilhelm Kübel, Editio Coloniensis, 36/1 (Münster: Aschendorff, 1993)
———, *Super Dionysium De divinis nominibus*, ed. by Paul Simon, Editio Coloniensis, 37/1 (Münster: Aschendorff, 1972)
———, *Super Dionysium De ecclestica hierarchia*, ed. by Maria Burger, Editio Coloniensis, 36/2 (Münster: Aschendorff, 1999)
———, *Super I Sententiarum (distinctiones 1–3)*, ed. by Maria Burger, Editio Coloniensis, 29/1 (Münster: Aschendorff, 2015)
Anselmus Cantuariensis, *De libertate arbitrii*, ed. by Francisus Salesius Schmitt, Anselmi Opera Omnia, 1 (Stuttgart-Bad Cannstatt: Friedrich Frommann, 1946)
Aristoteles Latinus, *De generatione et corruptione: Translatio vetus*, ed. by Joanna Judycka, Aristoteles Latinus, 9/1 (Leiden: Brill, 1986)
Augustine, *De duabus animabus contra Manichaeos*, ed. by Joseph Zycha, CSEL, 25/1 (Vienna: Verlag der Österreichischen Akademie der Wissenschaften), pp. 51–80
Avicenna, *Metaphysics of the Healing (al-Shifāʾ: Ilāhiyyāt)*, trans. by Michael Marmura (Provo, UT: Brigham Young University Press, 2005)
Avicenna Latinus, *Liber de philosophia prima sive Scientia divina, V–X*, ed. by Simone Van Riet (Leuven: Peeters, 1980)
Bacon, Roger [ps.], *Quaestiones super quatuor libros Physicorum Aristotelis*, ed. by Ferdinand Marie Delorme and Robert Steele, Opera hactenus inedita Rogeri Baconi, 8 (Oxford: Clarendon Press, 1928)
Bonaventura, *Commentaria in quatuor libros Sententiarum, in secundum librum sententiarum*, ed. by PP. Collegii S. Bonaventurae, Opera omnia, 2 (Quaracchi: Collegium S. Bonaventurae, 1885)
Johannes Damascenus, *De fide orthodoxa: Versions of Burgundio and Cerbanus*, ed. by Eligius Buytaert, Franciscan Institue Publications, 8 (St Bonaventure, NY: Franciscan Institute, 1955)
*Liber de causis* [Arabic], ed. by Richard C. Taylor, in 'The "Liber de causis" (*Kalām fī maḥḍ al-khair*): A Study of Medieval Neoplatonism' (unpublished doctoral thesis, University of Toronto, 1981)

*Liber de causis* [Latin], ed. by Winfried Fauser, in Albertus Magnus, *De causis et processu universitatis a prima causa*, ed. by Winfried Fauser, Editio Coloniensis, 17/2 (Münster: Aschendorff, 1993), pp. 67–191

Moses Maimonides, *Dalālat al-ḥā'irīn (Le guide des égarés): Traité de théologie et de philosophie*, ed. by Salomon Munk, 3 vols (Paris: A. Franck, 1856–66)

———, *Dux seu Director dubitantium aut perplexorum*, ed. by Agostino Giustiniani (Paris: J. B. Ascensius, 1520)

Petrus Lombardus, *Sententiae in IV libris distinctae*, ed. by PP. Collegium S. Bonaventurae, 2 vols (Quaracchi: Editiones Collegii S. Bonaventurae, 1971)

Philippi Cancellarii, *Summa de bono*, ed. by Nikolaus Wicki, Corpus philosophorum Medii Aevi: Opera philosophica Mediae Aetatis selecta, 2 (Bern: Francke, 1985)

Proclus, *The Elements of Theology: A Revised Text with Translation, Introduction, and Commentary*, ed. by Eric R. Dodds, 2nd ed. (Oxford: Clarendon, 1963)

Pseudo-Dionysius Areopagita, *De coelesti hierarchia*, ed. by Günter Heil and Adolf Martin Ritter, Corpus Dionysiacum, 2 (Berlin: De Gruyter, 1991)

———, *De divinis nominibus*, ed. by Beate Regina Suchla, Corpus Dionysiacum, 1 (Berlin: De Gruyter, 1990)

Pseudo-Dionysius Latinus, *De divinis nominibus*, trans. by Johannes Sarracenus, in Philippe Chevalier, *Dionysiaca: Receuil donnant l'ensemble des traditions latines des ouvrages attribués au Denys l'Aréopage*, vol. 1 (Bruges: Désclée de Brouwer, 1937)

*Robert Kilwardby, Quaestiones in librum secundum Sententiarum*, ed. by Gerhard Leibold (Munich: Verlag der Bayerischen Akademie der Wissenschaften, 1992)

Thomas Aquinas, *De substantiis separatis*, ed. by Commissio Leonina, Opera omnia iussu Leonis XIII P.M. edita, 40/D–E (Rome: Ad Sanctae Sabinae, 1969)

William of Auxerre, *Summa aurea*, ed. by Jean Ribaillier, 7 vols (Rome: Quaracchi, 1980–87)

## Secondary Works

Adamson, Peter, *The Arabic Plotinus: A Philosophical Study of the 'Theology of Aristotle'* (London: Duckworth, 2002)

Adelmann, Frederick J., 'The Theory of Will in John Damascene', in *The Quest for the Absolute*, ed. by Frederick J. Adelmann (Chestnut Hill: Boston College, 1966), pp. 22–46

Alarcón, Enrique, 'S. Alberto Magno y la *Epistola Aristotelis de Principio Universi Esse*', in *Actas del I Congreso Nacional de Filosofía Medieval* (Zaragoza: Ibercaja, 1992), pp. 181–92

Anzulewicz, Henryk, 'Albertus Magnus als Vermittler zwischen Aristoteles und Platon', *Acta Mediaevalia*, 18 (2005), 63–87

———, 'Die aristotelisch-platonische Synthese Alberts des Großen', in *Język, Rozumienie, Komunikacja*, ed. by Mikołaj Domaradzki, Emanuel Kulczycki, and Michał Wendland (Poznań: Wydawnictwo Naukowe Instytutu Filozofii UAM, 2011), pp. 291–302

———, 'Die Denkstruktur des Albertus Magnus: Ihre Dekodierung und ihre Relevanz für die Begrifflichkeit und Terminologie', in *L'élaboration du vocabulaire philosophique au Moyen Âge: Actes du Colloque international de Louvain-la-Neuve et Leuven 12–14 septembre 1998 organisé par la Société Internationale pour l'Étude de la Philosophie Médiévale*, ed. by Jacqueline Hamesse and Carlos Steel (Turnhout: Brepols, 2000), pp. 369–96

———, 'Einleitung', in Albertus Magnus, *Buch über die Ursachen und den Hervorgang von allem aus der ersten Ursache (Liber de causis et processu universitatis a prima causa)*, ed. and trans. by Henryk Anzulewicz, Maria Burger, Silvia Donati, Ruth Meyer, and Hannes Möhle (Hamburg: Felix Meiner, 2006), pp. xiii–xxxvi

———, 'Die Emanationslehre des Albertus Magnus: Genese, Gestalt und Bedeutung', in *Via Alberti: Texte – Quellen – Interpretationen*, ed. by Ludger Honnefelder, Hannes Möhle, and Susana Bullido del Barrio (Münster: Aschendorff, 2009), pp. 219–42

———, 'Der *Liber de causis* als Quelle der Intellektlehre des Albertus Magnus', in *Pseudo-Aristotelian Texts in Medieval Thought*, ed. by Monica Brinzei, Daniel Coman, Ioana Curut, and Andrei Marinca (Turnhout: Brepols, 2023), pp. 153-82

———, 'Die platonische Tradition bei Albertus Magnus: Eine Hinführung', in *The Platonic Tradition in the Middle Ages: A Doxographic Approach*, ed. by Stephen Gersh and Martin Hoenen (Berlin: De Gruyter, 2002), pp. 207–77

———, 'Pseudo-Dionysius Areopagita und das Strukturprinzip des Denkens von Albert dem Großen', in *Die Dionysius-Rezeption im Mittelalter. Internationales Kolloquium in Sofia vom 8. Bis 11. April 1999 unter Schirmherrschaft der Société Internationale pour l'Étude de la Philosophie Médiévale*, ed. by Tzotcho Boiadjiev, Georgi Kapriev, and Andreas Speer (Turnhout: Brepols, 2000), pp. 251–95

———, and Katja Krause, 'From Content to Method: the *Liber de causis* in Albert the Great', in *Reading Proclus and the Book of Causes*, vol. 1, ed. by Dragos Calma (Leiden: Brill, 2019), pp. 180–208

Armstrong, Arthur Hilary, '"Emanation" in Plotinus', *Mind*, 46 (1937), 61–66

Bächli-Hinz, Andreas, *Monotheismus und neuplatonische Philosophie: Eine Untersuchung zum pseudo-aristotelischen Liber de causis und dessen Rezeption durch Albert den Großen* (Sankt Augustin: Academia, 2004)

Baldner, Steven, 'Albertus Magnus on Creation: Why Philosophy Is Inadequate', *American Catholic Philosophical Quarterly*, 88 (2014), 63–79

Baumgarten, Alexander, 'L'interpretation de la proposition 90 du *Liber de Causis* chez Albert le Grand et saint Thomas d'Aquin', *Chôra*, 1 (2003), 161–71

Beierwaltes, Werner, 'Der Kommentar zum "Liber de causis" als neuplatonisches Element in der Philosophie des Thomas von Aquin', *Philosophische Rundschau*, 11 (1964), 192–215

Bonin, Thérèse, *Creation as Emanation: The Origin of Diversity in Albert the Great's 'On the Causes and the Procession of the Universe'* (Notre Dame, IN: Notre Dame University Press, 2001)

———, 'The Emanative Psychology of Albertus Magnus', *Topoi*, 19 (2000), 45–57

Booth, Edward, 'Conciliazioni ontologiche delle tradizioni platonica e aristotelica in Sant'Alberto e San Tommaso', in *Sant'Alberto Magno, l'uomo e il pensatore* (Milan: Massimo, 1982), pp. 60–81

Cerami, Cristina, *Génération et substance: Aristote et Averroès entre physique et métaphysique* (Berlin: De Gruyter, 2015)

Chase, Michael, 'Discussions on the Eternity of the World in Antiquity and Contemporary Cosmology', *ΣΧΟΛΗ*, 7 (2013), 20–68

D'Ancona, Cristina, 'The *Liber de causis*', in *Interpreting Proclus: From Antiquity to the Renaissance*, ed. by Stephen Gersh (Cambridge: Cambridge University Press, 2014), pp. 137–61

———, 'Nota sulla traduzione latina del *Libro di Aristotele sull'esposizione del bene puro* e sul titolo *Liber de causis*', in *Scientia, fides, theologia: Studi di filosofia medievale in onore di Gianfranco Fioravanti*, ed. by Stefano Perfetti (Pisa: Edizioni ETS, 2011), pp. 89–101

———, 'Un "quindicesimo libro" per la Metafisica di Aristotele: Il *Liber de Causis* in alcuni commenti del XIII secolo', *Cultura e Scuola*, 109 (1989), 88–96

———, *Recherches sur le 'Liber de causis'* (Paris: Vrin, 1995)

———, and Richard C. Taylor, 'Le *Liber de causis*', in *Dictionnaire des Philosophes Antiques*, ed. by Richard Goulet, vol. 1 Supplément (Paris: Éditions du CNRS, 2003), pp. 599–647

Davidson, Herbert, *Proofs for Eternity, Creation and the Existence of God in Medieval Islamic and Jewish Philosophy* (Oxford: Oxford University Press, 1987)

de Libera, Alain, *Albert le Grand et la philosophie* (Paris: Vrin, 1990)

———, 'Albert le Grand et le platonisme: De la doctrine des idées à la théorie des trois états de l'universel', in *On Proclus and His Influence in Medieval Philosophy*, ed. by Egbert P. Bos and Piet A. Meijer (Leiden: Brill, 1992), pp. 89–119

———, 'Albert le Grand et Thomas d'Aquin interprètes du *Liber de causis*', *Revue des sciences philosophiques et théologiques*, 74 (1990), 347–78

———, 'Épicurisme, stoïcisme, péripatétisme: L'histoire de la philosophie vue par les latins (XII^e^–XIII^e^ siècle)', in *Perspectives arabes et médiévales sur la tradition scientifique et philosophique grecque*, ed. by Ahmad Hasnawi, Abdelali Elamrani-Jamal, and Maorun Aouad (Leuven: Peeters, 1997), pp. 343–64

———, *Métaphysique et noétique: Albert le Grand* (Paris: Vrin, 2005)

———, *Raison et foi: Archéologie d'une crise d'Albert le Grand à Jean-Paul II* (Paris, Éditions du Seuil, 2003)

Di Segni, Diana, 'Early Quotations from Maimonides's *Guide of the Perplexed* in the Latin Middle Ages', in *Interpreting Maimonides: Critical Essays*, ed. by Charles Manekin and Daniel Davies (Cambridge: Cambridge University Press, 2018), pp. 190–207

Dörrie, Heinrich, 'Emanation: Ein unphilosophisches Wort im spätantiken Denken', in *Parusia: Studien zur Philosophie und Problemgeschichte des Platonismus*, ed. by Kurt Flasch (Frankfurt am Main: Minerva, 1965), pp. 119–41

Drouin, Fernand, 'Le libre arbitre dans l'organisme psychologique selon Albert le Grand', *Études d'histoire littéraire et doctrinale du XIII^e^ siècle*, 1, no. 2 (1932), 91–120

Ducharme, Leonard, '*Esse* chez saint Albert le Grand: Introduction à la métaphysique de ses premiers écrits', *Revue de l'Université d'Ottawa*, 27 (1957), 209*–252*

Duhem, Pierre, *Le système du monde: Histoire des doctrines cosmologiques de Platon à Copernic*, vol. 5 (Paris: A. Hermann, 1917)

Ehret, Charles, 'The Flow of Powers: Emanation in the Psychologies of Avicenna, Albert the Great, and Aquinas', *Oxford Studies in Medieval Philosophy*, 5 (2017), 87–121

Fidora, Alexander, 'From Arabic into Latin into Hebrew: Aristotelian Psychology and its Contribution to the Rationalisation of Theological Traditions', in *Philosophical Psychology in Medieval Arabic and Latin Aristotelianism of the 13th Century*, ed. by Luis Xavier López-Farjeat and Jörg Tellkamp (Paris: Vrin, 2013), pp. 17–39

Frede, Michael, 'John of Damascus on Human Action, the Will, and Human Freedom', in *Byzantine Philosophy and Its Ancient Sources*, ed. by Katerina Ierodiakonou (Oxford: Clarendon Press, 2002), pp. 63–95

Freudenthal, Gad, 'The Astrologization of the Aristotelian Cosmos: Celestial Influences on the Sublunary World in Aristotle, Alexander of Aphrodisias, and Averroes', in *A Companion to Aristotle's Cosmology: Collected Papers on 'De Caelo'*, ed. by Alan C. Bowen and Christian Wildberg (Leiden: Brill, 2009), pp. 239–81

Gauthier, René-Antoine, 'Saint Maxime le Confesseur et la psychologie de l'acte humain', *Recherches de théologie ancienne et médiévale*, 21 (1954), 51–100

Grabmann, Martin, 'Die Lehre des heiligen Albertus Magnus vom Grunde der Vielheit der Dinge und der lateinische Averroismus', in *Mittelalterliches Geistesleben: Abhandlungen zur Geschichte der Scholastik und Mystik*, vol. 2 (Munich: M. Hueber, 1936), pp. 287–312

Hankey, Wayne, '*Ab uno simplici non est nisi unum*: The Place of Natural and Necessary Emanation in Aquinas' Doctrine of Creation', in *Divine Creation in Ancient, Medieval, and Early Modern Thought*, ed. by Michael Treschow, Willemien Otten, and Walter Hannam (Leiden: Brill, 2007), pp. 309–33

Hasnawi, Ahmad, 'Fayḍ (épanchement, émanation)', in *Encyclopédie philosophique universelle II: Les notions philosophiques – Dictionnaire*, ed. by André Jacob and Sylvain Auroux, vol. 1 (Paris: Presses universitaires de France, 1990), pp. 966–72

Hasse, Dag Nikolaus, 'Avicenna's "Giver of Forms" in Latin Philosophy, Especially in the Works of Albertus Magnus', in *The Arabic, Hebrew and Latin Reception of Avicenna's Metaphysics*, ed. by Dag Nikolaus Hasse and Amos Bertolacci (Berlin: De Gruyter, 2012), pp. 225–49

Hedwig, Klaus, *Sphaera lucis: Studien zur Intelligibilität des Seienden im Kontext der mittelalterlichen Lichtspekulation* (Münster: Aschendorff, 1980)

Hissette, Roland, 'Albert le Grand et Thomas d'Aquin dans la censure parisienne du 7 mars 1277', *Miscellanea Mediaevalia*, 15 (1982), 226–46

———, *Enquête sur les 219 articles condamnés à Paris le 7 mars 1277* (Louvain: Publications universitaires, 1977)

Hoffmann, Tobias, 'Voluntariness, Choice, and Will in Albert and Aquinas', *Documenti e studi sulla tradizione filosofica medievale*, 17 (2006), 71–92

Imbach, Ruedi, 'Notule sur le commentaire du *Liber de causis* de Siger de Brabant et ses rapports avec Thomas d'Aquin', *Freiburger Zeitschrift für Philosophie und Theologie*, 43 (1996), 304–23

Janssens, Jules, 'Creation and Emanation in Ibn Sīnā', *Documenti e studi sulla tradizione filosofica medievale*, 8 (1997), 455–77

Kaiser, Rudolf, 'Zur Frage der eigenen Anschauung Alberts des Grossen in seinen philosophischen Kommentaren', *Freiburger Zeitschrift für Philosophie und Theologie*, 9 (1962), 53–62

Lizzini, Olga, 'Flusso, preparazione appropriata e *inchoatio formae*: brevi osservazioni su Avicenna e Alberto Magno', in *Medioevo e filosofia per Alfonso Maierù*, ed. by Massimiliano Lenzi, Cesare A. Musatti, and Luisa Valente (Rome: Viella, 2013), pp. 129–50

———, *Fluxus (fayḍ): Indagine sui fondamenti della metafisicae e della fisica di Avicenna* (Bari: Edizioni di Pagina, 2011)

Lottin, Odon, 'Libre arbitre et liberté depuis s. Anselme jusqu'a la fin du XIII^e^ siècle', in Odon Lottin, *Psychologie et morale aux XII^e^ et XIII^e^ siècles*, 2nd ed., vol. 1 (Gembloux: J. Duculot, 1957), pp. 11–389

———, 'La psychologie de l'acte humain chez saint Jean Damascène et les théologiens du XIII^e^ siècle occidental', in *Psychologie et morale aux XII^e^ et XIII^e^ siècles*, 2nd ed., vol. 1 (Gembloux: J. Duculot, 1957), pp. 393–424

Malik, Josef, 'Albert der Große und das Willensproblem: Zum 15. November 1280', *Theologie und Glaube*, 70 (1980), 371–97

McCluskey, Colleen, 'Human Action and Human Freedom: Four Theories of *Liberum arbitrium* in the Early Thirteenth Century' (unpublished doctoral thesis, University of Iowa, 1997)

———, 'The Roots of Ethical Voluntarism', *Vivarium*, 39 (2001), 185–208

———, 'Worthy Constraints in Albertus Magnus's Theory of Action', *Journal of the History of Philosophy*, 39 (2001), 491–533

Michaud-Quantin, Pierre, *La psychologie de l'activité chez Albert le Grand* (Paris: Vrin, 1966)

Milazzo, Sébastien, 'Introduction' and 'Commentaire', in Albertus Magnus, *Le traité du flux: Tractatus de fluxu causatorum a causa prima et causarum ordine*, trans. by Sébastien Milazzo (Paris: Les Belles Lettres, 2013), pp. xi–lxxxi and 67–290

Miteva, Evelina, 'I Want to Break Free: Albert the Great's Naturalistic Account of Freedom of Choice and Its Limitations', *Bochumer Jahrbuch*, 21 (2018), 11–28

Molina, Enrique, 'Movens-Efficiens-Agens: La fundamentación causal en San Alberto Magno', *Cuadernos de Filosofía*, 8 (1998), 279–345

Moulin, Isabelle, 'Éduction et émanation chez Albert le Grand: des commentaires sur Denys le Pseudo-Aréopagite au *De causis et processu universitatis a prima causa*', in *Via Alberti: Texte – Quellen – Interpretationen*, ed. by Ludger Honnefelder, Hannes Möhle, and Susana Bullido del Barrio (Münster: Aschendorff, 2009), pp. 243–64

———, and David Twetten, 'Causality and Emanation in Albert', in *A Companion to Albert the Great: Theology, Philosophy, and the Sciences*, ed. by Irven M. Resnick (Leiden: Brill, 2012), pp. 694–721

Nardi, Bruno, 'La posizione di Alberto Magno di fronte all'averroismo', in *Studi di filosofia medievale* (Rome: Edizioni di Storia e Letteratura, 1960), pp. 119–50

Palazzo, Alessandro, 'The Scientific Significance of Fate and Celestial Influences in some Mature Works by Albert the Great: *De fato, De somno et vigilia, De intellectu et intelligibili, Mineralia*', in *Per perscrutationem philosophicam: Neue Perspektiven der mittelalterlichen Forschung*, ed. by Alessandra Beccarisi, Ruedi Imbach, and Pasquale Porro (Hamburg: Felix Meiner, 2008), pp. 55–78

Piché, David, *La Condamnation parisienne de 1277: Nouvelle édition du texte latin, traduction, introduction et commentaire par David Piché* (Paris: Vrin, 1999)

Porreca, David, 'Albert the Great and Hermes Trismegistus: An Update', *Mediaeval Studies*, 72 (2010), 245–81

Porro, Pasquale, 'Prima rerum creatarum est esse: Henri de Gand, Gilles de Rome et la quatrième proposition du *De causis*', in *L'aristotélisme exposé: Aspects du débat philosophique entre Henri de Gand et Gilles de Rome*, ed. by Valérie Cordonier and Tiziana Suarez-Nani (Fribourg: Academic Press Fribourg, 2014), pp. 55–81

———, 'The University of Paris in the Thirteenth Century: Proclus and the *Liber de causis*', in *Interpreting Proclus: From Antiquity to the Renaissance*, ed. by Stephen Gersh (Cambridge: Cambridge University Press, 2014), pp. 264–98

Principe, Walter, *Philip the Chancellor's Theology of the Hypostatic Union* (Toronto: Pontifical Institute of Mediaeval Studies, 1975)

Rigo, Caterina, 'Zur Rezeption des Moses Maimonides im Werk des Albertus', in *Albertus Magnus zum Gedenken nach 800 Jahren: Neue Zugänge, Aspekte und Perspektiven*, ed. by Walter Senner (Berlin: Akademie, 2001), pp. 29–66

Roques, René, *L'univers Dionysien: Structure hiérarchique du monde selon le Pseudo-Denys* (Paris: Éditions du Cerf, 1983)

Saffrey, Henri Dominique, 'Introduction', in Thomas Aquinas, *Super Librum de causis expositio*, ed. by Henri Dominique Saffrey (Fribourg: Société philosophique, 1954), pp. xiii–lxxiii

Schönberger, Rolf, 'Rationale Spontaneität: Die Theorie des Willens bei Albertus Magnus', in *Albertus Magnus: Zum Gedenken nach 800 Jahren; Neue Zugänge, Aspekte und Perspektiven*, ed. by Walter Senner (Berlin: Akademie, 2001), pp. 221–34

Schwartz, Yossef, 'Celestial Motion, Immaterial Causality and the Latin Encounter with Arabic Aristotelian Cosmology', in *Albertus Magnus und der Ursprung der Universitätsidee: Die Begegnung der Wissenschaftskulturen im 13. Jahrhundert und die Entdeckung des Konzepts der Bildung durch Wissenschaft*, ed. by Ludger Honnefelder (Berlin: Berlin University Press, 2011), pp. 277–98 and 500–11

———, 'Divine Space and the Space of the Divine: On the Scholastic Rejection of Arab Cosmology', in *Représentations et conceptions de l'espace dans la culture médiévale: Colloque Fribourgeois 2009*, ed. Tiziana Suarez-Nani and Martin Rohde (Berlin: De Gruyter, 2011), pp. 89–120

Seeskin, Kenneth, *Maimonides on the Origin of the World* (Cambridge: Cambridge University Press, 2005)

Siedler, Dionys, *Intellektualismus und Voluntarismus bei Albertus Magnus* (Münster: Aschendorff, 1941)

———, and Paul Simon, 'Prolegomena', in Albert the Great, *Summa theologiae sive De mirabili scientia dei I: Liber 1, Pars 1, Quaestiones 1–50A*, ed. by Dionysius Siedler, Editio Coloniensis, 34/1 (Münster: Aschendorff, 1978), pp. v–xxvi

Spiering, Jamie, 'An Innovative Approach to *Liberum arbitrium* in the Thirteenth Century: Philip the Chancellor, Albert the Great and Thomas Aquinas' (unpublished doctoral thesis, The Catholic University of America, 2010)

———, '"*Liber est causa sui*": Thomas Aquinas and the Maxim "The Free is the Cause of Itself"', *Review of Metaphysics*, 65 (2011), 351–76

Sturlese, Loris, *Storia della filosofia tedesca nel medioevo: Il secolo XIII* (Florence: Olschki, 1996)

Sweeney, Leo, '*Esse Primum Creatum* in Albert the Great's *Liber de causis et processu universitatis*', *The Thomist*, 44 (1980), 599–646

Taylor, Richard C., 'Contextualizing the *Kalām fī maḥḍ al-khair / Liber de causis*', in *Reading Proclus and the Book of Causes*, vol. 2: *Translations and Acculturations*, ed. by Dragos Calma (Leiden: Brill, 2020), pp. 211–32

Trouillard, Jean, 'Procession néoplatonicienne et création judéo-chrétienne', in *Néoplatonisme: Mélanges offerts à Jean Trouillard*, ed. by Jacqueline Bonnamour (Fontenay-aux-Roses: École normale supérieure, 1981), pp. 1–30

Twetten, David, 'Albert's Arguments for the Existence of God and the Primary Causes', in *A Companion to Albert the Great: Theology, Philosophy, and the Sciences*, ed. by Irven M. Resnick (Leiden: Brill, 2012), pp. 658–87

———, 'Albert the Great, Double Truth, and Celestial Causality', *Documenti e studi sulla tradizione filosofica medievale*, 12 (2001), 275–358

———, 'Albert the Great's Early Conflations of Philosophy and Theology on the Issue of Universal Causality', in *Medieval Masters: Essays in Memory of Msgr. E. A. Synan*, ed. by Rollen E. Houser (Houston: Center for Thomistic Studies, 1999), pp. 25–62

———, 'Aristotelian Cosmology and Causality in Classical Arabic Philosophy and its Greek Background', in *Ideas in Motion in Baghdad and Beyond: Philosophical and Theological Exchanges between Christians and Muslims in the Third/Ninth and Fourth/Tenth Centuries*, ed. by Damien Janos (Leiden: Brill, 2015), pp. 312–433

———, 'Whose Prime Mover Is More (un)Aristotelian, Broadie's, Berti's or Averroes'?', in *La philosophie arabe à l'étude: Sens, limites et défis d'une discipline moderne*, ed. by Jean-Baptiste Brenet and Olga L. Lizzini (Paris: Vrin, 2019), pp. 347–92

Vargas, Rosa, 'Albert on Being and Beings: The Doctrine of *Esse*', in *A Companion to Albert the Great: Theology, Philosophy, and the Sciences*, ed. by Irven M. Resnick (Leiden: Brill, 2012), pp. 627–48

Wielockx, Robert, 'Zur "Summa Theologiae" des Albertus Magnus', *Ephemerides Theologicae Lovanienses*, 66 (1990), 78–110

Zavattero, Irene, 'La βούλησις nella psicologia dell'agire morale della prima metà del XIII secolo', in *Il desiderio nel Medioevo*, ed. by Alessandro Pallazzo (Rome: Edizioni di Storia e Letteratura, 2014), pp. 133–50

———, '*Voluntas est duplex*: La dottrina della volontà dell'anonimo Commento di Parigi sull'*Ethica nova e vetus* (1235–1240)', *Medioevo*, 40 (2015), 65–96

# Index of Sources

## Ancient Greek Authors

## Christian Authors

## Muslim Authors

## Jewish Authors

# Index of Subjects and Names